Introduction to Biostatistics

A Guide to Design, Analysis, and Discovery

Introduction to Biostatistics

A Guide to Design, Analysis, and Discovery

Ronald N. Forthofer
Longmont, Colorado

Eun Sul Lee
School of Public Health
The University of Texas
Health Sciences Center at Houston
Houston, Texas

Academic Press
San Diego New York Boston London Sydney Tokyo Toronto

This book is printed on acid-free paper. ♾

Academic Press
A Division of Harcourt Brace & Company
525 B Street, Suite 1900, San Diego, California 92101-4495

United Kingdom Edition published by
Academic Press Limited
24-28 Oval Road, London NW1 7DX

Library of Congress Cataloging-in-Publication Data

Introduction to Biostatistics: A Guide to Design, Analysis, and Discovery / edited by Ronald M. Forthofer, Eun Sul Lee
p. cm.
Includes bibliographical references and index.
ISBN 0-12-262270-7
1. Medicine--Research--Statistical methods. 2. Biometry.
I. Forthofer, Ron N., date. II. Lee, Eun Sul.
R853.S7B54 1995
574'.01'5195--dc20 94-24912
CIP

PRINTED IN THE UNITED STATES OF AMERICA
03 04 05 EB 10 9 8 7 6 5 4 3

To Mary and Chong Mahn
Without their love and support,
this book would not have been possible.

Contents

2 Data and Numbers

3 Sampling

4 Descriptive Tools

5 Probability and Life Tables

6 Probability Distributions

7 Interval Estimation

8 Designed Experiments

9 Tests of Hypotheses

10 Nonparametric Tests

11 Analysis of Categorical Data

12 Analysis of Survival Data

13 Tests of Hypotheses Based on the Normal Distribution

14 Analysis of Variance

15 Linear and Logistic Regression

APPENDIX A SAS and Stata Commands

APPENDIX B Statistical Tables

APPENDIX C Selected Governmental Biostatistical Data

APPENDIX D Solutions to Selected Exercises

Preface

This introductory biostatistics textbook encourages readers to consider the full context of the problem being examined. The context includes what the data actually represent, why and how the data were collected, whether or not one can generalize from the sample to the target population, and what problems occur when the data are incomplete due to people refusing to participate in the study or due to the researcher failing to obtain all the relevant data from some sample subjects. Although many introductory biostatistical textbooks do a very good job in presenting statistical tests and estimators, they are limited in their presentations of the context. In addition, most textbooks do not emphasize the relevance of biostatistics to people's lives and well being. We have written this textbook to address these deficiencies and to provide a good introduction to statistical methods. We address the context as well as the importance of research design, particularly in controlling for confounding variables and in dealing with reversion to the mean. We focus on these issues in Chapters 1 to 3 and Chapter 8 and raise them again in examples and exercises throughout the book.

This textbook also differs from the other texts in that it uses real data for most of the exercises and examples in the book. For example, real data on the relation between prenatal care and birthweight, instead of data from tossing dice or dealing cards, are used in the definition of probability and in the demonstration of the rules of probability. We then show how these rules are applied to the life table, a major tool used by health analysts. Another major difference between this and other texts is Chapter 12 on the analysis of the follow-up life table. The follow-up life table can be used to summarize survival data and is one of the more important tools used in clinical trials.

We also include material on tolerance and prediction intervals, topics generally ignored in other texts. We demonstrate in which situations these intervals should be used and how they provide different information than that provided by confidence intervals. Two other topics, usually not mentioned in other introductory texts, introduced here are multiple regression and logistic regression, two of the more useful methods of analysis in statistics and epidemiology.

We do not assume that the reader has prior knowledge of statistical methods, but we do assume that the reader is not rendered unconscious by the sight of a formula. In dealing with a formula, we first try to explain the concept underlying the formula. We then show how the formula is a translation of the concept into something that can be measured. The emphasis is on when and how to apply the formula, not on its derivation. We also show how the calculation can be quickly performed using a statistical package. The package shown in the text is MINITAB. Comparable commands for two other packages, *Stata* and *SAS*, are shown in the Appendix.

The textbook is designed for a two-quarter course for graduate students and for a two-semester course for undergraduate students. If used for a one-semester course, possible deletions include sections on the following topics: the geometric mean, the life table, the Poisson distribution, the distribution-free approach to intervals, the confidence interval and test of hypothesis for the correlation coefficient, the Kruskal–Wallis test, the trend test for r by 2 contingency tables, the two-way ANOVA and the linear model representation of the ANOVA.

We wish to acknowledge especially useful suggestions and comments provided by Joel A. Harrison and Mary Forthofer. Others who made valuable contributions include Herbert Gautschi, Irene Easling, Anna Baron, Mary Grace Kovar, and the students at the University of Texas School of Public Health Satellite Program in El Paso who reviewed parts or all of the text. Any problems in the text are the responsibility of the authors, not of the reviewers.

CHAPTER 1

Introduction

B*iostatistics* is the application of statistical methods to the biological and life sciences. Statistical methods include procedures for: (1) collecting data, (2) presenting and summarizing data, and (3) drawing inferences from sample data to a population. These methods are particularly useful in studies involving humans because the processes under investigation are often very complex. Because of this complexity, a large number of measurements on the study subjects are usually made to aid the discovery process; however, this complexity and abundance of data often mask the underlying processes. It is in these situations that the systematic methods found in Statistics help create order out of the seeming chaos. Some areas of application are:

1. A collection of vital statistics, for example, mortality rates, used to *inform* about and to *monitor* the health status of the population.
2. Clinical trials to *determine* whether or not a new hypertension medication performs better than the standard treatment for mild to moderate essential hypertension.

3. Surveys to *estimate* the proportion of low-income women of child-bearing age with iron-deficiency anemia.
4. Studies to *examine* whether or not exposure to electromagnetic fields is a risk factor for leukemia.

Biostatistics aids administrators, legislators, and researchers in answering questions. The questions of interest are explicit in examples 2 and 4 above: Is the new drug more effective than the standard and is exposure to the electromagnetic field a risk factor? In examples 1 and 3 the values or estimates obtained are measurements at a point in time which could be used with measures at other time points to determine whether or not a policy change, for example, a 10 percent increase in Medicaid funding in each state, had an effect.

I. DATA: THE KEY COMPONENT OF A STUDY

In this textbook, much of the material relates to methods to be used in the analysis of data. It is necessary to become familiar with these methods and their use as this knowledge will enable one to: (1) better understand reports of studies, and (2) better design and carry out studies. Readers, however, must not let the large number of methods of analysis and the associated calculations presented in this book overwhelm them. More important than the methods used in the analysis is the use of the correct study design and the correct definition and measurement of the study variables. *The key to a good study is good data*! The following examples demonstrate the importance of the data.

Sometimes because of an incomplete understanding of the data or of possible problems with the data, the conclusion from a study may be problematic. For example, consider a study to examine whether or not circumcision status is associated with cancer of the cervix. One issue the researcher must decide is *how to determine the circumcision status*. The easiest way is to ask the male if he had been circumcised; however, Lilienfeld and Graham (1) found that 34 percent of 192 consecutive male patients they studied gave incorrect answers about their circumcision status. Most of the incorrect responses were due to the men not knowing they had been circumcised. Hence the use of a direct question instead of an examination may lead to an incorrect conclusion about the relation between circumcision status and cancer of the cervix.

In the preceding example, reliance on the study subject's memory or knowledge could be a mistake. Yaffe and Shapiro (2) provide another example of potential problems when the study subjects' responses are used. They examined the accuracy of subjects' reports of health care utilization and expenditures for 7 months compared with that shown in their medical and insurance records for two geographical areas. In the Baltimore area, which provided data from approximately 375 households, subjects reported only 73 percent of the identified physician office visits and only 54 percent of the clinic visits. The results for Washington County, Maryland, based on about 315 households, showed 84 percent accuracy for physician office visits but only 39 percent accuracy for clinic visits. Hence the reported utilization of health services by subjects can greatly underestimate the actual utilization and, perhaps more importantly, the accuracy can vary by type of utilization and by population subgroups.

An example of how a wrong conclusion could be reached because of a failure to understand how data are collected comes from Norris and Shipley (3). Figure 1.1 shows the infant mortality rates, calculated conventionally as the ratio of the number of infant deaths to the number of live births during the same period multiplied by 1000, for different racial groups in California and the United States in 1967.

Norris and Shipley questioned the accuracy of the rate for American Indians in California because it was much lower than the corresponding American Indian rate in the U.S., and even lower than the rates of the Chinese- and Japanese-Americans in California. Therefore they used a cohort method to recalculate the infant mortality rates. The cohort rate is based on following all the children that were born in California during a year and observing how many of those infants died before they reached 1 year of age. Some deaths were missed, for example, infants that died out of California, but it was estimated that almost 97 percent of the infant deaths of the cohort were captured in the California death records.

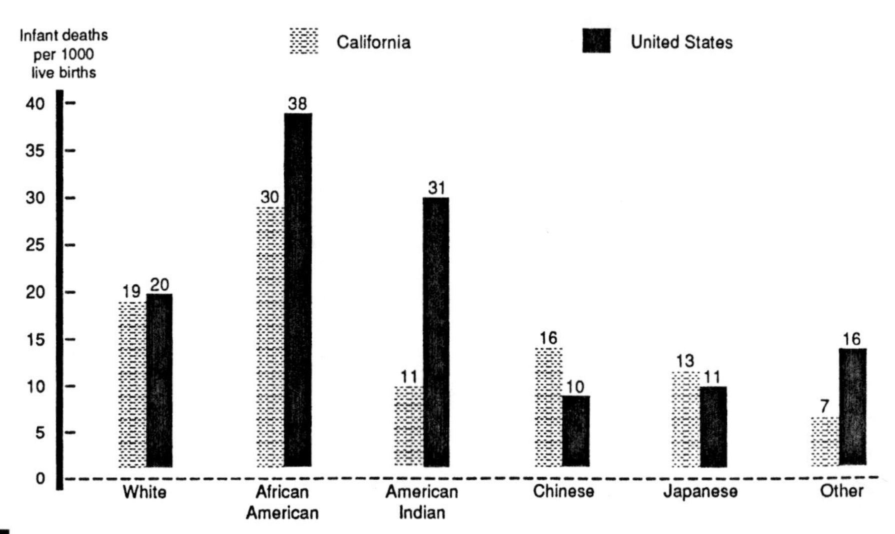

FIGURE 1.1 Infant mortality rates per 1000 live births by race for California and the United States in 1967.

Norris and Shipley used 3 years of data in their reexamination of the infant mortality to provide better stability for the rates. Figure 1.2 shows the conventional and the cohort rates for the 1965–1967 period by race. The use of data from 3 years has not changed the conventional rates much. The conventional rate for American Indians in California is still much lower than the rate for American Indians in the U.S., although now it is slightly above the Chinese- and Japanese-American rates. The cohort rate for American Indians, however, is now much closer to the corresponding rate found in the United States. The rates for the Chinese- and Japanese-Americans and other races have also increased substantially when the cohort method of calculation is used. What is the explanation for this discrepancy in results between these methods of calculating infant mortality rates?

Norris and Shipley attributed much of the difference to how the birth and death certificates, used in the conventional method, were completed. They found that the birth certificate is typically filled out by hospital staff who deal mostly with the mother; hence, the birth certificate usually reflects the race of the mother. The funeral director is responsible for completing the death record and usually deals with the father who may be of a different racial group than the mother. Hence, the racial identification of an infant can vary between the birth and death records—a mismatch of the numerator (death) and the denominator (birth) in the calculation of the infant death rate. The cohort method is not affected by this possible difference because it uses only the child's race from the birth certificate.

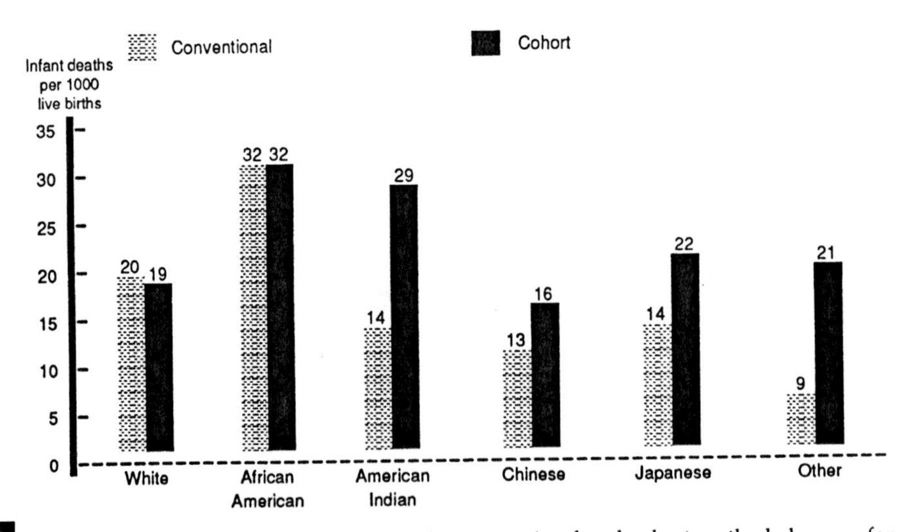

FIGURE 1.2 Infant mortality rates per 1000 live births by conventional and cohort methods by race for California, 1965–1967.

Beginning with the 1989 data year, the National Center for Health Statistics (NCHS) (4, page 53) uses primarily the race of the mother taken from the birth certificate in tabulating data on births. This change should remove the problem caused by having parents from two racial groups in the use of the conventional method of calculating infant mortality rates.

As can be seen, data rarely speak clearly and usually require an interpreter. The interpreter—someone like Norris and Shipley in the earlier example—is someone who is familiar with the subject matter, who understands what the data are supposed to represent, and who knows how the data were collected.

II. REPLICATION: PART OF THE SCIENTIFIC METHOD

Even though most of the examples and problems in this book refer to the analysis of data from a single study, the reader must remember that one study rarely tells the complete story.

Statistical analysis of data may demonstrate that there is a high probability of an *association* between two variables; however, a single study rarely provides proof that such an association exists. Results must be *replicated* by additional studies that eliminate other factors that could have accounted for the relationship observed between the study variables. For example, many studies have examined the role of cigarette smoking in lung cancer and other diseases. Proponents of smoking argue that these studies do not prove that smoking is the cause of lung cancer; however, through the large number of studies, which almost always have found an association between smoking and lung cancer in a wide variety of situations, it has become clear that smoking greatly increases the risk of developing lung cancer.

Another example of the use of replication is provided by the Food and Drug Administration (FDA). The FDA requires a pharmaceutical company to present data from a number of drug trials before it considers the drug. The FDA believes that a single trial does not provide sufficient evidence of the drug's efficacy and safety.

III. CONTENTS

The following chapters continue the theme of combining substantive knowledge with statistical methods. Where possible, we also demonstrate how the figures and calculations being considered can be created or performed on the computer. We believe the computer can be an asset as it removes the burden of the calculations and provides more time for the student to deal with the big picture, that is, the concepts. The computer also provides the capability to experiment, that is, to try different analyses on a set of data.

Chapters 2, 3, and 8 deal with data and data collection methods. Chapter 2 discusses types of data, methods for collecting data, and possible problems with these methods. Chapter 3 covers procedures for collecting observational data via sample surveys, and Chapter 8 discusses the collection of data from designed experiments.

The basic descriptive tools for presenting and summarizing data employed in biostatistics are introduced in Chapter 4. These tools include both numerical and graphical methods and they provide the first step in the attempt to understand the data.

Chapter 5 provides an introduction to probability with illustrations of its use in life tables. In Chapter 6, several probability distributions are introduced and applications of these distributions provided.

The remaining chapters deal with the third component of statistical methods, inference to a population from information about a sample. Chapter 7 presents point and interval estimation of parameters in the population based on the sample data. Chapter 9 presents the concept of hypothesis testing and the related terminology. Chapters 10, 11, and 12 demonstrate the use of procedures that make few assumptions about the data in the testing of hypotheses. Chapter 10 deals with continuous data, and Chapter 11 focuses on methods for examining the relationship between two and three discrete (or categorical) variables. In Chapter 12 some of the procedures introduced in Chapters 10 and 11 are extended and used in the analysis of survival data.

Chapter 13 demonstrates the use of a particular distribution, the normal distribution, in testing hypotheses about means from one and two populations. Chapter 14 presents one-way and two-way analyses of variance, extensions of the material in Chapter 13. These methods of analysis examine the relationship between a continuous response variable and one or more discrete predictor variables. The linear model and its use are also introduced in Chapter 14. Chapter 15 shows the use of the linear model in simple and multiple regression analyses, methods for examining the relationship between a continuous response variable and one or more continuous predictor variables. It also introduces logistic regression analysis, a method for examining the relation between a response variable with two outcomes and one or more predictive variables.

Following these chapters are several appendices. Appendix A presents SAS and Stata statements, either of which can be used in place of the MINITAB statements shown in the test. Appendix B contains several statistical tables that are referenced in the text. Appendix C lists major sources of health data. Appendix D presents solutions to selected exercises.

EXERCISES

1.1. Provide an example from your area of interest in which data collection is problematic or data are misused and discuss the nature of the problem.

1.2. Since 1972 the National Institute on Drug Abuse has periodically conducted surveys in the homes of adolescents on their use of cigarettes, alcohol, and marijuana. In the early surveys, respondents answered the questions aloud. Since 1979 private answer sheets were provided for the alcohol questions. Why do you think the agency made this change? What effect, if any, do you think this change might have had on the proportion of adolescents who reported consuming alcohol during the last month? Would you believe the reported values for the early surveys?

1.3. The infant mortality rate for Pennsylvania for the period 1983–1985 was 10.9 per 1000 live births compared with a rate of 12.5 for Louisiana. Is it appropriate to conclude that Pennsylvania had a better record than Louisiana relative to infant mortality? What other variable(s) might be important to consider here? The infant mortality rate was 9.4 for whites and 20.9 for African Americans in Pennsylvania. This is contrasted with rates of 9.1 and 18.1 for whites and African-Americans, respectively, in Louisiana [rates from (5, Table 15)]. Hence the race-specific rates were lower in Louisiana than in Pennsylvania, yet the overall rate was higher in Louisiana. Explain how this situation could arise.

REFERENCES

1. Lilienfeld, A. M., and Graham, S. (1958). Validity of determining circumcision status by questionnaire as related to epidemiological studies of cancer of the cervix. *J. Natl. Cancer Inst.* (U.S.) **21,** 713–720.
2. Yaffe, R., and Shapiro, S. (1979). Reporting accuracy of health care utilization and expenditures in a household survey as compared with provider records and insurance claims records. Presented at the Spring Meeting of the Biometric Society, Eastern North American Region, April, 1979.
3. Norris, F. D., and Shipley, P. W. (1971). A closer look at race differentials in California's infant mortality, 1965–67. *HSMHA Health Rep.* **86,** 810–814.
4. National Center for Health Statistics (1991). Monthly Vital Statistics Report "Advance Report of Final Natality Statistics, 1989," Vol. 40, No. 8, Supplement.
5. National Center for Health Statistics (1988). "Health, United States, 1987," DHHS Publ. No. 88-1232. Public Health Service, Hyattsville, MD.

CHAPTER 2

Data and Numbers

Appropriate use of statistical procedures requires that we understand the data and the process that generated them. This chapter focuses on data, specifically: (1) the linkage between numbers and phenomena, (2) types of variables, (3) data reliability and validity, and (4) ways data quality can be compromised.

I. DATA: NUMERICAL REPRESENTATION

Any record, descriptive account, or symbolic representation of an attribute, event, or process may constitute a data point. Data are usually measured on a numerical scale or classified into categories that are numerically coded. Three examples are:

1. Blood pressure (diastolic) is measured for all middle and high school students in a school district to learn what percentage of students have a diastolic blood pressure reading greater than 90 mm Hg (data = blood pressure reading).

2. All employees of a large company are asked to report their weight every month to evaluate the effects of a weight control program (data = self-reported weight measurement).
3. The question "Have you ever driven a car while intoxicated?" was asked of all licensed drivers in a large university to build the case for an educational program [data = yes (coded as 1) or no (coded as 0)].

We try to understand the real world, for example, blood pressure, weight, and the prevalence of drunken driving, through data recorded as or converted to numbers. This numerical representation and the understanding of the reality, however, do not occur automatically. It is easy for problems to occur in the conceptualization and measurement processes, which make the data irrelevant or imprecise. Referring to the earlier examples, blood pressure may be measured inaccurately by inexperienced school teachers, those employees who do not measure their weight regularly each month may report inaccurate values, and some drivers may be hesitant to report drunken driving. Therefore, we must not draw any conclusions from the data before we ascertain whether or not any problems exist in the data and, if so, their possible effects. Guarding against misuse of data is as important as learning how to make effective use of data. Repeated exposure to misuses of data may lead people to distrust data altogether. Even a century ago, Bernard Shaw (1) described people's attitudes toward statistical data as follows:

> The man in the street . . . All he knows is that "you can prove anything by figures," though he forgets this the moment figures are used to prove anything he wants to believe.

The situation is certainly far worse today as we are constantly exposed to numbers purported to be important in advertisements, news reporting, and election campaigns. We need to learn to use numbers carefully and to examine critically the meaning of the numbers to distinguish fact from fiction.

II. OBSERVATIONS AND VARIABLES

In statistics, we observe or measure characteristics, called *variables*, of study subjects, called *observational units*. For each study subject, the numerical values assigned to the variables are called *observations*. For example, in a study of hypertension among school children, the investigator measures systolic and diastolic blood pressures for each pupil; *systolic and diastolic blood pressure are the variables, the blood pressure readings are the observations, and the pupils are the observational units*. We usually observe more than one variable on each unit, for example, in a study of hypertension among 500 schoolchildren, we may record the pupil's age, height, and weight in addition to the two kinds of blood pressure readings. In this case we have a data set of 500 students with observations recorded on each of five variables for each student or observational unit.

III. SCALES USED WITH VARIABLES

Four scales are used with variables: nominal, ordinal, interval, and ratio. The scales are defined in terms of the information conveyed by the numerical values assigned to the variable. The distinction between the scales is not of crucial importance. These scale types have frequently been used in the literature, and we are presenting them to be sure that the reader understands the terms.

In some cases the numbers are simply indicators of a category. For example, when considering gender, 1 may be used to indicate that the person is female and 2 to indicate that the person is male. When the numbers merely indicate to which category a person belongs, a *nominal scale* is being used. Hence gender is measured on a nominal scale. It makes no difference what numerical values are used to represent females and males.

In other cases the numbers represent an ordering or ranking of the observational units on some variable. For example, from a worker's job description or work location, it may be possible to estimate the exposure to asbestos in the workplace, with 1 representing low, 2 representing medium, and 3 representing high exposure. In this case, the exposure to asbestos variable is measured on the *ordinal scale*. Values of 10, 50, and 100 could have been used instead of 1, 2, and 3 for representing the categories of low, medium, and high. The only requirement is that the order is maintained.

Other variables are measured on a scale of equal units, for example, temperature in degrees Celsius (*interval scale*) or height in centimeters (*ratio scale*). There is a subtle distinction between interval and ratio scales, and it is that a ratio scale has a zero value, which means there is none of the quantity being measured. For example, zero height means there is no height, but zero degrees Celsius does not mean there is no heat. When a variable is measured on a ratio scale, the ratio of two numbers is meaningful. For example, a boy 140 centimeters (cm) tall is 70 cm taller and also twice as tall as a boy 70 cm tall. In contrast, temperature in degrees Celsius is an interval variable, but not a ratio variable because an oven at 300 degrees is not twice as hot as one at 150 degrees. This distinction between interval and ratio scales is of little importance in statistics and both are measured on a scale continuously marked off in units.

These different scales give rise to three types of data: nominal (categorical), ordinal (ordered) data, and continuous (interval or ratio) data. The scale used often depends more on the method of measurement or the use made of it than on the property measured. The same property can be measured on different scales; for example, age can be measured in years (ratio scale); placed into young, middle-aged, and elderly age groups (ordinal scale); or classified as economically productive (ages 16 to 64) and dependent (under 16 and over 64) age groups (nominal scale). It is possible to convert a higher-level scale (ratio or interval) into a lower-level scale (ordinal and nominal scales), but not to convert from a lower level to a higher level. One final point is that all recorded measurements themselves are discrete. Age, for example, can be measured in years, months, or even hours, but it is still measured in discrete steps. It is possible to talk about a continuous variable, yet actual measurements are limited by the measuring instruments.

IV. RELIABILITY AND VALIDITY

Data are collected by direct observation or measurement and from responses to questions. For example, height, weight, and blood pressure of schoolchildren are directly measured in a health examination. The investigator is concerned about accurate measurement. The measurement of height and weight sounds easy, but the measurement process must be well defined and used consistently. For example, height is to be measured without shoes and weight measured before a meal. Therefore, to understand any measurement we need to know the operational definition, that is, the actual procedures used in the measurement. In measuring blood pressure, the investigator must specify what instrument is to be used, how much training will be given to the measurers, at what time of the day the blood pressure should be measured, in what position it will be measured (sitting or standing), and how many times it should be measured for each pupil.

There are two issues in specifying operational definitions: reliability and validity.

Reliability requires that the operational definition should be sufficiently precise so that all persons using the procedure or repeated use of the procedure by the same person will have the same or approximately the same results. If the procedures for measuring height and weight of students are reliable, then the values measured by two observers, say, the teacher and the nurse, will be the same. If the person reading the blood pressure is hard of hearing, the diastolic blood pressure values, recorded at the point of complete cessation of the Korotkoff sounds or, if no cessation, at the point of muffling, may not be reliable.

Validity is concerned with the appropriateness of the operational definition, that is, whether or not the procedure measures what it is supposed to measure. For example, if a biased scale is used, the measured weight is not valid, even though the repeated measurements give the same results. Another example of a measurement that may not be valid is the blood pressure reading obtained when the wrong size cuff is used. In addition, the person reading the blood pressures may have a digit preference which also threatens validity. The data shown in Figure 2.1 from Forthofer (2) suggest that there may have been a digit preference in the blood pressure data for children and adolescents in the second National Health and Nutrition Examination Survey (NHANES II). This survey, conducted by the NCHS in 1976–1980, provides representative health and nutrition data for the noninstitutionalized U.S. population. In this survey, the blood pressure values ending in zero have a much greater frequency of occurrence than the other values.

The reliability and validity issues are not only of concern for data obtained from measurements, but also for data obtained from questionnaires. In fact, the concern may be greater because of the larger number of

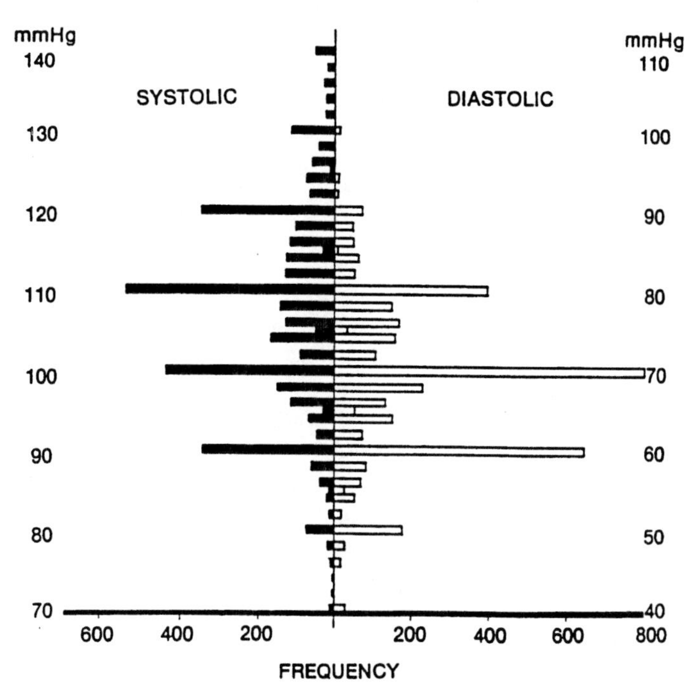

FIGURE 2.1 Blood pressure values (first reading) for 4053 children and adolescents in NHANES II. From Forthofer (2).

ways that problems threatening data accuracy can be introduced with questionnaires (3–5). One problem is that the question may be misinterpreted, and thus a wrong or irrelevant response may be elicited. For example, in a mail survey, a question used the phrase "place of death" instead of instructing the respondent to provide the county and state where a relative had died. One person responded that the deceased died in bed. Such problems can be avoided or greatly reduced if careful thought goes into the design of questionnaires and into the preparation of instructions for the interviewers and the respondents. Even when there are no obvious faults in the question, however, a different phrasing may elicit a different response. For example, age can be ascertained by asking age at the last birthday or date of birth. It is known that the question about the date of birth tends to obtain the more accurate age.

Another problem often encountered is that many people are uncomfortable in appearing to be out of step with society. As a result, these people may provide a socially acceptable but false answer about their feelings on an issue. A similar problem is that many people are reluctant to provide accurate information regarding personal matters, and often the respondent refuses to answer or intentionally distorts the response. Some issues are particularly sensitive, for example, questions about whether a woman has had an abortion or whether a person has attempted suicide. The responses, if any are obtained, to these sensitive questions are of questionable accuracy. The following section addresses one way of obtaining data on sensitive issues that should be accurate.

V. RANDOMIZED RESPONSE TECHNIQUE

There is a statistical technique that allows investigators to ask sensitive questions, for example, about drug use or driving under the influence of alcohol, in a way that should elicit an honest response. It is designed to protect the privacy of individuals and yet provide valid information. This technique is called randomized response (6,7) and has been used in surveys about abortions, drinking and driving, drug use, and cheating on examinations.

In this technique, a sensitive question is paired with a nonthreatening question and the respondent is told to answer only one of the questions. The respondent uses a chance mechanism, for example, the toss of a coin, to determine which question is to be answered and only the respondent knows which question was answered. The interviewer records the response without knowing which question was answered. It appears that these answers are of little value, but the following example demonstrates that they can be of use.

In the drinking and driving situation, the sensitive question is, "Have you driven a car while intoxicated during the last 6 months?" This question is paired with an unrelated, nonthreatening question such as, "Were you born in either September or October?" Each respondent is asked to toss a coin and not to reveal the outcome; those with heads are asked to answer the sensitive question and those with tails to answer the nonthreatening question. The interviewer records the yes or no response without knowing which question is being answered. Because only the respondent knows which question has been answered, there is less reason to answer dishonestly.

Suppose 36 people were questioned and 12 gave yes answers. At first glance, this information does not seem very useful because we do not know which question was answered; however, Figure 2.2 shows how we can use this information to estimate the proportion of the respondents who had been driving while intoxicated during the past 6 months.

As a fair coin was tossed by each respondent, we expect that half the respondents answered the drunk driving question and half answered the birthday question. We also expect that one-sixth (2 of 12 months) of those who answered the birthday question will give a yes response. Hence the number of yes responses from the birthday question should be 3 [(36/2) ×

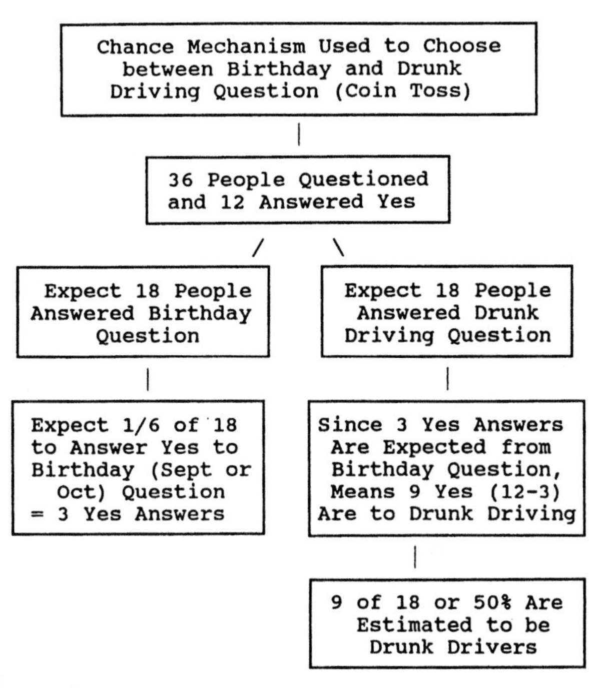

FIGURE 2.2 Use of randomized response information.

(1/6)]; the expected number of yes responses to the drinking and driving question is 9 (the 12 yes answers minus the 3 yes answers from the birthday question). Then the estimated proportion of drunk drivers is 50 percent (9/18).

There is no way of proving that the respondents answered honestly, but they are more likely to tell the truth when the randomized response method was used rather than the conventional direct question. Note that the data gathered by the randomized response technique cannot be used without understanding the process by which the data were obtained. Individual responses are not informative but the aggregated responses can provide useful information at the group level. Of course, we need to include a sufficiently large number of respondents in the survey to make the estimate reliable.

VI. COMMON DATA PROBLEMS

Examination of data can sometimes provide evidence of poor quality. Some clues to poor quality include many missing values, impossible or unlikely values, inconsistencies, irregular patterns, and suspicious regularity. Data with too many missing values will be less useful in the analysis and may indicate that something went wrong with the data collection process. Sometimes data contain extreme values that are seemingly unreasonable. For example, a person's age of 120 would be suspicious and 200 would be impossible. Missing values are often coded as 99 or 999 in the data file and these may be mistakenly interpreted as valid ages. The detection of numerous extreme ages in a data set would cast doubt on the process by which the data were collected and recorded and, hence, on all other observations, even if they appear reasonable.

Inconsistencies are often present in the data set. For example, a college graduate's age of 15 may appear inconsistent with the usual progress in school, but it is difficult to attribute this to an error. Some inconsistencies are obvious errors. An example can be found in the history of the U.S. population census. In an attempt to study community mental health, Edward Jarvis (1803–1884) discovered that there were numerous inconsistencies in the 1840 population census reports; for example, in many towns in the North, the numbers of African-American "insane and idiots" were larger than the total numbers of African-Americans in those towns. He published the results in medical journals and demanded that the federal government take remedial action. This demand led to a series of statistical reforms in the 1850 population census (8).

A careful inspection of data sometimes reveals irregular patterns. For example, ages reported in the 1945 census of Turkey have a much greater frequency of multiples of 5 than numbers ending in 4 or 6 and more even-numbered ages than odd-numbered ages (9), as shown in Figure 2.3. This tendency of digit preference in age reporting is quite common. Even in the U.S. census we can find a slight clumping or heaping at age 65 when most of the social benefit programs for the elderly begin. The same phenomenon of digit preference is often found in laboratory measurements, as was shown above with the blood pressure measurements in NHANES II.

Large and consistent differences in the values of a variable may indicate that there was a change in the measurement process that should be investigated. An example of large differences is found in data used in the Report of the Second Task Force on Blood Pressure Control in Children, 1987 (10). Systolic blood pressure values for 5-year-old boys averaged 103.5 mm Hg in a Pittsburgh study compared with 85.6 mm Hg in a Houston study. These averages were based on 61 and 181 boys aged 5 in the Pittsburgh and Houston studies, respectively. Hence these differences were not due to small sample sizes. Similar differences were seen for 5-year-old girls and for 3- and 4-year-old boys and girls as well. There are large differences between other studies also used by this task force, but the differences are smaller for older children. These incredibly large differences between the Pittsburgh and Houston studies were likely due to a

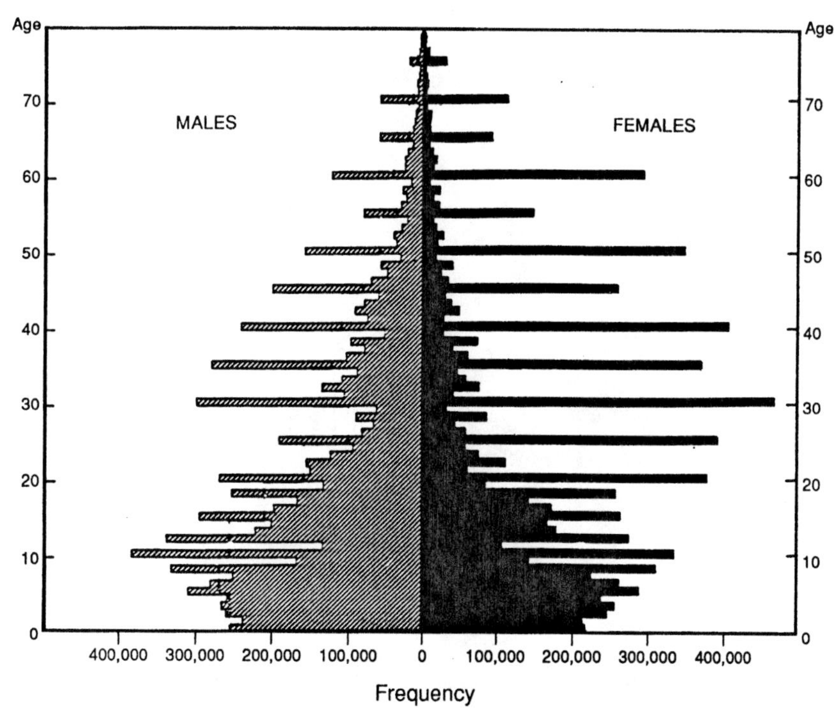

FIGURE 2.3 Population of Turkey, 1945, by sex and by single years of age.

difference in the measurement process. In the Houston study, the children were at the clinic at least 30 minutes before the blood pressure was measured, whereas they had a much shorter wait in the Pittsburgh study. Because the measurement processes differed, the values obtained do not reflect the same variable across these two studies. The use of data from these two studies without any adjustment for the difference in the measurement process is questionable.

The use of data from laboratories is another area in which it is crucial to monitor constantly the measurement process, that is, the equipment and the personnel who use the equipment. In large multicenter trials that use different laboratories, or even a single laboratory, referent samples are routinely sent to the laboratories to determine if the measurement processes are under control. This enables any problems to be detected quickly and prevents subjects from being either unduly alarmed or wrongly comforted. It also prevents false values from being entered into the data set.

The Centers for Disease Control (CDC) has an interlaboratory program, and data from it demonstrate the need for monitoring. The CDC distributes samples to about 100 laboratories throughout the United States. The April 1980 results of measuring lead concentration in blood are shown in Figure 2.4 (11). According to the author of the article, the best estimate of the blood lead concentration in the distributed sample was 41 micrograms per deciliter (μg/dl) but the average reported by all participating laboratories was 44 μg/dl. The large variability from the value of 41 shown in Figure 2.4 is a reason for concern, particularly because the usual value in human blood lies between 15 and 20 μg/dl.

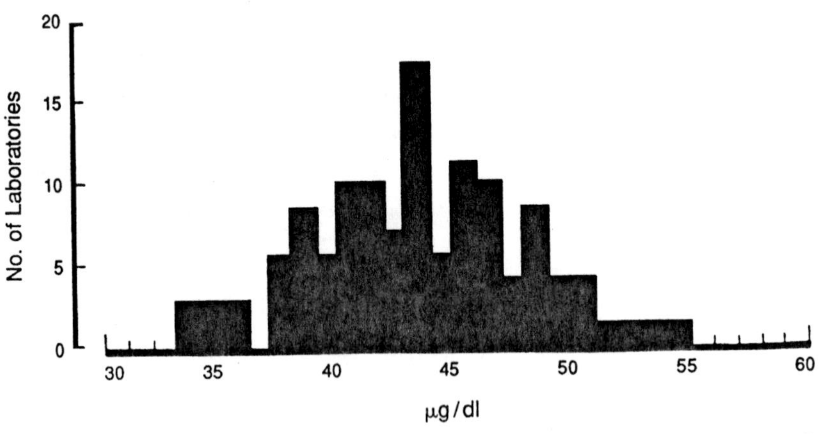

FIGURE 2.4 Distribution of measurements of blood lead concentration by separate laboratories, Centers for Disease Control.

Of course the lack of inconsistencies and irregularities does not mean that there are no problems with the data. Too much consistency and regularity sometimes are grounds for a special inquiry into the causes. Scientific frauds have been uncovered in some investigations in which the investigator discarded data that did not conform with theory. Abbe Gregor Mendel, the 19th-century monk who pioneered modern gene theory by breeding and crossbreeding pea plants, came up with such perfect results that later investigators concluded he probably tailored his data to fit predetermined theories. Another possible fabrication of data in science is the case of Sir Cyril Burt, a British pioneer of applied psychology. In his frequently cited studies of intelligence and its relation to heredity, he reported the same correlation in three studies of twins with different sample sizes (0.771 for twins reared apart and 0.944 for twins reared together). The consistency of his results eventually raised concern as it is highly unlikely that the exact same correlations would be found in studies of humans with different sample sizes. Science historians generally agree that his analyses were creations of his imagination with little or no data to support them (12).

VII. CONCLUDING REMARKS

Data are a numerical representation of a phenomenon. By assigning numerical values to occurrences of the phenomenon, we are thus able to describe and analyze it. The assignment of the numerical values requires an understanding of the phenomenon and careful measurement. In the measurement process, some unexpected problems may be introduced and the data then contain the intended numerical facts as well as the unintended fictions. Therefore we cannot use data blindly. The meaning of data and its implications have been explored in a number of examples in this chapter. In the next chapter, we consider some ways data are obtained.

EXERCISES

2.1. Identify the scale used for each of the following variables:
   a. Calories consumed during the day
   b. Marital status
   c. Perceived health status reported as poor, fair, good, or excellent
   d. Blood type
   e. IQ score

2.2. A person's level of education can be measured in several ways. It could be recorded as years of education or it could be treated as an ordinal variable, for example, less than high school, high school graduate, and so on. Is it always better to use years of education than the ordinal variable measurement of education? Explain your answer.

2.3. In a health interview survey, a large number of questions are asked. For the following items, discuss: (1) how the variable should be defined operationally, (2) whether nonresponse is likely to be high or low, and (3) whether reliability is likely to be high or low. Explain your answers.
   a. Weight
   b. Height
   c. Family income
   d. Unemployment
   e. Number of stays in mental hospitals

2.4. The pulse is usually reported as the number of heartbeats per minute, but the actual measurement can be done in several different ways, for example:
   a. Count for 60 seconds
   b. Count for 30 seconds and multiply the count by 2
   c. Count for 20 seconds and multiply the count by 3
   d. Count for 15 seconds and multiply the count by 4

   Which procedure would you recommend to be used in clinics, considering accuracy and practicality?

2.5. The first U.S. census was taken in 1790 under the direction of Thomas Jefferson. The task of counting the people was given to 16 federal marshals who in turn hired enumerators to complete the task in 9 months. In October of 1791, all of the census reports had been turned in except the one from South Carolina, which was not received until March 3, 1792. As can be expected, the marshals encountered many obstacles and the counting was incomplete. The first census revealed a population of 3,929,326. This result was viewed as an undercount as is indicated in the following excerpt from a letter written by Jefferson:

> I enclose you also a copy of our census, written in black ink, so far as we have actual returns, and supplied by conjecture in red ink, where we have no returns; but the conjectures are known to be very near the truth. Making very small allowance for omissions, which we know to have been very great, we are certainly above four millions, (13)

   Discuss what types of obstacles they might have encountered and what might have led Jefferson to believe there was an undercounting of the people.

2.6. The NCHS matched a sample of death certificates in 1960 with the 1960 population census records to assess the quality of data and reported the following results (14):

Percentage Agreement and Disagreement in Age Reporting, 1960

	White		Nonwhite		
	Male	*Female*	*Male*	*Female*	*Total*
Agreement	74.5%	67.9%	44.7%	36.9%	68.8%
Disagreement					
1-year difference	16.6	18.8	20.8	20.2	17.8
2+-year difference	8.9	13.3	34.5	42.9	13.4

Do you think that age reported in the death certificate is more accurate than that reported in the census? How do you explain the differential agreement by gender and race? How do you think these disagreements affect the age-specific death rates calculated by single years and those computed by 5-year age groups?

2.7. Discuss possible reasons for the digit preference in the 1945 population census of Turkey that is shown in Figure 2.3. Why was the digit preference problem more prominent among females than among males? How would you improve the quality of age reporting in census or surveys? How do you think the digit preference affects the age-specific rates calculated by single years of age and those computed by 5-year age groups?

REFERENCES

1. Shaw, B. (1909). Preface on doctors. "The Doctor's Dilemma," p. lxix. Brentano's, New York.
2. Forthofer, R. N. (1991). Blood pressure standards in children. Paper presented at the American Statistical Association meeting, August, 1991. See Appendix C for details of the National Health and Nutrition Examination Survey.
3. Suchman, L., and Jordan, B. (1990). Interactional troubles in face-to-face survey interviews. *J. Am. Stat. Assoc.* **85,** 232–241.
4. Marquis, K. H., Marquis, M. S., and Polich, J. M. (1986). Response bias and reliability in sensitive topic surveys. *J. Am. Stat. Assoc.* **81,** 381–389.
5. Juster, F. T. (1986). Response errors in the measurement of time use. *J. Am. Stat. Assoc.* **81,** 390–402.
6. Campbell, C., and Joiner, B. L. (1973). How to get the answer without being sure you've asked the question. *Am. Stat.* **27,** 229–231.
7. Warner, S. L. (1965). Randomized response: A survey technique for eliminating evasive answer bias. *J. Am. Stat. Assoc.* **60,** 63–69.
8. Regan, O. G. (1973). Statistical reforms accelerated by sixth census errors. *J. Am. Stat. Assoc.* **68,** 540–546. See Appendix D for details of the population census.
9. United Nations (1955). "Methods of Appraisal of Quality of Basic Data for Population Estimates," Popul. Stud. No. 23, p. 34. United Nations Dept. of Economic and Social Affairs, New York.

10. NHLBI Task Force on Blood Pressure Control in Children (1987). The report of the second task force on blood pressure control in children, 1987. *Pediatrics* **79,** 1–25.
11. Hunter, J. S. (1980). The national system of scientific measurement. *Science* **210,** 869–874.
12. Gould, S. J. (1981). "The Mismeasure of Man." Norton, New York.
13. Washington, H. A., ed. (1853). "The Writings of Thomas Jefferson," **III,** 287. Taylor & Maury, Washington, D.C.
14. National Center for Health Statistics (1968). "Vital and Health Statistics," Ser. 2. Public Health Service, Washington, U.S. Government Printing Office.

CHAPTER 3

Sampling

In meeting a set of data we must first check the credentials of the data, that is, what the data represent and how the data were collected. In Chapter 2 we discussed the linkage between concepts and numbers, that is, what the data represent. As far as data collection is concerned, two basic methods are used to obtain data, the sample survey and the designed experiment. In this chapter we examine the sample survey; in Chapter 8, we consider the designed experiment.

I. WHAT AND WHY SAMPLING

Sampling means selecting a few units from all the possible observational units in the population. The idea of sampling is not new to us because we all use some form of sampling in our daily life. For example, in buying fruit from the produce section, we examine (sample) several pieces of the fruit before deciding whether or not to make a purchase. If we examine only the fruit at the top of the basket, we sometimes make a wrong decision about the quality of the fruit. It turns out that certain sampling methods tend to cause fewer wrong decisions (introduce less bias) than others.

For practical purposes, any data set is a sample. Even if a complete census is attempted, there are missing observations. This means that we must pay attention to the intended as well as the unintended sampling when evaluating a sample. This also suggests that we cannot evaluate a sample by looking at the sample itself, but we need to know what sampling method was used and how well it was executed. We are interested in the process of selection as well as the sample obtained.

Sampling is used extensively today for many reasons. In many situations a sample produces information about the population more accurate than that provided by a census. Two reasons for obtaining more accurate information from a sample are the following. As was mentioned in Chapter 2, a census often turns out to be incomplete and the impact of the missing information is most often unknown. Additionally, in obtaining a sample, fewer interviewers are required and it is likely that they will be better trained than the huge team of interviewers required to perform a census.

Even more pragmatically, collecting data from a sample is cheaper and faster than attempting a complete census. In addition, in many situations a census is impractical or even impossible. The following three examples illustrate situations in which sampling was used and reasons for the use of samples.

1. Even in the U.S. population census, many data items are collected from a sample of households. In the 1990 census, for example, only a few basic demographic data items—gender, age, race, and marital status—were asked of each individual in all households in the short form of the questionnaire. Many questions about socioeconomic characteristics such as education, income, and occupation are included in the long form, which was distributed to about 17 percent of U.S. households. In small towns, a larger proportion of households received the long form to ensure reliable estimates. Conversely, in large cities, proportionately fewer households received the long form. Use of sampling not only reduced the cost of the census, but also shortened the data collection burden and time.

2. Pharmaceutical companies routinely sample a small fraction of their products to examine the quality and the chemical contents. On the basis of this examination, a decision is made whether to accept the entire lot and ship it or reject the lot and change the manufacturing process. In this case the sample is destroyed to check the quality; a company cannot afford to inspect the entire lot.

3. Health departments of large urban areas monitor ambient air quality. As the health department cannot afford to monitor the air everywhere in its coverage area, a sample of sites are selected and the values of several different pollutants are continuously recorded.

II. SAMPLING AND SELECTION BIAS

A smart shopper is conscious of the possible variability in the quality of fruit between the top and bottom of the fruit basket. The smart shopper looks at pieces of fruit throughout the basket, even though it is more convenient to look at the pieces on top, before making a purchase. In the same way, a researcher is aware of the possible variability among observational units in the population. A good researcher takes steps to ensure that the process for selecting units from the population deals with this possible variability. The failure to do so means that the selected sample may not adequately represent the population.

Selecting a sample of units because of convenience also poses a problem for a researcher just as it did for the shopper. The opinions of people interviewed during lunch time on downtown street corners, although convenient to obtain, usually are not representative of the residents of the city. Those who never go to the center of the city during lunch time are not represented in the sample and they may have different opinions from those who go to the city center.

Before performing any sampling, it is important to define clearly the population of interest. Similarly, when we are given a set of data, we need to know what group the sample represents, that is, from what population the data were collected. The definition of population is often implicit and assumed to be known, but we should ask what the population was before using the data or accepting the information. When we read an election poll, we should know whether the population was all adults or all registered voters to interpret the results appropriately. In practice, the population is defined by specifying the *sampling frame*, the list of units from which the sample was selected. Ideally, the sampling frame should include all units of the defined population. But, as we shall see, it is often difficult to obtain the sampling frame and we need to rely on a variety of alternative approaches.

The failure to include all units contained in the defined population in the sampling frame leads to selecting a biased sample. A biased sample is not representative of the population. The average of a variable obtained from a biased sample is likely to be consistently different from the corresponding value in the population. *Selection bias* is the consistent divergence of a sample value (*statistic*) from the corresponding population value (*parameter*) because of an improper selection process. Even with a complete sampling frame, selection bias can occur if proper selection rules were not followed. Two basic sources of selection bias are the use of an incomplete sampling frame and the use of improper selection procedures. The following example illustrates the importance of the sampling frame.

In the 1936 presidential election, the *Literary Digest* confidently predicted that the Republican nominee, Alfred M. Landon, would defeat the

Democratic incumbent, Franklin D. Roosevelt (1). This prediction was based on 2.3 million returns out of 10 million survey ballots mailed to the magazine's subscribers, telephone customers, and persons on other mailing lists. The prediction was wrong and there were several causes of this mistake. One of the key causes was the use of an incomplete sampling frame of eligible voters, which resulted in a biased sample. The *Literary Digest*'s mailing list overrepresented people with high incomes. A second problem was the low response rate of 23 percent, which meant there was the possibility of a large *nonresponse bias*, a type of selection bias. When there is nonresponse, it means that the respondents were self-selected and hence might not adequately represent the sampling frame.

The Report of the Second Task Force on Blood Pressure Control in Children provides another example of the possibility of selection bias in data (2). This task force used existing data from several studies, only one of which could be considered representative of the U.S. noninstitutionalized population. In this convenience sample, more than 70 percent of the data came from Texas, Louisiana, and South Carolina, with little data from the Northeast or West. Data from England were also used for newborns and children up to 3 years of age. The representativeness of these data for use in the creation of blood pressure standards for U.S. children is questionable. Unlike the *Literary Digest* survey, in which the errors in the sampling were shown to lead to the wrong conclusion, it is not clear that the blood pressure standards are wrong. All we can point to is the use of convenience sampling and, with it, the likely introduction of selection bias by the Second Task Force.

Telephone surveys may provide another example of failure of the sampling frame to include all members of the target population. If the target population is all the resident households in a geographical area, a survey conducted using the telephone will miss a portion of the resident households. Even though more than 90 percent of the households in the United States have telephones, the percentage varies with race and socioeconomic status. The telephone directory was used frequently in the past as the sampling frame, but it excluded households without telephones as well as households with unlisted numbers. A technique called *random digit dialing* (RDD) has been developed to deal with the unlisted number problem in an efficient manner (3). As the name implies, telephone numbers are basically selected at random from the prefixes—the first 3 digits—thought to contain residential numbers, not from a telephone directory. But the concern about the possible selection bias resulting from missing households without telephones and people who do not have a stable place of residence remains.

To avoid or minimize selection bias, every sample needs to be selected on the basis of a carefully drawn sample design. The design defines the population the sample is supposed to represent, identifies the sampling frame from which the sample is to be selected, and specifies the procedural rules for selecting units. The sample data are then evaluated based on the sample design and the way the design was actually executed. The next section introduces the key to modern sampling methods, the introduction of randomness into the sample selection process.

III. IMPORTANCE OF PROBABILITY SAMPLING

We are familiar with the use of a random mechanism to remove possible biases. For example, to start a football game, a coin toss—a random mechanism—is used to decide which team receives the opening kickoff. A random or chance mechanism is also used to select a sample in an attempt to remove biases. Any sample selected using a random mechanism that results in *known chances* of selection of the observational units is called a *random* or *probability sample*. This definition requires only that the chances of selection are known. It does not require that the chances of the observational units being selected into the sample are equal.

Knowledge of the chance of selection is the basis for the statistical inference from the sample to the population. A sample selected with unknown chances of selection cannot be linked appropriately to the population from which the sample was drawn. This point will become clearer when we study probability in Chapter 5 and probability distributions in Chapter 6.

IV. SIMPLE RANDOM SAMPLING

The simplest probability sample is a *simple random sample* (SRS). In a SRS, each unit in the sampling frame has the same chance of being included in the sample as any other unit. Use of a SRS removes the possibility of any bias, conscious or unconscious, on the part of the researcher in selecting the sample from the sampling frame.

One method of drawing a SRS is to place numbered slips of paper in an urn, mix them up thoroughly, and then have a neutral party pick out the slips. This is basically the method the Selective Service officials attempted to use in the 1970 draft lottery. Figure 3.1 shows the lottery results (4). It appears that the process did not work as intended as the months at the end of the year, which were put into the container last and were not mixed thoroughly, have much smaller lottery numbers than the earlier months. The unreliability of this traditional method of selecting a SRS has also been demonstrated empirically. Problems often result because it is difficult to mix the slips thoroughly enough to approximate a random selection.

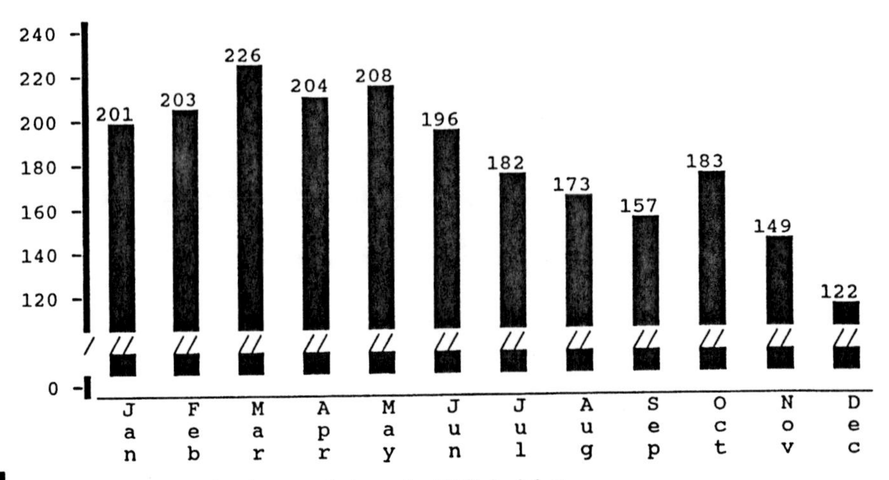

FIGURE 3.1 Average lottery number by month from the 1970 draft lottery.

A better method of selecting a SRS is to use a random number table or random numbers generated by a computer. If the population is relatively small, we can number all units sequentially. Next we locate a starting point in the random number table, Table B1 in Appendix B. We then begin reading random numbers in some systematic fashion, for example, across a row or down a column or diagonally, but the direction of reading should be decided ahead of the time. The units in the sampling frame whose unique numbers match the random numbers that have been read are selected into the sample.

For example, suppose that we have 50 students in a classroom and they are sequentially labeled from 00 to 49 by row, starting at the left end of the first row. We wish to select a SRS of 10 students. We decide to use the upper left-hand corner of the table as our starting point and we go across the row. By reading the two-digit numbers from the first row of the random digit table, the following 10 numbers are obtained:

17, 17, 47, 59, 08, 43, 30, 67, 70, 61.

As four numbers are greater than 49, they cannot be used and we must draw additional numbers until we have 10 random numbers smaller than 50. In addition, the number 17 occurred twice. If we were to use the value of the variable of interest for student 18 twice, the sample would be called a *sample with replacement*. As there is no good practical reason for including the same element twice in the sample, we should draw another number that has not been selected previously. A sample that does not allow duplicate selections is called a *sample without replacement*. We usually sample without replacement, but this distinction is moot when selecting a sample from large populations because the chance of selecting a unit more than once would be very small. The next five valid numbers are 07, 44, 48, 36, and 47. The students whose labels match the 10 valid numbers drawn are selected as the sample.

Another way of dealing with this problem of drawing invalid numbers is to subtract 50 from values greater than or equal to 50 in the first set of 10 random numbers. For example, 59, 67, 70, and 61 become 09, 17, 20, and 11. We now select the students with labels 09, 17, 20, and 11. This procedure is based on the premise that each student is represented by two numbers differing by 50 in value. For example, the first student will be selected if either 00 or 50 were read; the second would be selected if either 01 or 51 were read and so on, until the last student would be selected if 49 or 99 were read. Note that even with the subtraction of 50, we again have another 17. We would still have to draw other random numbers until we had 10 distinct values.

In using this second procedure (subtracting 50), each unit (student) in the sampling frame had the same number (two) of labels associated with it. If there are 30 students in a class, we can label them in three cycles, 1 through 30, 31 through 60, and 61 through 90, but we cannot assign 91 through 99 and 00 to any student. If we assigned these last 10 values to some of the students, some students would have 3 labels associated with them whereas other students would have 4 labels. The students would have unequal chances of being selected. By not using the last 10 values, each student has 3 labels (numbers). The first student is assigned the numbers 01, 31, and 61; the second student is assigned the numbers 02, 32, and 62 and so on for the other students.

Let us take another sample of 10 students from the original group of 50 students. We now are going to use two-digit numbers from the beginning of the third row. The set of 10 numbers are the following:

24, 04, 13, 38, 00, 09, 97(47), 63(13), 67(17), 85(35).

The value 50 was subtracted from numbers greater than 49 and the values in parentheses are the result of the subtraction and they indicate which students are to be selected. Because the fourteenth student is selected twice, that is, the number 13 appears twice, additional numbers have to be selected until there are 10 distinct values.

In this example, we used two-digit random numbers because we could not provide distinct labels for all 50 students with only a single digit. The number of digits to be used is dependent on the size of the population under consideration. For example, when we have 570 units in the population, we need to use three digits. A population which contains 7870 units would require four digit random numbers.

V. COMPLEX SAMPLING DESIGNS

The idea of simple random sampling is essential in statistical thinking and most methods of analysis assume that the data were collected using a SRS; however, when we attempt to use a SRS in the collection of data, we often encounter difficulties. Suppose we wanted a SRS of 500 adults from a large city. First, a sampling frame is not readily available. Developing a list of all adults in the city is very costly and should be considered impractical. Even though we are able to select a SRS of 500 adults from a reasonably complete list, it would be expensive to send interviewers to sample persons scattered all over the city. A solution to these practical difficulties is to sample people based on geographical areas, for example, census tracts. Most survey agencies and researchers use a *multistage sample design* in this situation. First, a random sample of census tracts is selected, then blocks within each selected tract are randomly selected. Within the selected blocks a list of households can be prepared and a sample of households can be selected systematically from the list, say, every third household. Finally, within each of the selected households, an adult may be randomly chosen.

In the above sampling design, elementary units (individuals) in the population are grouped into *clusters*, for example, groups of households, of blocks and of tracts that usually are close together. The clusters are then sampled, which reduces the travel time and cost of the sampling.

In addition to the use of clusters, *stratification* is often used in complex sample designs. In a *stratified random sample* design, the units in the sampling frame are first divided into groups, called *strata* and a separate SRS is taken in each stratum to form the total sample. The strata are formed to keep similar units together, for example, a female stratum and a male stratum. In this design, units need not have equal chances of being selected and some strata may be deliberately oversampled. For example, in NHANES I, the elderly, persons in poverty areas, and women of childbearing age were oversampled to provide sufficient numbers of these groups for in-depth analysis (5). If a SRS had been used, it is likely that too few people in these groups would have been selected to allow any in-depth analysis.

Another advantage of stratification is that it can reduce the variability of sample statistics over that of a SRS, thus reducing the sample size required for analysis. This reduction in variability occurs when the units in a stratum are similar, but there is variation across strata. Another way of saying this is that the reduction occurs when the variable used to form the strata is related to the variable being measured. Let us consider a small example that illustrates this point.

In this example, we wish to estimate the average weight of persons in the population. The population contains six persons, three females and three males. The weights of the females in the population are 110, 120, and

130 pounds and the weights of the males are 160, 170, and 180 pounds. We form our estimate of the population average weight by taking a sample of size 2 without replacement.

If we use a SRS, the smallest possible estimate is 115 pounds [= (110 + 120)/2] and the largest possible estimate is 175 [= (170 + 180)/2]. As an alternative, we could use a stratified random sample where the strata are formed on the basis of gender. If one person is randomly selected from each stratum, the smallest estimate is 135 pounds [= (110 + 160)/2] and the largest estimate is 155 pounds [= (130 + 180)/2]. The estimates from the stratified sample approach have less variation, that is, have greater precision, than those from the SRS approach.

A stratified random sample is often taken in the early stages in multistage sampling. For instance, in the earlier example of multistage sampling, the list of census tracts at the first stage of sampling could have been stratified by the degree of minority population concentration or by the median years of education in the tract. A separate SRS from each stratum could have then been selected. Similarly, stratification can be applied at the block and household levels.

The sample design can be more complicated than illustrated in earlier examples. The additional complications are introduced for a variety of reasons, for example, to control costs, to save time, to take known sources of variation into account, and to improve precision of sample estimates. In these more complex sample designs, the selection probabilities are unequal; the sampling unit may be a cluster of households, not a person, and hence persons in the sample are related to other persons by virtue of belonging to the same cluster. The data collected from these more complex sample designs require different analyses than data from a SRS; however, it is beyond the scope of this textbook to deal with the more complicated analysis of data from these complex designs. Books on the analysis of data from a complex survey are available (6,7), although it is best to consult a statistician when dealing with data from a complex sample.

VI. PROBLEMS CAUSED BY UNINTENDED SAMPLING

In analyzing data it is imperative to understand the sample design as well as how the design was actually executed in the field. Deviations from the intended sample design are reflected in the data. Even in a well-designed survey, it is usually not possible to collect data from all the units sampled because there is almost always some nonresponse. Hence, the respondents, a subset of the sampled persons, are self-selected from the sampled persons through some procedure that is usually unknown to the designer of the study. As the respondents are no longer a random sample of the study population, there is concern that the data may be unusable because of nonresponse bias.

If the percentage of nonresponse is small, say, less than 5 to 10 percent, there is usually little concern because the bias, if any, is also likely to be small. If the nonresponse is on the order of 20 to 30 percent, the possibility of a substantial bias exists. For example, assume that we wish to estimate the proportion of people without health insurance in our community. We select a SRS and find that 20 percent of the respondents were without health insurance. However, one-fourth of those selected to be in the sample did not respond. If we knew the proportion of those without health insurance among the nonrespondents, it would be easy to combine this value with that of the respondents to obtain the total sample estimate. The proportions of those without health insurance among the respondents and nonrespondents would be weighted by the corresponding proportion of respondents and nonrespondents in the sample.

For example, if none of these nonrespondents had health insurance, the total sample estimate would be 40 percent [= (20% * 0.75) + (100% * 0.25)], twice as large as the rate for the respondents only. If all of the nonrespondents had health insurance, then the total sample estimate becomes 15 percent [= (20% * 0.75) + (0% * 0.25)]. Hence, although 20 percent of the respondents were without health insurance, the total sample estimate can range from 15 to 40 percent when one-fourth of the sample are nonrespondents.

For nonresponse bias to occur, the nonrespondents must differ from the respondents with respect to the variable of interest. In the example, it may be that many of the nonrespondents were unemployed homeless whereas few of the respondents were unemployed or homeless. In this case, the respondents and nonrespondents would likely differ with respect to health insurance coverage. If they do differ, there would be a large nonresponse bias. With larger percentages of nonresponse, the likelihood of a substantial nonresponse bias is very high and this makes the use of the data questionable. Unfortunately, many large surveys have a high percentage of nonresponse or do not mention the level of nonresponse. Data from these surveys are problematic.

An example of a survey with poor response is the Nationwide Food Consumption Survey conducted in 1987–1988 for the U.S. Department of Agriculture. This survey, conducted once per decade, was to be the basis for policy decisions regarding food assistance programs; however, only about one-third of the persons who were in the sample participated and, hence, the sample may not be representative of the U.S. population. An independent expert panel and the General Accounting Office of the U.S. Congress have concluded that information from this survey may be unusable (8).

There is no easy solution to the nonresponse problem. The best approach is a preventive one, that is, to exert every effort to obtain a high response rate. Even if you are unable to contact the sample person, perhaps a neighbor or family member can provide some basic demographic data about the person. If a sample person refuses to participate, again try to obtain some basic data about the person. If possible, try to obtain some information about the main topic of interest in the survey. The basic demographic data can be used to compare the respondents and nonrespondents. If there are no differences between the two groups on the demographic variables, that does not necessarily guarantee the absence of nonresponse bias. It does, however, eliminate the demographic variables as a cause of the potential nonresponse bias. If there is a difference, it may be possible to take those differences into account and create an adjusted estimator. The following calculations show one of many possible adjustment methods.

Suppose we found that there was a difference in the gender distribution between the respondents and nonrespondents. Sixty percent of the respondent group were females and 40 percent were males, whereas 30 percent of the nonrespondent group were females and 70 percent were males. If there were no difference in the proportions of females and males with health insurance, this difference in the gender distribution between the respondents and nonrespondents would be no problem. For this example, however, assume there was a difference. In the respondent group, 30 percent of the females were without health insurance compared with only 5 percent of the males. Figure 3.2 displays these percentages and the calculations involved in creating an adjusted rate.

The corresponding percentages with health insurance are unknown for the nonrespondent group. If, however, we assume that the female and male respondents' percentages with health insurance hold in the nonrespondent group, we can obtain an adjusted rate. The percentage of those without health insurance in the nonrespondent group under this assumption is found by weighting the proportions of females and males without health insurance by their proportions in the nonrespondent group, that is, $(30\% * 0.3) + (70\% * 0.05)$, which is 12.5 percent. We then use this value for the proportion of nonrespondents without health insurance and combine it with the proportion of respondents without health insurance to obtain a sex-adjusted estimate of the proportion of our community without health insurance. This adjusted estimate is 18.1 percent $[= (75\% * 0.20) + (25\% * 0.125)]$.

The adjusted rate does not differ much from the rate for the respondents only; however, this adjusted rate was based on the assumption that the proportions of females and males without health insurance were the same for respondents and nonrespondents. If this assumption is false, which we cannot easily check, then this adjusted estimate is incorrect. Whatever method of adjustment is employed, an assumption similar to the

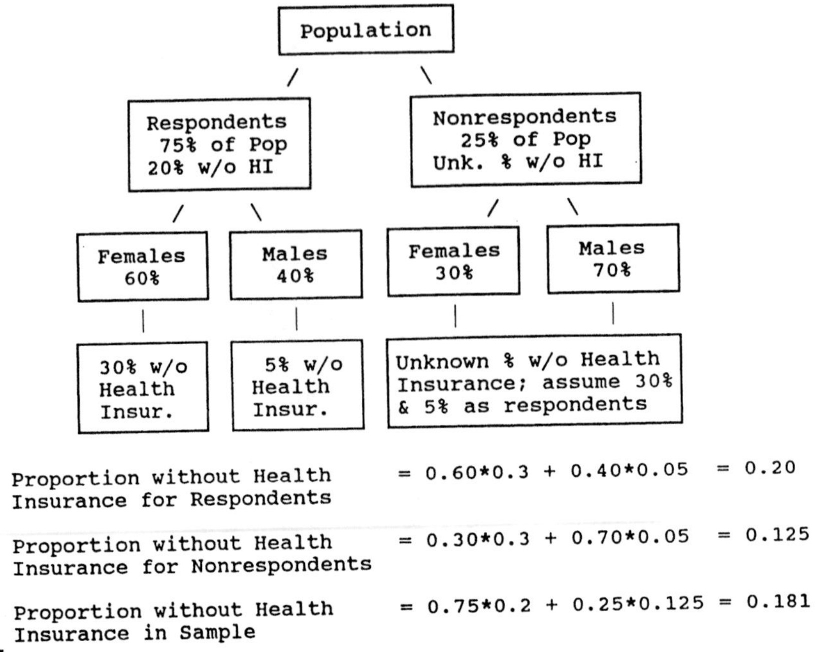

FIGURE 3.2 Display of the percentages for the health insurance example and calculation of the adjusted rate.

above must be made at some stage in the adjustment process (9). It is better to prevent nonresponse from occurring or to keep its rate of occurrence small.

The discussion so far has focused on *unit nonresponse*, that is, the observational unit did not participate in the survey. There is also *item nonresponse*, in which the sample person did not provide the requested information for some of the items in the survey. Just as there are no easy answers to unit nonresponse, item nonresponse or missing data also is a source of difficulty for the data analyst. Again if the percentage of item nonresponse is small, say less than 5 to 10 percent, it probably will not have much of an effect on the data analysis. In this case, the observations with the missing values may be deleted from the analysis. As the percentage of missing data increases, there is increasing concern about the representativeness of the sample persons remaining in the analysis. Because of the concern about the representativeness of the sample persons remaining, statisticians have developed methods for *imputing* or creating values for the missing data (9). By imputing values, it is no longer necessary to delete the sample persons with the missing data from the analysis. The imputation methods range from the very simple to the complex, depending on the amount of auxiliary data available.

As an example, suppose that in a survey to estimate the per capita expenditure for health care, we decided to substitute the respondents' sample average for those with a missing value on this variable. That is a reasonable imputation. If, however, we know the age of the sample persons, as age is highly related to health care expenditures, a better imputation would be to use the average expenditure from respondents in the same age group. There are other variables that could be used with age that would be even better than using age alone, for example, the combination of age and health insurance status. The sample mean from the respondents in the same age and health insurance group should be an even better estimate of the missing value than the mean from the age group or the overall mean. In using any imputation method, we must remember that the number of observations is really the number of sample persons with no missing data for the analysis performed, not the number of sample persons.

Other more complicated procedures are also available; however, none of these procedures guarantee that the value substituted for the missing data is correct. It is possible that the use of imputation procedures can lead to wrong conclusions being drawn from the data. Again, the best procedure for dealing with missing data is preventive, that is, make every effort to avoid missing data in the data collection process.

VII. CONCLUDING REMARKS

In this chapter we saw how to collect data using sample surveys and examined the principle of randomness related to the design of samples. We also presented some practical issues that cause more complicated sample designs to be used. Regardless of the complexity of the sample design, as long as we know the selection probability, we can infer from the sample to the population. The topic of probability is considered in detail in Chapter 5. In the next chapter, we consider ways to describe the sample data.

EXERCISES

3.1. Choose the most appropriate response from the choices listed after each question.

a. To determine whether a given set of data is a random sample from a defined population, one must ______.

— analyze the data
— know the procedure used to select the sample
— use a mathematical proof

b. A simple random sample is a sample chosen in such a way that every unit in the population has a(n) ______ chance of being selected into the sample.
  _ equal
  _ unequal
  _ known

c. In the random number table, Appendix Table B1, about _ percent of numbers are 9 or 2.
  _ 20
  _ 10
  _ unknown

d. Sampling with replacement from a large population gives virtually the same result as sampling without replacement.
  _ true
  _ false

e. In a stratified random sample, the selection probability for each element within a stratum is ______.
  _ equal
  _ unequal
  _ unknown

f. A probability sample is a sample chosen in such a way that each possible sample has a(n) ______ chance of being selected.
  _ equal
  _ unequal
  _ known
  _ unknown

3.2. If a population has 2000 members in it, how would you use Table B1 to select a simple random sample of size 25? Assume that the 2000 members in the population have been assigned numbers from 0 to 1999. Beginning with the first row in Table B1, select the 25 subjects for the sample.

3.3. In the following situations, do you consider the selected sample to be a simple random sample? Provide your reasoning for the answer.

a. A college administrator wishes to investigate students' attitudes concerning the college's health services program. A 10 percent random sample is to be selected by distributing questionnaires to students whose student ID number ends with a 5.

b. A medical researcher randomly selected five letters from the alphabet and abstracted data from the charts of patients whose surnames start with any of those five letters.

3.4. In NHANES II, 27 percent of the target sample did not undergo the health examination. In the examined sample, the weighted estimate of the percentage overweight was 25.7 percent [from Table 71 in NCHS (10)].

a. Assuming that these data were collected via a SRS, what is the range for the percentage overweight in the target sample?
b. Should any portion of the population be excluded in the measurement of overweight?

3.5. Discuss how sampling can be used in the following situations by defining: (1) the population, (2) the unit from which data will be obtained, (3) the unit to be used in sampling, and (4) the sample selection procedure.
a. A student is interested in estimating the total number of words in this book.
b. A city planner is interested in estimating the proportion of passenger cars that have only one occupant during rush hours.
c. A county public health officer is interested in estimating the proportion of dogs that have been vaccinated against rabies.

3.6. For each of the following situations discuss whether or not random sampling is used appropriately and why the use of random sampling is important.
a. A doctor selected every 20th file from medical charts arranged alphabetically to estimate the percentage of patients who have not had any clinic visits during the past 24 months.
b. A city public health veterinarian randomly selected 50 of 500 street corners and designated a resident at each corner to count the number of stray dogs for 1 week. He multiplied the number of stray dogs counted at the 50 corners by 10 as an estimate of the number of stray dogs in the city.
c. A hospital administrator reported to the board of directors that his extensive conversations with two randomly selected technicians revealed no evidence of support for a walkout by hospital technicians this year.

3.7. An epidemiologist wishes to estimate the average length of hospitalization for cancer patients discharged from the hospitals in her region of the country. There are 500 hospitals, with the number of beds ranging from 30 to 1200 in the region.
a. Discuss what difficulties the researcher might encounter in drawing a simple random sample.
b. Offer suggestions for drawing a random sample.

3.8. Discuss the advantages and disadvantages of the following sampling frames for a survey of the immunization levels of preschool children.
a. Telephone directory
b. List of children in kindergarten
c. List of registered voters

3.9. Discuss the interpretation of the following surveys:
a. A mail survey was conducted of 1000 U.S. executives and plant

managers. After a month, 112 responses had been received. The report of the survey results stated that Japan, Germany, and South Korea were viewed as being better competitors than the United States in the world economy. Also, one-third of the managers did not believe their own operations were making competitive improvements.

b. A weekly magazine reported that most American workers are satisfied with the amount of paid vacation they are allowed to take. This conclusion was based on the results of a telephone poll of 522 full-time employees (margin of error is plus or minus 4 percent; "not sure" omitted). The question asked was, "Should you have more time off or is the amount of vacation you have fair?"

More time off	33%
Current amount fair	62%

REFERENCES

1. Bryson, M. C. (1976). The Literary Digest poll: Making of a statistical myth. *Am. Stat.* **30,** 184–185.
2. NHLBI Task Force on Blood Pressure Control in Children (1987). The report of the second task force on blood pressure control in children, 1987. *Pediatrics* **79,** 1–25.
3. Waksberg, J. (1978). Sampling methods for random digit dialing. *J. Am. Stat. Assoc.* **73,** 40–46.
4. Fienberg, S. E. (1971). Randomization and social affairs: The 1970 draft lottery. *Science* **171,** 255–261.
5. National Center for Health Statistics (1973). Plan and Operation of the Health and Nutrition Examination Survey, United States, 1971–73. "Vital and Health Statistics," Ser. 1, No. 10a, DHEW Publ. No. (HSM) 73-1310. Public Health Service, Washington, U.S. Government Printing Office.
6. Wolter, K. M. (1985). "Introduction to Variance Estimation." Springer-Verlag, Berlin.
7. Lee, E. S., Forthofer, R. N., and Lorimor, R. J. (1989). "Analyzing Complex Survey Data." Sage Publ., Newbury Park, CA.
8. General Accounting Office (1991). "Nutrition Monitoring: Mismanagement of Nutrition Survey Has Resulted in Questionable Data," GAO/RCED-91-117. GAO, Washington, DC.
9. Kalton, G. (1983). "Compensating for Missing Survey Data," Res. Rep. Ser., Institute for Social Research, University of Michigan, Ann Arbor.
10. National Center for Health Statistics (1992). "Health, United States, 1991 and Prevention Profile," DHHS Publ. No. 92-1232. Public Health Service, Hyattsville, MD.

CHAPTER 4

Descriptive Tools

This chapter focuses on the summarization of data that were obtained from a simple random sampling process. Numerical and pictorial procedures are useful in the summarization of data. Both sets of tools are introduced in this chapter along with computer procedures based on the MINITAB package. Appendix A contains the comparable statements for SAS and Stata, two other statistical software packages.

I. USE OF THE COMPUTER: MINITAB

A dietary data set selected from a larger study by McPherson *et al.* (1) is introduced here to illustrate the use of various descriptive tools. Students in grades 5 through 8 in two suburban Houston schools were requested to keep food records for three randomly selected days, two weekdays and one weekend day, during a 2-week period. Calories, protein, total fat, and vitamin A consumed on the first day are shown in Table 4.1 for the 33 boys who participated in the study. These data will be explored by using various descriptive tools.

TABLE 4.1 Dietary Intake from Food Records for 33 Boys Enrolled in Two Middle Schools outside of Houston[a]

ID	Day of week	Grade	Calories	Protein (g)	Total fat (g)	Vitamin A (IU)	ID	Day of week	Grade	Calories	Protein (g)	Total fat (g)	Vitamin A (IU)
10	3	8	1823	83	63	4,876	50	1	7	1292	32	59	1,379
11	4	8	2007	64	62	6,202	51	5	7	3049	88	121	4,535
13	4	8	1053	23	33	964	101	7	6	3277	113	133	5,242
14	5	8	4322	128	202	6,761	105	7	6	2039	73	99	3,747
16	6	8	1753	84	83	1,704	107	6	6	2000	54	55	2,246
17	3	8	2685	105	103	2,671	118	3	6	1781	69	84	10,451
26	5	7	2340	157	73	4,288	120	2	6	2748	84	123	2,687
27	4	7	3532	172	227	12,812	127	7	5	2348	75	88	8,675
30	3	7	2842	135	121	4,450	130	3	5	2773	136	91	8,516
32	2	7	2074	44	69	820	137	3	5	2310	71	82	1,459
33	2	7	1505	97	25	9,490	139	2	5	2594	98	82	5,874
39	3	7	2330	60	87	4,315	141	2	5	1898	99	98	1,921
40	1	7	2436	86	115	6,754	145	7	5	2400	93	68	8,631
41	6	7	3076	89	121	8,034	148	1	5	2011	45	28	12,493
44	6	7	1843	94	69	9,710	149	1	5	1645	70	78	1,826
46	3	7	2301	62	74	4,248	150	3	5	1723	45	43	5,703
47	1	7	2546	72	124	2,284							

[a] A CDC grant funded the data collection; data used with permission of R. S. McPherson.

Before considering the descriptive tools, we introduce MINITAB, a computer package for statistical analysis, which facilitates the description of the data. The MINITAB statements in each section provide examples, not detailed instructions, of the commands. In-depth instructions are available in the "MINITAB User Guide" (2) and the online MINITAB Help commands. The MINITAB Quick Reference Card (3) provides an overview of commands. The first step is to enter data into the MINITAB worksheet. Three ways of entering the data in Table 4.1 are shown in Boxes 4.1 and 4.2.

MINITAB BOX 4.1

The first method of data entry uses the SET command to enter the values of calories into a column, denoted by the letter *c* followed by a number. We entered everything to the right of the >'s in the following:

```
MTB > set cl
DATA> 1823 2007 1053 4322 1753 2685 2340 3532 2842 2074 1505
DATA> 2330 2436 3076 1843 2301 2546 1292 3049 3277 2039 2000
DATA> 1781 2748 2348 2773 2310 2594 1898 2400 2011 1645 1723
DATA> end
```

There is a space between each caloric value and a carriage return at the end of each of these lines. The word *end* in the last line indicates that all of the data for cl had been entered. It is useful to label the column by using the NAME command (limited to eight characters).

```
MTB > name cl 'calories'
```

The READ command is another way of entering data from the keyboard. It is useful when there are several columns of data to be entered, whereas the SET command is more appropriate when there are only a few columns of data to enter. The first three rows of the day of the week, grade, protein, total fat, and vitamin A data are entered to demonstrate the use of the READ command. The values of these five variables are stored in columns c2, c3, c4, c5, and c6.

```
MTB > read c2-c6
DATA> 3 8 83 63 4876
DATA> 4 8 64 62 6220
DATA> 4 8 23 33 964
DATA> end
      3 ROWS READ
```

It is useful to examine whether or not we have entered the data correctly. This can be done by looking at what we have entered or by using the PRINT command, which shows what values are in the columns.

```
MTB > print c2-c6
 ROW   C2    C3   C4   C5   C6
   1    3     8   83   63 4876
   2    4     8   64   62 6220
   3    4     8   23   33  964
```

The second value (row) in c6 is supposed to be 6202, not 6220. This can be corrected by reentering c6 or by using the LET command as shown now:

```
MTB > let c6(2)= 6202
```

In c6(2), the 2 in the parentheses indicates that we are referring to the second element (row) in c6 and we are setting its value to 6202. We can use the PRINT command to see if we have been successful in making the correction.

```
 MTB > print c6
C6     4876    6202   964
```

(The correction was made.)

MINITAB BOX 4.2

We can also read data from a file that has already been created instead of entering the data at the keyboard. This method also uses the READ command followed by the name of the file containing the data in single quotes and the column numbers in which the data will be stored.

```
MTB > read 'bookch4.dat' c1-c6
     33 ROWS READ
 ROW       C1     C2     C3    C4    C5    C6
   1     4876   1823     63    83     3     8
   2     6202   2007     62    64     4     8
   3      964   1053     33    23     4     8
   4     6761   4322    202   128     5     8
   .   .   .
```

The order of the data in these six columns is different from that shown in Table 4.1. That poses no problem as we simply label the columns to reflect their contents.

```
MTB > name c1 'vit A' c2 'calories' c3 'tot fat'

MTB > name c4 'protein' c5 'day' c6 'grade'
```

II. TABULAR AND GRAPHICAL PRESENTATION

One- and two-way frequency tables and several types of figures—line graph, bar chart, histogram, stem-and-leaf plot, scatter plot, and box plot—that aid in the description of data are introduced in this and subsequent sections.

A. Frequency Tables

A *one-way frequency table* shows the results of the *tabulation* of the observations at each level of a variable. For example, Table 4.2 shows the frequencies of the days of the week when the first measurements were made for

TABLE 4.2 Frequency of Days of the Week of the First Measurement

Day	Number of boys	Percentage
1 (Sunday)	5	15.2
2 (Monday)	5	15.2
3 (Tuesday)	9	27.3
4 (Wednesday)	3	9.1
5 (Thursday)	3	9.1
6 (Friday)	4	12.1
7 (Saturday)	4	12.1
Total	33	100.1

the 33 boys from Table 4.1. Over one-quarter of the observations were made on Tuesday followed by Sunday and Monday with five observations each. Note that the total number of boys is 33 as it must be. The sum of the percents should be 100.0, although a small allowance is made for rounding. Note also that the title of the table contains sufficient information to allow the reader to understand the table.

Two-way frequency tables, formed by the *crosstabulation* of two variables, are usually more interesting than one-way tables because they show the relationship between the variables. The variables can be nominal, ordinal, or continuous. Usually when continuous variables are used, their values are grouped into categories. Table 4.3 shows the relationship between day of the week and caloric intake where caloric intake has been grouped into below 2500 calories and 2500 calories and above. The value of 2500 calories was chosen because it is approximately the average intake of boys ages 12 to 15. Day 5, Thursday, appears to be different than the other days as it is the only day with a majority of its values greater than or equal to 2500

TABLE 4.3 Crosstabulation of Day of the Week and Caloric Intake with Row Percentages in Parentheses

	Calories				
Day of week	<2500		≥ 2500		Total
1	4	(80%)	1	(20%)	5
2	3	(60%)	2	(40%)	5
3	6	(67%)	3	(33%)	9
4	2	(67%)	1	(33%)	3
5	1	(33%)	2	(67%)	3
6	3	(75%)	1	(25%)	4
7	3	(75%)	1	(25%)	4
Total	22	(67%)	11	(33%)	33

calories. However, there are so few observations for most of the levels of the day variable that the difference between day 5 and the other days may be due to sampling variation; that is, this difference may be an artifact of the sample that was selected.

One way of reducing the number of levels with only a few observations is to combine levels. For example, a natural grouping is weekdays, combining Monday through Friday, and weekends, combining Saturday and Sun-

MINITAB BOX 4.3

The frequencies in Table 4.2 can be easily obtained by using the following command.

```
MTB > table c5
 ROWS: C5
          COUNT
   1        5
   2        5
   3        9
   4        3
   5        3
   6        4
   7        4
  ALL      33
```

Before creating Table 4.3, we require one additional step. In Table 4.3, we had categorized the calories into below 2500 and greater than or equal to 2500 calories. We use the CODE command in MINITAB to accomplish this recoding.

```
MTB > code (1000:2499) 0 (2500:4500) 1 c2 c7
```

This statement assigns a value of 0 in column c7 for boys who consumed from 1000 to 2499 calories on their first recording day and a value of 1 in c7 for boys who consumed 2500 calories or more. The ranges shown are from 1000 to 2499 and from 2500 to 4500 because no boy had a value less than 1000 or greater than 4500. The caloric values come from column c2 and the recoded values are stored in c7. Now the two-way table can be created.

```
MTB > table c5 c7
 ROWS: C5       COLUMNS: C7
              0         1        ALL
 1            4         1         5
 2            3         2         5
 3            6         3         9
 4            2         1         3
 5            1         2         3
 6            3         1         4
 7            3         1         4
 ALL         22        11        33
  CELL CONTENTS --
                    COUNT
```

TABLE 4.4 Health Expenditures as a Percentage of Gross Domestic Product over Time

Year	Great Britain	United States	West Germany
1960	3.9	5.2	4.7
1965	4.1	6.0	5.1
1970	4.5	7.4	5.5
1975	5.5	8.4	7.8
1980	5.8	9.2	7.9
1985	6.0	10.6	8.2
1987	6.1	11.2	8.2

Source: Table 104 in "Health, United States, 1990" (4).

day. There are more observations for the weekday and weekend categories and, if there are differences now, they are more likely to be real. In forming the groups, we should not allow the data to guide us. We should use our knowledge of the subject matter, and not use the data, in selecting the categorization. If we use the data to guide us, it is easy to obtain apparent differences that are not real but only artifacts of the data. Box 4.3 shows how to create tables from the data we already entered.

Other data besides frequencies can be presented in tabular format. For example, Table 4.4 shows the health expenditures of three nations as a percentage of gross domestic product (GDP) over time. Health expenditures as a percentage of GDP are increasing much more rapidly in the United States than in either Great Britain or West Germany.

B. Line Graphs

A *line graph* shows the value of a variable over time. The values of the variable are given on the vertical axis; the horizontal axis is the time variable. Figure 4.1 shows three line graphs for the data shown in Table 4.4.

These line graphs also show the rapid increase in health expenditures in the United States compared with those of two other countries with national health plans. The trends are immediately clear in the line graphs, whereas one has to study Table 4.4 before the same trends are recognized.

It is possible to give different impressions about the data by shortening or lengthening the horizontal and vertical axes or by including only a portion of an axis. In creating and studying line graphs, one must be aware of the scales used for the horizontal and vertical axes. For example, with numbers that are extremely variable over time, a logarithmic transformation (discussed later) of the variable on the vertical axis is frequently used to allow the line graph to fit on a page.

It is also possible to represent different variables in the same figure as Figure 4.2 shows. The right vertical axis is used for lead emissions and the

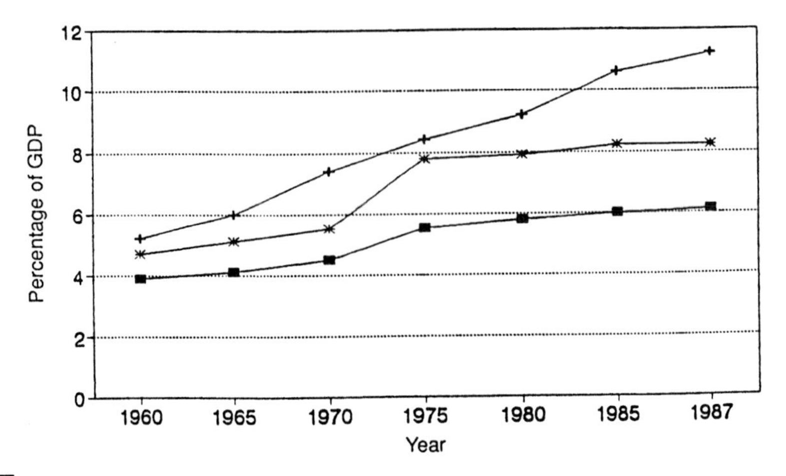

FIGURE 4.1 Line graph: Health expenditures as percentage of GDP. ■, Great Britain; +, United States; *, West Germany. *Source*: Table 104 in "Health, United States, 1990" (4).

left vertical axis for sulfur oxides emissions. Both pollutants are decreasing, but the decrease in lead emissions is quite dramatic, from approximately 200×10^3 metric tons in 1970 to only about 8×10^3 metric tons in 1988. During this same period, sulfur oxides emissions decreased from about 28×10^6 to about 21×10^6 metric tons. The decrease in the lead emissions is partially related to the use of unleaded gasoline, which was introduced during the 1970s.

Box 4.4 shows how to create plots and Box 4.5 shows the result.

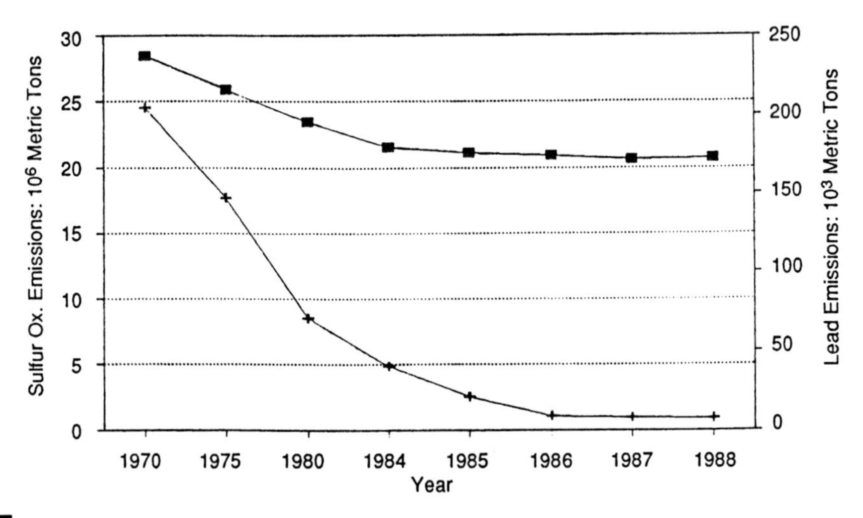

FIGURE 4.2 Line graphs of sulfur oxides and lead emissions in the United States. ■, Sulfur oxides; +, lead. *Source*: Table 64 in "Health, United States, 1990" (4).

MINITAB BOX 4.4

MINITAB has a number of commands for creating plots. For example, the command PLOT is used to create a single line graph, whereas the command MPLOT can be used to create multiple line graphs in a figure. MINITAB also has high-resolution graphics available for a number of its plotting commands. To invoke the high-resolution graphics, add the letter G in front of the command, for example, GPLOT or GMPLOT instead of PLOT or MPLOT. When PLOT or MPLOT is used, the points are shown, but no connecting lines are drawn. When GPLOT or GMPLOT is used, MINITAB draws the line(s) connecting the points.

The following shows the use of MINITAB in the creation of the line graphs in Figure 4.1. The first group of statements shows the data entry. The values of the percentage of GDP are entered in columns c8 to c10; c11 contains the years. The second group of statements creates three lines in a single graph, using the MPLOT command. The commands of height and width allow the user to indicate how many lines should be used for the figure (height) and how many columns wide the figure should be (width). The MPLOT command ends with a semicolon, indicating that a subcommand will follow. Several subcommands, closed by semicolons, are used to provide details about what is contained in the figure. There is a period at the end of the last subcommand; the period tells MINITAB that all the information for the command has now been entered.

```
MTB > set c8
DATA> 3.9 4.1 4.5 5.5 5.8 6.0 6.1
DATA> set c9
DATA> 5.2 6.0 7.4 8.4 9.2 10.6 11.2
DATA> set c10
DATA> 4.7 5.1 5.5 7.8 7.9 8.2 8.2
DATA> set c11
DATA> 60 65 70 75 80 85 87
DATA> end
MTB > height 35
MTB > width 55
MTB > mplot c8 c11, c9 c11, c10 c11;
SUBC> title='Health Expenditures as % of GDP over Time';
SUBC> footnote='A=Great Brit, B=U.S. and C=West Germ';
SUBC> ylabel='% of GDP';
SUBC> xlabel='Year'.
```

The plot is shown in Box 4.5. In this plot the points must be connected by hand to produce the line graphs. The following MINITAB statements create a high-resolution graph with three lines connecting the points in the previous graph.

```
MTB> gmplot c8 c11, c9 c11, c10 c11;
SUBC> lines c8 c11;
SUBC> lines c9 c11;
SUBC> lines c10 c11;
SUBC> footnote 'A=Great Brit, B=U.S., and C=West Germ';
SUBC> ylabel '% of GDP';
SUBC> xlabel 'Year'.
```

MINITAB BOX 4.5

Health Expenditures as % of GDP over Time (A = Great Britain, B = United States, C = West Germany)

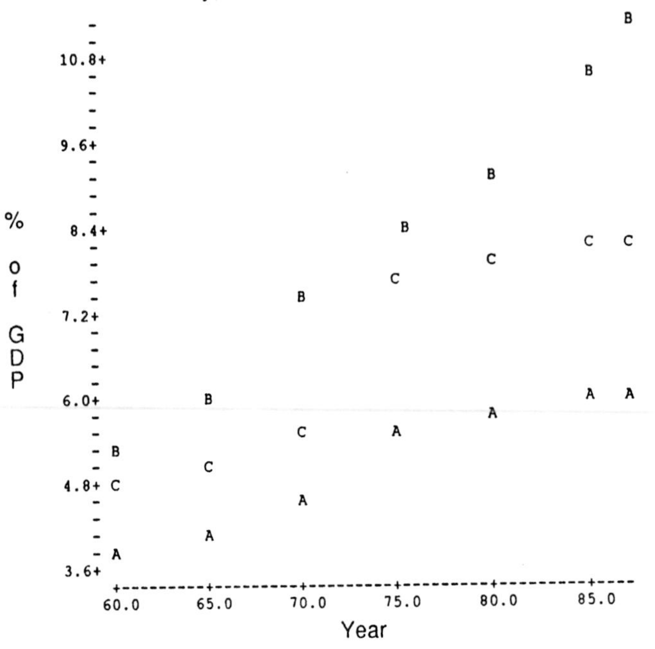

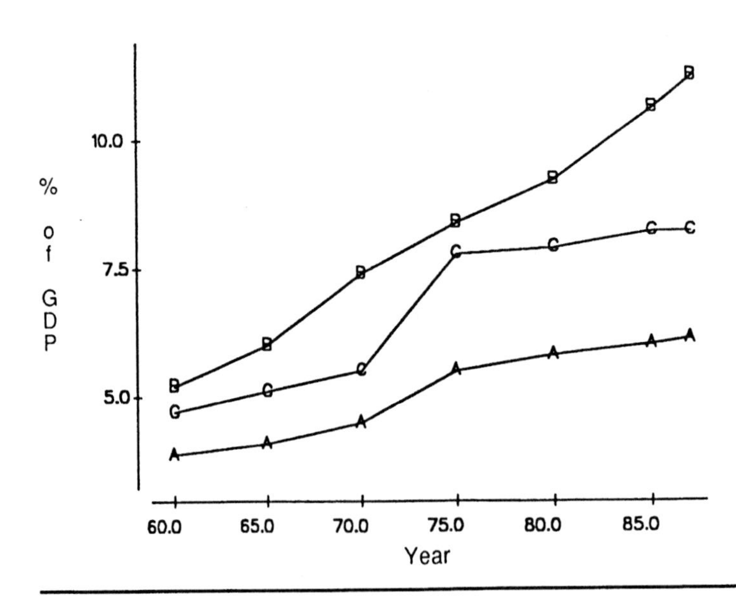

C. Bar Charts

A *bar chart* provides a picture of data that could also often be reasonably presented in a tabular format. Bar charts can be created for nominal, ordinal, or continuous data, although they are most frequently used with nominal data. If used with continuous data, the chart could be called a *histogram* (see below) instead of a bar chart. The bar chart can show the number or proportion of people by levels of a nominal or ordinal variable. For example, the numbers of people enrolled in health maintenance organizations (HMOs) in the United States by year (ordinal variable) are shown in Figure 4.3.

This bar chart makes it very clear that there has been explosive growth in HMO enrollment, particularly between 1982 and 1986. The actual enrollments by year are 7.45, 10.81, 25.73, and 33.03 million. The numbers also document this growth, but it is more dramatic in the visual presentation.

In bar charts, the length of the bar shows the number of observations or the value of the variable of interest for the levels of the nominal or ordinal variable. The widths of the bar are the same for all the levels of the nominal or ordinal variable; the width has no meaning. The levels of the nominal or ordinal variable are usually separated by several spaces which makes it easier to view the data. The bars are usually presented vertically but they could also be horizontal.

More complicated data can also be presented in bar chart format. Figure 4.4 shows death rates for selected causes for persons 45 to 64 years of age by race/ethnicity (the nominal variable) in the United States in 1988. This bar chart, a segmented bar chart, presents a large amount of information that is quickly understandable. African-Americans have the highest mortality rates in this age group and they also have the highest rates of

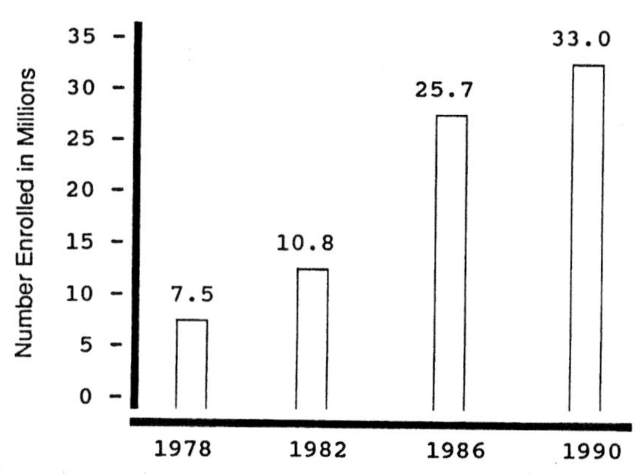

FIGURE 4.3 Bar chart of the number of persons (in millions) enrolled in health maintenance organizations by year. *Source*: Table 126 in "Health, United States, 1990" (4).

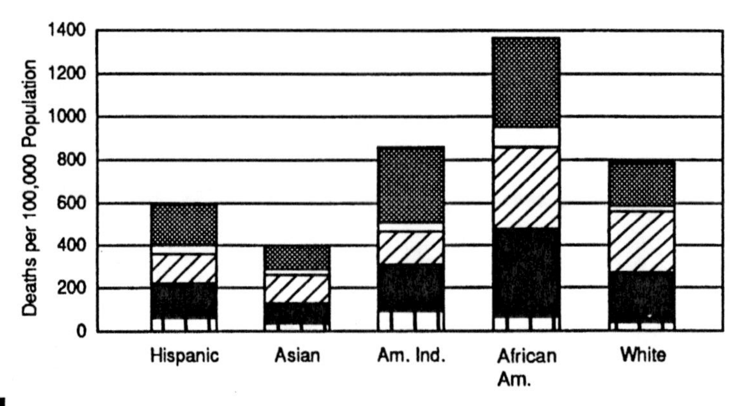

FIGURE 4.4 Mortality rates for persons aged 45 to 64 for selected causes by race/ethnicity in the United States in 1988. [legend], Injuries; [legend], cerebrovascular disease; [legend], heart disease; [legend], malignant neoplasms; [legend], all other causes. *Source*: Figure 9 in "Health, United States, 1990" (4).

mortality from heart disease and malignant neoplasms. Asians and Pacific Islanders have the lowest mortality rates for this age group.

Figure 4.5, a three-dimensional bar chart, shows infant mortality rates by race and year for Harris County, Texas, during the period 1980–1986. Levels of one nominal variable (race) and one ordinal variable (year) are used in the creation of this bar chart. From this figure it is clear that the

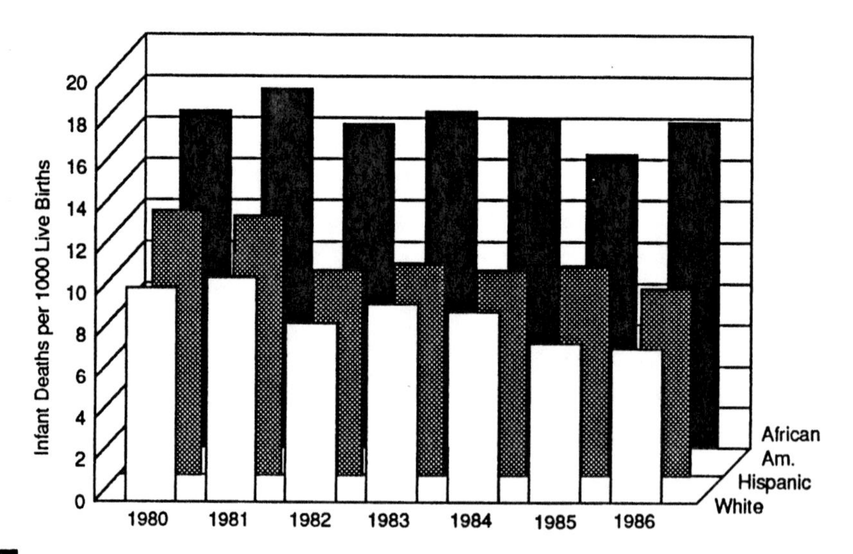

FIGURE 4.5 Infant mortality rates by race for Harris County, Texas, by year. *Source*: Figure B-19 in Harris County Health Department (5).

infant mortality rates for whites and Hispanics are decreasing over time, whereas the rate for African-Americans is remaining almost constant. It is also easy to see that the rate for African-Americans is considerably higher than that of the other two groups.

These last two figures demonstrate that bar charts can be quite effective in presenting information on several variables.

D. Histograms

As was mentioned earlier, a histogram is similar to a bar chart but is used with interval/ratio variables. The values of the variable are grouped into intervals which are usually of equal width. Rectangles are drawn above each interval and the area of the rectangle represents the number of observations in that interval. If all the intervals are of equal width, then the height of the interval, as well as its area, represents the frequency of the interval. In contrast to bar charts, there are no spaces between the rectangles unless there are no observations in some interval.

We demonstrate the creation of a histogram for the data in Table 4.5 which contains systolic blood pressure values that could be seen in typical 12-year-old U.S. boys.

Before creating the histogram, however, we create a one-way table which will facilitate the creation of the histogram. Table 4.6 gives the frequency of each blood pressure value. Note that a large proportion of the blood pressure values appear to end in zero: 43 of the 100 observations end in zero. All the values are also even numbers, with the exception of 11 values that end with a 5. This suggests that the persons who recorded the blood pressure values may have had a preference for numbers ending in 0 or 5. This type of finding is not unusual in blood pressure studies or in the reporting of age, as was seen in Chapter 2. In spite of this possible digit

TABLE 4.5 Systolic Blood Pressure (mm Hg) Values for 100 Typical 12-Year-Old U.S. Boys

130	100	125	92	98	108	104	100	100	102	120	110	100
112	110	110	100	128	122	110	120	108	94	130	110	104
120	118	84	115	102	100	112	104	100	120	110	110	106
130	120	108	104	106	114	96	112	114	100	112	80	100
110	126	95	100	100	94	102	95	140	124	98	110	90
80	102	116	102	90	116	110	128	140	90	104	130	104
105	80	116	106	100	95	105	90	108	88	105	112	134
116	108	108	100	105	110	90	95	125				

preference, we are going to create some histograms based on these values. The histogram provides a visual summarization of the values shown in Tables 4.5 and 4.6.

Three questions must be answered before we can draw the histogram for these data:

1. How many intervals should there be?
2. How large should the intervals be?
3. Where should the intervals be located?

Tarter and Kronmal (6) discuss these three questions in some depth. There are no hard and fast answers to these questions; only guidelines are provided.

The number of intervals is related to the number of observations. Generally 5 to 15 intervals would be used, with a smaller number of intervals used for smaller sample sizes. There is a trade-off between many small intervals, which allow for greater detail with few observations in any category, and a few large intervals, with little detail and many observations in the categories.

Once the number of intervals, call this number k, is decided, the size of the intervals can be determined. One way of choosing the interval size is to calculate the difference between the maximum and minimum observed values and divide this difference by $k - 1$. This is a reasonable approach unless there are some relatively large or small values. In this case, exclude these unusual values from the difference calculation. This approach will yield larger first and last intervals but the other intervals will be the same size.

The location of the intervals is also arbitrary. Most researchers either begin the interval with a round number or have the midpoint of the interval be a round number.

TABLE 4.6 Frequency of Individual Systolic Blood Pressures (mm Hg)

Value	Frequency	Value	Frequency	Value	Frequency
80	**3**	102	5	118	1
84	1	104	6	**120**	**5**
88	1	105	4	122	1
90	**5**	106	3	124	1
92	1	108	6	125	2
94	2	**110**	**11**	126	1
95	4	112	5	128	2
96	1	114	2	**130**	**4**
98	2	115	1	134	1
100	**13**	116	4	**140**	**2**

Let us create several histograms for these 100 systolic blood pressure values to see the effect of our choices. First we try 11 intervals. As the maximum and minimum observed values are 140 and 80, respectively, the difference is 60. Dividing 60 by 10 yields 6 as the interval size. As in Figure 4.6, we could choose the lower boundary of the first interval to be the minimum observed value (80 mm Hg), a most reasonable choice for these data, or the minimum value could be located in the first interval as in Figure 4.7. These two figures look different although they have the same number of intervals and the intervals are of the same size. The starting points of the first interval cause the difference in appearance.

Figure 4.8 shows the effect of using 10 intervals instead of 11. An interval of size 7 (60 divided by 9 is approximately 7) and a starting value of 73.5 are used.

The histograms in Figures 4.6, 4.7, and 4.8 use the same data but give different impressions about the data. We see that the shapes of the histograms are dissimilar because of the decisions we made; however, the histograms say basically the same thing about the distribution of the sample data even though their shapes are different. All three histograms show that the blood pressures are tightly clustered, with most of the values between approximately 98 and 114, and that there are a few small values around 80 and a few large values around 140. The lesson of these histograms is not to become enamored of the shape of the histogram but to look at what the histogram is saying about the data.

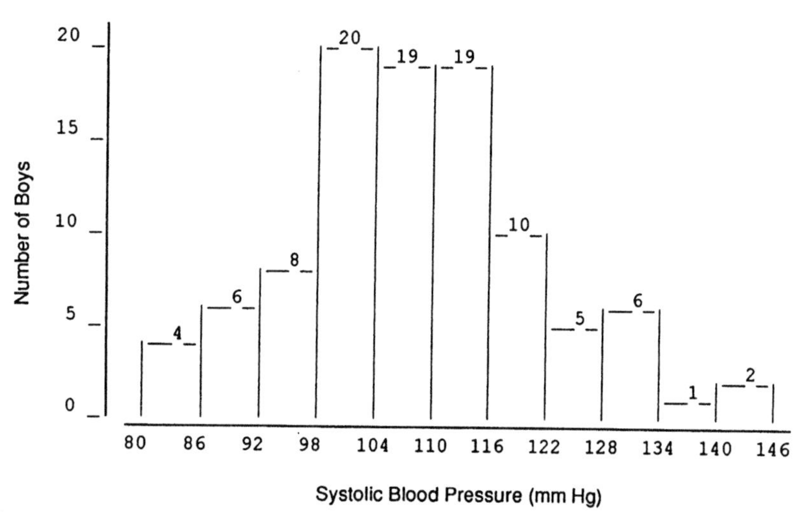

FIGURE 4.6 Histogram of 100 systolic blood pressure values using 11 intervals of size 6 starting at 80. Interval boundaries are at the bottom of the figure.

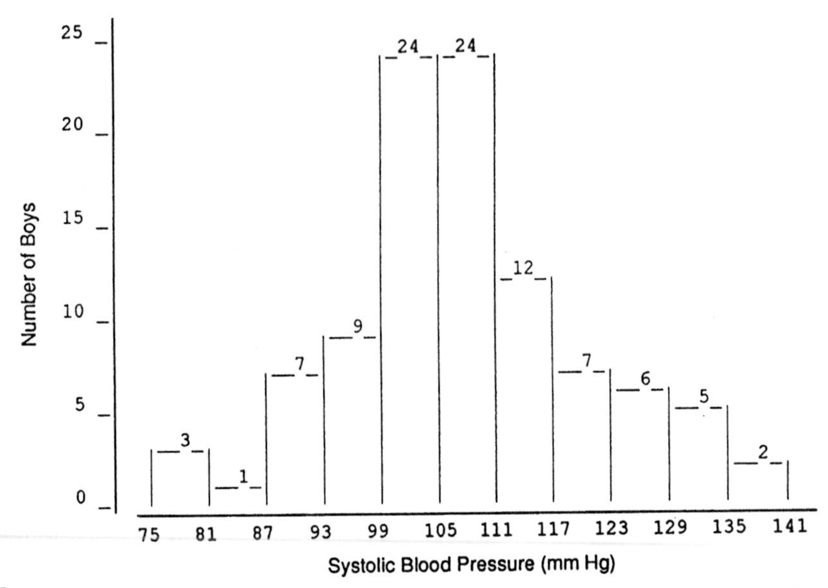

FIGURE 4.7 Histogram of 100 systolic blood pressure values using 11 intervals of size 6 starting at 75. Interval boundaries are at the bottom of the figure.

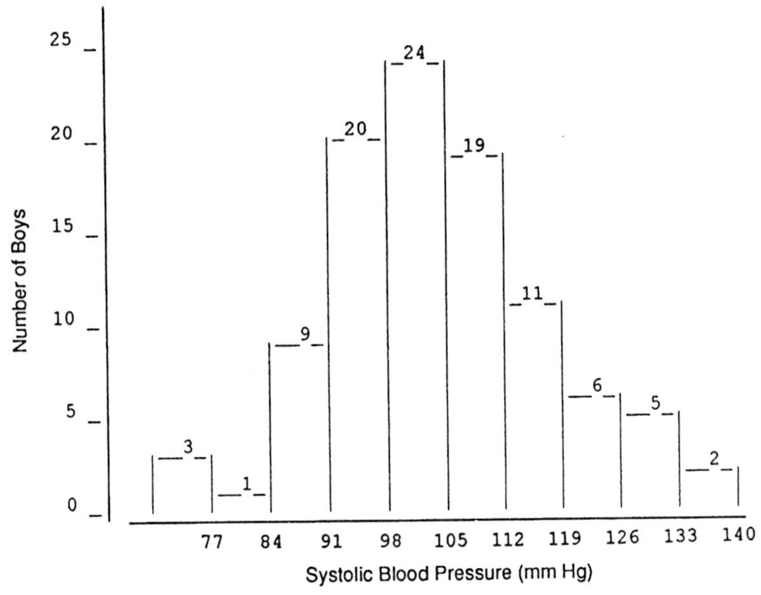

FIGURE 4.8 Histogram of 100 systolic blood pressure values using 10 intervals of size 7 starting at 73.5. Interval midpoints are at the bottom of the figure.

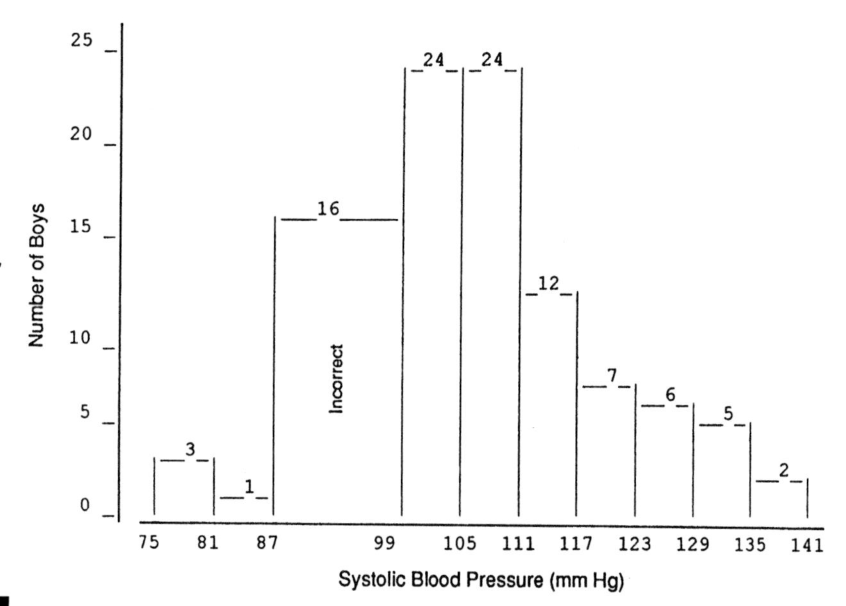

FIGURE 4.9 An incorrect histogram of the 100 systolic blood pressures. Interval boundaries are at the bottom of the figure.

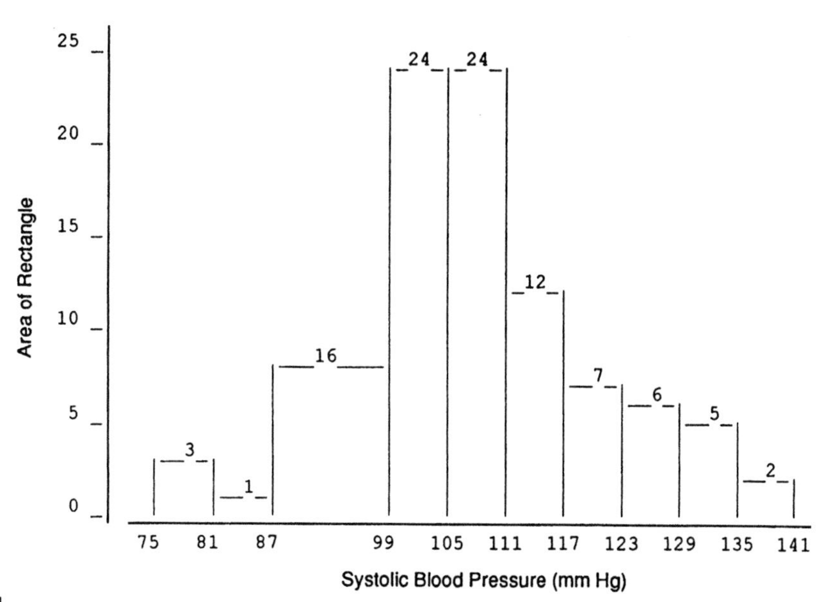

FIGURE 4.10 A correct histogram of the 100 systolic blood pressures. Interval boundaries are at the bottom of the figure.

MINITAB BOX 4.6

The data in Table 4.5 are entered in c8, although not shown here. The following shows the MINITAB creation of a histogram shown in Figure 4.8.

```
MTB > hist c8;
SUBC> increment 7;
SUBC> start 77.
Histogram of C8    N = 100
 Midpoint   Count
    77.00       3   ***
    84.00       1   *
    91.00       9   *********
    98.00      20   ********************
   105.00      24   ************************
   112.00      19   *******************
   119.00      11   ***********
   126.00       6   ******
   133.00       5   *****
   140.00       2   **
```

MINITAB plots histograms sideways in contrast to the conventional manner of presentation. As is demonstrated in the HISTOGRAM (abbreviated as HIST) statement, there are optional subcommands that can be used to customize the histogram. The size of the interval can be specified; for example, it is of size 7 (because INCREMENT was set to 7) in this example. As an additional instruction is still to follow, the INCREMENT 7 subcommand is followed by a semicolon. The value of the first interval's midpoint is next specified in the START subcommand and it is set at 77. Because there are no more subcommands that we wish to specify, START 77 is followed by a period, indicating the end of the HISTOGRAM command.

The first column in the output gives the midpoint of the interval and is followed by a column showing the number of observations in the interval. The next several columns show the histogram using asterisks to represent each value in the interval. When we use GHIST instead of HIST, a high-resolution graph can be obtained as shown.

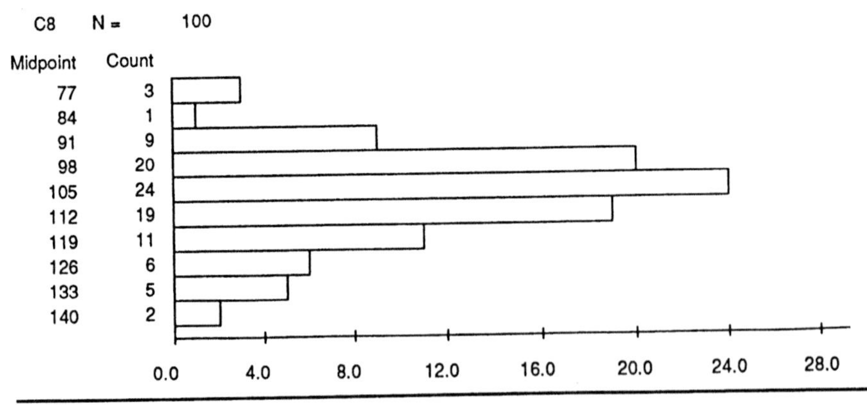

Before leaving histograms, we must examine unequal size intervals and how to deal with them. Suppose that in Figure 4.7 the third and fourth intervals were combined, that is, the third interval now extends from 87 to 99 mm Hg, and all the other intervals remained the same. As was mentioned above, the area of the rectangle reflects the frequency of the interval. Therefore, because this interval is twice as wide as the other intervals, to have the appropriate area within the rectangle it is necessary to divide its height by 2. Figure 4.9 is incorrect because it fails to adjust for the greater width of the third interval. Figure 4.10 is more appropriate than Figure 4.9 because it has taken the greater width into account. Box 4.6 shows how MINITAB is used to create the histograms for the systolic blood pressure data.

E. Stem-and-Leaf Plot

The *stem-and-leaf* plot looks similar to a histogram except that the stem-and-leaf plot shows the data values instead of using asterisks or bars to represent the height of an interval. In the blood pressure example, the stem could be the tens units and the leaves then would be the units. We use MINITAB to create a stem-and-leaf plot for the systolic blood pressure data as shown in Box 4.7.

MINITAB BOX 4.7

```
MTB > stem c8
Stem-and-leaf of C8          N = 100
Leaf Unit = 1.0
    4     8 0004
    5     8 8
   13     9 00000244
   20     9 5555688
   44    10 000000000000022222444444
  (13)   10 5555666888888
   43    11 000000000002222244
   25    11 566668
   19    12 0000024
   12    12 55688
    7    13 00004
    2    13
    2    14 00
```

This looks like a histogram except that we now know the values of all the observations. The first column in this output shows a cumulative count of all the observations from the top and from the bottom to the interval in which the median value is found. The *median* is the value such that 50 percent of the values are less than it and 50 percent are greater than it. The number of observations in the interval containing the median is shown in parentheses. The second column is the stem and the subsequent columns contain the leaves. For example, in the first row we read a stem of 8 and leaves of 0, 0, 0, and 4. As the stem represents units of 10 in this case and the leaf unit is 1, these four numbers are 80, 80, 80, and 84. The second row has the same stem and a leaf of 8 and, thus, represents a blood pressure value of 88. Note that the first number in the second row is 5, which is the cumulative count of observations in the first two rows. In the third row the stem is 9 and there are 8 leaves. These blood pressures are 90, 90, 90, 90, 90, 92, 94, and 94. Because there are 8 values in the third row, the cumulative count is now 13. As the interval from 105 to 109 has 44 observations less than it, 43 observations greater than it, and 13 observations in it, the median is in this interval and its value is 106.

Note in Box 4.7 that the interval size of 5 units was used for the pressure values. We can choose the interval size as shown in Box 4.8. This stem-and-leaf plot still shows the same values as the first stem-and-leaf plot but they are grouped differently.

Another characteristic of the data that can be seen from histograms or stem-and-leaf plots is whether or not the data are symmetrically distributed. Data are symmetrically distributed when the half of the distribution above the median matches the distribution below the median. Data could also come from a *skewed* or asymmetric distribution. Data from a skewed

MINITAB BOX 4.8

We can choose the interval size by using the subcommand INCREMENT with the STEM command. An inverval of size 10 is used below.

```
MTB > stem c8;
SUBC> increment 10.
 Stem-and-leaf of C8          N = 100
Leaf Unit = 1.0
    5    8 00048
   20    9 000002445555688
  (37)  10 0000000000000222224444445555666888888
   43   11 000000000002222244566668
   19   12 000002455688
    7   13 00004
    2   14 00
```

distribution typically have extreme values in one end of the distribution but no extreme values in the other end of the distribution. When there is a long tail to the right, or to the bottom if the histogram is presented sideways, data are said to be *positively skewed*. If there are some extremely small values without corresponding extremely large values, the distribution is said to be *negatively skewed*.

For the blood pressure values shown in MINITAB Box 4.7, the sample data appear to be slightly asymmetric as the data above the median are not grouped as tightly as those below the median.

The stem-and-leaf plot in Box 4.9 uses the vitamin A values from Table 4.1, data that are positively skewed. MINITAB tells us that the leaf unit is 100 which means that the stem unit is 1000. Hence the two values in the first row represent one observation in the 800s and one value in the 900s. From Table 4.1 we see the actual values are 820 and 964. The single observation in the fourth row has a value of three thousand seven hundred and something. The actual value is 3747. The median is in the 4000s and there are observations as large as 12,000. The distance from the median to the larger values is much greater than that to the smallest values. These data have a long tail to the right, that is, the vitamin A data are positively skewed. This statement is more informative than simply saying that the data are asymmetric.

The stem-and-leaf plot works best with relatively small sample sizes. With large sample sizes, having the exact numerical value of every observation can be overwhelming.

MINITAB BOX 4.9

```
MTB > stem c1
Stem-and-leaf of C1      N = 33
Leaf Unit = 100
    2     0 89
    7     1 34789
   11     2 2266
   12     3 7
   (6)    4 223458
   15     5 278
   12     6 277
    9     7
    9     8 0566
    5     9 47
    3    10 4
    2    11
    2    12 48
```

F. Scatter Plot

The two-dimensional *scatter plot* is analogous to the two-way frequency table in that it facilitates the examination of the relationship between two variables. Unlike the two-way table, the two-dimensional scatter plot is most effectively used when the variables are continuous. Just as it is possible to have higher-dimensional frequency tables, it is possible to have higher-dimensional scatter plots, but they become more difficult to comprehend.

One way of pictorially examining the relationship between grams of protein and grams of total fat shown in Table 4.1 is to use a scatter plot. It is easy to create two-dimensional scatter plots with MINITAB as can be seen in Box 4.10. c3 is the column containing the total fat values and c4 contains the protein values. Each asterisk represents a boy's protein and total fat values. For example, the asterisk in the lower left-hand corner of the plot represents the boy in Table 4.1 whose diet included 23 grams (g) of protein and 33 g of total fat. The asterisk in the upper right-hand corner represents the boy whose food intake contained 172 g of protein and 227 g of total fat. Numerical values in the plot represent the frequency of the point. For example, the leftmost 2 in the plot refers to two boys whose diets contained about 70 g of protein and 80 g of total fat. In this case the exact values were 69 and 84 for one of the boys and 70 and 78 for the other boy.

The plot shows that there is a strong tendency for boys with large protein values to also have large values of total fat and that small values of

MINITAB BOX 4.10

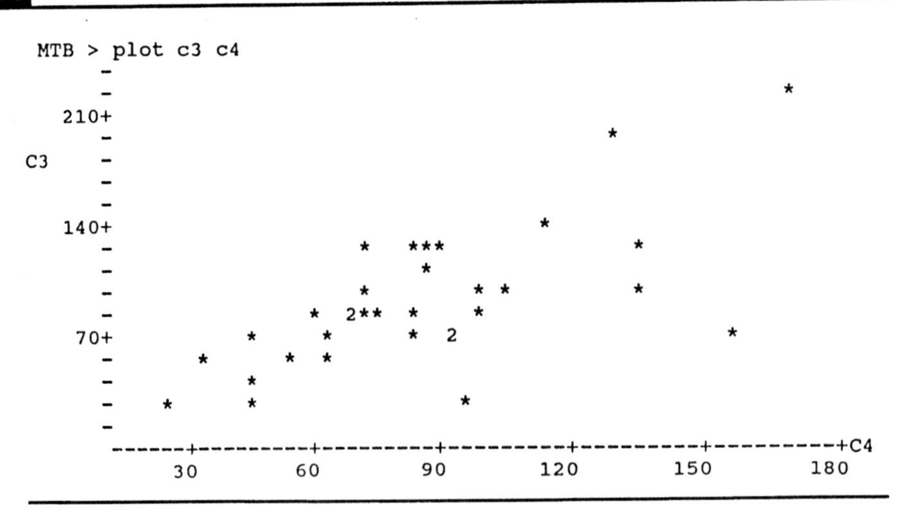

protein generally correspond with small values of total fat. The relationship is not perfect as can be seen from the point in the lower right-hand corner which corresponds to a boy with a large protein value (157 g) and a small total fat value (73 g). It is much easier to see this tendency for a positive association between protein and total fat from this scatter plot than by looking at the values in Table 4.1. The association is said to be positive because boys with large values of protein also tend to have large values of total fat and boys with small values of protein also tend to have small total fat values. When there is a positive association, the points in the plot tend to larger values on the vertical axis as we move from the left to the right in the graph. The association would be negative if boys with large values of one variable tended to have small values of the other variable and conversely.

Scatter plots are most effective for small to moderate sample sizes. When there are thousands of observations, it is often difficult to obtain a sense of the relationship. In these cases, it may be useful to calculate the sample average of the variable on the vertical axis (C3 in the scatter plot above) separately for each value of the variable on the horizontal axis (C4 above). If there are 20 distinct values of the variable on the horizontal axis, call it X, there would be 20 distinct sample means calculated for the variable on the vertical axis, call it Y. Then we plot the sample means of Y versus the values of X. This approach can aid the visualization of the relationship between Y and X even though it does not use all the available information.

This completes the presentation of the pictorial tools in common use with the exception of the box plot, which is shown later in this chapter. The following material introduces the more frequently used statistics that aid us in describing and summarizing data.

III. MEASURES OF CENTRAL TENDENCY

Simple descriptive statistics can be useful in data editing as well as in aiding our understanding of the data. The minimum and the maximum values of a variable are useful statistics when editing the data. Are the observed minimum and maximum observed values reasonable or even possible? For the boys shown in Table 4.1, the minimum intake on the first day is 1053 calories and the maximum is 4322 calories. These values are somewhat unusual given that the average number of calories consumed by 12- to 15-year-old boys is about 2500, but they are not impossible. The boy whose intake was 1053 calories would not show much growth if he were to continue with this level of intake, but we all have unusual days. We consider other ways of identifying unusual values in later sections.

A. Mean, Median, and Mode

In terms of describing data, people usually think of the typical or average value. For example, the average caloric intake for boys was useful in determining whether or not the maximum and minimum values were reasonable. There are three frequently used measures of central tendency: the mean, the median, and the mode.

The *sample mean* ($\bar{x}$) is the sum of all the observed values of a variable divided by the number of observations. The *median*, previously mentioned, is defined to be the middle value, that is, the value such that 50 percent of the observed values fall above it and 50 percent fall below it. It can also be called the 50th percentile, where the ith percentile represents a value such that i percent of the observations are less than it. The mode is the most frequently occurring value.

Let us calculate these statistics for the caloric intake variable. The sample mean is the arithmetic average, that is,

$$\frac{1823 + 2007 + 1053 + \ldots + 1723}{33} = \frac{76{,}356}{33} = 2313.8$$

We can also represent the mean succintly using symbols. We use X as the symbol for the variable under study, in this case, the caloric intake of the boys based on their first food record. We use x, with subscripts to distinguish the boys' caloric intake from one another, to represent the observed value of the variable. For example, the first boy's intake is represented by x_1 and its value is 1823 calories. The second boy's intake is x_2 and his intake is 2007 calories. In the same way, x_3 is 1053, . . . , and x_{33} is 1723. Then the sum of the caloric intakes can be represented by

$$x_1 + x_2 + \cdots + x_{33} = \sum_{i=1}^{33} x_i.$$

The symbol Σ means summation. The value of i beneath Σ gives the subscript of the first x to be included in the summation process. The value above Σ gives the subscript of the last x to be included in the summation. The value of i increases in steps of one from the beginning value to the ending value. Thus, all the observations with subscripts ranging from the beginning value to the ending value are included in the sum. The formula for the sample mean variable, $\bar{x}$ (pronounced x-bar), is

$$\bar{x} = \frac{\sum_{i=1}^{n} x_i}{n}$$

In this case, the sample mean, $\bar{x}$, is 2313.8.

If we have the data for the entire population, not just for a sample of observations from the population, the mean is denoted by the Greek letter

μ (pronounced mu). Values that come from samples are *statistics* and values that come from the population are *parameters*. For example, the sample statistic $\bar{x}$ is an estimator of the population parameter μ. The population mean is defined as

$$\mu = \frac{\sum_{i=1}^{N} x_i}{N}$$

where N is the population size.

In calculating the median, it is useful to have the data sorted from the lowest to the highest value as that assists in finding the middle value. Table 4.7 shows the sorted caloric intake values for the 33 boys. For a sample of size n, the sample median is the value such that half ($n/2$) of the sample values are less than it and $n/2$ are greater than it. When the sample size is odd, the sample median is the $[(n + 1)/2]$th largest value. For a sample of size 33, the median is thus the 17th largest value. The value 17 comes from $(33 + 1)/2$. When the sample size is even, there is no observed sample value such that one-half of the sample falls below it and one-half falls above it. By convention, we use the average of the two middle sample values as the median, that is, the average of the $(n/2)$th and $[(n/2) + 1]$th largest values. For these data, the sample median is 2310, the 17th largest value.

The *mode* is the most frequently occurring value. When all the values occur the same number of times, we usually say that there is no unique mode. When two values occur the same number of times and more than any other values, the distribution is said to be bimodal. If three values occur the same number of times and more than any other value, the distribution could be called trimodal. Usually one would not go beyond trimodal in labeling a distribution.

For these data, none of the values occurs more than once and hence there is no unique modal value. This result is not unexpected when dealing

TABLE 4.7 Sorted Caloric Intakes for 33 Boys

1053	2007	2546
1292	2011	2594
1505	2039	2685
1645	2074	2748
1723	2301	2773
1753	2310	2842
1781	2330	3049
1823	2340	3076
1843	2348	3277
1898	2400	3532
2000	2436	4322

with continuous data as it is unlikely that two people have exactly the same values of a continuous variable.

B. Use of the Measures of Central Tendency

Now that we understand how these three measures of central tendency are defined and found, when are they used? Note that in calculating the mean, we summed the observations. Hence we can only calculate a mean when we can perform arithmetic operations on the data. For example, we cannot calculate the mean day for these data because we cannot perform meaningful arithmetic operations on nominal data. Therefore, the mean should be used only when we are working with continuous data, although sometimes we find it being used with ordinal data as well. The median does not require us to sum observations, and thus it can be used with continuous and ordinal data, but it also cannot be used with nominal data. The mode can be used with all types of data because it simply says which level of the variable occurs most frequently. Day 3, Tuesday, is the modal value for the days of the week as it occurs 9 times, more than any other day.

The mean is affected by extreme values, whereas the median is not. Hence, if we are studying a variable such as income which has some extremely large values, that is, it is is positively skewed, the mean will reflect these large values and move away from the center of the data. The median is unaffected, and it remains at the center of the data. For data that are symmetrically distributed or approximately so, the mean and median will be the same or very close to each other. The calories and vitamin A variables demonstrate situations in which the mean is close to the center (calories) as well as when it has moved away from the center (vitamin A). As was mentioned above, the caloric intake ranged from 1053 to 4322 for the 33 observations. The sample mean was 2313.8 and the sample median was 2310. These values are very similar as there were no extremely high caloric intakes. The corresponding observations for vitamin A show some relatively high intakes. Two values are above 12,000 international units (IU), and these values have caused the mean of 5326.3 IU to be larger than the median of 4535 IU. The median might be the more appropriate measure of central tendency in this case because it is unaffected by the two relatively large values.

C. The Geometric Mean

Another measure of central tendency is used when the numbers reflect population counts that are extremely variable. For example, in a laboratory setting, the growth in the number of bacteria per area is examined over time. The number of microbes per area does not change by the same amount from one period to the next, but the change is proportional to the number of microbes that were present during the previous period. Another way of saying this is that the growth is multiplicative, not additive. The areas under study may also have used different media, and the microbes may not do well in some of the media, whereas in other media the growth is explosive. Hence we may have counts in the hundred or thousands for some cultures and in the millions or billions for other cultures.

The arithmetic mean would not be close to the center of the values in this situation because of the effect of the extremely large values. The median could be used in this situation; however, another measure that is used in these situations is the geometric mean. The sample *geometric mean* for n observations is the nth root of the product of the values, that is,

$$\overline{x_g} = \sqrt[n]{x_1 * x_2 * \cdots * x_n}$$

Note that because the nth root is used in its calculation, the geometric mean cannot be used when a value is negative or zero.

This definition of the geometric mean is completely analogous to the definition of the arithmetic mean. The arithmetic mean is the value such that if we add it to itself $n - 1$ times, it equals the sum of all the observations. It is found by summing the observations and then dividing the sum by n, the sample size. Because in the situation above we are dealing with data resulting from a multiplicative process, our measure of central tendency should reflect this. The geometric mean is the value such that if we multiply it by itself $n - 1$ times, it equals the product of all the observations. It is found by multiplying the observations and then taking the nth root of the product.

When n is 2, there is little difficulty in finding the geometric mean as the product of the two observed values is usually not large and we know that the second root is the square root. For larger values of n, however, the product of the observed values may become very large and we may lose some accuracy in calculating it, even when we use a computer. Fortunately there is another way of calculating the product of the observations that does not cause any accuracy to be lost.

We can transform the observations to a logarithmic scale. Use of the logarithmic scale provides for accurate calculation of the geometric mean. After finding the logarithm of the geometric mean, we transform the value back to the original scale and have the value of the geometric mean. In this section, we use logarithms to the base 10 although other bases could be used.

To understand what we mean by logarithm, consider some positive number y. The base 10 logarithm of y is x where x satisfies the relation that 10^x equals y. For example, the base 10 logarithm of 10, often written as $\log_{10}(10)$, is 1 because 10^1 is 10. The value of $\log_{10}(100)$ is 2 because 10^2 equals 100. The value of $\log_{10}(1000)$ is 3 because 10^3, equal to $10 * 10 * 10$, is

1000. Therefore, base 10 logarithms of numbers between 10 and 100 will be between 1 and 2, base 10 logarithms of numbers between 100 and 1000 will be between 2 and 3, and so on. There are tables as well as keys on calculators and commands in MINITAB that can be used to find logarithms of positive numbers. Figure 4.11 shows a plot of base 10 logarithms of positive numbers up to 40.

The logarithms have negative values for numbers between 0 and 1. For example, using the definition of logarithms, the base 10 logarithm of 0.1 (= 1/10 = 10^{-1}) is −1.

A key property of the logarithmic transformation is that the level of the mathematical operation performed on the arithmetic scale is reduced a level when the logarithmic scale is used. For example, a product on the arithmetic scale becomes a sum on the logarithmic scale. Therefore the logarithm of the product of n values is

$$\log(x_1 * x_2 * \cdots * x_n) = \sum_{i=1}^{n} \log x_i.$$

In addition, taking the nth root of a product on the arithmetic scale becomes division by n on the logarithmic scale, that is, finding the mean logarithm. In symbols, this is

$$\sqrt[n]{x_1 * x_2 * \cdots * x_n} = \frac{\sum_{i=1}^{n} \log_{10} x_i}{n} = \overline{\log_{10} x}.$$

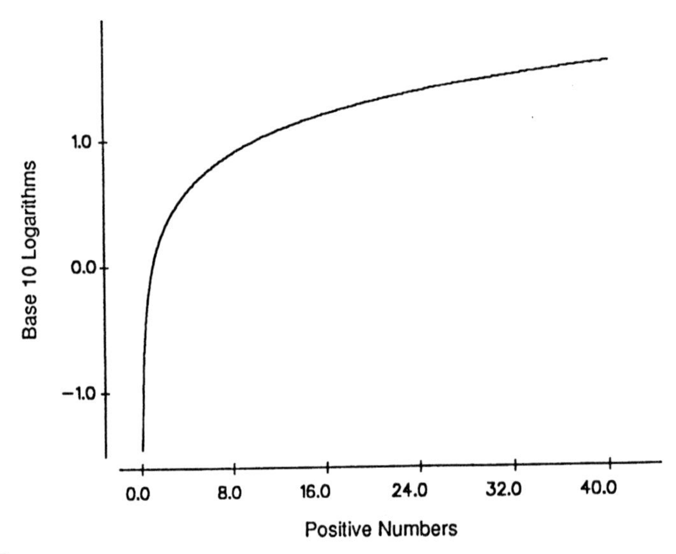

FIGURE 4.11 Plot of base 10 logarithms of positive numbers up to 40.

We now have the logarithm of the geometric mean, and to obtain the geometric mean, we must take the antilogarithm of the mean logarithm, that is,

$$\bar{x}_g = \text{antilog}\ (\overline{\log_{10} x})$$

Suppose that the number of microbes observed from six different areas are 100, 100, 1000, 1000, 10,000, and 1,000,000. The geometric mean is found by taking the logarithm of each observation and then finding the mean logarithm. The corresponding base 10 logarithms are 2, 2, 3, 3, 4, and 6 and their mean is 3.33. The geometric mean is the antilog of 3.33, which is 2154.43. The arithmetic mean of these observations is 168,700, a value much larger than the geometric mean and also much larger than five of the six values. The usual mean does not provide a good measure of central tendency in this case. The value of the median is the average of the two middle values, 1000 and 1000, giving a median of 1000 which is of the same order of magnitude as the geometric mean.

The geometric mean has also been used in the estimation of population counts, for example, of mosquitos, through the use of capture procedures over several time points or areas. These counts can be quite variable by time or area; hence, the geometric mean is the preferred measure of central tendency in this situation too.

These are the more common measures of central tendency employed in the description of data. The value of the central tendency does not completely describe the data, however. For example, consider the nine observations of calories made on day 3:

1823 2685 2842 2330 2301 1781 2773 2310 1723.

Suppose that the four smallest observations were decreased by 1000 calories and the four largest were increased by 1000 calories. The values would now be the following:

823 3685 3842 3330 1301 781 3773 2310 723.

The means and medians of these two data sets are the same, yet the sets are very different. The sample mean of 2285.3 and the sample median of 2310 capture the essence of the first data set. In the second data set, however, the measures of central tendency are less informative as only one value is close to the mean and median. Therefore, some additional characteristics of the data must be used to provide for a more complete summary and description of the data and to distinguish between dissimilar data sets. The next section deals with this additional characteristic, the variability of the data.

IV. MEASURES OF VARIABILITY

The observations in the second set above varied much more than those in the first set, but the means were the same. Hence to provide for a more complete description of the data, we need to include a measure of its variability. A number of measures or values—the range, the interquartile range, selected percentiles, the variance, the standard deviation, and the coefficient of variation—are used to describe the variability in data.

A. Range and Percentiles

The range is defined as the maximum value minus the minimum value. It is simple to calculate and it provides some idea of the spread of the data. For the first data set above, the range is the difference between 2842 and 1723, which is 1119. In the second data set the range is found by subtracting 723 from 3842, which yields 3119. The large difference in the two ranges points to a difference between the two data sets. Although the range can be informative, the range has two major deficiencies: (1) it ignores most of the data since only two observations are used in its definition; and (2) its value depends indirectly on sample size. The range will either remain the same or increase as more observations are added to a data set; it cannot decrease. A better measure of variability would use more of the information in the data by using more of the data points in its definition and would not be dependent on sample size.

The *interquartile range*, the difference between the 75th and 25th percentiles (also called the third and first quartiles) uses more information from the data than does the range. In addition, the interquartile (or semi-quartile) range can either increase or decrease as the sample size increases. The interquartile range is a measure of the spread of the middle 50 percent of the values. Finding the value of the interquartile range requires that the first and third quartiles be specified and there are several reasonable ways of calculating them. We use the following procedure to calculate the 25th percentile for a sample of size n:

1. If $(n + 1)/4$ is an integer, then the 25th percentile is the value of the $(n + 1)/4$th smallest observation.

2. If $(n + 1)/4$ is not an integer, then the 25th percentile is a value between two observations. For example, if n is 22, $(n + 1)/4$ is $(22 + 1)/4 = 5.75$. The 25th percentile then is a value three-fourths of the way between the 5th and 6th smallest observations. To find it, we sum the 5th smallest observation and 0.75 of the difference between the 6th and 5th smallest observations.

The sample size is 9 for the two data sets above. According to our procedure, we first calculate $(9 + 1)/4$, which is 2.5. Hence the 25th percentile is a value halfway between the second and third smallest observations. When the value is halfway between two observations, the above formula reduces to the average of the two observations. Therefore, for the first data set, the 25th percentile is (1781 + 1823)/2, which is 1802. For the second data set, the value is (781 + 823)/2 = 802. The 75th percentile is found in the same way except that we use $3*(n + 1)/4$ in place of $(n + 1)/4$. The 75th percentiles are 2729 and 3729 for the first and second data sets, respectively. Hence the interquartile ranges are 2729 − 1802 = 927 and 3729 − 802 = 2927 in the two data sets. These values show that there is relatively little difference in caloric intake in the middle 50 percent of boys in the first data set, whereas there is a tremendous spread, more than three times as large as that in the first data set, in caloric intake in the middle 50 percent of the boys in the second set. The interquartile range in the second set suggests that there is tremendous variability in the data and that the measures of central tendency are of far less interest than in the first data set.

The values of five selected percentiles—10th, 25th, 50th, 75th and 90th—when considered together provide good descriptions of the central tendency and the spread of the data. When the sample size is very small, however, the calculation of the extreme percentiles is problematic. For example, when n is 5, it is difficult to determine how the 10th percentile should be calculated. Because of this difficulty and also because of the instability of the extreme percentiles for small samples, we calculate them only when the sample size is reasonably large, say larger than 30. Therefore we calculate these percentiles for the caloric intake of all 33 boys in Table 4.1 instead of using only the nine observations from day 3. In calculating the first and ninth deciles (10th and 90th percentiles), we use the same idea that was used with the quartiles.

Table 4.7 shows the sorted caloric intakes. The 50th percentile, the median, was already found to be 2310 calories. The 25th percentile is based on $(n + 1)/4$, which equals 8.5. Therefore the 25th percentile is the average of the 8th and 9th smallest observations. In this case, the 8th and 9th smallest values are 1823 and 1843, respectively. Thus the first quartile value is 1833 calories. The third quartile, the 75th percentile, is based on $3*(n + 1)/4$, which is 25.5. Therefore the third quartile is the average of the 25th and 26th smallest values.

For these data the third quartile is (2685 + 2748)/2, which is 2716.5 calories. The first decile, the 10th percentile, is based on $(n + 1)/10$, which is 3.4 for these data. Therefore the first decile is the sum of the third smallest value, 1505 calories, plus 0.4 times the difference between the fourth and third smallest values. The first decile is then 1505 plus 0.4 multiplied by 1645–1505, which is 1561 calories. The ninth decile is calculated in a similar fashion and found to be 3196.6 calories.

The values of the five percentiles are thus 1561, 1833, 2310, 2716.5, and 3196.6 calories. These values tell us that an intake of less than 1561 calories is somewhat unusual, with only 10 percent of the sample having an intake less than that. Intakes greater than 3197 calories are also somewhat unusual, with only 10 percent of the sample having values greater than that. The middle 50 percent of the sample have intakes between 1833 and 2717 calories and 50 percent of the sample have intakes greater than 2310 calories. These five numbers provide a good summary of the central tendency and the variability in the caloric intakes of the sample. Other values, for example, the 5th and 95th percentiles or the minimum and maximum, are sometimes used in place of the 10th and 90th percentiles. The next measure of variability to be discussed is the variance but, before considering it, we discuss the box plot because of its relationship to the five percentiles.

B. Box Plot

The *box plot* graphically gives the approximate location of the quartiles, including the median, and the extreme values. The box plot can also reveal whether or not the data are symmetrically distributed. The figure in Box 4.11 is a box plot of the vitamin A data from Table 4.1.

The lower and upper ends (*hinges*) of the box mark approximate locations of the first and third quartiles, respectively, and the plus symbol gives the approximate location of the median. The first quartile is thus approximately 2250 IU, the median is approximately 4500 IU, and the third quartile is about 8000 IU. The dashes (*whiskers*) indicate how far the data extend beyond the hinges. The difference between the upper and lower hinges is approximately the interquartile range, about 5750 IU. If any values are very unusual, they are indicated by either an asterisk or a zero. Unusual points are identified in relation to the interquartile range. Values from [1.5 to 3 times the interquartile range] less (greater) than the lower (upper) hinge are represented by an asterisk. Values more than 3 times the interquartile less (greater) than the lower (upper) hinge are represented by a zero. There are no very unusual values for the vitamin A data according to these criteria. The distance from the median to the third quartile is much

MINITAB BOX 4.11

The vitamin A data from Table 4.4 are entered into c1.

```
MTB > boxp c1

               ------------------------
        ------I           +           I------------------
               ------------------------

+---------+---------+---------+---------+---------+------C1
0       2500      5000      7500     10,000    12,500
```

greater than the corresponding distance to the first quartile, and the whiskers in the upper tail extend much further than the whiskers in the lower tail of the distribution. This picture makes it very clear that the vitamin A data are very skewed.

Box 4.12 shows how to create box plots for subgroups.

MINITAB BOX 4.12

The systolic blood pressure data are treated as coming from two groups to demonstrate the BY subcommand, which is an option of the BOXPLOT statement. The first group contains the first 50 values and the second group contains the other set of 50 values. This is easily accomplished in MINITAB as the following commands show. First a column is created in which 50 ones are followed by 50 twos. This column, c12, is used with the BY subcommand in the BOXPLOT statement as shown.

```
MTB > set c12
DATA> 50(1) 50(2)
DATA> end
```

The number before the parentheses tells how many of the values in the parentheses are to be created, in this case 50 ones and 50 twos.

```
 MTB > boxp c8;
 SUBC> by c12.
C12
                      ---------------
1          -------------I      +  I------------
                      ---------------
                  ----------------
2       -------------I      +     I--------------------    *
                  ----------------
   -------+----------+----------+----------+----------+----------+C8
         84         96        108        120        132        144
```

The use of the BY subcommand makes it possible to visually compare two or more distributions. The first 50 blood pressure values appear to be more symmetrically distributed than the second set, to have less variability and no extreme values, and to have a slightly higher median value.

C. Variance and Standard Deviation

The variance and its square root, the *standard deviation* are the two most frequently used measures of variability and both use all the data in their calculations. The *variance* measures the variability in the data from the mean of the data. The population variance, denoted by σ^2, for a population of size N is defined as

$$\sigma^2 = \frac{\sum_{i=1}^{N} (x_i - \mu)^2}{N}.$$

For a sample of size n, the sample variance, s^2, an estimator of σ^2, is defined by

$$s^2 = \frac{\sum_{i=1}^{n} (x_i - \bar{x})^2}{n - 1}.$$

The population variance could be interpreted as the average squared difference from the population mean and the sample variance has almost the same interpretation about the sample mean.

The variance uses the sum of the squared differences from the mean and the sample variance uses $n - 1$ in its denominator. Why were the squared differences chosen for use instead of the differences themselves? Perhaps Table 4.8 will clarify this. The sum of calories minus the mean, which would be zero except for rounding, must be zero because the positive differences cancel the negative differences.

Additionally, why is $n - 1$ used instead of n in the denominator of the sample variance? It can be shown mathematically that the use of n results in an estimator of the population variance that on the average slightly underestimates it. The following gives some feel for the use of $n - 1$.

In the formula for the sample variance, the population mean is estimated by the sample mean. This estimation of the population mean reduces the number of independent observations to $n - 1$ instead of n, as is shown next.

TABLE 4.8 Differences and Squared Differences from the Mean for the Nine Observations of Day 3 Caloric Intake

	Calories	Sample mean	Calories − mean	(Calories − mean)²
	1,823	2,285.33	−462.33	213,749
	2,685	2,285.33	399.67	159,736
	2,842	2,285.33	556.67	309,881
	2,330	2,285.33	44.67	1,995
	2,301	2,285.33	15.67	246
	1,781	2,285.33	−504.33	254,349
	2,773	2,285.33	487.67	237,822
	2,310	2,285.33	24.67	609
	1,723	2,285.33	−562.33	316,215
Total	20,568	20,567.97	0.03	1,494,602

For example, you are told that there are three observations and that two of the values along with the sample mean are known. Can you find the value of the other observation? If you can, this means that there are only two independent observations, not three, once the sample mean is calculated. Suppose that the two values are 6 and 10 and the sample mean is 9. As the mean of the three observations is 9, this indicates that the sum of the values is 27 and that the unknown value is 27 − (6 + 10), which is 11. In this sample of size 3, given knowledge of the sample mean, only two of the observations are independent or free to vary. Hence once a parameter, in this case the population mean, is estimated from the data, it reduces the number of independent observations (*degrees of freedom*) by one. To account for this reduction in the number of independent observations, $n - 1$ is used in the denominator of the sample variance.

For the nine caloric values in Table 4.8, the nine values from the first of the two data sets given above, the value of the *sample variance* is 1,494,602/(9 − 1), which is 186,825.3. This number is large, but is hard to interpret as it is in squared units. Because of this, the square root of the sample variance, called the *sample standard deviation,* is also often used as a measure of variability. The sample standard deviation, s, on the average slightly underestimates the population standard deviation σ. For these data, the value of the standard deviation is $\sqrt{186{,}825.3}$, which is 432.2. The sample variance and standard deviation for the nine values in the second data set are 1,937,325.2 and 1391.9, respectively, values much larger than the corresponding statistics for the first nine values. These statistics reflect, as they must, the much greater variation in the second data set than in the first data set.

The above calculations showed how the variance changed with nonconstant changes in the data. How does the value of the variance change when (1) a constant is added to (subtracted from) all the observations in the data set and (2) all the observations are multiplied (divided) by a constant?

The answer to the first question is that there is no change in the value of the variance as can be seen from the following. If all the observations are increased by a constant, say by 10 units, the mean is also increased by the same amount. Therefore, the constants simply cancel each other out in the squared differences, that is,

$$[(x_i + 10) - (\mu + 10)]^2 = (x_i - \mu)^2$$

and thus there is no change in the sum of the squared differences or in the variance.

When all the observations are multiplied by a constant, the variance is multiplied by the square of the constant as can be seen from the following. If all the observations are multiplied by a constant, say by 10, the mean is also multiplied by the same amount. Therefore in the squared differences we have

$$[(x_i * 10) - (\mu * 10)]^2 = [(x_i - \mu) * 10]^2 = (x_i - \mu)^2 * 10^2$$

and the sum of the squared differences, and thus the variance, is multiplied by the constant squared. This means that the standard deviation is multiplied by the constant. These two properties will be used in Chapter 6.

In later chapters, the variance and the standard deviation are shown to be the most appropriate measures of variation when the data come from a *normal distribution*, as knowledge of them and the mean is all that is necessary to completely describe the data. The normal distribution is the bell-shaped distribution often used in the grading of courses; it is the most widely used distribution in statistics. The interquartile range and the five percentiles are useful statistics for characterizing the variation in data regardless of the distribution from which the data are selected, but they are not as informative as the mean and variance are when the data come from a normal distribution.

One last measure of variation is the *coefficient of variation*, defined as 100 percent times the ratio of the standard deviation to the mean. In symbols this is $(\sigma/\mu) * 100$ percent, and it is estimated by $(s/\bar{x}) * 100$ percent. The coefficient of variation is a relative measure of variation, because dividing by the mean directly takes the magnitude of the values into account. Large values of the coefficient suggest that the data are quite variable.

The coefficient of variation has several uses. One use is in comparison of the precision of different studies. If another experiment has a coefficient of variation much smaller than that in your study of the same substance, this suggests that there may be room for improvement in your study procedures. Another use is in determination of whether or not there is so much variability in the data that the measure of central tendency is of little value. For example, the National Center for Health Statistics (NCHS) does not publish sample means for variables if the estimated coefficient of variation is greater than 30 percent.

Let us calculate the estimated coefficients of variation for our two sets of nine observations. For the first set, s was 432.2, and for the second set, 1391.9. The sample mean was 2285.3 in both sets which leads to coefficients of variation of 18.9 percent [= (432.2/2285.3) * 100%] and 60.9 percent in sets 1 and 2, respectively. These values reinforce our feeling that the mean provided useful information in the first set, but was of less value in describing the data in the second set.

Box 4.13 shows how to calculate descriptive statistics.

Box 4.14 shows how to sort data for the calculation of the median and other percentiles. The sorted data facilitate the calculation of the percentiles because the data are ordered by size. For example, in the text we stated that the first decile (the 10th percentile) of the 33 calorie values is found by taking the third smallest calorie value plus 0.4 times the differ-

MINITAB BOX 4.13

We use the DESCRIBE (or DESC for short) command to calculate the descriptive statistics for the data stored in columns and to read the output.

```
MTB > desc c1-c4
                N      MEAN    MEDIAN    TRMEAN     STDEV    SEMEAN
vit A          33      5326      4535      5127      3364       586
calories       33      2314      2310      2281       668       116
tot fat        33     90.39     83.00     86.24     42.84      7.46
protein        33     84.85     84.00     83.31     33.89      5.90

              MIN       MAX        Q1        Q3
vit A         820     12812      2265      8275
calories     1053      4322      1833      2717
tot fat     25.00    227.00     65.50    118.00
protein     23.00    172.00     63.00     98.50
```

Most of the column headings are clear. STDEV stands for standard deviation and the first and third quartiles are indicated by Q1 and Q3. SEMEAN will be discussed later, but it stands for the standard error of the mean and is found by dividing STDEV by the square root of *N*. TRMEAN stands for trimmed mean and it attempts to remove the effect of the extreme observations from the calculation of the mean. It does this by removing the smallest and largest 5 percent of the values and then calculates the mean of the remaining observations. With 33 values, 5 percent of 33 is 1.65 and this is rounded to 2. Hence the two smallest and two largest values are deleted and the mean is calculated on the remaining 29 values.

ence between the third and fourth smallest values. From c9, the third and fourth smallest values are easily seen to be 1505 and 1645 and the first decile is then 1505 + 0.4 * 140, which equals 1561.

The coefficient of variation (CV) is not part of the output from Box 4.13 and thus additional MINITAB commands are required to find its value, as

MINITAB BOX 4.14

The SORT command is used to rearrange the data. In the following, c9 will receive the sorted calorie values.

```
MTB > sort 'calories' c9
MTB > print c9
  C9
1053 1292 1505 1645 1723 1753 1781 1823 1843 1898 2000
2007 2011 2039 2074 2301 2310 2330 2340 2348 2400 2436
2546 2594 2685 2748 2773 2842 3049 3076 3277 3532 4322
```

is shown in Box 4.15. Thus the CV for calories for these 33 boys is almost 29 percent, denoting a large variation in these values.

MINITAB BOX 4.15

In the following, k1 is the name of the variable that contains the standard deviation of the data stored in column c2 and k2 is the name of the variable that contains the mean. MINITAB uses the letter *k* followed by a number to identify a single value.

```
MTB > stdev c2 k1
    ST.DEV. =        667.89
MTB > mean c2 k2
    MEAN     =        2313.8
```

The CV is 100 percent times the ratio of k1 to k2, and k3 will contain that value.

```
MTB > let k3=(k1/k2)*100
MTB > print k3
K3          28.8655
```

MINITAB BOX 4.16

First we enter the data used in the example for calculating the geometric mean.

```
MTB > set c10
DATA> 100 100 1000 1000 10000 1000000
DATA> end
```

We next require the logarithms and we use those with 10 as the base; the LET command is used again.

```
MTB > let c11=logten(c10)
MTB > print c11
C11
   2.00000    2.00000    3.00000    3.00000    4.00000    6.00000
```

The next step is to find the mean of the logarithms and then take the antilogarithm of the mean. This value is the geometric mean.

```
MTB > mean c11 k4
   MEAN     =        3.3333
MTB > antilog k4 k5
     ANSWER =        2154.43
```

Thus the geometric mean is 2154.43 for these data, the same value reported earlier in this chapter.

Box 4.16 illustrates the calculation of the geometric mean.

As can be seen from this material, MINITAB is easy to use and it can greatly reduce the burden of the calculations. Other software could be used instead of MINITAB. Regardless of the software used, you are encouraged to use the computer in carrying out the calculations to allow time for thinking about what you are analyzing and why.

V. RATES: CRUDE, SPECIFIC, AND ADJUSTED

The rates of diseases and vital rates, which include death rates in general, infant mortality rates, feto-infant, neonatal, and postneonatal mortality rates, and birth rates, are frequently used measures in public health. These rates are useful in determining the health status of a population, in monitoring the health status over time, in comparing the health status of populations, and in assessing the impact of policy changes.

For example, the infant mortality rate is often used in comparing the performance of health systems in different countries. In 1988, the United States had an infant mortality rate higher than that of 22 other nations. The U.S. rate was 10.0 infant deaths under 1 year of age per 1000 live births compared with a low rate of 4.8 for Japan. Most of the Western European nations and some Pacific Rim nations (Japan, Singapore, and Hong Kong) had lower rates than the United States. Canada's health system is often touted as a model for the United States because of its lower cost. How does Canada's infant mortality rate compare with that of the United States? Canada's infant mortality rate in 1988 was 7.2, almost 30 percent lower than the U.S. rate. The progress in reducing infant mortality has been most impressive as can be seen from the U.S. rate for 1967 of 22.4 shown in Figure 1.1 and the 1988 rate of 10.0.

As can be seen from the following definition, a rate is basically a mean multiplied by a constant. A rate is defined as the product of two parts: (1) the number of persons who have experienced the event of interest divided by the population size; and (2) a standard population size. For example, according to the data compiled by the Harris County Health Department, there were 15,585 deaths in an estimated population of 2,942,550 in Harris County, Texas, in 1986. The corresponding death rate per 100,000 is found by taking (15,585/2,942,550) * 100,000, and it equals 529.6 deaths per 100,000 population. This is considerably lower than the corresponding rate for the United States of 873.2 deaths per 100,000. This difference will be explored in a later section.

As is often the case with rates, however, there is a problem in determining the value of the denominator, that is, the 1986 Harris County population. What is meant by the 1986 population size? Is it as of January 1, July 1, December 31, or some other date? Convention is that the population size in the middle of the period (mid-1986) is used. An additional problem is that census data were available for 1980 but not for 1986 which introduces some uncertainty in the value used. In this case, the Harris

County Health Department used an estimate of the 1986 population based on projections from the Texas Department of Health. The uncertainty in the value of the denominator of the rate should be of little concern given the magnitude of the numbers involved in this situation.

Death rates are usually expressed per 1000 or per 100,000 population. As was mentioned above, infant mortality rates are expressed per 1000 live births with the exception of feto-infant mortality rates. Feto-infant mortality rates are based on the number of late fetal deaths plus infant deaths under 1 year per 1000 live births plus late fetal deaths. Neonatal mortality rates are based on deaths of infants who were less than 28 days old, and postneonatal rates are based on deaths of infants between 28 and 365 days of age. This split of infant deaths is useful because often the neonatal deaths may be the result of genetic factors, whereas the postneonatal deaths may have more to do with the environment. The birth rate is defined as 1000 times the ratio of the number of live births to the population size.

Note that as the infant mortality rate example in Chapter 1 showed, the children whose deaths are used in the conventional method of calculating this rate may have been born in 1987, not 1988. Hence the numerator, the number of deaths, comes from both 1987 and 1988 births, whereas the denominator is based solely on 1988 births. This should cause no problem unless something happened that caused the mortality experience or the number of births to differ greatly between the two years. One way of dealing with this possibility of a difference between the years is to combine several years of data. Often health agencies pool data over 3 years to provide protection against the instability of small numbers and to reduce the possible, but unlikely, effect of very different birth or mortality experiences across the years.

A. Crude and Specific Rates

Rates may be either crude or specific. *Crude rates* use the total number of events in their definition, whereas specific rates apply to subgroups in the population. For example, there may be age-, gender-, or race-specific death rates. For an age-specific death rate, only the deaths of individuals in the specific age group are used in the numerator and the denominator is the total number of individuals in the specific age group. *Specific rates* are used because they supply more information and also allow for more appropriate comparisons of groups.

For example, perhaps the difference seen above in the Harris County and U.S. death rates for 1986 is related to age. The age-specific rates, shown in Table 4.9, provide a better description of the mortality experience than the crude rates of 529.6 and 873.2 for Harris County and the United States, respectively. This table shows that those less than 25 years old in

TABLE 4.9 Age-Specific Mortality Rates per 100,000 for Harris County, Texas, and the United States as well as U.S. Deaths and Population for 1986

	Deaths per 100,000 in		United States	
Age	Harris County	United States[a]	Deaths	Population
			(in 1000s)	
0–4	250.2	255.4	46.4	18,152
5–14	19.6	26.0	8.8	33,860
15–24	99.8	102.3	39.9	39,021
25–34	146.8	132.1	56.5	42,779
35–44	218.5	212.9	70.4	33,070
45–54	464.7	504.8	115.2	22,815
55–64	1320.2	1255.1	279.0	22,232
65–74	2832.8	2801.4	485.5	17,332
≥75	8101.1	8470.9	1002.6	11,836
Total			2104.3	241,097

[a] Rates may not exactly equal the ratio of deaths to population because of rounding.
Sources: "Health, United States, 1990," Tables 1 and 23 (4) and "The Health Status of Harris County Residents, 1980–1986," Tables 3.F and Appendix (5).

Harris County have lower mortality rates than the corresponding U.S. groups, but from then on the results are mixed. Without knowledge of the age distributions, it is difficult to conclude whether or not the age variable is responsible for the difference in the crude rates.

As shown above, one problem with the use of specific rates is that they are not easily summarized. They do provide more information than the crude rate which gives a single value for a population, but sometimes it is difficult to draw a conclusion based on the examination of the specific rates. However, because of the strong linkage between mortality and age, age often must be taken into account in the comparison of two or more populations. One way of adjusting for age or other variables while avoiding the problem of many specific rates is to use adjusted rates.

B. Adjusted Rates

Adjusted rates are weighted rates as is shown below. There are direct and indirect methods of adjustment; the choice of which method to use depends on what data are available. The *direct method* requires that we have the specific rates for each population and a standard population. Table 4.9

provides the age-specific death rates for both populations of interest. The *standard population* provides a referent for purposes of comparison. To provide more stable values, the standard population is usually larger than the population(s) under study. The choice of a standard population is subjective. For example, in comparing the rates between states, often the U.S. population would be used as the standard. In comparing counties of a state, the state population often would be used as the standard. For comparing rates over time, the population at a previous time point could be used as the standard. Another alternative might be to pool the populations of the areas or times under study and use the pooled population as the standard. In performing the age adjustment here, we have decided to use the 1986 U.S. population shown in Table 4.9 as the standard. The age-adjusted rate for Harris County differs from its crude rate, reflecting the effect of using the 1986 U.S. age distribution.

The adjustment process consists of applying the Harris County age-specific mortality rates to the standard population's age distribution and then summing the expected number of deaths over the age categories. Another way of saying this is that each age category's mortality rate is weighted by that age category's share of the standard population. Table 4.10 shows the calculation of age-adjusted death rate for Harris County by the direct standardization method. Hence the direct age-adjusted death rate for Harris County using the United States as the standard population is 860.9 deaths per 100,000 population, quite a contrast to the crude rate of 529.6 and very close to the U.S. rate of 873.2. The difference in crude rates between Harris County and the United States can be accounted for by the

TABLE 4.10 Direct Method of Adjusting the 1986 Harris County Death Rate Using 1986 U.S. Population as the Standard

	Harris County Population				
Age	Number	Proportion	Specific rates (1)	U.S. population proportion (2)	Expected deaths per 100,000 (1) * (2)
0–4	253,776	0.0862	250.2	0.0753	18.84
5–14	469,446	0.1595	19.6	0.1404	2.75
15–24	489,053	0.1662	99.8	0.1618	16.15
25–34	640,813	0.2178	146.8	0.1774	26.04
35–44	444,366	0.1510	218.5	0.1372	29.98
45–54	275,007	0.0935	464.7	0.0946	43.96
55–64	190,352	0.0647	1320.2	0.0922	121.72
65–74	111,870	0.0380	2832.8	0.0719	203.68
≥75	67,867	0.0231	8101.1	0.0491	397.76
Total	2,294,550	1.0000		0.9999	860.88

difference in the age distributions. As shown in Table 4.10, Harris County had proportionately far fewer persons over 55 years of age than did the United States and this contributed to its much lower crude death rate. After adjustment for the age distribution, there was little difference between the Harris County and U.S. death rates in 1986.

The *indirect method* is an alternative to be used when we do not have the data required for the direct method or when the specific rates may be unstable because they were based on small numbers. The indirect method requires the specific rates for the standard population and the age (or, e.g., gender or race) distribution for the population to be adjusted. It is more likely that these data will be available than the age-specific death rates in the population to be adjusted. The first step in calculating the indirect age-adjusted death rate is to multiply the age-specific death rates of the standard (U.S.) population by the corresponding age distribution of the population to be adjusted (Harris County). Table 4.11 shows this calculation for the Harris County data; in this example we ignore the availability of the Harris County age-specific death rates. The observed crude death rate for Harris County is 529.6 and the expected rate when the U.S. age-specific mortality rates are applied is 534.6. The ratio of observed to expected death rates, the standardized mortality ratio, is 0.99 which indicates that Harris County's death rate is very similar to that of the United States once age is taken into account. To find the indirect age-adjusted death rate for Harris County, we now multiply the crude rate for the standard population, the United States, by the value 0.99. Thus the indirect age-adjusted mortality rate for Harris County is 0.99 × 873.2, which equals 864.5 deaths per 100,000 population.

TABLE 4.11 The First Step in the Calculation of the 1986 Indirect Age-Adjusted Death Rate for Harris County Using the Estimated 1986 U.S. Population as the Standard

Age	Harris County age distribution	*	U.S. age-specific death rates per 100,000	=	Expected deaths per 100,000
0–4	0.0862		255.4		22.0
5–14	0.1595		26.0		4.1
15–24	0.1662		102.3		17.0
25–34	0.2178		132.1		28.8
35–44	0.1510		212.9		32.1
45–54	0.0935		504.8		47.2
55–64	0.0647		1255.1		81.2
65–74	0.0380		2801.4		106.5
≥75	0.0231		8470.9		195.7
Total	1.0000				534.6

In this case, both the direct and indirect age-adjusted death rates for Harris County are very similar to one another and to the U.S. crude rate. The difference in the crude death rates between the United States and Harris County disappeared once the age distributions were taken into account.

The calculation of adjusted rates can be easily done with MINITAB. For example, the calculation of the indirect adjusted rate above is demonstrated in Box 4.17.

MINITAB BOX 4.17

Column c1 contains the Harris County age distribution, and c2 contains the U.S. age-specific death rates. The constant k1 contains the crude death rate for Harris County, k2 contains the crude death rate for the United States, and k3 contains the standardized mortality ratio.

```
MTB > set c1
DATA> .0862 .1595 .1662 .2178 .1510 .0935 .0647 .0380 .0231
DATA> set c2
DATA> 255.4 26.0 102.3 132.1 212.9 504.8 1255.1 2801.4 8470.9
DATA> end

MTB > mult c1 c2 c3
MTB > sum c3 k3
   SUM = 534.62
MTB > let k1=529.6
MTB > let k2=873.2
MTB > let k4=(k1/k3)*k2
MTB > print k4
K4        865.003
```

Note that MINITAB's value for the indirect age-adjusted mortality rate is 865.003, slightly larger than the value of 864.5 found above. This difference is due to our use of only two decimal places for the standardized mortality rate instead of the more precise value used by MINITAB.

It is possible to adjust for more than one variable at a time; for example, age and gender are often used together. Gender is frequently used because the mortality experiences are often quite different for females and males.

VI. CORRELATION COEFFICIENTS

Earlier in the chapter, we presented a scatter plot of protein and total fat, and we concluded that there was a strong, although imperfect, positive association between protein and fat. Although this statement is informative, it is imprecise. To be more precise, a numerical value that reflects the strength of the association is needed. *Correlation coefficients* do just that; that is, they reflect the strength of association.

A. Pearson Correlation Coefficient

The most widely used measure of association between two variables, X and Y, is the *Pearson correlation coefficient* denoted by ρ (rho) for the population and by r for the sample. This measure is named after Karl Pearson, a leading British statistician of the late 19th and early 20th century, for his role in the development of the formula for the correlation coefficient.

We want the correlation coefficient to be large, approaching $+1$ as a limit, as the values of the X, Y pair show an increasing tendency to be large or small together. When the values of the X, Y pair tend to be opposite in magnitude, that is, a large value of X with a small value of Y or vice versa, the measure should be large negatively, approaching -1 as the limit. If there is no overall tendency of the values of the X, Y pairs, the measure should be close to 0.

By large or small, we mean in relation to its mean. Because of the above requirements for the correlation coefficient, one simple function that may be of interest here is the product of $x_i - \bar{x}$ and $y_i - \bar{y}$. Let us focus on the sign of the differences, temporarily ignoring the magnitude. The possibilities are the following:

$x_i - \bar{x}$	$y_i - \bar{y}$	*Product*
+	+	+
−	−	+
+	−	−
−	+	−

The product of the differences does what we want; that is, it is positive when the X, Y pairs are large or small together and negative when one variable is large and the other variable is small. The sum of the products of the differences over all the sample pairs should give some indication of whether there is a positive, negative, or no association in the data. If all the products are positive (negative), the sum will be a large positive (negative) value. If there is no overall tendency, the positive terms in the sum will tend to cancel out with the negative terms in the sum, driving the value of the sum toward zero.

The value of the sum of the products depends on the magnitude of the data. As we want the maximum value of our measure to be 1, we must do something to remove the dependence of the measure on the magnitude of the values of the variables. If we divide the measure by something reflecting the variability in the X and Y variables, this should remove this dependence. The actual formula for r, reflecting these ideas, is

$$r = \frac{\sum_{i=1}^{n} (x_i - \bar{x}) * (y_i - \bar{y})}{\sqrt{\sum_{i=1}^{n} (x_i - \bar{x})^2 * \sum_{i=1}^{n} (y_i - \bar{y})^2}}$$

Dividing the numerator and denominator of this formula by $n - 1$ enables us to rewrite the formula in terms of familiar statistics, that is,

$$r = \frac{\sum_{i=1}^{n} (x_i - \bar{x}) * (y_i - \bar{y})/(n - 1)}{\sqrt{s_x^2 * s_y^2}}.$$

In this version, we used the formula for the sample variance, that is, $s_x^2 = \sum(x_i - \bar{x})^2/(n - 1)$. The sample variance can also be expressed as $s^2 = \sum(x_i - \bar{x}) * (x_i - \bar{x})/(n - 1)$. Hence the sample variance could be said to measure how the X variable varies with itself. The numerator looks very similar to this, and it measures how the variables X and Y covary.

The denominator, $\sqrt{s_x^2 * s_y^2}$, standardizes r so that it varies from -1 to $+1$. For example, if $Y = X$, then the numerator becomes $\sum(x_i - \bar{x})^2/(n - 1)$, that is, s_x^2, which is the same as the denominator and their ratio is $+1$.

We can use MINITAB to find r for the protein and total fat data from Table 4.1, as shown in Box 4.18. This value shows that protein and total fat have a strong positive association with one another.

MINITAB BOX 4.18

```
MTB > corr c3 c4
Correlation of C3 and C4 = 0.648
```

The following example shows that ρ is not a general-purpose measure of association, but that it measures linear association, that is, the tendency of the x_i, y_i pairs to lie on a straight line. The values of Y and X are the following:

Y	4	1	0	1	4
X	−2	−1	0	1	2

The sample mean of Y is 2 and the sample mean of X is 0. The pieces required to calculate r are the following:

	Y	X	(Y − 2)	*	(X − 0)	=	Product	$(Y - 2)^2$	$(X - 0)^2$
	4	−2	2	*	−2		−4	4	4
	1	−1	−1	*	−1		1	1	1
	0	0	−2	*	0		0	4	0
	1	1	−1	*	1		−1	1	1
	4	2	2	*	2		4	4	4
Total	10	0	0		0		0	14	10

The estimated Pearson correlation coefficient, r, is then $0/(\sqrt{14 * 10}) = 0$. There is no linear association between Y and X. Note, however, that the first column (values of Y) and the last column (X^2) are the same. Hence there is a perfect quadratic (squared) relationship between Y and X that was not found by the Pearson correlation coefficient. The scatter plot in Box 4.19 graphically shows this relationship.

Thus, even if r is 0, it does not mean that the two variables are unrelated; it means that there is no linear relationship between the two variables. The use of a scatter plot first, followed by the calculation of r, may find the existence of a nonlinear association that could be missed when r alone is used.

MINITAB BOX 4.19

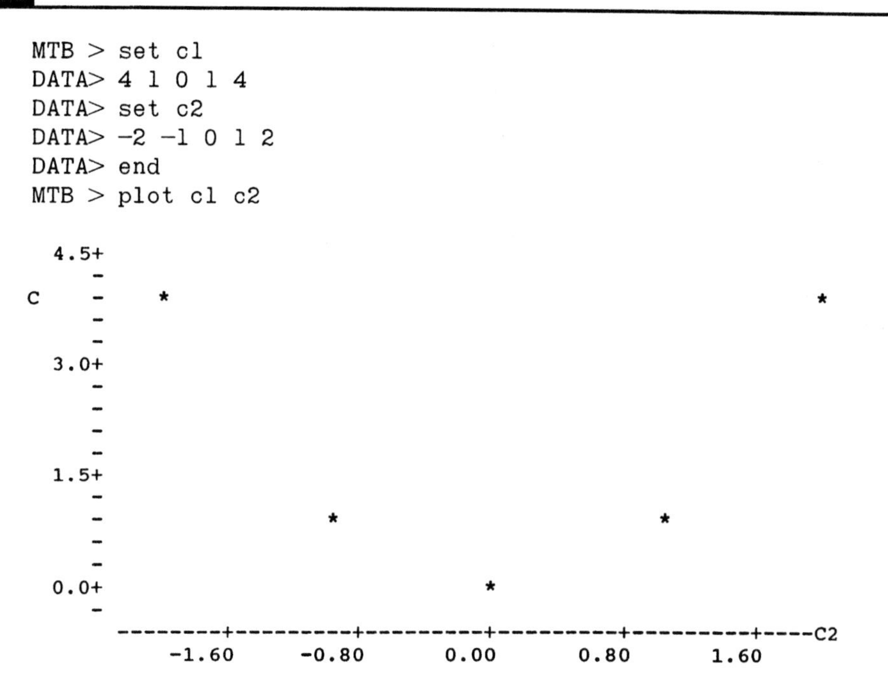

```
MTB > set c1
DATA> 4 1 0 1 4
DATA> set c2
DATA> -2 -1 0 1 2
DATA> end
MTB > plot c1 c2
```


Connecting these points gives the parabola shape associated with a quadratic relationship.

B. Spearman Rank Correlation Coefficient

The Pearson correlation coefficient was designed to be used with jointly normally distributed variables; however, it is used, sometimes incorrectly, with all types of data in practice. Instead of using the Pearson correlation coefficient with nonnormally distributed variables, it may be better to use a modification suggested by Spearman, an influential British psychometrician, in 1904. Spearman suggested ranking the values of Y and also ranking the values of X. These ranks are then used instead of the actual values of Y and X in the formula for the sample Pearson correlation coefficient. The result of this calculation is the sample Spearman rank correlation coefficient, denoted by r_S. This calculation is demonstrated in Box 4.20 for the protein and total fat data. Hence r_S is 0.573, slightly less than the Pearson value of 0.648.

MINITAB BOX 4.20

Protein data are in c3 and total fat data in c4. The ranks of c3 and c4 are put into c5 and c6, respectively.

```
MTB > rank c3 c5
MTB > rank c4 c6
MTB > corr c5 c6
Correlation of C5 and C6 = 0.573
```

In addition to being used with nonnormal continuous data, the Spearman rank correlation coefficient can also be used with ordinal data. When ordinal data are used, ties (two or more subjects having exactly the same value of a variable) are likely to occur. In the case of ties, the tied observations receive the same average rank. For example, if three observations of X are tied for the third smallest value, the ranks involved are 3, 4, and 5. The average of these three ranks is 4, and that is the rank that each of the three observations would be assigned. The occurrence of ties causes no problem in the calculation of the Spearman correlation coefficient when the Pearson formula is used with the ranks.

VII. CONCLUDING REMARKS

In this chapter we have presented tables, graphs, and plots as well as a few key statistics. The pictures and the statistics together enable one to describe single variables and the relationship between two variables for the sample data. Although the description of the sample data and the provision of estimates of population parameters are important, sometimes we wish to go beyond that, for example, to give a range of likely values for the population parameters or to determine whether or not it is likely that two populations under study have the same mean. To do this requires the use of probability distributions, a topic presented in Chapter 6. Before studying probability distributions, however, it is useful to understand probability, the topic covered in the next chapter.

EXERCISES

4.1. Create a bar chart of the following data on serum cholesterol for non-Hispanic whites based on Table II-42 in "Nutrition Monitoring in the United States" (7).

Gender	*Age*	*N*	*Mean serum cholesterol*[a] *(mg/dl)*[1]
Male	40–49	572	223.5
	50–59	575	228.9
	60–69	1354	226.2
	70–74	427	215.8
Female	40–49	615	218.5
	50–59	649	243.6
	60–69	1487	249.0
	70–74	533	248.3

[a] These data are from the Second National Health and Nutrition Examination Survey of noninstitutionalized persons conducted during the period 1976–1980 (8).

A high value of serum cholesterol is thought to be a risk factor for heart disease. The National Cholesterol Education Program (NCEP) of the National Institutes of Health in 1987 stated that the recommended value for serum cholesterol is below 200 mg/dl and a value between 200 and 240 is considered to be borderline. A value above 240 may indicate a problem and NCEP recommended that a lipoprotein analysis should be performed. On the basis of these data, it appears that many non-Hispanic whites, particularly women, have serum cholesterol values that are too high.

a. Give some possible reasons why non-Hispanic white men have higher mortality from heart and cerebrovascular diseases when it appears from these data that non-Hispanic white women should have the higher rates.
b. Provide a possible explanation why the serum cholesterol values for older men are lower than those for the younger men and the reverse is true for women.

4.2. Create line graphs for the following expenditures for the Food Stamps Program in New York State during the 1980s.
What, if any, tendencies in the expenditures (both actual and inflation-adjusted) do you see? Which expenditure data do you think should be used in describing the New York State Food Stamps Program? Explain your choice.

Year	*Actual expenditures (in millions of dollars)*	*Inflation-adjusted expenditures*[a]
1980	745.3	745.3
1981	901.2	814.1
1982	835.7	717.3
1983	930.9	766.8
1984	904.4	709.3
1985	939.4	712.2
1986	926.5	685.3
1987	901.8	638.7
1988	909.1	613.4
1989	964.7	616.4

[a] Expenditures adjusted for inflation using the consumer price index for the Northeast Region with 1980 as the base.
Source: Table 3.2 in the Division of Nutritional Sciences, Cornell University (9).

4.3. Use line graphs to represent the short-stay hospital occupancy rates shown here.

		Hospital Ownership		
Year	*Federal*	*Nonprofit*	*Proprietary*	*State/local*
1960	82.5	76.6	65.4	71.6
1970	77.5	80.1	72.2	73.2
1975	77.6	77.4	65.9	69.7
1980	77.8	78.2	65.2	70.7
1985	74.3	67.2	52.1	62.8
1989	71.0	68.8	51.7	64.8

Source: Table 105 in National Center for Health Statistics (10).

Discuss the trends, if any, in these data.

4.4. The following data on hazardous government jobs appeared as a bar chart in the USA Snapshots section of *USA Today* on April 30, 1992. The variable shown was the number of assaults suffered by federal officers based on 1990 FBI figures. The least number of assaults suffered were by the Internal Revenue Service (3 assaults), the Bureau of Indian Affairs (5 assaults), and the Postal Inspectors (6 assaults). The most assaults were suffered by the Immigration and Naturalization Service with 409, followed by U.S. Attorneys with 269 and the Bureau of Prisons with 185 assaults. What additional information do you need to conclude anything about which federal officers have the more hazardous (from the perspective of assaults) jobs?

4.5. A study was performed to determine which of three drugs was more effective in the treatment of a health problem. The responses of subjects who received each of three drugs (A, B, and C) were provided by Cochran (11). The following table shows the pattern of response for the 46 subjects; 1 indicates a favorable response and 0 an unfavorable response.

Response to			
A	*B*	*C*	*Frequency*
1	1	1	6
1	1	0	16
1	0	1	2
1	0	0	4
0	1	1	2
0	1	0	4
0	0	1	6
0	0	0	6
Total			46

a. Give an example of a type of health problem that would be appropriate for this study.
b. Create a two-way frequency table showing the relationship between drugs A and C. Does it appear that the responses to these drugs are related?
c. Create a bar chart that shows the number of subjects with a favorable response by drug.

4.6. Using the data shown in Table 4.1, calculate the coefficient of variation for vitamin A. Do you think that any measure of central tendency adequately describes these data? Explain your answer.

4.7. Lee (12) presented survival times in months from diagnosis for 71 patients with either acute myeloblastic leukemia (AML) or acute lymphoblastic leukemia (ALL).

AML patients:
18 31 31 31 36 01 09 39 20 04 45 36 12 08 01 15 24 02 33 29 07 00 01 02 12 09 01 01 09 05 27 01 13 01 05 01 03 04 01 18 01 02 01 08 03 04 14 03 13 13 01

ALL patients:
16 25 01 22 12 12 74 01 16 09 21 09 64 35 01 07 03 01 01 22

a. Calculate the sample mean and median for both AML and ALL patients separately. Which measure do you believe is more appropriate to use with these data? Explain.
b. Create box plots, histograms, and stem-and-leaf plots to show the distributions of the survival times for AML and ALL patients. Which type of figure is more informative for these data? Which type of patient has the longer survival time after diagnosis?

c. Give examples of additional variables that are needed to interpret appropriately these survival times.

4.8. Is it possible to calculate the mean occupancy rate for the short-stay hospitals in 1960 given the data provided in Exercise 4.3? If it is, calculate it. If not, state why it cannot be calculated.

4.9. Provide an appropriate summarization of the following data on the results of inspections of food establishments (e.g., food processing plants, food warehouses, and grocery stores) conducted by the Division of Food Inspection Services of the New York State Department of Agriculture and Markets.

	Number inspected		*Approximate number failed*	
Year	*Upstate*	*NYC & LI*[a]	*Upstate*	*NYC & LI*[a]
1980	19,599	23,676	2,548	5,209
1982	17,183	22,767	3,093	6,830
1984	13,731	18,677	2,746	6,350
1986	10,915	15,948	2,292	6,379
1988	13,614	15,070	3,267	6,179
1990	12,609	16,285	3,026	6,677

[a] New York City and Long Island.
Source: Table 2.5 in the Division of Nutritional Sciences, Cornell University (9).

Do you think that there were more or fewer cases of foodborne illness in New York State in 1990 than in 1980?

4.10. Diagnosis-related groups (DRGs) are used in the payment for the health care of Medicare-funded patients. In the creation of the DRGs, suppose that the lengths of stay for patients in one of the proposed groups were the following:

1 1 2 2 2 2 2 2 3 3 3 4 4 4 5 5 5 5 6 6 6 7 7 8 8 8 9 9 10 12 13 15 15 17 17 18 19 19 20 23 26 29 31 34 36 43 49 52 67 96

Calculate the mean, standard deviation, coefficient of variation, and five key percentiles for these data. Are these data skewed? Do the patients in this DRG appear to have homogeneous lengths of stay? Which measures, if any, should be used in the description of these data? Explain your answer.

4.11. The following data represent bacteria counts measured in water with levels of 0, 1, and 3 percent sodium chloride.

a. Calculate the mean and coefficient of variation for these data.
b. Calculate the median and geometric mean.
c. Comment on which measure of central tendency is appropriate for these data.

Level of sodium chloride (%)	*Count (number/ml)*
0	10^7, 10^6, 10^8, 10^9, 10^8, 10^{10}
1	10^4, 10^4, 10^5, 10^6
3	10^3, 10^4, 10^4, 10^3, 10^5

4.12. In Harris County, Texas, in 1986, there were 24,346 live births to whites, 11,365 to African-Americans, 14,849 to Hispanics, and 2093 to other groups. There were 187 infant deaths among whites, 125 of which were to infants less than 28 days old. The corresponding numbers were 183 and 121 for African-Americans, 137 and 95 for Hispanics, and 12 and 7 for the other groups. Based on these numbers, calculate the neonatal and postneonatal mortality rates for these four groups. Comment on any rate that appears to be unusual.

4.13. Of the estimated 1,488,939 male residents of Harris County, Texas, in 1986, there were 8672 deaths. Of the 1,453,611 female residents, there were 6913 deaths. The estimated 1986 U.S. population was approximately 48.7 percent male and 51.3 percent female.
   a. Calculate the crude death rate and sex-specific death rates for Harris County in 1986.
   b. Do you believe that a sex-adjusted death rate will be very different from the crude death rate? Provide the reason for your belief.
   c. Calculate a sex-adjusted death rate for Harris County in 1986.

4.14. The Pearson correlation coefficient between protein and total fat for the data in Table 4.1 was 0.648. This suggests a strong linear relationship between these two variables; however, this relationship may be reflecting the amount of food consumed (calories). One way of adjusting for the calories is to create new variables by dividing the protein and total fat variables by the calories consumed.
   a. Create a scatter plot of protein per calories by total fat per calories.
   b. Calculate the Pearson correlation coefficient for the new variables protein/calories and total fat/calories. Which measure of correlation do you think best characterizes the strength of the relationship?

4.15. Data from NHANES II showed that 39.7 percent of persons ages 20 to 74 had hypertension (systolic blood pressure ≥ 140 mm Hg, diastolic blood pressure ≥ 90 mm Hg, or taking blood pressure medication) and that 30.3 percent had borderline high or high values of serum cholesterol. The sources of these percentages are Tables 69 and 70 in "Health, United States and Prevention Profile, 1991" (10). Based on these data, can we correctly conclude that 70 percent

(39.7 + 30.3) of the noninstitutionalized U.S. population ages 20 to 74 had either hypertension or borderline high or high values of serum cholesterol? Provide the rationale for your answer.

REFERENCES

1. McPherson, R. S., Nichaman, M. Z., Kohl, H. W., Reed, D. B., and Labarthe, D. R. (1990). Intake and food sources of dietary fat among schoolchildren in The Woodlands, Texas. *Pediatrics* **86**(4), 520–526.
2. MINITAB Inc. (1993). "MINITAB Reference Manual, Release 9 for Windows." MINITAB Inc., State College, PA.
3. MINITAB Inc. (1989). "MINITAB Quick Reference Card, Release 9 for Windows." MINITAB Inc., State College, PA.
4. National Center for Health Statistics (1991). "Health, United States, 1990," DHHS Publ. No. 91-1232. Public Health Service, Hyattsville, MD.
5. Canfield, M., ed. (1990). "The Health Status of Harris County Residents: Births, Deaths and Selected Measures of Public Health, 1980–86." Harris County Health Department, Houston, Texas.
6. Tarter, M. E., and Kronmal, R. A. (1976). An introduction to the implementation and theory of nonparametric density estimation. *Amer. Stat.* **30**, 105–112.
7. Life Sciences Research Office, Federation of American Societies for Experimental Biology (1989). "Nutrition Monitoring in the United States: An Update Report on Nutrition Monitoring," DHHS Publ. No. (PHS) 89-1255. U.S. Department of Agriculture and the U.S. Department of Health and Human Services, Public Health Service, Washington, U.S. Government Printing Office.
8. National Center for Health Statistics, McDowell, A., Engel, A., Massey, J. T., and Maurer, K. (1981). Plan and operation of the second National Health and Nutrition Examination Survey, 1976–80. "Vital and Health Statistics," Ser. 1, No. 15, DHHS Publ. No. (PHS) 81-1317. Public Health Service, Washington, U.S. Government Printing Office.
9. Division of Nutritional Sciences, Cornell University in cooperation with the Nutrition Surveillance Program of the New York State Department of Health (1992). "New York State Nutrition: State of the State, 1992." NYSDH, New York.
10. National Center for Health Statistics (1992). "Health, United States, 1991 and Prevention Profile," DHHS Publ. No. 92-1232. Public Health Service, Hyattsville, MD.
11. Cochran, W. G. (1955). The comparison of percentages in matched samples. *Biometrika* **37**, 356–366.
12. Lee, E. T. (1980). "Statistical Methods for Survival Data Analysis." Wadsworth, Belmont, CA.

CHAPTER 5

Probability and Life Tables

As was mentioned in Chapter 4, we often wish to do more than simply describe or summarize the data by graphs or descriptive statistics. For example, we may want to determine whether or not two drugs or treatments are equally effective and safe, or whether the age-adjusted death rates for two areas are the same. To answer these questions, we require knowledge of probability, the topic of this chapter.

I. A DEFINITION OF PROBABILITY

We have all encountered the use of probability, for example, in the weather forecast. The forecast usually involves an estimate of the probability of rain, as in the statement that the probability of rain tomorrow is 20 percent. As its use in the weather forecast demonstrates, *probability* is a numerical assessment of the likelihood of the occurrence of an outcome of a random variable. In the weather forecast, weather is the random variable and rain is one of its possible outcomes.

Before considering the numerical assessment of likelihood, we should consider random variables. There are both discrete and continuous random variables. A *discrete* (nominal, categorical, or ordinal) *random variable* is a quantity that reflects an attribute or characteristic that takes on different values with specified probabilities. A *continuous* (interval or ratio) *random variable* is a quantity that reflects an attribute or characteristic that falls within an interval with specified probabilities.

Hypertension status is a discrete random variable when the values or levels of this variable are defined as its presence (can be defined as systolic blood pressure greater than 140 mm Hg, diastolic blood pressure greater than 90 mm Hg, or taking antihypertensive medication) or absence. Other examples of discrete random variables include racial status, the number of children in a family, and type of health insurance. Examples of continuous random variables include height, blood pressure, and amount of lead emissions as these are usually measured.

We define the probability of the occurrence of an outcome or interval of a random variable as its relative frequency in an infinite number of trials or in a population. A probability is a population parameter. An observed proportion (relative frequency) from a sample is a statistic which can be used to estimate a probability. We use the data in Table 5.1 to demonstrate the calculation of the probability of different racial categories in the United States in 1990. As shown in Table 5.1, in the U.S. population census there are four major racial groups and a fifth category that combines all other races. Hispanics are counted mostly in the White and Other categories.

The probability of a person selected at random being white was 0.803 (= 199,686,070/248,709,873), or 80.3 percent. The corresponding probabilities of being African-American, American Indian, Asian, and other were 0.121 (= 29,986,060/248,709,873), 0.008, 0.029, and 0.039, respectively. These five probabilities sum to 1.000, or 100.0 percent, as shown in Table 5.1 (1).

As probability is the number of occurrences of an outcome divided by the total number of occurrences of all possible outcomes of the variable

TABLE 5.1 Racial Composition of the 1990 U.S. Population

Racial category	Number	Percent
White	199,686,070	80.3
African-American	29,986,060	12.1
American Indian, Eskimo, or Aleut	1,959,234	0.8
Asian or Pacific Islander	7,273,662	2.9
Other races	9,804,847	3.9
Total population	248,709,873	100.0

Source: The 1990 Census of Population and Housing, the United States, STF1A (1).

under study, this means that a probability cannot be larger than 1.00 or 100 percent in value. By the same reasoning, a probability cannot be smaller than 0.00 or 0 percent in value. Therefore, the only valid values for probabilities range from 0 to 1 or 0 to 100 percent. Additionally, use of the relative frequency definition means that the sum of the probabilities of all the possible outcomes of a random variable must be 1.00 or 100 percent. If a probability falls outside the range 0 to 1, or if the sum of the probabilities of all the possible outcomes of a variable do not sum to 1 (with allowance for rounding), a mistake has been made.

For many variables in the health field, the probability of an outcome is estimated from a large number of observations and may change over time. For example, the probabilities of the different racial groups in the United States in 2000 will be different from the 1990 probabilities. As an additional example of changing probabilities, the estimates of the age-adjusted probabilities of hypertension among U.S. adult males increased from 0.414 in 1960–1962 to 0.440 in 1971–1974 to 0.453 in 1976–1980 (2). This change in the values of a probability contrasts with the lack of change in the probabilities associated with physical phenomena such as tossing a coin or a pair of dice. For example, when a fair coin is tossed, the probability of a head is assumed to be 0.5 or 50 percent, and it does not change.

The listing of the probabilities of all possible outcomes of a discrete variable is its *probability distribution*. For example, the probability distribution of the racial composition of the U.S. population in 1990 is shown in the last column of Table 5.1. More will be said about probability distributions and their use in the next chapter.

II. RULES FOR CALCULATING PROBABILITIES

A few basic rules govern the calculation of probabilities of compound outcomes or events. We use the data in Table 5.2 to help us discover these rules. Entries in Table 5.2 (3) are the number of live births by birth weight

TABLE 5.2 Number of Live Births by Trimester of First Prenatal Care and Birth Weight for Harris County, Texas, in 1986 (Excluding 1180 Births with Unknown Trimester or Birth Weight)

	Trimester prenatal care began				
Birth weight	First	Second	Third	No care	Total
≤5.5 lb ≈ 2500 g	2,412	754	141	234	3,541
5.6–7.7 lb ≈ 2500–3500 g	20,274	5480	1458	1014	28,226
>7.7 lb ≈ 3500 g	15,250	3271	738	447	19,706
Total	37,936	9505	2337	1695	51,473

Source: "The Health Status of Harris County Residents, 1980–1986," Table 1.S (3).

and the trimester in which prenatal care was begun for women in Harris County, Texas in 1986. For example, the entry in the third row, second column, 3271, is the number of live births to women who had begun their prenatal care during their second trimester and whose babies' birth weights were greater than 7.7 lb.

A. Addition Rule for Probabilities

The data in Table 5.2 can be used to determine whether a relationship exists between the timing of the beginning of prenatal care and birth weight. Before examining this issue, however, let us calculate a few additional probabilities. For example, the probability of a woman in Harris County in 1986 having a low-birth-weight baby (less than or equal to 5.5 lb) was 0.069 (= 3541/51,473). This value is very close to the 1986 value of 0.068 for the United States (2). Let us now consider a slightly more complex example. The probability of late prenatal (third trimester) *or* no prenatal care is simply the sum of their individual probabilities, that is, 2337/51,473 + 1695/51,473, which is 0.078 (= 4032/51,473). This value is slightly greater than the corresponding 1986 U.S. value of 0.060 (2). In these calculations of probabilities, we are considering births in Harris County in 1986 as our population. If the intended population were Texas or the United States, then the above values would be sample estimates, that is, observed proportions, of the probabilities. A sample consisting of births in Harris County should not, however, be used to draw inferences about births in Texas or the United States because the Harris County births are likely not to be representative of either of these two larger units.

So far, these probabilities have focused on row or column totals (marginal totals), not on the numbers in the interior of the table (cell entries). Entries in the interior of the table deal with the intersection of outcomes or events. For example, the outcome of a woman bearing a live infant weighing 5.5 lb or less and having begun her prenatal care during the first trimester is the intersection of those two individual outcomes. The probability of this intersection, that is, of these two outcomes occurring together, is easily found to be 0.047 (= 2412/51,473).

Above we found the probability of a baby weighing 5.5 lb or less by using the row total of 3541 and dividing it by the grand total of 51,473. Note that we can also express this probability in terms of the probability of the intersection of a birth weight of 5.5 lb or less with each of the prenatal care levels, that is,

$$\Pr\{\leq 5.5 \text{ lb}\} = \Pr\{\leq 5.5 \text{ lb \& 1st trim.}\} + \Pr\{\leq 5.5 \text{ lb \& 2nd trim.}\}$$
$$+ \Pr\{\leq 5.5 \text{ lb \& 3rd trim.}\} + \Pr\{\leq 5.5 \text{ lb \& no care}\}$$
$$= \frac{2412}{51{,}473} + \frac{754}{51{,}473} + \frac{141}{51{,}473} + \frac{234}{51{,}473} = \frac{3541}{51{,}473}.$$

This can be expressed in symbols. Let A represent the outcome of a birth weight of 5.5 lb or less and B_i, $i = 1$ to 4, represent the four prenatal care levels. Then we have

$$\Pr\{A\} = \Pr\{A \text{ and } B_1\} + \Pr\{A \text{ and } B_2\} + \Pr\{A \text{ and } B_3\} + \Pr\{A \text{ and } B_4\}$$

which, using the summation symbol, is

$$\Pr\{A\} = \sum_i \Pr\{A \text{ and } B_i\}.$$

Suppose now that we want to find, for a woman who had a live birth, the probability that either the birth weight was 5.5 lb or less *or* the woman began her prenatal care during the first trimester. It is tempting to add the two individual probabilities—of a birth weight of 5.5 lb or less and of prenatal care beginning during the first trimester—as we had done above. If, however, we added the entries in the first row (birth weights of 5.5 lb or less) to those in the first column (prenatal care begun during the first trimester), the entry in the intersection of the first row and column would be included twice. Therefore, we have to subtract this intersection from the sum of the two individual probabilities to obtain the correct answer. The calculation is

$$\begin{aligned}\Pr\{\leq 5.5 \text{ lb or 1st trim.}\} &= \Pr\{\leq 5.5 \text{ lb}\} + \Pr\{\text{1st trim.}\} \\ &\quad - \Pr\{\leq 5.5 \text{ lb and 1st trim.}\} \\ &= \frac{3541 + 37{,}936 - 2412}{51{,}473} = 0.759.\end{aligned}$$

This can be succinctly stated in symbols. Let A represent the outcome of a live birth of 5.5 lb or less and B represent the outcome of the initiation of prenatal care during the first trimester. The intersection of these two outcomes is represented by A and B. In symbols, the rule is

$$\Pr\{A \text{ or } B\} = \Pr\{A\} + \Pr\{B\} - \Pr\{A \text{ and } B\}.$$

This rule also was used in the earlier example of late or no prenatal care, but, in that case, the outcomes were disjoint; that is, there was no overlap or intersection. Hence the probability of the intersection was zero.

As the sum of the probabilities of all possible outcomes is one, if there are only two outcomes, say A and not A (represented by $\overline{A}$), we also have the following relationship:

$$\Pr\{A\} = 1 - \Pr\{\overline{A}\}$$

B. Conditional Probabilities

Suppose we change the wording slightly in the above example. We now want to find the probability of a woman bearing a live infant weighing 5.5 lb or less (event A) *conditional on or given* that her prenatal care was begun during the first trimester (event B). The word *conditional* limits our view in that we now focus on the 37,936 women who began their prenatal care during the first trimester. Thus, the probability of a woman bearing a live infant weighing 5.5 lb or less, given that she began her prenatal care during the first trimester, is 0.064 (= 2412/37,936). Dividing both the numerator and denominator of this calculation by 51,473 (the total number of women) does not change the value of 0.064, but it allows us to define this *conditional probability* (the probability of A conditional on the occurrence of B) in terms of other probabilities. The numerator divided by the total number of women (2412/51,473) is the probability of the intersection of A and B, and the denominator divided by the total number of women (37,936/51,473) is the probability of B. In symbols, this is expressed as

$$\Pr\{A \mid B\} = \frac{\Pr\{A \text{ and } B\}}{\Pr\{B\}}$$

where $\Pr\{A \mid B\}$ represents the probability of A given that B has occurred.

Conditional probabilities often are of greater interest than the unconditional probabilities we have been dealing with as will be shown below. Before doing that, note that we can use the conditional probability formula to find the probability of the intersection, that is,

$$\Pr\{A \text{ and } B\} = \Pr\{A \mid B\} * \Pr\{B\}.$$

Thus, if we know the probability of A conditional on the occurrence of B, and we also know the probability of B, we can find the probability of the intersection of A and B. Note that we can also express the probability of the intersection as

$$\Pr\{A \text{ and } B\} = \Pr\{B \mid A\} * \Pr\{A\}.$$

Table 5.3 repeats the data from Table 5.2 along with three different sets of probabilities. The first set of probabilities is conditional on the birth weight; that is, it uses the row totals as the denominators in the calculations. The second set is conditional on the trimester that prenatal care was begun; that is, it uses the column totals in the denominator. The third set of probabilities is the unconditional set, that is, those based on the total of 51,473 live births. The probabilities in the Total column are the probabilities of the different birth weight categories, that is, the probability distribution of the birth weight variable, and those beneath the Total row are the probabilities of the different trimester categories, that is, the probability distribution of the prenatal care variable. As mentioned above, these probabilities are based on the population of births in Harris County, Texas in 1986.

Let us consider the entries in the row 1, column 1 cell. The first two entries below the frequency of the cell are conditional probabilities. The value 0.681 (= 2412/3541) is the probability based on the row total, that is,

TABLE 5.3 Number and Probabilities of Live Births by Trimester of First Prenatal Care and Birth Weight for Harris County, Texas, in 1986 (Excluding 1180 Births with Unknown Trimester or Birth Weight)

Birth weight		Trimester prenatal care began: First	Second	Third	No care	Total
≤5.5 lb ≈ 2500 g		2,412	754	141	234	3,541
	R[a]	0.681	0.213	0.040	0.066	
	C	**0.064**	**0.079**	**0.060**	**0.138**	**0.069**
	U	0.047	0.015	0.003	0.005	
5.6–7.7 lb ≈ 2500–3500 g		20,274	5480	1458	1014	28,226
	R	0.718	0.194	0.052	0.036	
	C	**0.534**	**0.577**	**0.624**	**0.598**	**0.548**
	U	0.394	0.106	0.028	0.020	
>7.7 lb ≥ 3500 g		15,250	3271	738	447	19,706
	R	0.774	0.166	0.037	0.023	
	C	**0.402**	**0.344**	**0.316**	**0.264**	**0.383**
	U	0.296	0.064	0.014	0.009	
Total		37,936	9505	2337	1695	51,473
	R	0.737	0.185	0.045	0.033	1.000

[a] R, row; C, column; and U, unconditional.

the probability of a woman having begun her prenatal care during the first trimester given that the baby's birth weight was 5.5 lb or less. The value 0.064 (= 2412/37,936) is the probability based on the column total, that is, the probability of a birth weight of 5.5 lb or less given that the woman had begun her prenatal care during the first trimester. The last value, 0.047 (= 2412/51,473), is the unconditional probability; it is based on the grand total of 51,473 live births. It is the probability of the intersection of a birth weight of 5.5 lb or less with prenatal care having been begun during the first trimester.

As Table 5.3 shows, at least three different probabilities, or observed proportions if the data are a sample, can be calculated for the entries in the two-way table. The choice of which probability (row, column, or unconditional) to use depends on the purpose of the investigation. In this case, the data may have been tabulated to determine whether or not the timing of the initiation of the prenatal care had any effect on the birth weight of the infant. If this is the purpose of the study, the column-based probabilities may be the more appropriate to use and report. The column-based calculations give the probabilities of the different birth weight categories conditional on when the prenatal care was begun. The row-based calculations give the probability of trimester prenatal care being initiated given the birth weight category; however, these row-based probabilities are of no interest because birth weight cannot affect the timing of prenatal care. The unconditional probabilities are less informative in this situation as they also reflect the row and column totals. For example, compare the unconditional probabilities in the first and third columns in the first row: 0.047 and 0.003. Even though we have seen that there is little difference in the corresponding column-based probabilities of 0.064 and 0.060, these unconditional values are very different. The value of 0.047 is larger mainly because there are 37,936 live births in the first column compared with only 2337 live births in the third column. The unconditional probabilities may, however, be useful in planning and allocating resources for maternal and child health services programs.

Using the column-based values, women who began their prenatal care during the first trimester had a probability of bearing a low-birth-weight baby of 0.064. This value is compared to 0.079, the probability of bearing a low-birth-weight baby for those who began their prenatal care during their second trimester, to 0.060 for those who began their prenatal care during the third trimester, and to 0.138 for those who received no prenatal care. There is little difference in the probabilities of bearing a low-birth-weight baby among women who received prenatal care; however, the probability of bearing a low-birth-weight baby is about twice as large for women who received no prenatal care compared with women who received prenatal care. The effect of prenatal care is most clearly evident in the probability of bearing a baby weighing more than 7.7 lb. In this category, the probabilities are 0.402, 0.344, 0.316, and 0.264 for the first, second, third trimesters, and no prenatal care, respectively.

Based on the trend in the probabilities of a birth weight greater than 7.7 lb, one might conclude that there is an effect of prenatal care. To do so, however, is inappropriate without further information. First, although these births can be viewed as constituting a population, that is, all the live births in Harris County in 1986, they could also be viewed as a sample in time, one year selected from many, or in place, one county selected from many. From the perspective that these births are a sample, there is sampling variation to be taken into account and this is covered in Chapter 11. Second, and more important, these data do not represent a true experiment. Chapter 8 presents more on experiments but, briefly, the women were not randomly assigned to the different prenatal care groups, that is, to the first, second, or third trimester groups or to the no prenatal care group. Thus the women in these groups may differ on variables related to birth weight, for example, smoking habits, amount of weight gained, and dietary behavior. Without further examination of these other factors, it is wrong to conclude that the variation in the probabilities of birth weights is due to the time when prenatal care was begun.

TABLE 5.4 Probabilities of Birth Weight Level Conditional on Trimester of First Prenatal Care for Harris County, Texas, in 1986 (Excluding 1180 Births with Unknown Trimester or Birth Weight)

	Trimester prenatal care began				
Birth weight	First	Second	Third	No care	Total
≤5.5 lb ≈ 2500 g	0.064	0.079	0.060	0.138	0.069
5.6–7.7 lb ≈ 2500–3500 g	0.534	0.577	0.624	0.598	0.548
>7.7 lb ≥ 3500 g	0.402	0.344	0.316	0.264	0.383
Total	1.000	1.000	1.000	1.000	1.000

C. Independent Events

Suppose that we were satisfied that there were no additional factors of interest in the examination of prenatal care and birth weight. Only the data in Table 5.2 were to be used to determine whether or not there was a relationship between when prenatal care was initiated and birth weight. Table 5.4 shows the column-based probabilities, that is, those conditional on which trimester care was begun or whether care was received, and these are the probabilities to be used in the study.

If there is no relationship between the prenatal care variable and the birth weight variable, that is, these two variables are independent, what values should the column-based probabilities have? If these variables are *independent*, this means that the birth weight probability distribution is the same in each of the columns. The last column in Table 5.4 gives the birth weight probability distribution, and this is the distribution that will be in each of the columns if the birth weight and prenatal care variables are independent. Table 5.5 shows the birth weight probability distribution for the situation when these two variables are independent.

TABLE 5.5 Probabilities Conditional on Trimester under the Assumption of Independence of Birth Weight Level and Trimester of First Prenatal Care for Harris County, Texas, in 1986 (Excluding 1180 Births with Unknown Trimester or Birth Weight)

	Trimester prenatal care began				
Birth weight	First	Second	Third	No care	Total
≤5.5 lb ≈ 2500 g	0.069	0.069	0.069	0.069	0.069
5.6–7.7 lb ≈ 2500–3500 g	0.548	0.548	0.548	0.548	0.548
>7.7 lb ≥ 3500 g	0.383	0.383	0.383	0.383	0.383
Total	1.000	1.000	1.000	1.000	1.000

The entries in Table 5.5 are conditional probabilities, for example, of a birth weight of 5.5 lb or less (A) given that prenatal care began during the first trimester (B) under the assumption of independence. Hence, under the assumption of independence of A and B, the probability of A given B is equal to the probability of A. In symbols, this is

$$\Pr\{A \mid B\} = \Pr\{A\}$$

when A and B are independent. Combining this formula with the formula for the probability of the intersection, that is,

$$\Pr\{A \text{ and } B\} = \Pr\{A \mid B\} * \Pr\{B\}$$

yields

$$\Pr\{A \text{ and } B\} = \Pr\{A\} * \Pr\{B\}$$

when A and B are independent.

When considering diseases, it is unlikely that the disease status of one person is independent of that of another person for many infectious diseases; however, it is likely that the disease status of one person is independent of that of another for many chronic diseases. For example, let π be the probability that a person has Alzheimer's disease. One person's Alzheimer status should be independent of another's status. Therefore, the probability of persons A and B both having Alzheimer's disease is the product of the probabilities of either having the disease, that is, $\Pr\{A \text{ and } B\} = \pi * \pi$.

Establishing the dependence (a relationship exists) or independence (no relationship) of variables is what much of health research is about. For example, in the disease context, Is disease status related to some variable? If there is a relationship (dependency), the variable is said to be a risk factor for the disease. The identification of risk factors leads to strategies for preventing or reducing the occurrence of the disease.

Some additional uses of conditional probabilities and the concept of independence are introduced in the next section.

III. DEFINITIONS FROM EPIDEMIOLOGY

Many quantities used in epidemiology are defined in terms of probabilities, particularly conditional probabilities. Several of these useful quantities are defined in this section and used in the next section to illustrate Bayes' rule.

A. Prevalence and Incidence

Prevalence of a disease is the probability of having the disease. It is the number of people with the disease divided by the number of people in the population. The observed proportion of those with the disease in a sample is the sample estimate of prevalence. When the midyear population is used for the denominator, it is possible that the numerator contains persons not included in the denominator. For example, persons with the disease that move into the area in the second half of the year are not counted in the denominator, but they are counted in the numerator. When prevalence or other quantities use midperiod population values, they are not really probabilities or proportions, although this distinction usually is unimportant.

Incidence of a disease is the probability that a person without the disease will develop the disease during some specified interval of time. It is the number of new cases of the disease that occur during the specified time interval divided by the number of people in the population who do not already have the disease.

Prevalence provides an idea of the current magnitude of the disease problem and incidence informs as to whether the disease problem is getting worse or not.

Data on AIDS from Harris County excluding Houston will be used to demonstrate the calculation of prevalence and incidence. In 1986, the population of Harris County excluding Houston was estimated to be 1,004,947. According to Table 8.6 in "The Health Status of Harris County Residents" (3), 132 cases of AIDS had been reported to the Harris County Health Department by the end of 1986, and of those, 79 were diagnosed in 1986. There is no information on the number of individuals who had died from AIDS, but we shall assume that 60 percent of those diagnosed prior to 1986 had died by the end of 1986. Thus, of the 132 reported cases, we are assuming that 32 individuals [= 0.60 * (132–79)] had died, leaving 100 persons with AIDS at the end of 1986.

The prevalence of AIDS at the end of 1986 then was 0.0000995 (= 100/1,004,947). Prevalence and incidence are often converted to rates, for example, the number of cases per 1000 or 100,000 population. In this case, the prevalence rate is 9.95 cases per 100,000 population. The incidence is the probability of new cases during some period. We shall calculate the incidence for 1986. There were 79 new cases diagnosed in 1986; the eligible population is the number of people without the disease. Therefore the eligible population is 1,004,947 minus the number of people who had AIDS prior to 1986. There were 53 cases diagnosed prior to 1986, and of these, some had already died prior to 1986. We shall assume that 30 percent of these individuals had died prior to 1986; that is, 16 individuals are assumed to have died prior to 1986. Therefore we must subtract 37 (= 53 − 16) from 1,004,947 in the denominator of the incidence calculation. The 1986 incidence of AIDS in Harris County excluding Houston was 0.0000716 (= 72/1,004,910), or 7.16 cases per 100,000 population. The incidence is almost as large as the prevalence, suggesting that the disease problem is worsening. In this situation, the subtraction of the number of cases from the denominator had little practical importance; however, it must be done and can be important in many other situations.

B. Sensitivity, Specificity, and Predicted Value Positive and Negative

Laboratory test results are part of the diagnostic process for determining if a patient has some disease. Unfortunately in many cases, a positive test result, that is, the existence of an unusual value, does not guarantee that a patient has the disease. Nor does a negative test result, the existence of a typical value, guarantee the absence of the disease. To provide some information on the accuracy of testing procedures, their developers use two conditional probabilities, sensitivity and specificity.

The *sensitivity* of a test (symptom) is the probability that there was a positive result (the symptom was present) given that the person has the disease. The *specificity* of a test (symptom) is the probability that there was a negative result (the symptom was absent) given that the person does not have the disease. Note that one minus sensitivity is the false-negative rate and one minus specificity is the false-positive rate. Thus, large values of sensitivity and specificity imply small false-negative and false-positive rates.

Sensitivity and specificity are probabilities of the test result conditional on the disease status. These are values that the developer of the test has estimated during extensive testing in hospitals and clinics. As a potential patient, however, we are more interested in the probability of disease status conditional on the test result. Names given to two conditional probabilities that address the patient's concern are predicted value positive and predicted value negative. *Predicted value positive* is the probability of disease given a positive test result, and *predicted value negative* is the probability of no disease given a negative test result.

These four quantities can be expressed succinctly in symbols. Let T^+ represent a positive test result and T^- represent a negative result. The presence of disease is indicated by D^+ and its absence is indicated by D^-. These four quantities can be expressed as conditional probabilities:

Sensitivity	$\Pr\{T^+ \mid D^+\}$
Specificity	$\Pr\{T^- \mid D^-\}$
Predicted value positive	$\Pr\{D^+ \mid T^+\}$
Predicted value negative	$\Pr\{D^- \mid T^-\}$

All four of these probabilities should be large for a screening test to be useful to the screener and to the screenee. Discussions of these and related issues are plentiful in the epidemiological literature (4).

It is possible to estimate these probabilities. One way is to select a large sample of the population and subject the sample to a screening or diagnostic test as well as to a standard clinical evaluation. The standard clinical

TABLE 5.6 Disease Status by Test Results for a Large Sample from the Population

Disease status	Test result		Total
	Positive	Negative	
Presence	a	b	$a + b$
Absence	c	d	$c + d$
Total	$a + c$	$b + d$	$a + b + c + d$

evaluation is assumed to provide the true disease status. Then the sample persons can be classified into one of the four cells in the 2 × 2 table shown below. For example, hypertension status is first screened by the sphygmomanometer in the community and by a comprehensive clinical evaluation in the clinic; or persons are screened for mental disorders first by the DIS (Diagnostic Interview Schedule) and then by a comprehensive psychiatric evaluation. The results from a two-stage diagnostic procedure would look like Table 5.6. Sensitivity is estimated by $a/(a + b)$, specificity is estimated by $d/(c + d)$, predicted value positive is estimated by $a/(a + c)$, and predicted value negative is estimated by $d/(b + d)$. Similarly, the false-positive rate is estimated by $c/(a + c)$ and the false-negative rate by $b/(b + d)$.

For many diseases of interest, the prevalence is so low that there would be few persons with the disease in the sample. This means that the estimates of sensitivity and the predicted value positive would be problematic. Therefore, some alternate sample design must be used to estimate these conditional probabilities. When a large number of people are screened by a test in a community and a sample of persons with positive test results and those with negative test results are subjected to clinical evaluations, the predicted value positive and the predicted value negative can be directly calculated from the results of clinical evaluations, and sensitivity and specificity can be indirectly estimated. Conversely, when sensitivity and specificity are directly estimated by applying the test to persons with the disease and persons without the disease in the clinic setting, the predicted value positive and the predicted value negative can be indirectly estimated if the prevalence rate of disease is known. These indirect estimation procedures are explained in the next section.

IV. BAYES' THEOREM

We wish to find the predicted value positive and predicted value negative using the known values for disease prevalence, sensitivity and specificity. Let us focus on predicted value positive, that is, $\Pr\{D^+ \mid T^+\}$, and see how it can be expressed in terms of sensitivity, $\Pr\{T^+ \mid D^+\}$, specificity, $\Pr\{T^- \mid D^-\}$, and disease prevalence, $\Pr\{D^+\}$.

We begin with the definition of the predicted value positive, which is

$$\Pr\{D^+ \mid T^+\} = \frac{\Pr\{D^+ \text{ and } T^+\}}{\Pr\{T^+\}}. \tag{1}$$

Recall that the probability of the intersection of D^+ and T^+ can also be expressed as

$$\Pr\{D^+ \text{ and } T^+\} = \Pr\{T^+ \mid D^+\} * \Pr\{D^+\}.$$

On substitution of this expression for the probability of the intersection in (1), we have

$$\Pr\{D^+ \mid T^+\} = \frac{\Pr\{T^+ \mid D^+\} * \Pr\{D^+\}}{\Pr\{T^+\}} \tag{2}$$

which shows that predicted value positive can be obtained by dividing the product of sensitivity and prevalence by $\Pr\{T^+\}$.

Recall that the probability of an event can be expressed as the sum of the probabilities of the intersection of that event with all possible outcomes of another variable, that is,

$$\Pr\{A\} = \sum_i \Pr\{A \text{ and } B_i\}.$$

By use of the relationship between the probability of the intersection and conditional probabilities, this in turn can be reexpressed as

$$\Pr\{A\} = \sum_i \Pr\{A \mid B_i\} * \Pr\{B_i\}.$$

We use this formula to reexpress the probability of a positive test result. T^+ is substituted for A, D^+ replaces B_1, and D^- replaces B_2 in this formula, and this gives

$$\Pr\{T^+\} = \Pr\{T^+ \mid D^+\} * \Pr\{D^+\} + \Pr\{T^+ \mid D^-\} * \Pr\{D^-\}.$$

The first component in this sum is the product of sensitivity and disease prevalence and the second component is the product of (1 − specificity) and (1 − disease prevalence). Therefore, predicted value positive (PVP) is

$$\text{PVP} = \frac{\Pr\{T^+ \mid D^+\} * \Pr\{D^+\}}{\Pr\{T^+ \mid D^+\} * \Pr\{D^+\} + \Pr\{T^+ \mid D^-\} * \Pr\{D^-\}}.$$

Predicted value negative (PVN) follows immediately:

$$\text{PVN} = \frac{\Pr\{T^- \mid D^-\} * \Pr\{D^-\}}{\Pr\{T^- \mid D^-\} * \Pr\{D^-\} + \Pr\{T^- \mid D^+\} * \Pr\{D^+\}}.$$

These two formulas are special cases of the theorem discovered by Reverend Thomas Bayes (1702–1761). In terms of the events A and B_i, Bayes' theorem is

$$\Pr\{B_i \mid A\} = \frac{\Pr\{A \mid B_i\} * \Pr\{B_i\}}{\sum_i \Pr\{A \mid B_i\} * \Pr\{B_i\}}.$$

As an example, consider the use of the count of blood vessels in breast tumors. A high density of blood vessels indicates a patient who is at high risk of having cancer spread to other organs (5). The use of the count of blood vessels appears to be worthwhile in women with very small tumors and no lymph node involvement, the node-negative case. Suppose that during the development stage of this procedure, its sensitivity was estimated to be 0.85; that is, of the women who had cancer spread to other organs, 85 percent had a high count of blood vessels in their breast tumors. The specificity of the test was estimated to be 0.90; that is, of the women for whom there was no spread of cancer, 90 percent had a low count of blood vessels in their tumors. Assume that the prevalence of cancer spread from breast cancers is 0.02. Given these assumed values, what is the predicted value positive (PVP) of counting the number of blood vessels in the small tumors?

Use the formula from above:

$$\begin{aligned}\text{PVP} &= \frac{\text{prevalence} \times \text{sensitivity}}{[\text{prevalence} \times \text{sensitivity}] + [(1 - \text{prevalence}) \times (1 - \text{specificity})]}\\ &= (0.02 * 0.85)/[(0.02 * 0.85) + (1 - 0.02) * (1 - 0.90)]\\ &= 0.017/0.115 = 0.148.\end{aligned}$$

Using the assumed values above for sensitivity, specificity, and prevalence, there is approximately a 15 percent chance of having cancer spread from a small breast tumor given a high density of blood vessels in the tumor. This value may be too low for the test to be useful. If the true values for specificity or prevalence are higher than the values assumed above, then the PVP will also be higher. For example, if the prevalence is 0.04 instead of 0.02, then the PVP is 0.262 instead of 0.148.

V. PROBABILITY IN SAMPLING

One probability-related issue in sampling alluded to in Chapter 3 is reexamined here using conditional probability. Simple random sampling was presented as giving all the units in the population the same chance of being selected into the sample. The equal probability of selection is clear in sampling with replacement as the total number of units in the population remains constant during the sampling. In sampling without replacement, however, once a subject is selected, it is removed from the population, and the number of units in the population is decreased by one unit. Does this decrease in the denominator as a unit is selected invalidate the equal probability of selection for subsequent units? The following example addresses this matter.

Suppose that a class has 30 students and a SRS of 5 students is to be selected without allowing duplicate selections. The probability of selection for the first draw will be 1/30 and that for the student selected second will be 1/29, as one student was already selected. This line of thinking seems to suggest that random sampling without replacement is not an equal probability sampling model. Is anything wrong in our thinking?

We have to realize that the selection probability of 1/29 for the second draw is a conditional probability. The student selected in the second draw is available for selection only if the student were not selected in the first draw. The probability of not being selected in the first draw is 29/30. Thus, the event of being selected during the second draw is the intersection of the events of not being selected during the first draw (B) and being selected during the second draw (A). Using the rule for the probability of the intersection, that is, $\Pr\{A \text{ and } B\} = \Pr\{A \mid B\} * \Pr\{B\}$, the probability of this intersection is (1/29) * (29/30), which yields 1/30. The same argument can be made for subsequent draws as is shown in Table 5.7.

The demonstration in Table 5.7 indicates that the probability of being selected in any draw is 1/30 and hence the equal probability of selection also holds for sampling without replacement. Now we can state that the probability for a particular student to be included in the sample will be 5/30, as the student can be drawn in any one of the five draws. In general, a SRS of size n from a population of size N will give a selection probability of n/N to each unit of the population, regardless whether sampling is done with replacement or without replacement.

The selection probability in a SRS without replacement can be examined by considering all possible samples that can be drawn. Consider a situation where a SRS of size 3 is drawn without replacement from a

TABLE 5.7 Calculation of Inclusion Probabilities in Drawing a SRS of 5 from 30 without Replacement

Order of draw	Conditional probability (1)	Probability not selected in previous draws (2)	Product of (1) and (2)
1	1/30	1	1/30
2	1/29	29/30	1/30
3	1/28	(29/30)(28/29) = 28/30	1/30
4	1/27	(29/30)(28/29)(27/28) = 27/30	1/30
5	1/26	(29/30)(28/29)(27/28)(26/27) = 26/30	1/30

population containing 5 units (labeled as A, B, C, D, and E). There are 10 possible ways of selecting a sample of 3 as listed below. As we used a random selection mechanism, any one of the 10 possible samples is equally likely to be chosen with a probability of 1/10.

		Elements in the population				
	Sample	*A*	*B*	*C*	*D*	*E*
1	ABC	x	x	x		
2	ABD	x	x		x	
3	ABE	x	x			x
4	ACD	x		x	x	
5	ACE	x		x		x
6	ADE	x			x	x
7	BCD		x	x	x	
8	BCE		x	x		x
9	BDE		x		x	x
10	CDE			x	x	x

From the above configuration, we can easily see that each element of the universe is represented in 6 of the 10 possible samples, suggesting that the probability of a particular element being selected into any sample will be 6/10, which is consistent with $n/N = 3/5$. The statement that each of the possible samples is equally likely implies that each unit in the population has the same probability of being included in the sample.

VI. ESTIMATING PROBABILITIES BY SIMULATION

Our approach to finding probabilities has been to enumerate all possible outcomes and to base calculation of probabilities on this enumeration. This approach works well with simple phenomena, but it is difficult to use with complex events. Another way of assessing probabilities is to simulate the random phenomenon by using repeated sampling. With the wide availability of microcomputers, the simulation approach has become a powerful tool to approach many statistical problems.

For example, consider the following question. How likely is it that two students in a class of 30 will share the same birthday? The answer is not immediately apparent, but the chance does not appear to be very high. Let us find an answer by simulation. First, we assume that the birthdays of 30 students are independent. Second, any of the 365 dates, ignoring February 29th, is equally likely to be a student's birthday. This situation is then equivalent to selecting a random sample of 30 dates from the 365 days using the sampling with replacement procedure. As described in Chapter 3, we can use the random number table in Appendix B. For example, we can read 30 three-digit numbers between 1 and 365 from the table and check to see if any duplicate numbers are selected. We can repeat the operation many times and see how many of the trials produced duplicates. As this manual simulation would require considerable time, we can use MINITAB as shown in Box 5.1. Table 5.8 shows the results of the MINITAB simulation. Eight of these ten trials have duplicates, which suggests that there is a 80 percent probability of finding at least one common birthday among 30 students. Not shown are the results of 10 additional trials in which 5 of the 10 had duplicates. Combining these two sets of 10 trials, the probability of finding common birthdays among 30 students is estimated to be 65 percent [= (8 + 5)/20]. As we increase the number of trials, the estimated probability should approach the true value of 70.6 percent.

Let us consider another example. Population and family planning program planners in Asian countries have been dealing with the effects of the preference for a son on population growth. If all couples continue to have children until they have two sons, what is the average number of children they would have? To build a probability model for this situation, we assume that genders of successive children are independent and the chance of a son is 1/2. To simulate the number of children a couple has, we select single digits from the random number table, considering odd numbers as boys and even numbers as girls. Random numbers are read until the second odd number is encountered, and the number of values required to obtain two odd values is noted. Table 5.9 shows the results for 20 trials

MINITAB BOX 5.1

A set of sequential numbers 1 through 365 is entered into c1 and a sample of 30 numbers are randomly selected with replacement from c1 and stored in c11. To check for duplicates, the results were sorted in an ascending order. This operation is repeated 10 times and the results are stored in c11–c20. The results are printed as shown in Table 5.8.

```
MTB > set c1
DATA> 1:365
DATA> end
MTB > sample 30 c1 c11;
SUBC> replace.
MTB > sort c11 c11
MTB > sample 30 c1 c12;
SUBC> replace.
MTB > sort c12 c12
    ....................
MTB > sample 30 c1 c20;
SUBC> replace.
MTB > sort c20 c20.
MTB > print c11-c20
```

TABLE 5.8 Simulation via MINITAB to Find the Probability of Common Birthdays among 30 Students

Row	C11	C12	C13	C14	C15	C16	C17	C18	C19	C20
1	4	2	3	44	8	3	7	5	8	12
2	10	30	10*	52	21	4	47	7	18	19
3	21	46	10*	72	24	22	48	7	27	31
4	47	67	15	85	76	23	54	18	45	48
5	48	97	23	106	91	27	80	23	50	65
6	64	100	26	116	100	42	82	37	66	80
7	65	105	35	120	113	57	93	54	90	82
8	78	106	41	123	124	64	119	59	91	103
9	93	106	53	132	143*	72	123	64	94	116
10	95	109	73	143	143*	104	137	89	97	169
11	101	133	78	151	147	107	138	109	104	175
12	115	140	86	180	150	119	140	120	132	182
13	154	145	87	181	155	132	162	138	149	193
14	165	158	163	188	166	152	179	143	153	195
15	167	191	166	208	172	167	185	173	180	208
16	185	209*	176	231	200	210	191	201	187	217
17	193	209*	186	248	205	229	199	209*	188	247
18	220	220	200	249	241	230	203	209*	189	249
19	232	223	209	255	243	233	213	215	193	261
20	242	229	220	259*	248	236	232	223	196	262*
21	257	241	251	259*	250	253	238	224	242	262*
22	282	249	260	267	263	307	252	231	250	305
23	284	268	264	270	281	321	259	239	324	307
24	285	286	265	285	283	326	267	259	333	309
25	288	317	283	286	307	327	272	274	338	321
26	299	323	295	288	310	334	287	335	354	326
27	309	335*	297	296	311	336	295	342	360*	328
28	346	335*	300	310	326	343*	308	352	360*	330
29	347	336	352	327	335	343*	313	357	360*	347
30	357	356	355	352	336	362	363	358	360*	356

TABLE 5.9 Simulation of Childbearing until the Second Son Is Born

Trial	Digits	Number of digits	Trial	Digits	Number of digits
1	19	2	11	37	2
2	2239	4	12	367	3
3	503	3	13	6471	4
4	4057	4	14	509	3
5	56287	5	15	940001	6
6	13	2	16	927	3
7	96409	5	17	277	3
8	125	3	18	544264882425	12
9	31	2	19	3629	4
10	425448285	9	20	045467	6
			Total		85

Average = 85/20 = 4.25

(couples). The average number of children based on this very small simulation is estimated to be 4.25 (= 85/20). Additional trials would provide an estimate closer to the true value of 4 children.

VII. PROBABILITY AND THE LIFE TABLE

Perhaps the oldest probability model that has been applied to a problem related to health is the life table. The basic idea was conceived by John Graunt (1620–1674) and the first life table, published in 1693, was constructed by Edmund Halley (1656–1742). Later Daniel Bernoulli (1700–1782) extended the model to determine how many years would be added to the average life span if small pox were eliminated as a cause of death. Now the life table is used in a variety of fields, for example, in life insurance calculations, in clinical research, and in the analysis of processes involving attrition, aging, and wearing out of industrial products.

We present the life table here to show an additional application of the probability rules described above. Table 5.10 is the abridged life table for the total U.S. population in 1990 (6). It is based on information from all death certificates filed in the 50 states and the District of Columbia. It is called an abridged life table because it uses age groupings instead of single years of age. Other types of life tables are available from the National Center for Health Statistics. A brief history and sources for life tables for the United States can be found in Appendix C.

One use of the life table is to summarize the life experience of the population. A direct way of creating a life table is to follow a large cohort, say 100,000 infants born on the same day, until the last member of this cohort dies. For each person, the exact length of life can be obtained by counting the number of days elapsed from the date of birth. This yields 100,000 observations of the length of life. The random variable is the length of life in years or even in days. We can display the distribution of this random variable and calculate the mean, median, first and third quartiles, and minimum and maximum. As most people die at older ages, we expect that the distribution is skewed to the left and hence the median length of life is larger than the mean length of life. The mean length of life is the life expectancy. We can tabulate the data using the following age intervals: 0–1, 1–5, 5–10, 10–15, . . . , 80–85, and 85 or over. All the intervals are the same length, 5 years, except for the first two and the last interval. The first interval is of a special interest, as quite a few infants die. From this tabulation, we can also calculate the relative frequency distribution by dividing the frequencies by 100,000. These relative frequencies give the probability of dying in each age interval. This probability distribution can be used to answer many practical questions regarding life expectancy. For instance,

TABLE 5.10 Abridged Life Table for the Total U.S. Population, 1990

Age interval: Period of life between two exact ages stated in years (1)	Proportion dying: Proportion of persons alive at beginning of age interval dying during interval (2)	Of 100,000 born alive: Number living at beginning of age interval (3)	Of 100,000 born alive: Number dying during age interval (4)	Stationary population: In the age interval (5)	Stationary population: In this and all subsequent age intervals (6)	Average remaining lifetime: Average number of years of life remaining at beginning of age interval (7)
x to $x + n$	${}_nq_x$	l_x	${}_nd_x$	${}_nL_x$	T_x	e_x
0–1	0.0093	100,000	927	99,210	7,535,219	75.4
1–5	0.0018	99,073	183	395,863	7,436,009	75.1
5–10	0.0011	98,890	110	494,150	7,040,146	71.2
10–15	0.0013	98,780	127	493,654	6,545,996	66.3
15–20	0.0044	98,653	430	492,290	6,052,342	61.3
20–25	0.0055	98,223	539	489,794	5,560,052	56.6
25–30	0.0062	97,684	607	486,901	5,070,258	51.9
30–35	0.0077	97,077	743	483,571	4,583,357	47.2
35–40	0.0099	96,334	952	479,425	4,099,786	42.6
40–45	0.0126	95,382	1,203	474,117	3,620,361	38.0
45–50	0.0187	94,179	1,759	466,820	3,146,244	33.4
50–55	0.0290	92,420	2,685	455,809	2,679,424	29.0
55–60	0.0457	89,735	4,101	439,012	2,223,615	24.8
60–65	0.0706	85,634	6,044	413,879	1,784,603	20.8
65–70	0.1029	79,590	8,186	378,369	1,370,724	17.2
70–75	0.1519	71,404	10,847	330,846	992,355	13.9
75–80	0.2211	60,557	13,389	270,129	661,509	10.9
80–85	0.3239	47,168	15,276	197,857	391,380	8.3
≥85	1.0000	31,892	31,892	193,523	193,523	6.1

Source: National Center for Health Statistics (6).

what is a 20-year-old person's probability of surviving to the retirement age of 65?

Acquiring such data poses a problem, however. It would take more than 100 years to collect. Moreover, information obtained from such data may be of some historical interest, but are not useful in answering current life expectancy questions, as current life expectancy may be different from that of earlier times. To solve this problem, we have to find ways to use current mortality information to construct a life table. The logical current mortality data for this purpose are the age-specific death rates. For the time being, we assume that age-specific death rates measure the probability of dying in each age interval. Note that these rates are conditional probabilities. The death rate for the age group 5 to 10 years is computed on the condition that its members survived the previous age intervals.

As studied in Chapter 4, the age-specific death rate is calculated by dividing the number of deaths in a particular age group by the midyear population in that age group. This is not exactly a proportion, whereas a probability is. Therefore the first step in constructing a life table is to convert the age-specific death rates to the form of a probability. One possible conversion is based on the assumption that the deaths were occurring evenly throughout the interval. Under this assumption, we expect that one-half of the deaths occurred during the first half of the interval. Thus, the number of persons at the beginning of an interval is the sum of the midyear population and one-half of the deaths that occurred during the interval. Then the conditional probability of dying during the interval is the number of deaths divided by the number of persons at the beginning of the interval. Actual conversions use more complicated procedures for different age groups, but we are not concerned about these details.

A. The First Four Columns in the Life Table

With this background, we are now ready to examine Table 5.10. The first column shows the age intervals between two exact ages. For instance, 5–10 indicates the 5-year interval between the fifth and tenth birthdays. This age grouping is slightly different from those of under 5, 5–9, 10–14, and so on, used in the census publications. In the life table, age is considered as a continuous variable, whereas in the census, counting of people by age (ignoring the fractional year) is emphasized.

The second column shows the proportion of the persons alive at the beginning of the interval who will die before reaching the end of the interval. It is labeled as ${}_nq_x$, where the first subscript on the left denotes the length of the interval and the second subscript on the right denotes the exact age at the beginning of the interval. The first entry in the second column, ${}_1q_0$, is 0.0093, which is the probability of infants dying during the first year of life. The second entry is ${}_4q_1$, which equals 0.0018. It is the conditional probability of dying during the interval between ages 1 and 5 provided the child survived the first year of life. The rest of the entries in this column are conditional probabilities of dying in a given interval for those who survived the preceding intervals. These conditional probabilities are estimated from the current age-specific death rates. Note that the last entry of column 2 is 1.0000, indicating everybody dies some time after age 85.

Thus we have a series of conditional probabilities of dying. Given these conditional probabilities of dying, we can also find the conditional probabilities of surviving. The probability of surviving the first year of life is

$$1 - {}_1q_0 = 1 - 0.0093 = 0.9907.$$

Likewise, the conditional probability of surviving the interval between exact ages 1 and 5, provided the infants had survived the first year of life, is

$$1 - {}_4q_1 = 1 - 0.0018 = 0.9982.$$

Surviving the first 5 years of life is the intersection of surviving the 0–1 interval and the 1–5 interval. The probability of this intersection can be obtained as the product of the probability of surviving the 0–1 interval and the conditional probability of surviving the 1–5 interval given survival during the 0–1 interval, that is,

$$\Pr\{\text{surviving the intervals 0–1 and 1–5}\} = (1 - {}_1q_0) * (1 - {}_4q_1)$$
$$= (1 - 0.0093) * (1 - 0.0018) = (0.9907) * (0.9982) = 0.9889.$$

Similarly, the probability of surviving the first 10 years of life, the first three intervals, is

$$(1 - {}_1q_0) * (1 - {}_4q_1) * (1 - {}_5q_5).$$

Using this approach, we can calculate the survival probabilities from birth to the beginning of any subsequent age intervals. These survival probabilities are reflected in the third column, the number alive, l_x, at the beginning of the interval that begins at x years of age, out of a cohort of 100,000. Note that the entries in this column may differ slightly from the product of the survival probabilities and 100,000 because, although only four digits to the right of the decimal point are shown in the second column, more digits are used in the calculations. The first entry in this column, l_0, called the *radix*, is the size of the birth cohort. The second entry, the number alive at the beginning of the interval beginning at 1 year of age, l_1, is found by taking the product of the number alive at the beginning of the previous interval and the probability of surviving that interval, that is,

$$l_1 = l_0 * (1 - {}_1q_0) = l_0 - (l_0 * {}_1q_0) = l_0 - {}_1d_0.$$

This quantity, l_1, is equivalent to taking the number alive at the beginning of the previous period minus the number that died during that period, ${}_1d_0$. The numbers that died during each interval are shown in the fourth column, which is labeled as ${}_nd_x$.

The number who died during the 4-year age interval from 1 to 5 is ${}_4d_1$. This is found by taking the product of the number alive at the beginning of this interval, l_1, and the probability of dying during the interval, ${}_4q_1$, that is, ${}_4d_1 = l_1 * {}_4q_1$. The number alive at the beginning of the interval of 5 to 10 years of age, l_5, can be found by subtracting the number who died during the previous age interval, ${}_4d_1$, from the number alive at the beginning of the previous interval, l_1, that is, $l_5 = l_1 - {}_4d_1$. Repeating this operation yields the rest of the entries in the third and fourth columns. The fourth column can also be obtained directly from the third column. For example,

$$_1d_0 = l_0 - l_1, \qquad _4d_1 = l_1 - l_5, \qquad \text{etc.}$$

Note that the last entry in the third column is the same as the last entry in the fourth column, because all the survivors at age 85 will die subsequently. Note further that the l_x value in each row is a cumulative total of $_nd_x$ values in that and all subsequent rows.

Dividing the entries in the third and fourth columns by 100,000, we obtain the probabilities of surviving from birth to the beginning of the current interval and dying during the current interval, respectively. Note that the entries in the fourth column sum to 100,000, meaning that the probability of dying sums to one. As we expected, the distribution is negatively skewed, with the larger probabilities of dying at older ages.

B. Some Uses of the Life Table

The last three columns are discussed in a following section. Before doing that, we wish to show how the first four columns, particularly the third column, can be used to answer some questions regarding life expectancy.

For example, what is the probability of surviving from one age to a subsequent age, say from age 5 to age 20? This is a conditional probability, conditional on the survival to age 5. The intersection of the events of surviving to age 20 and surviving to age 5 is surviving to age 20. Thus the probability of this intersection is the probability of surviving from birth to age 20. This is the number alive at the beginning of the interval 20–25 divided by the number alive at the beginning, that is, l_{20}/l_0. The probability of surviving from birth to age 5 is l_5/l_0. Therefore, the conditional survival probability from age 5 to age 20 is found by dividing the probability of the intersection by the probability of surviving to age 5, that is,

$$\left(\frac{l_{20}}{l_0}\right) \Big/ \left(\frac{l_5}{l_0}\right) = \frac{l_{20}}{l_5} = \frac{98{,}223}{98{,}890} = 0.9933.$$

The survival probabilities from any age to an older age can be calculated in a similar fashion.

We know the conditional probability of dying in any single interval; however, we may be interested in the probability of dying during a period formed by the first two or more consecutive intervals. For example, what is the probability of dying during the first 5 years of life? This probability can be found by subtracting the probability of surviving the first 5 years from 1, that is,

$$1 - [1 - {}_1q_0) * (1 - {}_4q_1)] = 1 - \left(\frac{l_1}{l_0} * \frac{l_5}{l_1}\right) = 1 - \frac{l_5}{l_0}$$

$$= 1 - \frac{98{,}890}{100{,}000} = 1 - 0.9889 = 0.0111.$$

This is simply 1 minus the ratio of the number alive at the beginning of the final interval of interest and 100,000.

A similar question relates to the probability of dying during a period formed by two or more consecutive intervals given that one had already survived several intervals. For example, what is the probability that a 30-year-old person will die between the ages of 50 and 60? This conditional probability is found by dividing the probability of the intersection of the event of dying between the ages of 50 and 60 and the event surviving until 30 by the probability of the event of surviving until 30 years of age. The intersection of dying between 50 and 60 and surviving until 30 is dying between 50 and 60. The probability of dying between 50 and 60 is the number of persons dying, l_{50} minus l_{60}, divided by the total number, l_0. The probability of surviving until age 30 is simply l_{30} divided by l_0. Therefore, the probability of dying between 50 and 60 given survival until 30 is

$$\left(\frac{l_{50} - l_{60}}{l_0}\right) \bigg/ \left(\frac{l_{30}}{l_0}\right) = \frac{l_{50} - l_{60}}{l_{30}} = \frac{92{,}420 - 85{,}634}{97{,}077} = 0.0699.$$

Another slightly more complicated question concerns the joint survival of persons. Suppose that a 40-year-old person has a 5-year-old child. What will be the probability that both the parent and child survive 25 more years until the parent's retirement? If we assume that the survival of the parent and that of the child are independent, we can calculate the desired probability by multiplying the individual survival probabilities. Applying the rule for the probability of surviving from one age to a subsequent age from the first question, this is

$$\frac{l_{65}}{l_{40}} * \frac{l_{30}}{l_5} = \frac{79{,}590}{95{,}382} * \frac{97{,}077}{98{,}890} = 0.8344 * 0.9817 = 0.8191.$$

The probability that both the parent and the child will die during the 25 years is

$$\left(1 - \frac{l_{65}}{l_{40}}\right) * \left(1 - \frac{l_{30}}{l_5}\right) = (1 - 0.8344) * (1 - 0.9817) = 0.0030.$$

The probability that the parent will die but the child will survive during the 25 years is

$$\left(1 - \frac{l_{65}}{l_{40}}\right) * \left(\frac{l_{30}}{l_5}\right) = (1 - 0.8344) * (0.9817) = 0.1626.$$

The probability that the parent will survive but the child will die during the 25 years is

$$\left(\frac{l_{65}}{l_{40}}\right) * \left(1 - \frac{l_{30}}{l_5}\right) = (0.8344) * (1 - 0.9817) = 0.0153.$$

These four probabilities sum to 1, because those four events represent all the possible outcomes in considering the life and death of two persons.

C. Expected Values in the Life Table

The most widely used expected or average value in the life table is the mean length of life, which is known as the life expectancy. This is found by summing all the ages of deaths and dividing by 100,000. This is the same as multiplying the age of death by the probability of death at that age and summing over all ages. As we have age groups, not individual ages, we can approximate life expectancy by using the midpoints of the age intervals shown in column 1. As shown in Box 5.2, these midpoints are multiplied by the probabilites of dying in that interval (column 4 divided by 100,000). The midpoint for the last open inverval is arbitrarily entered as 92.5, assuming that the length of interval is 15 years. The sum of these products approximates the life expectancy. The approximate mean turns out to be 75.8 years, which is slightly larger than 75.4 shown as the first entry of column 7 in the life table.

MINITAB BOX 5.2

Column c1 contains the midpoints of the age intervals and c2 contains the number of deaths during the age interval.

```
MTB > set c1
DATA> .5 3 7.5 12.5 17.5 22.5 27.5 32.5 37.5 42.5
DATA> 47.5 52.5 57.5 62.5 67.5 72.5 77.5 82.5 92.5
DATA> set c2
DATA> 927 183 110 127 430 539 607 743 952 1203 1759
DATA> 2685 4101 6044 8186 10847 13389 15276 31892
DATA> end
MTB > let c3=c2/100000
MTB > let k1=sum(c1*c3)
MTB > print k1
K1        75.7571
```

The quartiles are approximated by interpolation as shown below. To find the median, the second quartile, we must find the value such that 50 percent of the values fall below it. By examining column 3 in the life table, we find that 60,557 persons are alive at the beginning of the age interval 75–80 whereas only 47,168 are alive at the beginning of the interval 80–85. As 50,000 is between 60,557 and 47,168, we know that the median is somewhere between 75 and 80 years of age. If we assume that the 13,389 (= 60,557 − 47,168) deaths are uniformly distributed over this age interval, we can find the median by interpolation. We add a proportion of the 5 years, the length of the interval, to the age at the beginning of the interval, 75 years. The proportion is the ratio of the difference between 60,557 and 50,000 to the 13,389 deaths that occurred in the interval. The calculation is

$$\text{median} = 75 + 5 * \left(\frac{60{,}557 - 50{,}000}{13{,}389}\right) = 78.94.$$

The corresponding calculations for the first and third quartiles are

$$Q_1 = 65 + 5 * \left(\frac{79{,}590 - 75{,}000}{8186}\right) = 67.80$$

and

$$Q_3 = 85 + 15 * \left(\frac{31{,}892 - 25{,}000}{31{,}892}\right) = 88.24.$$

As expected, the mean is smaller than the median. Perhaps, it is more enlightening to know that one-half of a birth cohort will live to age 79 than to know that an average length of life is about 75 years.

The above calculations of the mean and quartiles are based on the assumption that deaths were distributed evenly within each interval. This assumption is realistic for most intervals but it is not for the intervals at both ends of the distribution. For instance, vital statistics show that more deaths occur during the first week of life than in any other week during the first year of life. Therefore, the use of the midpoint for the first year of life in the calculation of the mean should have inflated the mean slightly, as seen above. The last three columns in the life table are based on additional information which removes the need to assume that the deaths are distributed uniformly throughout the interval.

D. Columns 5, 6, and 7 in the Life Table

The fifth column of the life table, denoted by ${}_nL_x$, shows the person-years lived during each interval. For instance, the first entry in the fifth column is 99,210, which is the total number of person-years of life contributed by 100,000 infants during the first year of life. This value consists of 99,073 years contributed by the infants that survived the full year plus 137 years contributed by the 927 infants who died during the year. The value of 137 years is based on actual mortality data coupled with mathematical smoothing. It cannot be found from the first four columns in the table. The value of 137 years is much less than the 400 to 500 years of life expected if the deaths had been distributed uniformly during the year. This value also suggests that most of the deaths occurred during the first half of the interval. The second entry in the fifth column is much larger than the first entry, mainly reflecting that the length of the second interval is greater than the length of the first interval. Each person surviving this second interval contributed 4 person-years of life.

In the life table, the fifth column is labeled as the "stationary population in the age interval." The label *stationary population* is based on a model of the long-term process of birth and death. If we assume 100,000 infants are born every year for 100 years, with each birth cohort subject to the same probabilities of dying specified in the second column of the life table, then we expect that 100,000 people will be dying at the indicated ages every year. This means that the number of people in each age group will be the numbers shown in the fifth column. This hypothetical population will maintain the same size, as the number of births is the same as the number of deaths and it also keeps the same age distribution. That is, the size and structure of population are invariant, and hence this is called a stationary population.

The sixth column of the life table, denoted by T_x, shows cumulative totals of ${}_nL_x$ values starting from the last age interval. The T_x value in each interval indicates the number of person-years remaining in that and all subsequent age intervals. For example, the T_{80} value of 391,380 is the sum of ${}_5L_{80}$ (= 197,857) and ${}_{15}L_{85}$ (= 193,523).

The last column of the life table, denoted by e_x, shows the life expectancies at various ages, which are calculated by $e_x = T_x/l_x$. The first entry in the last column is the life expectancy for newborn infants, and all subsequent entries are conditional life expectancies. Conditional life expectancies are more useful information than the expectancies figured for newborn infants. For instance, those who survived to age 85 are expected to live 6.1 years more ($e_{85} = 6.1$) (the last entry of the last column), whereas newborn infants are expected to live 1.93 years beyond age 85 (T_{85}/l_0 = 193,523/100,000 = 1.93).

On the basis of T_x values, more complicated conditional life expectancies can be calculated. For instance, suppose that a 30-year-old person was killed in an industrial accident and had been expected to retire at age 65 if still alive. For how many years of unearned income should that person's heirs be compensated? The family may request a compensation for 35 years; however, based on the life table, the company argues for a smaller number of years. The total number of years of life remaining during the interval from 30 to 65 is T_{30} minus T_{65}, and there are l_{30} persons remaining at age 30 to live those years. Therefore, the average number of years of life remaining is found by

$$\frac{T_{30} - T_{65}}{l_{30}} = \frac{4{,}583{,}357 - 1{,}370{,}724}{97{,}077} = 33.1 \text{ years.}$$

Finally, the notion of stationary population can be used to make certain inferences for population planning and manpower planning. The birth rate of the stationary population can be obtained by dividing 100,000 by the total years of life lived by the stationary population, or

$$\frac{l_0}{T_0} = \frac{100{,}000}{7{,}535{,}219} = \frac{1}{75.4} = 0.013$$

or 13 per 1000 population. The death rate should be the same. But note that the birth rate equals the reciprocal of the life expectancy at birth ($1/e_0$). In other words, the birth rate (replacement rate) and death rate (attrition rate) are determined entirely by the life expectancy under the stationary population assumption.

VIII. CONCLUDING REMARKS

Probability has been defined as the relative frequency of an event in an infinite number of trials or in a population. Its use has been demonstrated in a number of examples and a number of rules for the calculation of probabilities have been presented. The use of probabilities and the rules for calculating probabilities have been applied to the life table, a basic tool in public health research.

Now that we have an understanding of probability, we shall examine particular probability distributions in the next chapter.

EXERCISES

5.1. Choose the most appropriate answer.

a. Which of the following is not a probability model?
 __ the life table
 __ a sampling distribution
 __ the random digit table

b. If you get 10 straight heads in tossing a fair coin, a tail is ______ on the next toss.
 __ more likely
 __ less likely
 __ neither more likely nor less likely

c. In the U.S. life table, the distribution of the length of life (or age at death) is
 __ skewed to the left
 __ skewed to the right
 __ symmetric

d. A test with high sensitivity is very good at
 __ screening out patients who do not have the disease
 __ detecting patients with the disease
 __ determining the probability of the disease

e. In the U.S. life table the life expectancy (mean) is ______ the median length of life.
 __ the same as
 __ greater than
 __ less than

f. ${}_4q_1$ is called a ______ because an infant cannot die in this interval unless it survived the first year of life.

_ personal probability
_ marginal probability
_ conditional probability

g. In the U.S. life table, the mean length of life for those who died during ages 0–1 is
_ about 1/2 year
_ more than 1/2 year
_ less than 1/2 year

5.2. The following table gives estimates of the probabilities that a randomly chosen adult in the United States falls into each of six gender-by-education categories [based on relative frequencies from NHANES II (7)]. The three education categories used are (1) less than 12 years, (2) high school graduate, and (3) more than high school graduation.

	Category of education		
Gender	*1*	*2*	*3*
Female	0.166	0.194	0.164
Male	0.149	0.140	0.187

a. What is the estimate of the probability that an adult is a high school graduate?
b. What is the estimate of the probability that an adult is a female?
c. From the NHANES II data, it is also estimated that the probability that a female is taking a vitamin supplement is 0.426. What is the estimate of the probability that the adult is a female and taking a vitamin supplement?
d. From the NHANES II, it is also estimated that the probability of adults taking a vitamin supplement is 0.372. What is the estimate of the probability that a male is taking a vitamin supplement?

5.3. Suppose that the failure rate for a brand of smoke detector is 1 in 2000. For safety, two of these smoke detectors are installed in a laboratory.
a. What is the probability that smoke is not detected in the laboratory when smoke is present in the laboratory?
b. What is the probability that both detectors sound an alarm when smoke is present in the laboratory?
c. What is the probability that one of the detectors sounds the alarm and the other fails to sound the alarm when smoke is present in the laboratory?

5.4. Suppose that the probability of conception for a married woman in any month is 0.2. What is the probability of conception in 2 months?

5.5. A new contraceptive device is said to have only a 1 in 100 chance of failure. Assume that the probability of conception for a given month, without using any contraceptive, is 20 percent. What is the probability of having at least one unwanted pregnancy if a woman were to use this device for 10 years? *Hint*: This would be the complement of the probability of avoiding pregnancy for 10 years or 120 months. The probability of conception for any month with the use of the new contraceptive device would be $0.2 * (1 - 0.99)$. This and related issues are examined by Keyfitz (8).

5.6. In a community, 5500 adults were screened for hypertension by the use of a standard sphygmomanometer and 640 were found to have a diastolic blood pressure of 90 mm Hg or higher. A random sample of 100 adults from those with diastolic blood pressure of 90 mm Hg or higher and another random sample of 100 adults from those with blood pressure less than 90 mm Hg were subjected to more intensive clinical evaluation for hypertension, and 73 and 13 of the respective samples were confirmed as being hypertensive.

a. What is an estimate of the probability that an adult having blood pressure greater than or equal to 90 at the initial screening will actually be hypertensive (predicted value positive)?
b. What is an estimate of the probability that an adult having blood pressure less than 90 at the initial screening will not actually be hypertensive (predicted value negative)?
c. What is an estimate of the probability that an adult in this community is truly hypertensive (prevalence rate of hypertension)?
d. What is an estimate of the probability that a hypertensive person will be found to have blood pressure greater than or equal to 90 at the initial screening (sensitivity)?
e. What is an estimate of the probability that a person without hypertension will have blood pressure less than 90 at the initial screening (specificity)?

5.7. How likely is it to find two students in a class of 23 sharing a birthday? Simulate using the random number table in Appendix B or MINITAB.

5.8. What is the average number of children per family if every couple were to have children until a son is born? Simulate using the random number table or MINITAB.

5.9. Calculate the following probabilities from the 1990 U.S. Abridged Life Table.

a. What is the probability that a 35-year-old person will survive to retirement at age 65?
b. What is the probability that a 20-year-old person will die between ages 55 and 65?

5.10. Calculate the following expected values from the 1990 U.S. Abridged Life Table.
   a. How many years is a newborn expected to live before his fifth birthday?
   b. How many years is a 20-year-old person expected to live after retirement at age 65? Repeat the calculation for a 60-year-old person. How would you explain the difference?
   c. A 35-year-old person is killed in a factory accident. How many years would the person have been expected to live before retirement at age 65 if the accident had not occurred?

REFERENCES

1. U.S. Bureau of the Census (1991). The 1990 Census of Population and Housing, Summary Tape File 1A. On CD-ROM Technical Documentation/prepared by the Bureau of the Census.—Washington: The Bureau, 1991.
2. National Center for Health Statistics (1992). "Health, United States, 1991 and Prevention Profile," DHHS Publ. No. 92-1232. Public Health Service, Hyattsville, MD.
3. Canfield, M., ed. (1990). "The Health Status of Harris County Residents: Births, Deaths and Selected Measures of Public Health, 1980–1986." Harris County Health Department, Houston, TX.
4. Weiss, N. S. (1986). "Clinical Epidemiology: The Study of Outcome of Disease." Oxford Univ. Press, New York.
5. Weidner, N., Folkman, J., Pozza, F., Bevilacqua, P., Allred, E. N., Moore, D. H., Meli, S. and Gasparini, G. (1992). "Tumor angiogenesis: A new significant and independent prognostic indicator in early-stage breast carcinoma." *J. Nat. Cancer Inst.* **84,** 1875–1887.
6. National Center for Health Statistics (1993). Advance report of final mortality statistics, 1990. *Mon. Vital Stat. Rep.* **41,** No. 7, Suppl.
7. National Center for Health Statistics (1982). "Second National Health and Nutrition Examination Survey (NHANES II)," tabulation of data for adults 18 years of age and over by the authors.
8. Keyfitz, N. (1971). How birth control affects births. *Soc. Biol.* **18,** 109–121.

CHAPTER 6

Probability Distributions

This chapter introduces three probability distributions: the binomial and the Poisson for discrete random variables, and the normal for continuous random variables. For a discrete random variable, its probability distribution is a listing of the probabilities of its possible outcomes or a formula for finding the probabilities. For a continuous random variable, its probability distribution is usually expressed as a formula that can be used to find the probability that the variable will fall in a specified interval. Knowledge of the probability distribution (1) allows us to summarize and describe data through the use of a few numbers; and (2) helps to place results of experiments in perspective, that is, it allows us to determine whether or not the result is consistent with our ideas. We begin the presentation of probability distributions with the binomial distribution.

I. THE BINOMIAL DISTRIBUTION

As its name suggests, the *binomial distribution* refers to random variables with two outcomes. Three examples of random variables with two outcomes are (1) hypertension status—a person does or does not have hypertension, (2) exposure to benzene—a worker was or was not exposed to benzene in the workplace, and (3) health insurance coverage—a person does or does not have health insurance. The random variable of interest in the binomial setting is the number of occurrences of the event under study, for example, the number of adults in a sample of size *n* who have hypertension, or who have been exposed to benzene, or who have health insurance. For the binomial distribution to apply, the status of each subject must be independent of that of the other subjects. For example, in the hypertension question, we are assuming that each person's hypertension status is unaffected by any other person's status.

We consider a simple example to demonstrate the calculation of binomial probabilities. Suppose that four adults (labeled A, B, C, and D) have been randomly selected and asked whether or not they have hypertension. The random variable of interest in this example is the number of persons who respond yes to the question about hypertension. The possible outcomes of this variable are 0, 1, 2, 3, and 4.

The outcomes (0, 1, 2, 3, or 4) translate to estimates of the proportion of persons who answer yes (0.00, 0.25, 0.50, 0.75, and 1.00, respectively). Any of these outcomes could occur when we draw a random sample of four adults. As a demonstration, let us draw 10 random samples of size 4 from a population in which the proportion of adults who answer yes to the hypertension question is 0.25. We are using the value of 0.25 instead of the value of 0.397 mentioned in Exercise 4.15 because many people are unaware that they have hypertension. We can use a random number table in performing this demonstration or we can use MINITAB as shown in Box 6.1.

MINITAB BOX 6.1

The command to be used is RANDOM, which tells MINITAB to draw samples and store the results in a column. We supply the number of samples to be drawn, 10 in this example, and a column to receive the results, c1. The subcommand identifies the distribution from which the samples are drawn, in this case the binomial. The binomial distribution is characterized by two parameters, the sample size and the population proportion having the characteristic of interest. In this case, the sample size is 4 and the population proportion is 0.25.

```
 MTB > random 10 cl;
 SUBC> binom 4 .25.
 MTB > print cl
C1   1  1  0  2  2  1  1  1  1  0
```

The printed values represent the number of people who answered yes in each of the 10 random samples of size 4. Two samples had zero yes responses, six samples had one yes response, and in two samples there were two yes responses. These results translate to two estimates having the value of 0.00, six estimates having the value of 0.25, and two estimates having the value of 0.50.

Hence the sample estimate does not necessarily equal the population parameter and the estimates can vary considerably. In practice a single sample is selected, and in making an inference from this one sample to the population, this sample-to-sample variability must be taken into account. The probability distribution does this. Now let us calculate the binomial probability distribution for a sample of size 4.

Suppose that in the population, the proportion of people that would respond yes to this question is π. The probability of each of the outcomes can be found in terms of π by listing all the possible outcomes. Table 6.1 provides this listing.

As each person is independent of all the other persons, the probability of the joint occurrence of any outcome is simply the product of the probabilities associated with each person's outcome. That is, the probability of four yes responses is $\pi * \pi * \pi * \pi$, which is π^4. In the same way, the probability of three yes responses is $4 * \pi^3 * (1 - \pi)$ as there are four occurrences of three yes responses. The probability of two yes responses is $6 * \pi^2 * (1 - \pi)^2$, the probability of one yes response is $4 * \pi * (1 - \pi)^3$, and the probability of zero yes responses is $(1 - \pi)^4$. If we know the value of π, we can calculate the numerical value of these probabilities.

Suppose π is the previously mentioned value of 0.25. Then the probability of each outcome is as follows:

$$\Pr\{4 \text{ yes responses}\} = 1 * (0.25)^4 * (0.75)^0 = 0.0039 = \Pr\{0 \text{ no responses}\}$$
$$\Pr\{3 \text{ yes responses}\} = 4 * (0.25)^3 * (0.75)^1 = 0.0469 = \Pr\{1 \text{ no response}\}$$

TABLE 6.1 Possible Outcomes and Their Probabilities of Occurrence

Person					
A	**B**	**C**	**D**	**Probability of occurrence**	
y[a]	y	y	y	$\pi * \pi * \pi * \pi$	$= \pi^4 * (1-\pi)^0$
y	y	y	n	$\pi * \pi * \pi * (1-\pi)$	$= \pi^3 * (1-\pi)^1$
y	y	n	y	$\pi * \pi * (1-\pi) * \pi$	$= \pi^3 * (1-\pi)^1$
y	n	y	y	$\pi * (1-\pi) * \pi * \pi$	$= \pi^3 * (1-\pi)^1$
n	y	y	y	$(1-\pi) * \pi * \pi * \pi$	$= \pi^3 * (1-\pi)^1$
y	y	n	n	$\pi * \pi * (1-\pi) * (1-\pi)$	$= \pi^2 * (1-\pi)^2$
y	n	y	n	$\pi * (1-\pi) * \pi * (1-\pi)$	$= \pi^2 * (1-\pi)^2$
y	n	n	y	$\pi * (1-\pi) * (1-\pi) * \pi$	$= \pi^2 * (1-\pi)^2$
n	y	y	n	$(1-\pi) * \pi * \pi * (1-\pi)$	$= \pi^2 * (1-\pi)^2$
n	y	n	y	$(1-\pi) * \pi * (1-\pi) * \pi$	$= \pi^2 * (1-\pi)^2$
n	n	y	y	$(1-\pi) * (1-\pi) * \pi * \pi$	$= \pi^2 * (1-\pi)^2$
y	n	n	n	$\pi * (1-\pi) * (1-\pi) * (1-\pi)$	$= \pi^1 * (1-\pi)^3$
n	y	n	n	$(1-\pi) * \pi * (1-\pi) * (1-\pi)$	$= \pi^1 * (1-\pi)^3$
n	n	y	n	$(1-\pi) * (1-\pi) * \pi * (1-\pi)$	$= \pi^1 * (1-\pi)^3$
n	n	n	y	$(1-\pi) * (1-\pi) * (1-\pi) * \pi$	$= \pi^1 * (1-\pi)^3$
n	n	n	n	$(1-\pi) * (1-\pi) * (1-\pi) * (1-\pi)$	$= \pi^0 * (1-\pi)^4$

[a] y indicates a yes response and n indicates a no response.

$\Pr\{2 \text{ yes responses}\} = 6 * (0.25)^2 * (0.75)^2 = 0.2109 = \Pr\{2 \text{ no responses}\}$
$\Pr\{1 \text{ yes response}\} = 4 * (0.25)^1 * (0.75)^3 = 0.4219 = \Pr\{3 \text{ no responses}\}$
$\Pr\{0 \text{ yes responses}\} = 1 * (0.25)^0 * (0.75)^4 = 0.3164 = \Pr\{4 \text{ no responses}\}$

The sum of these probabilities is one as it must be because these are all the possible outcomes. If the probabilities do not sum to one (with allowance for rounding), a mistake has been made. Figure 6.1 shows a plot of the binomial distribution for n equal to 4 and π equal to 0.25.

Are these probabilities reasonable? Because the probability of a yes response is assumed to be 0.25 in the population, in a sample of size 4, the probability of one yes response should be the largest. It is also reasonable that the probabilities of zero and two yes responses are the next largest as these values are closest to one yes response. The probability of four yes responses is the smallest, as is to be expected. Figure 6.1 shows the rapid decrease in the probabilities as the number of yes responses moves away from the expected response of one.

In the calculation of the probabilities, several patterns are visible. The exponent of the probability of a yes response matches the number of yes responses being considered, and the exponent of the probability of a no response also matches the number of no responses being considered. The sum of the exponents is always the number of persons in the sample. These patterns are easy to capture in a formula which eliminates the need to enumerate the possible outcomes. The formula may appear complicated, but it is really not all that difficult to use. The formula, also referred to as the *probability mass function* for the binomial distribution, is

$$\Pr\{X = x\} = \binom{n}{x} * \pi^x * (1 - \pi)^{n-x} \qquad \text{where } \binom{n}{x} = {}_nC_x = \frac{n!}{x! * (n - x)!},$$

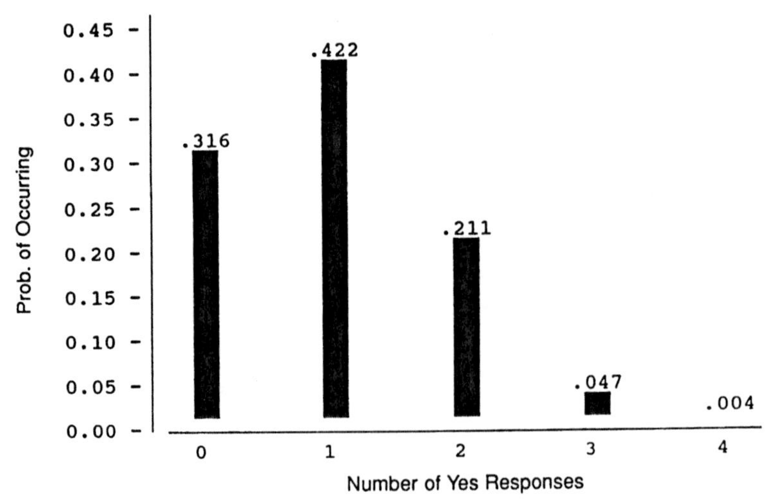

FIGURE 6.1 Bar chart showing the binomial distribution for $n = 4$ and $\pi = 0.25$.

$k! = k * (k - 1) * (k - 2) * \cdots * 1$, and $0!$ is defined to be 1. The symbol $k!$ is called *k factorial*, and ${}_nC_x$ is read as n combination x, which gives the number of ways that x elements can be selected from n elements without regard to order. In this formula, n is the number of persons or elements selected and x is the value of the random variable which goes from 0 to n. Another representation of this formula is

$$B(x; n, \pi) = \binom{n}{x} * \pi^x * (1 - \pi)^{n-x} = B(n - x; n, 1 - \pi)$$

where B represents binomial. The equality of $B(x; n, \pi)$ and $B(n - x; n, 1 - \pi)$ is a symbolic way of saying that the probability of x yes responses from n persons, given that π is the probability of a yes response, equals the probability of $n - x$ no responses.

The hypertension situation can be used to demonstrate the use of the formula. To find the probability that $X = 3$, we have

$$\Pr\{X = 3\} = \binom{4}{3} * (0.25)^3 * (0.75)^1 = \left(\frac{4!}{3! * 1!}\right) * 0.015625 * 0.75$$

$$= \left(\frac{4 * 3 * 2 * 1}{3 * 2 * 1}\right) * 0.01172 = 4 * 0.01172 = 0.0469.$$

This is the same value we found by listing all the outcomes and the associated probabilities. There are easier ways of finding binomial probabilities as is shown next.

There is a recursive relationship between the binomial probabilities that makes it easier to find them than to use the binomial formula for each different value of X. The relationship is

$$\Pr\{X = x + 1\} = \left(\frac{n - x}{x + 1}\right) * \left(\frac{\pi}{1 - \pi}\right) * Pr\{X = x\}$$

for x ranging from 0 to $n - 1$. For example, the probability that X equals 1 in terms of the probability that X equals 0 is

$$\Pr\{X = 1\} = \left(\frac{4 - 0}{0 + 1}\right) * \left(\frac{0.25}{0.75}\right) * 0.3164 = 4 * \left(\frac{1}{3}\right) * 0.3164 = 0.4219$$

which is the same value we calculated above.

A still easier method is to use Appendix Table B2, a table of binomial probabilities for n ranging from 2 to 20 and π beginning at 0.01 and ranging from 0.05 to 0.50 in steps of 0.05. There is no need to extend the table to values of π larger than 0.50 because $B(x; n, \pi)$ equals $B(n - x; n, 1 - \pi)$. For example, if π were 0.75 and we wanted to find the probability that $X = 1$ for $n = 4$, $B(1; 4, 0.75)$, we find $B(3; 4, 0.25)$ in Table B2 and read the value of 0.0469. These probabilities are the same because when $n = 4$ and the probability of a yes response is 0.75, the occurrence of one yes response is the same as the occurrence of three no responses when the probability of a no response is 0.25.

Another way of obtaining binomial probabilities is to use MINITAB as shown in Box 6.2. MINITAB is particularly nice as it does not limit the values of π to being a multiple of 0.05 and n can be much larger than 20. More will be said about how large n can be in a later section.

MINITAB BOX 6.2

The command that can be used to obtain the probability distribution for the binomial is PDF, an abbreviation for *probability density function*, which is the name given to probability distributions for continuous variables and used by MINITAB for both discrete and continuous variables. The PDF command will give the probabilities associated with all the values specified after the PDF command or, if none are specified, for all possible values of X. A subcommand is required to specify which probability distribution is to be found. Because we are working with the binomial, we specify it and then provide the values of n and π that we are using. The semicolon at the end of the PDF line and the period after the value of π must be entered.

```
MTB > set cl
DATA> 0 1
DATA> end
MTB > pdf cl;
SUBC> binomial 4 0.25.
          K            P( X = K)
       0.00                0.3164
       1.00                0.4219
MTB > pdf;
SUBC> binom 4  0.25.
     BINOMIAL WITH N =   4  P = 0.250000
          K              P( X = K)
          0               0.3164
          1               0.4219
          2               0.2109
          3               0.0469
          4               0.0039
```

The probability mass function for the binomial gives $\Pr\{X = x\}$ for x ranging from 0 to n. Another function that is used frequently is the *cumulative distribution function* (cdf). This function gives the probability that X is less than or equal to x for all possible values of X. Table 6.2 shows both the probability mass function and the cumulative distribution function values for the binomial when n is 4 and π is 0.25. The entries in the cumulative distribution row are simply the sum of the probabilities in the row above it, the probability mass function row, for all values of X less than or equal to the value being considered (see Box 6.3 for MINITAB use). Cumulative distribution functions all have a general shape shown in Box 6.3. The value

TABLE 6.2 Probability Mass ($\Pr\{X = x\}$) and Cumulative ($\Pr\{X \leq x\}$) Distribution Functions for the Binomial When $n = 4$ and $\pi = 0.25$

x	0	1	2	3	4
$\Pr\{X = x\}$	0.3164	0.4219	0.2109	0.0469	0.0039
$\Pr\{X \leq x\}$	0.3164	0.7383	0.9492	0.9961	1.0000

of the function starts with a low value and then increases over the range of the X variable. The rate of increase in the function is what varies between different distributions. All the distributions eventually reach the value of one or approach it asymptotically.

MINITAB BOX 6.3

MINITAB produces these values by using the command CDF, the abbreviation for cumulative distribution function, command in the same way as the PDF command. The following plot shows the cdf for a binomial distribution when n is 4 and π is 0.25.

```
MTB > set cl
DATA> 0:4
DATA> end
MTB > cdf cl c2;
SUBC> binom 4 0.25.
MTB > plot c2 cl

          -
   1.00+                                           *         *
        -                               *
C2      -
        -
        -
   0.75+                     *
        -
        -
        -
        -
   0.50+
        -
        -
        -
        -   *
   0.25+
         --+---------+---------+---------+---------+---------+C1
          0.00      0.80      1.60      2.40      3.20      4.00
```

As seen above, if we know the data follow a binomial distribution, we can completely summarize the data through its two parameters, the sample size and the population proportion or an estimate of it. The sample estimate of the population proportion is the number of occurrences of the event in the sample divided by the sample size.

A. Mean and Variance of the Binomial Distribution

We can now calculate the mean and variance of the binomial distribution. The mean is found by summing the products of each outcome by its probability of occurrence, that is,

$$\mu = \sum_{x=0}^{n} x * \Pr\{X = x\}.$$

This appears to be different from the calculation of the sample mean in Chapter 4, but it is really the same because in Chapter 4 all the observations had the same probability of occurrence, $1/N$. Thus the formula for the population mean could be re-expressed as

$$\sum_{i=1}^{N} x_i/N = \sum_{i=1}^{N} x_i * (1/N) = \sum_{i=1}^{N} x_i * \Pr\{x_i\}.$$

The mean of the binomial variable, that is, the mean number of yes responses out of n responses, when n is 4 and π is 0.25, is

$$(0 * 0.3164) + (1 * 0.4219) + (2 * 0.2109) + (3 * 0.0469) + (4 * 0.0039) = 1.00 = n * \pi.$$

The expression of the binomial mean as $n * \pi$ makes sense because, if the probability of occurrence of an event is π, then in a sample of size n, we would expect $n * \pi$ occurrences of the event.

The variance of the binomial variable, the number of yes responses, can also be expressed conveniently in terms of π. From Chapter 4, the population variance was expressed as

$$\sigma^2 = \sum_{i=1}^{N} (x_i - \mu)^2/N.$$

In terms of the binomial, the X variable takes on the values from 0 to n, and we again replace the N in the divisor by the probability that X is equal to x. Thus, the formula becomes

$$\sigma^2 = \sum_{x=0}^{n} (x - n * \pi)^2 * \Pr\{X = x\}$$

which, with further algebraic manipulation, simplifies to $n * \pi * (1 - \pi)$. The variance is then $4 * 0.25 * (1 - 0.25)$, which is 0.75.

There is often interest in the variance of the proportion of yes responses, that is, in the variance of the number of yes responses divided by the sample size. This is the variance of the number of yes responses divided by a constant. From Chapter 4, we know that this is the variance of the number of yes responses divided by the square of the constant. Thus the variance of a proportion is $n * \pi * (1 - \pi)/n^2$, which becomes $\pi * (1 - \pi)/n$.

B. Example: Use of the Binomial Distribution

Let us consider a larger example now. In 1990, cesarean section (c-section) deliveries represented 23.5 percent of all deliveries in the United States, a tremendous increase since 1960 when the rate was only 5.5 percent. Concern has been expressed, for example, by the Public Citizen Health Research Group in its June 1992 Health Letter (1), that many unnecessary c-section deliveries are performed. Public Citizen believes unnecessary c-sections waste resources and increase maternal risks without achieving sufficient concomitant improvement in maternal and infant health. It is in this context that administrators at a local hospital are concerned as they believe that their hospital's c-section rate is even higher than the national average. Suppose as a first step in determining if this belief is correct, we select a random sample of deliveries from the hospital. Of the 62 delivery records pulled for 1990, we found 22 c-sections. Does this large proportion of c-section deliveries, 35.5 percent (= 22/62), mean that this hospital's rate is higher than the national average? The sample proportion of 35.5 percent is certainly larger than 23.5 percent, but our question refers to the population of deliveries in the hospital in 1990, not the sample. As we saw above, we cannot infer immediately from this sample without taking sample-to-sample variability into account. This is a situation where the binomial distribution can be used to address the question about the population based on the sample.

To put the sample rate into perspective, we need to answer the following question. How likely is a rate of 35.5 percent or higher in our sample if the rate of c-section deliveries is really 23.5 percent? Note that the question includes rates higher than 35.5 percent. We must include them because if the sum of their probabilities is large, we cannot conclude that a rate of 35.5 percent is inconsistent with the national rate regardless of how unlikely the rate of 35.5 percent is.

We can use the cdf for the binomial to find the answer to the above question. The cdf enables us to find the probability that a variable is less than a given value, in this case, less than the result we observed in our sample. Then we can subtract that probability from one to find how likely it is to obtain a rate as large or larger than our sample rate. The MINITAB calculation is shown in Box 6.4. Thus, the probability of 21 or fewer c-sections out of 62 deliveries, assuming that the national rate of 23.5 percent

MINITAB BOX 6.4

```
MTB > cdf 21;
SUBC> binom 62 0.235.
 K    P(X LESS OR = K)
21.00           0.9776
```

holds, is 0.9776. This means that the probability of 22 or more c-section deliveries is 1 − 0.9776 = 0.0224. The probability of having 22 or more c-sections is very small. It is unlikely that this hospital's c-section rate is the same as the national average; in fact, it appears to be higher. Further investigation is required to determine why the rate may be higher.

C. Shapes of the Binomial Distribution

The binomial distribution has two parameters, the sample size and the population proportion, that affect its appearance. So far we have seen the distribution of one binomial (Figure 6.1) which had a sample size of 4 and a population proportion of 0.25. Box 6.5 examines the effect of population proportion on the shape of the binomial distribution for a sample size of 10; the plots are shown in Box 6.6.

The plots in Box 6.6 would look like bar charts if a perpendicular line were drawn from the horizontal axis to the points above each outcome.

In the first plot with π equal to 0.10, the shape is quite asymmetric, with only a few of the outcomes having probabilities very different from zero. This plot has a long tail to the right. In the second plot with π equal to 0.20, the plot is less asymmetric.

The third binomial distribution, with π equal to 0.50, has a mean of 5 ($= n * \pi$). The plot is symmetric about its mean of 5, and it has the familiar bell shape. As π is 0.50, it is as likely to have one less occurrence as one more occurrence; that is, four occurrences of the event of interest are as likely as six occurrences, three as likely as seven, and so on, and the plot reflects this.

MINITAB BOX 6.5

Column c1 contains the integers from 0 to 10, the possible number of occurrences of the event of interest in a binomial situation when n = 10. Columns c2, c3, and c4 contain the binomial probabilities of each outcome for the population proportions of 0.1, 0.2, and 0.5, respectively.

```
MTB > set cl
DATA> 0:10
DATA> end
MTB > pdf cl c2;
SUBC> binom 10 .1.
MTB > pdf cl c3;
SUBC> binom 10 .2.
MTB > pdf cl c4;
SUBC> binom 10 .5.
```

(The binomial probabilities are plotted against the corresponding outcomes in Box 6.6)

MINITAB BOX 6.6

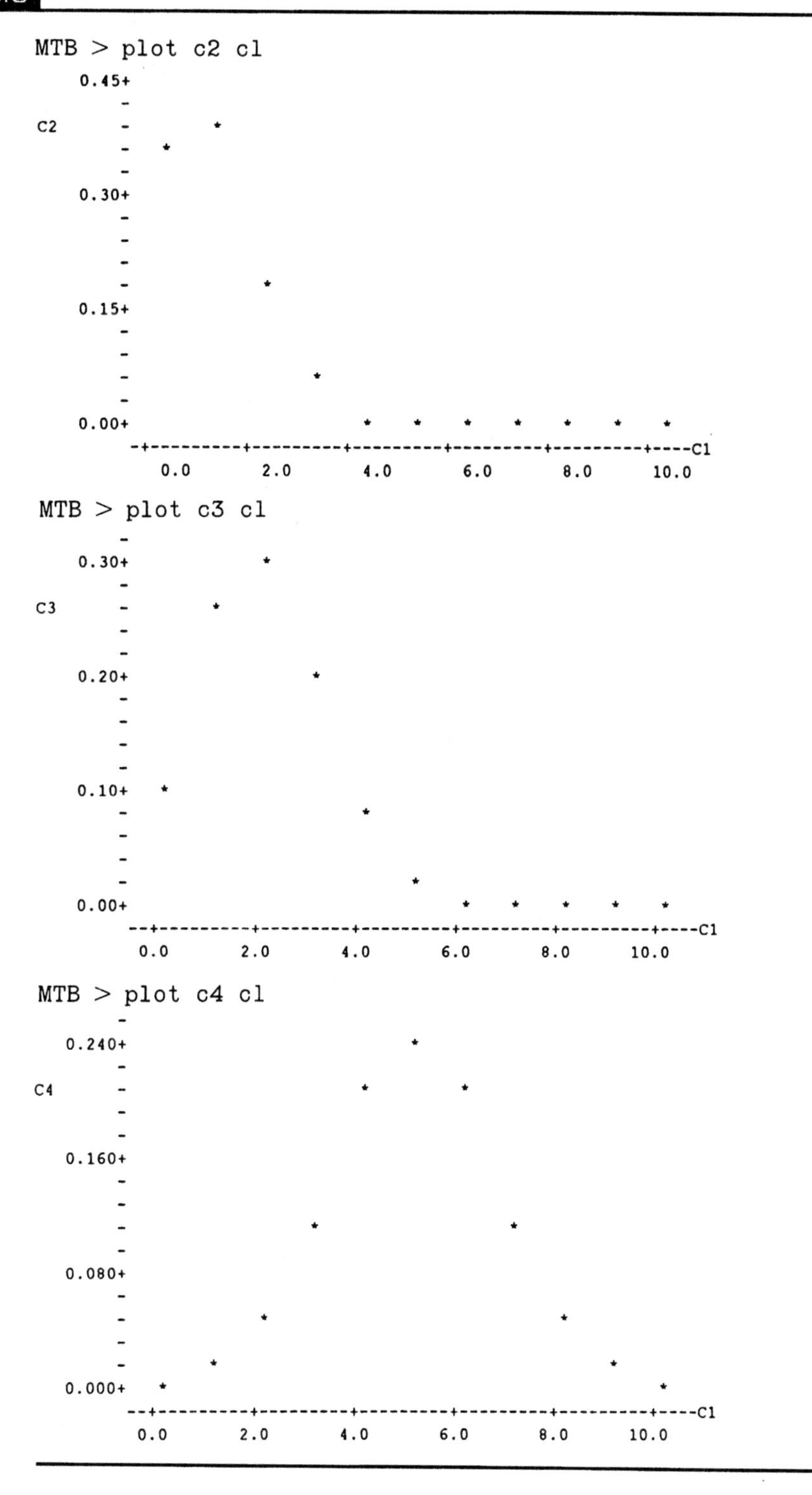

This completes the introduction to the binomial, although we shall say more about it later. The next section introduces the Poisson distribution, another widely used distribution.

II. THE POISSON DISTRIBUTION

The *Poisson distribution* is named for its discoverer, Siméon-Denis Poisson, a French mathematician from the late 18th and early 19th centuries. He is said to have once remarked that life is good for only two things: to do mathematics and to teach it (2, p. 569). The Poisson distribution is similar to the binomial in that it is also used with counts or the number of events. The Poisson is particularly useful when the events occur infrequently. It has been applied in the epidemiological study of many forms of cancer and other rare diseases over time. It has also been applied to the study of the number of elements in a small space when a large number of these small spaces are spread at random over a much larger space, for example, in the study of bacterial colonies on an agar plate.

Even though the Poisson and binomial distributions both are used with counts, the situations for their applications differ. The binomial is used when a sample of size n is selected and the numbers of events and non-events are determined from this sample. The Poisson is used when events occur at random in time or space, and the number of these events is noted. In the Poisson situation, *no* sample of size n has been selected.

The Poisson distribution arises from either of two models. In one model, quantities, for example, bacteria, are assumed to be distributed at random in some medium with a uniform density of λ (lambda) per unit area. The number of bacterial colonies found in a sample area of size A follows the Poisson distribution with a parameter μ equal to the product of λ and A.

In terms of the model over time, we assume that the probability of one event in a short interval of length t_1 is proportional to t_1, that is, Pr{exactly one event} is approximately $\lambda * t_1$. Another assumption is that t_1 is so short that the probability of more than one event during this interval is almost zero. We also assume that what happens in one time interval is independent of the happenings in another interval. Finally, we assume that λ is constant over time. Given these assumptions, the number of occurrences of the event in a time interval of length t follows the Poisson distribution with parameter μ where μ is the product of λ and t.

The Poisson probability mass function is

$$\Pr\{X = x\} = \frac{e^{-\mu} * \mu^x}{x!} \qquad \text{for } x = 0, 1, 2, \ldots$$

TABLE 6.3 Calculation of Poisson Probabilities, $\Pr\{X = x\} = e^{-\mu} * \mu^x/x!$, for $\mu = 1$ and 2

	$\mu = 1$			$\mu = 2$		
x	e^{-1}	*	$1^x/x! = \Pr\{X = x\}$	e^{-2}	*	$2^x/x! = \Pr\{X = x\}$
0	0.3679	*	1/1 = 0.3679	0.1353	*	1/1 = 0.1353
1	0.3679	*	1/1 = 0.3679	0.1353	*	2/1 = 0.2707
2	0.3679	*	1/2 = 0.1839	0.1353	*	4/2 = 0.2707
3	0.3679	*	1/6 = 0.0613	0.1353	*	8/6 = 0.1804
4	0.3679	*	1/24 = 0.0153	0.1353	*	16/24 = 0.0902
5	0.3679	*	1/120 = 0.0031	0.1353	*	32/120 = 0.0361
6	0.3679	*	1/720 = 0.0005	0.1353	*	64/720 = 0.0120
7	0.3679	*	1/5040 = 0.0001	0.1353	*	128/5,040 = 0.0034
8				0.1353	*	256/40,320 = 0.0009
9				0.1353	*	512/362,880 = 0.0002
Total			1.0000			0.9999

where e is a constant approximately equal to 2.71828 and μ is the parameter of the Poisson distribution. Usually μ is unknown and we must estimate it from the sample data. Before considering an example, we demonstrate in Table 6.3 the use of the probability mass function for the Poisson distribution to calculate the probabilities when $\mu = 1$ and $\mu = 2$. These probabilities are not difficult to calculate, particularly when μ is an integer. There is also a recursive relationship between the probability that $X = x + 1$ and the probability that $X = x$ that simplifies the calculations:

$$\Pr\{X = x + 1\} = \left(\frac{\mu}{x + 1}\right) * \Pr\{X = x\}$$

for x beginning at a value of 0. For example, for $\mu = 2$,

$$\Pr\{X = 3\} = (2/3) * \Pr\{X = 2\} = (2/3) * 0.2707 = 0.1804$$

which is the value shown in Table 6.3.

These probabilities are also found in Appendix Table B3 which gives the Poisson probabilities for values of μ beginning at 0.2 and increasing in increments of 0.2 up to 2.0, then in increments of 0.5 up to 7, and in increments of 1 up to 17. MINITAB can also provide the Poisson probabilities as shown in Boxes 6.7 and 6.8. Note that the Poisson distribution is totally determined by specifying the value of its one parameter, μ. The plots in Box 6.8 show the shape of the Poisson probability mass and cumulative distribution functions with $\mu = 2$.

The shape of the Poisson probability mass function with μ equal to 2 (the top figure in Box 6.8) is similar to the binomial mass function for a sample of size 10 and π equal to 0.2 shown above. The cdf (the bottom

MINITAB BOX 6.7

The following shows the use of the PDF and CDF commands in MINITAB for a mean value, μ, of 2.0. Column c1 contains the possible outcomes, c2 will contain the probability mass function, and c3 will contain the cumulative distribution function.

```
MTB > set c1
DATA> 0:10
DATA> end
MTB > pdf c1 c2;
SUBC> poisson 2.
MTB > cdf c1 c3;
SUBC> poisson 2.
```

(The Poisson probabilities are printed and plotted in Box 6.8.)

MINITAB BOX 6.8

```
MTB > print c1 c2 c3
  ROW C1        C2      C3
    1  0  0.135335 0.13534
    2  1  0.270671 0.40601
    3  2  0.270671 0.67668
    4  3  0.180447 0.85712
    5  4  0.090224 0.94735
    6  5  0.036089 0.98344
    7  6  0.012030 0.99547
    8  7  0.003437 0.99890
    9  8  0.000859 0.99976
   10  9  0.000191 0.99995
   11 10  0.000038 0.99999
MTB > plot c2 c1
    0.30+
        -         *    *
C2      -
        -
        -
    0.20+
        -                   *
        -
        -    *
        -
    0.10+                        *
        -
        -
        -                             *
        -                                  *
    0.00+                                       *    *    *    *
         --+---------+---------+---------+---------+---------+C1
          0.0       2.0       4.0       6.0       8.0      10.0
```

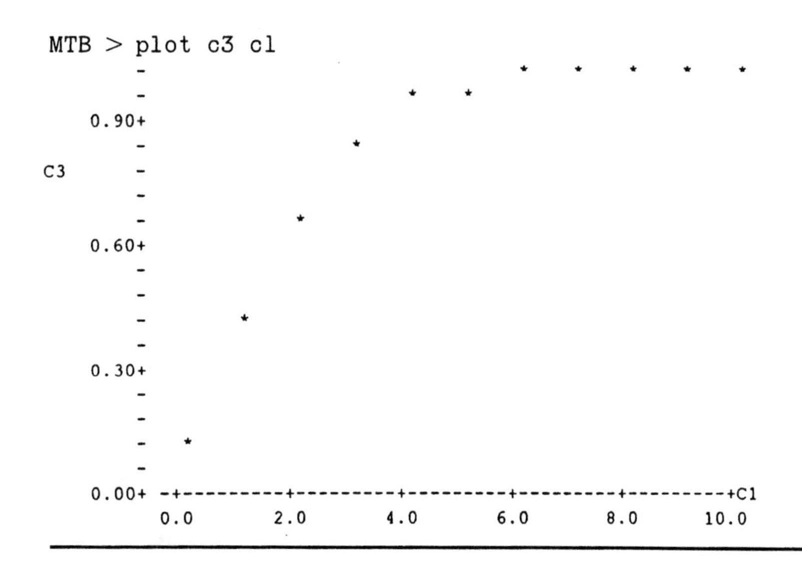

figure in Box 6.8) has the same general shape as that shown in the binomial example above, but the shape is easier to see here as there are more values for the X variable shown on the horizontal axis.

A. Mean and Variance of the Poisson Distribution

As discussed above, the mean is found by summing the products of each outcome by its probability of occurrence. For the Poisson distribution with parameter $\mu = 1$ (see Table 6.3), the mean is

$$\begin{aligned} \text{population mean} &= \sum_{x=0} x * \Pr\{X = x\} \\ &= 0 * 0.3679 + 1 * 0.3679 + 2 * 0.1839 + 3 * 0.0613 \\ &\quad + 4 * 0.0153 + 5 * 0.0031 + 6 * 0.0005 + 7 * 0.0001 \\ &= 1.0000 = \mu. \end{aligned}$$

The mean of the Poisson distribution is μ, which is also the parameter of the Poisson distribution. It turns out that the variance of the Poisson distribution is also μ.

B. Example 1: Finding Poisson Probabilities

A famous chemist and statistician, W. S. Gosset, worked for the Guinness Brewery in Dublin at the turn of the 20th century. Because Gosset did not wish the competitor breweries to learn of the potential application of his work for a brewery, he published his research under the pseudonym of Student. As part of his work, he studied the distribution of yeast cells over

TABLE 6.4 Observed Frequency of Yeast Cells in 400 Squares

	X						
	0	1	2	3	4	5	6
Frequency	103	143	98	42	8	4	2
Proportion	0.258	0.358	0.245	0.105	0.020	0.010	0.005
Poisson probability	0.267	0.352	0.233	0.103	0.034	0.009	0.003

400 squares of a hemacytometer, an instrument for the counting of cells (3). One of the four data sets he obtained is shown in Table 6.4.

Do these data follow a Poisson distribution? As was mentioned above, the Poisson distribution is determined by the mean value, which is unknown in this case. We can use the sample mean to estimate the population mean μ. The sample mean is the sum of all the observations divided by the number of observations, in this case 400. The sum of the number of cells is

$$103 * 0 + 143 * 1 + 98 * 2 + 42 * 3 + 8 * 4 + 4 * 5 + 2 * 6 = 529.$$

The sample mean is then 529/400 = 1.3225. Thus we can calculate the Poisson probabilities using the value of 1.3225 for the mean. As the value of 1.3225 for μ is not in Appendix Table B3, we must use some other means of obtaining the probabilities. We can calculate them using the recursive relationship shown above. We begin by finding the probability of squares with zero cells, $e^{-1.3225}$, which is 0.2665. The other probabilities are found from this value. We can also use MINITAB to calculate these probabilities as shown in Box 6.9. These values are also shown in Table 6.4 and they

MINITAB BOX 6.9

```
MTB > pdf;
SUBC> poiss 1.3225.
    POISSON WITH MEAN = 1.322
        K           P( X = K)
        0             0.2665
        1             0.3524
        2             0.2330
        3             0.1027
        4             0.0340
        5             0.0090
        6             0.0020
        7             0.0004
        8             0.0001
        9             0.0000
```

agree reasonably well with the actual proportions also shown in the table. Based on the visual agreement of the actual and theoretical proportions (from the Poisson), we cannot rule out the Poisson distribution as the distribution of the cell counts. The Poisson distribution agreed quite well for three of the four replications of the 400 cells that Gosset performed.

One reason for interest in the distribution of data is that knowledge of the distribution can be used in future occurrences of this situation. If future data do not follow the previously observed distribution, this can alert us to a change in the process for generating the data. It could also indicate, for example, that the blood cell counts of a patient under study differ from those expected in a healthy population or that there are more occurrences of some disease than was expected assuming that the disease occurrence follows a Poisson distribution with parameter μ. If there are more cases of the disease, it may indicate that there is some common source of infection, for example, some exposure in the workplace or in the environment.

A method of visual inspection of whether the data could come from a Poisson distribution is the *Poissonness plot*, presented by Hoaglin (4). The rationale for the plot is based on the Poisson probability mass distribution formula. If the data could come from a Poisson distribution, then a plot of the sum of the natural logarithm of the frequency of x and the natural logarithm of $x!$ against the value of x should be a straight line. We can use MINITAB with the data in Table 6.4 to create a Poissonness plot as shown in Box 6.10.

The plot appears to be approximately a straight line with the exception of a dip for $x = 4$. In Table 6.4, we see that the biggest discrepancy between the actual and theoretical proportions occurred when $x = 4$, confirmed by the Poissonness plot.

C. Example 2: Use of the Poisson Distribution

In 1986, 18 cases of pertussis were reported in Harris County, Texas, from its estimated 1986 population of 2,942,550. The reported national rate of pertussis was 1.2 cases per 100,000 population (5). Do the Harris County data appear to be consistent with the national rate?

The data are inconsistent if there are too many or too few cases of pertussis compared with the national rate. This concern about both too few as well as too many adds a complication lacking in the binomial example in which we were concerned only about too many occurrences. Our method of answering the question is as follows.

First calculate the pertussis rate in Harris County. If the rate is above the national rate, find the probability of at least as many cases occurring as were observed. If the rate is below the national rate, find the probability of the observed number of cases or fewer occurring. To account for both too few and too many in our calculations, we double the calculated probability.

MINITAB BOX 6.10

In the following MINITAB statements, the frequencies are entered in c1 and values of x are entered in c2. Column c3 contains the values of $x!$, c4 contains the natural logarithm of the frequencies, c5 has the natural logarithm of $x!$ and c6 is the sum of c4 and c5.

```
MTB > set cl
DATA> 103 143 98 42 8 4 2
DATA> set c2
DATA> 0:6
DATA> set c3
DATA> 1 1 2 6 24 120 720
DATA> end
MTB > let c4=loge(cl)
MTB > let c5=loge(c3)
MTB > let c6=c4+c5
MTB > plot c6 c2

       -                                                  *
C6     -
    7.0+
       -
       -
       -
       -                                        *
    6.0+
       -
       -                              *
       -
       -                   *                   *
    5.0+          *
       -
       -  *
         +---------+---------+---------+---------+---------+--C2
        0.0       1.2       2.4       3.6       4.8       6.0
```

Is the resultant probability large? If it is large, there is no evidence that the data are inconsistent with the national rate. If it is small, it is unlikely that the data are consistent with the national rate.

The rate of pertussis in Harris County was 0.61 cases per 100,000 population, less than the national rate. Therefore, we shall calculate the probability of 18 or fewer cases given the national rate of 1.2 cases per 100,000 population. The rate of 1.2 per 100,000 is multiplied by 29.4255 (the Harris County population of 2,942,550 divided by 100,000) to obtain the Poisson parameter for Harris County of 35.31. This value exceeds those listed in Table B3. Therefore we can either find the probability of zero cases and use the recursive formula shown above or use the computer. Box 6.11 calcu-

MINITAB BOX 6.11

The CDF command provides the probability that the variable is less than or equal to a specified value. In this case, we want the probability that a variable following the Poisson distribution with a mean of 35.31 is less than or equal to 18.

```
MTB > cdf 18;
SUBC> poiss 35.31.
       K  P( X LESS OR = K)
   18.00            0.0010
```

lates the probability of 18 or fewer cases. The probability of 18 or fewer cases is 0.001. Multiplying this value by 2 to account for the upper tail of the distribution gives a probability of 0.002, a very small value. It is therefore doubtful, as the probability is only 0.002, that the national rate of pertussis applies to Harris County.

This completes the introduction to the binomial and Poisson distributions. The following section introduces the normal probability distribution for continuous random variables.

III. THE NORMAL DISTRIBUTION

As was mentioned above, the probability distribution for a continuous random variable is usually expressed as a formula which can be used to find the probability that the continuous variable is within a specified interval. This differs from the probability distribution of a discrete variable which gives the probability of each possible outcome.

One reason why an interval is used with a continuous variable instead of considering each possible outcome is that there is really no interest in each distinct outcome. For example, when someone expresses an interest in knowing the probability that a male 45 to 54 years old weighs 160 pounds, exactly 160.000000000. . . pounds is not what is intended. What the person intends is related to the precision of the scale used, and the person may actually mean 159.5 to 160.5 pounds. With a less precise scale, 160 pounds may mean a value between 155 and 165 pounds. Hence the probability distribution of continuous random variables focuses on intervals rather than on exact values.

The probability density function for a continuous random variable *X* is a formula that allows one to find the probability of *X* being in an interval. Just as the probability mass function for a discrete random variable could be graphed, the probability density function can also be graphed. Its graph is a curve such that the area under the curve sums to one, and the area between two points, x_1 and x_2, is equal to the probability that the random variable X is between x_1 and x_2.

The *normal distribution* is also sometimes referred to as the *Gaussian distribution* after the German mathematician, Carl Gauss (1777–1855). Gauss, perhaps the greatest mathematician who ever lived, demonstrated the importance of the normal distribution, and today, it is the most widely used probability distribution in statistics. The normal distribution is so widely used because (1) it occurs naturally in many situations; (2) the sample means of many nonnormal distributions tend to follow it; and (3) it can serve as a good approximation to some nonnormal distributions.

The *normal probability density function* is

$$f(x) = \frac{1}{\sqrt{2\pi\sigma^2}} e^{-(x-\mu)^2/2\sigma^2}, \qquad -\infty < x < \infty$$

where μ is the mean and σ is the standard deviation of the normal distribution, and π is a constant approximately equal to 3.14159. The normal density function is bell-shaped as can be seen from the following plots from MINITAB.

Box 6.12 shows the standard normal density function, that is, the normal pdf with a mean of zero and a standard deviation of one, over the

MINITAB BOX 6.12

Column c2 contains the values from −3.5 to 3.5 in steps of 0.1 and c3 contains the pdf values for the standard normal distribution evaluated at the points in c2. When no subcommands are given with the PDF or CDF commands, MINITAB defaults to the standard normal distribution.

```
MTB > set c2
DATA> -3.5:3.5/.1
DATA> end
MTB > pdf c2 c3
MTB > plot c3 c2
C3      -                           *2*2*
        -                          2     2
        -                         *       *
    0.30+                        2         2
        -                       *           *
        -                      *             *
        -                      *             *
        -                    **               **
    0.15+                    *                 *
        -                  **                    **
        -                 **                      **
        -                2                          2
        -            **2*                            *2**
    0.00+      2*2*2**                                  **2*2*2
         --------+---------+---------+---------+---------+--C2
               -3.0      -1.5       0.0       1.5       3.0
```

range of -3.5 to $+3.5$. The area under the curve is one and the probability of X being between any two points is equal to the area under the curve between those two points.

Box 6.13 shows the effect of changing σ on the normal pdf. The area under both of these curves again is one, and both curves are bell-shaped. The standard normal distribution has smaller variability, evidenced by more of the area being closer to zero, as it must because its standard deviation is 50 percent of that of the other normal distribution. There is more area, or a greater probability of occurrence, under the second curve associated with values farther from the mean of zero than under the standard normal curve. The effect of increasing the standard deviation is to

MINITAB BOX 6.13

The standard normal pdf, its values are in c3, is plotted along with the pdf for a normal distribution with $\mu = 0$ and $\sigma = 2$ (its pdf is stored in column c4). The pdfs are shown over the range from -7 to 7 in increments of 0.1. Note that the GMPLOT command is used to obtain a high-resolution plot as the output from the MPLOT command was hard to follow. The plot is shown below.

```
MTB > set c2
DATA> -7:7/.1
DATA> end
MTB > pdf c2 c3
MTB > pdf c2 c4;
SUBC> normal 0 2.
MTB > gmplot c3 c2, c4 c2
```

C3: 0.45, 0.30, 0.15, 0.00

C2: -5.0, -2.5, 0.0, 2.5, 5.0

A = C3 vs. C2 B = C4 vs. C2

MINITAB BOX 6.14

Column c3 again contains the pdf for the standard normal distribution over the range −3.5 to 6.5 in increments of 0.1, and c4 will contain the pdf for a normal distribution with a mean of 3 and a standard deviation of 1.

```
MTB > set c2
DATA> -3.5:6.5/.1
DATA> end
MTB > pdf c2 c3
MTB > pdf c2 c4;
SUBC> normal 3 1.
MTB > gmplot c3 c2, c4 c2
```

```
      0.45 +
C3
      0.30 +
      0.15 +
      0.00 +
           +---------+---------+---------+---------+---------+-------- C2
         -4.0      -2.0       0.0       2.0       4.0       6.0
                   A = C3     vs. C2    B = C4    vs. C2
```

flatten the curve of the pdf, with a concomitant increase in the probability of more extreme values of X.

In Box 6.14, statements for graphing two additional normal probability density functions are shown and the resultant plots show the effect of changing the mean. Increasing the mean by 3 units has simply shifted the entire pdf curve 3 units to the right. Hence changing the mean shifts the curve to the right or left and changing the standard deviation increases or decreases the spread of the distribution.

A. Transforming Normally Distributed Data to the Standard Normal Distribution

As can be seen from the normal pdf formula and the plots, two parameters, the mean and the standard deviation, determine the location and spread of the normal curve. Hence there are many normal distributions, just as there are many binomial and Poisson distributions; however, it is not necessary to have many pages of normal tables for each different normal distribution because all the normal distributions can be transformed to the standard normal distribution. Thus only one normal table is needed.

Consider data from a normal distribution with a mean of μ and a standard deviation of σ. We wish to transform these data to the standard normal distribution which has a mean of zero and a standard deviation of one. The transformation has two steps. The first step is to subtract the mean, μ, from all the observations. In symbols, let y_i be equal to $x_i - \mu$. Then the mean of Y is μ_y, which equals

$$\mu_y = \Sigma \frac{x_i - \mu}{N} = \frac{\Sigma x_i - N * \mu}{N} = \frac{N * \mu - N * \mu}{N} = 0.$$

The second step is to divide y_i by its standard deviation. As we have subtracted a constant from the observations of X, the variance and standard deviation of Y are the same as those of X as was shown in Chapter 4. That is, the standard deviation of Y is also σ. In symbols, let z_i be equal to y_i/σ. What are the mean and standard deviation of Z? The mean is still zero but the standard deviation of Z is one. This is due to the second property of the variance shown in Chapter 4; that is, when all the observations are divided by a constant, the standard deviation is also divided by that constant. Therefore the standard deviation of Z is found by dividing σ, the standard deviation of Y, by the constant, σ. The value of this ratio is one.

Therefore any variable, X, which follows a normal distribution with a mean of μ and a standard deviation of σ can be transformed to the standard normal distribution by subtracting μ from all the observations and dividing all the observed deviations by σ. The variable Z, defined as $(X - \mu)/\sigma$, follows the standard normal distribution. A symbol for indicating that a variable follows a particular distribution or is "distributed as" is the asymptote, $\sim$; for example, $Z \sim N(0,1)$ means that Z follows a normal distribution with a mean of zero and a standard deviation of one. The observed value of a variable from a standard normal distribution tells how many standard deviations that value is from its mean of zero.

B. Calculation of Normal Probabilities

The cumulative distribution function of the standard normal distribution, denoted by $\Phi(z)$, represents the probability that the standard normal variable Z is less than or equal to the value z, that is, $\Pr\{Z \leq z\}$. It is also the area under the standard normal curve less than z as is depicted in Figure 6.2.

The shaded area represents the probability that a variable, Z, distributed as a $N(0,1)$ variable, is less than or equal to z. Table B4 presents the values of $\Phi(z)$ for values of z ranging from -3.79 to 3.79 in steps of 0.01.

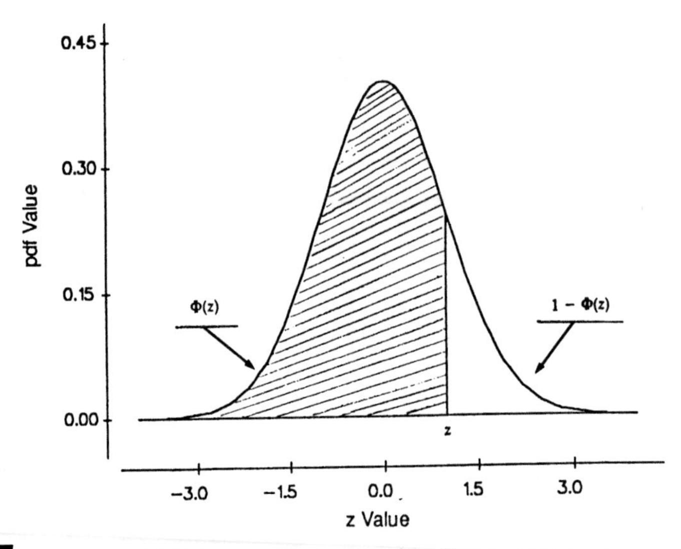

FIGURE 6.2 Depiction of $\Phi(z)$ for a positive value of z.

The unshaded area in Figure 6.2 represents the probability that Z, a $N(0,1)$ variable, is greater than some value z.

As can be seen from Figure 6.3, the probability that Z is less than or equal to a negative z is the same as the probability that Z is greater than the corresponding positive z. In symbols, this equivalence is expressed as $\Phi(-z) = 1 - \Phi(z)$.

Box 6.15 shows the cumulative distribution function for the standard

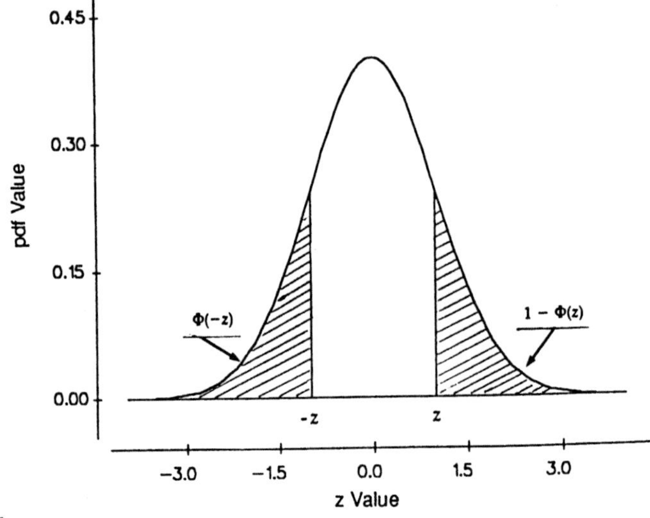

FIGURE 6.3 Equivalence of $\Phi(-z)$ and $1 - \Phi(z)$.

MINITAB BOX 6.15

Column c1 contains the values from −3.8 to 3.8 in increments of 0.1 and c2 will contain the corresponding values of the cdf over the range −3.8 to 3.8.

```
MTB > set c1
DATA> -3.8:3.8/0.1
DATA> end
MTB > cdf c1 c2
MTB > plot c2 c1

       1.05+
           -                                                 2*2*2*2*2*2*2
 C2        -                                              *2*
           -                                            *2
           -                                           2
       0.70+                                         **
           -                                        **
           -                                       2
           -                                     **
           -                                     *
       0.35+                                   **
           -                                  **
           -                                *2
           -                              *2
           -                          *2*2
       0.00+          2*2*2*2*2*2
             --------+---------+---------+---------+---------+---C1
                   -3.0      -1.5       0.0       1.5       3.0
```

normal distribution. The vertical axis gives the values of the probabilities corresponding to the values of z shown along the horizontal axis. The curve gradually increases from a probability of 0.0 for values of z around −3, to a probability of 0.5 when z is zero, and on to probabilities close to 1.0 for z values of 3 or larger.

Table 6.5 shows the values of the cdf, taken from c2, for z ranging from −3.8 to 3.8 in increments of 0.5. When z is −3.8, the value of the cdf is 0.000072; when z is −3.3, the value of the cdf is 0.000483; and, although not shown, when z is 0.0, we know that the cdf value is 0.5.

TABLE 6.5 Values of the Standard Normal cdf for Selected Values of z

z	$\Phi(z)$	z	$\Phi(z)$
−3.8	0.000072	0.2	0.579260
−3.3	0.000483	0.7	0.758036
−2.8	0.002555	1.2	0.884930
−2.3	0.010724	1.7	0.955435
−1.8	0.035930	2.2	0.986097
−1.3	0.096801	2.7	0.996533
−0.8	0.211855	3.2	0.999313
−0.3	0.382089	3.7	0.999892

1. Example 1: Probability of Being Greater Than a Value

Suppose that we wish to find the probability that an adult woman will have a diastolic blood pressure value greater than 95 mm Hg given that X, the diastolic blood pressure for adult women, follows the $N(80,10)$ distribution. Because the values in Table B4 are for variables that follow the $N(0,1)$ distribution, we first must transform the value of 95 to its corresponding Z value. To do this, we subtract the mean of 80 and divide by the standard deviation of 10. The value of 95 mm Hg therefore is

$$\frac{95 - 80}{10} = \frac{15}{10} = 1.5.$$

Thus the value of the Z variable corresponding to 95 mm Hg is 1.5, which means that 95 is 1.5 standard deviations above its mean of 80. We now want the probability that Z is greater than 1.5. Using Table B4, look for 1.5 under the z heading and then go across the columns until reaching the .00 column. The probability of a standard normal variable being less than 1.5 is 0.9332. Thus the probability of being greater than 1.5 is 0.0668 (= 1 − 0.9332).

The CDF command in MINITAB can be used to obtain the cumulative distribution function values for variables that follow a normal distribution. The CDF command provides for a wider coverage of values of z than Appendix Table B4 does, and the values do not have to be greater than or equal to zero. In MINITAB, we do not have to transform to the standard normal distribution as it does that for us. If no distribution is specified, MINITAB assumes that we are using the $N(0,1)$ distribution. Box 6.16 calculates the probability that X was greater than 95 mm Hg where $X \sim N(80,10)$.

MINITAB BOX 6.16

```
MTB > cdf 95;
SUBC> normal 80 10.
    95.0000     0.9332
```

As $\Pr\{X > 95\}$ is $1 - \Pr\{X \leq 95\}$, we subtract 0.9332 from 1.0 and obtain 0.0668, which is the same value we found above.

2. Example 2: Calculation of the Value of the *i*th Percentile

Table B4 can be used to answer a slightly different question as well. Suppose that we wish to find the 95th percentile of the diastolic blood pressure variable for adult women, that is, the value such that 95 percent of adult women had a diastolic blood pressure less than it. We look in the body of the table until we find 0.9500. We find the corresponding value in the z column and transform that value to the $N(80,10)$ distribution.

Examination of Table B4 shows the value of 0.9495 when z is 1.64 and of 0.9505 when z is 1.65. There is no value of 0.9500 in the table. As 0.9500 is exactly halfway between 0.9495 and 0.9505, we use the value of 1.645 for the corresponding z. We now must transform this value to the $N(80,10)$ distribution. This is easy to do because we know the relation between Z and X.

As $Z = (X - \mu)/\sigma$, on multiplication of both sides of the equation by σ, we have $\sigma * Z = X - \mu$. If we add μ to both sides of the equation, we have $\sigma * Z + \mu = X$. Therefore we must multiply the value of 1.645 by 10, the value of σ, and add 80, the value of μ, to it to find the value of the 95th percentile. This value is 96.45 (= 16.45 + 80) mm Hg.

MINITAB can also perform this calculation as shown in Box 6.17.

The percentiles of the standard normal distribution are used frequently; therefore, a shorthand notation for them has been developed. The ith percentile for the standard normal distribution is written as z_i, for example, $z_{0.95}$ is 1.645. From Table B4, we also see that $z_{0.90}$ is approximately 1.28 and $z_{0.975}$ is 1.96. By the symmetry of the normal distribution, we also know that $z_{0.10}$ is -1.28, $z_{0.05}$ is -1.645, and $z_{0.025}$ is -1.96.

The percentiles in theory could also be obtained from the graph of the cdf for the standard normal shown above. For example, if the 90th percentile was desired, find the value of 0.90 on the vertical axis and draw a line parallel to the horizontal axis from it to the graph. Next drop a line parallel to the vertical axis from that point down to the horizontal axis. The point where the line intersects the horizontal axis is the 90th percentile of the standard normal distribution.

MINITAB BOX 6.17

The INVCDF (inverse cumulative distribution function) command helps us find the percentile values.

```
MTB > invcdf 0.95;
SUBC> normal 80 10.
    0.9500   96.4485
```

The value of the 95th percentile is 96.4485, which agrees with the value found above. We could also use this command to find the percentiles of the standard normal distribution.

```
MTB > invcdf 0.95
    0.9500   1.6449
```

3. Example 3: Probability Calculation for an Interval

Suppose that we wished to find the proportion of women whose diastolic blood pressure was between 75 and 90 mm Hg. Before attempting to find the probabilities numerically, it is useful to make a drawing similar to Figure 6.4, which depicts the interval we are discussing, to aid our understanding of what is wanted. The figure also provides us with an idea of the probability's value. If the numerical value is not consistent with our idea of the value, perhaps we misused Appendix Table B4.

The first step in finding the proportion of women whose diastolic blood pressure is in this interval is to convert the values of 75 and 90 mm Hg to the $N(0,1)$ distribution. The value of 75 is transformed to $(75 - 80)/10$, which is -0.5, and 90 is converted to 1.0. We therefore must find the area under the standard normal curve between -0.5 and 1.0. Figure 6.5 shows this interval on the standard normal curve.

Note that the shaded areas in Figures 6.4 and 6.5 represent the same proportion of the area under each curve. Even though the Figures 6.4 and 6.5 look exactly the same, the vertical axes are different. In Figure 6.4, the vertical axis goes up to 0.045, whereas it goes up to 0.45 in Figure 6.5. If these two curves were plotted on the same graph, the $N(80,10)$ curve would be much flatter than the standard normal because of its much greater variability. When plotted separately, however, they look the same. This same appearance supports the use of the standard normal curve for all normal distributions.

One way of finding the area between -0.5 and 1.0 is to find the area under the curve less than or equal to 1.0 and to subtract from it the area

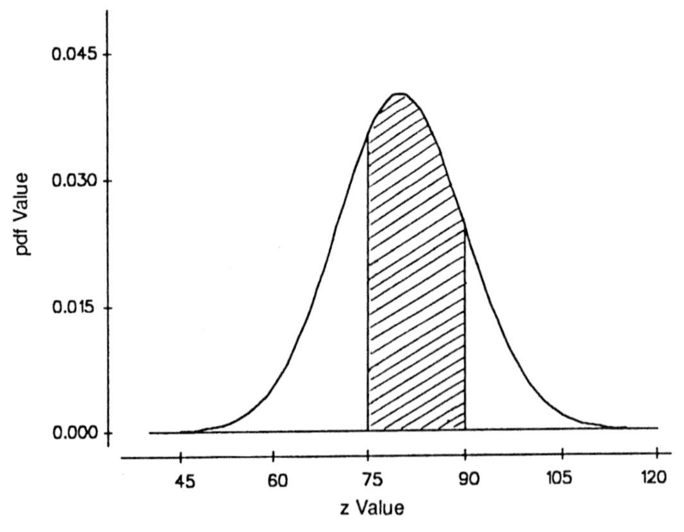

FIGURE 6.4 Area under the $N(80, 10)$ curve between 75 and 90 mm Hg.

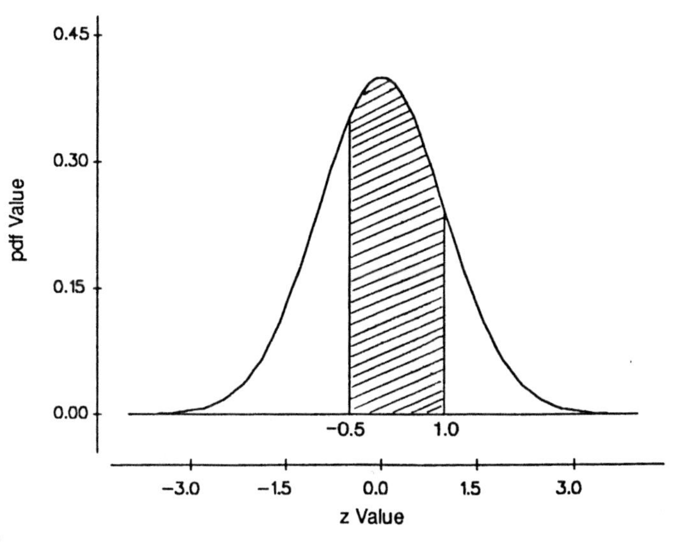

FIGURE 6.5 Area under the standard normal curve between −0.5 and 1.0.

under the curve less than or equal to −0.5. In symbols, this is

$$\Pr\{-0.5 \leq Z \leq 1.0\} = \Pr\{Z \leq 1.0\} - \Pr\{Z \leq -0.5\}.$$

From Table B4, we find that the area under the standard normal pdf curve less than or equal to 1.0 is 0.8413. The probability of a value less than or equal to −0.5 is 0.3085. Thus the proportion of women whose diastolic blood pressure is between 75 and 90 mm Hg is 0.5328 (= 0.8413 − 0.3085). Box 6.18 shows this calculation.

MINITAB BOX 6.18

We can do this calculation by using the CDF command in MINITAB as follows.

```
MTB > cdf 90;
SUBC> normal 80 10.
   90.0000     0.8413

MTB > cdf 75;
SUBC> normal 80 10.
   75.0000     0.3085
```

The difference between these two probabilities, 0.8413 and 0.3085, yields the value 0.5328, the same value as above.

C. The Normal Probability Plot

The normal probability plot provides a way of visually determining whether or not data might be normally distributed. This plot is based on the cdf of the standard normal distribution. Special graph paper, called *normal probability paper,* is used in the plotting of the points. The vertical axis of normal probability paper shows the values of the cdf of the standard normal. Table 6.5 showed some of the cdf values corresponding to z values of −3.8 to 3.7 in steps of 0.5 and we saw that the increase in value of the cdf was not constant per a constant increase in z. The vertical axis reflects this with very small changes in values of the cdf initially, then larger changes in the cdf values in the middle of plot, followed finally by very small changes in the cdf value. Numbers along the horizontal axis are in their natural units.

If a variable X is normally distributed, the plot of its cdf against X should be a straight line on normal probability paper. If the plot is not a straight line, it suggests that X is not normally distributed. As we do not know the distribution of X, we approximate its cdf in the following fashion.

We first sort the observed values of X from lowest to highest. Next we assign ranks to the observations from 1 for the lowest to n (the sample size) for the highest value. The ranks are divided by n and this gives an estimate of the cdf. This sample estimate is often called the *empirical distribution function*.

The points, determined by the values of the sample estimate of the cdf and the corresponding values of x, are plotted on normal probability paper. In practice, the ranks divided by the sample size are not used as the

TABLE 6.6 Values of Vitamin A and Their Ranks and Transformed Ranks for the 33 Boys in Table 4.1

Vitamin A (IU)	Rank	Transformed rank	Vitamin A (IU)	Rank	Transformed rank	Vitamin A (IU)	Rank	Transformed rank
820	1	0.0188	3747	12	0.3496	6754	23	0.6805
964	2	0.0489	4248	13	0.3797	6761	24	0.7105
1379	3	0.0789	4288	14	0.4098	8034	25	0.7406
1459	4	0.1090	4315	15	0.4398	8516	26	0.7707
1704	5	0.1391	4450	16	0.4699	8631	27	0.8008
1826	6	0.1692	4535	17	0.5000	8675	28	0.8308
1921	7	0.1992	4876	18	0.5301	9490	29	0.8609
2246	8	0.2293	5242	19	0.5602	9710	30	0.8910
2284	9	0.2594	5703	20	0.5902	10451	31	0.9211
2671	10	0.2895	5874	21	0.6203	12493	32	0.9511
2687	11	0.3195	6202	22	0.6504	12812	33	0.9812

estimate of the cdf. Instead, the transformation, (rank − 0.375)/(n + 0.25), is frequently used. One reason for this transformation is that the estimate of the cdf for the largest observation is now a value less than one, whereas the use of the ranks divided by n always results in a sample cdf value of one for the largest observation. A value less than one is desirable because it is highly unlikely that the selected sample actually contains the largest value in the population.

As an example, consider the vitamin A data for the 33 boys in Table 4.1. Table 6.6 shows the sorted values, their ranks, and the transformed ranks, which are plotted in Figure 6.6. The points in the plot do not appear to fall along a straight line. It is therefore doubtful that the vitamin A variable follows a normal distribution, a conclusion that we had previously reached in the discussion of symmetry in Chapter 4.

An alternative to normal probability paper is use of the computer as demonstrated in Box 6.19.

Let us now examine data from a normal distribution and see what its normality probability plot looks like. The example in Box 6.20 uses data from a $N(80,10)$ distribution. The plot looks like a straight line, but there are many points with the same normal scores. Box 6.21 shows a different plot of the same data, stretching the vertical axis and reducing the number of points with the same normal scores.

The points still appear to fall mostly on a straight line as they should. The smallest observed value of X is slightly larger than expected if the data

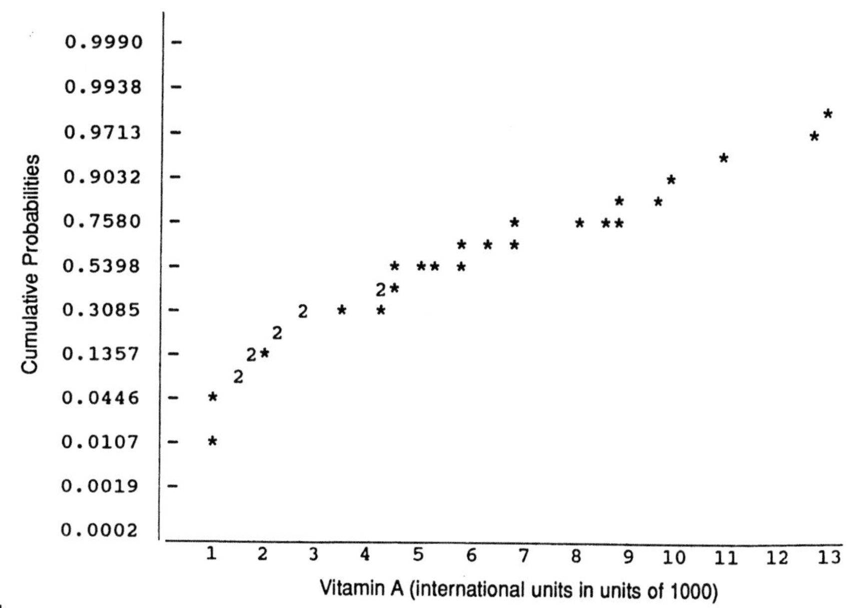

FIGURE 6.6 Normal probability plot of vitamin A data from Table 4.1.

MINITAB BOX 6.19

The command NSCORES in MINITAB transforms the data into normal scores in the following manner. First, ranks are assigned to the data and then the ranks are transformed in a manner similar to that shown above. These transformed ranks provide an estimate of the cdf, the proportion of values less than x for the X variable. The next step is to find the values of a standard normal variable that would have produced these same proportions. These are the normal scores. The plot of the normal scores versus the observed values of X should be linear if X is normally distributed. The following example uses the vitamin A data in c1.

```
MTB > nscores c1 c2
MTB > plot c2 c1

C2        -                                                        *
          -                                                       *
      1.5+                                                  *
          -                                               **
          -                                             2
          -                                        2    * *
          -                                 * 2 *
      0.0+                               2 *
          -                            * 3
          -                  2 2
          -                 **
          -                **
     -1.5+                 *
          -               *
          -              *
           +---------+---------+---------+---------+---------+--C1
           0       2500      5000      7500     10000     12500
```

This plot looks very similar to the normal probability plot shown above in Figure 6.6 as it must. They would be almost identical if the same scales had been used in the plotting.

MINITAB BOX 6.20

```
MTB > random 100 c1;
SUBC> normal 80 10.
MTB > desc c1
                 N     MEAN   MEDIAN   TRMEAN    STDEV   SEMEAN
C1             100   80.619   80.447   80.705    9.443    0.944

               MIN      MAX       Q1       Q3
C1          57.797  104.728   74.046   87.260

MTB > nscores c1 c2
MTB > plot c2 c1
```

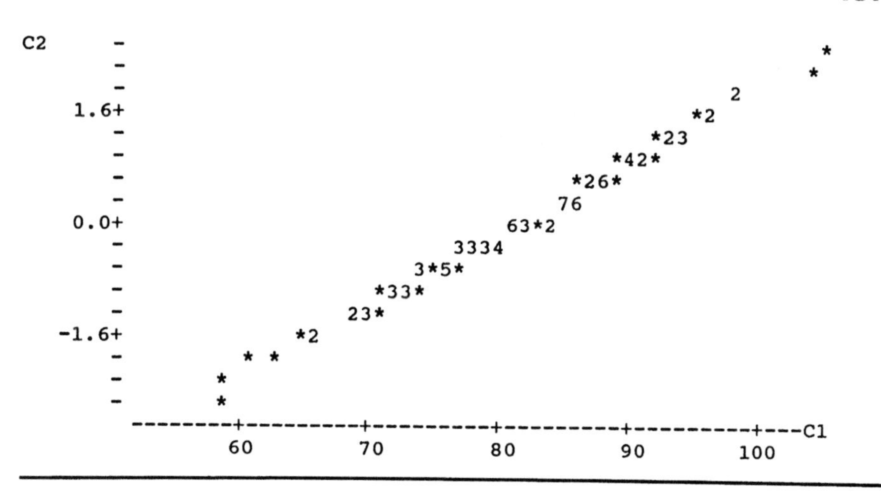

were perfectly normally distributed, but this deviation is relatively slight. Hence, based on this visual inspection, these data could come from a normal distribution.

It is difficult to determine visually whether or not data follow a normal distribution for small sample sizes unless the data deviate substantially from a normal distribution. As the sample size increases from 50 to 100, one can have more confidence in the visual determination.

IV. THE CENTRAL LIMIT THEOREM

As was mentioned above, one of the main reasons for the widespread use of the normal distribution is that the sample means of many nonnormal distributions tend to follow the normal distribution as the sample size increases. The formal statement of this is called the *central limit theorem*. Basically, for random samples of size n from some distribution with mean μ and standard deviation σ, the distribution of $\bar{x}$, the sample mean, is approximately $N(\mu, \sigma/\sqrt{n})$. This statement applies for any distribution as long as μ and σ are defined. The approximation to normality improves as n increases.

The proof of this theorem is beyond the scope of this book and also unnecessary for our understanding. We shall, however, demonstrate that it holds for a very nonnormal distribution, the Poisson distribution with mean one. First, Box 6.22 shows the probability mass function for this distribution. As can be seen from the plot in Box 6.22, the Poisson distribution with a mean of one is very nonnormal in appearance.

The following demonstration consists of drawing a large number of samples, say 100, from this distribution, calculating the mean for each sample, and examining the sampling distribution of the sample means. We do this for samples of size 5 in Box 6.23 and of size 40 in Boxes 6.24 and

MINITAB BOX 6.21

The HEIGHT command is used before the PLOT command to stretch out the plot to avoid some of the clumping. The usual plot of 17 lines is increased to 40 lines in the following.

```
MTB > height 40
MTB > plot c2 cl

C2      -                                                     *
        -
        -
    2.10+                                                    *
        -                                              *
        -                                              *
        -                                            *
        -                                           **
    1.40+                                          2
        -                                         2*
        -                                       *2
        -                                      2*
        -                                     22
    0.70+                                    4
        -                                  *22
        -                                  5
        -                                 5*
        -                               *22
    0.00+                             33
        -                            23
        -                          *32
        -                         32
        -                        4*
   -0.70+                      2**
        -                     22
        -                    2*
        -                   2*
        -                  3
   -1.40+                 2
        -              2
        -             *
        -           *
        -         *
   -2.10+       *
        -
        -
        -       *
        -
   -2.80+
         --------+---------+---------+---------+---------+---C1
                 60        70        80        90       100
```

The clumping has been reduced. The largest number of points with the same normal score value is 6 compared with 13 in the previous graph.

6.25. As was stated above, the mean of the means should be one and the sample estimate is 0.968. The standard deviation of the means is the standard deviation divided by the square root of the sample size. As the mean and variance of this Poisson distribution are both one, the standard deviation of the means should be $1/\sqrt{5}$, which is 0.4472. The sample estimate of 0.4519 is very close. Even for a sample of size 5, the sampling distribution of the sample means does not differ substantially from the bell shape.

Box 6.25 examines the normality of the sample means from Box 6.24. The normal scores plot in Box 6.25 as well as the box plot and histogram in

MINITAB BOX 6.22

```
MTB > set c1
DATA> 0:8
DATA> end
MTB > pdf c1 c2;
SUBC> poisson 1.
MTB > plot c2 c1
         -
    0.36+    *         *
         -
  C2     -
         -
         -
    0.24+
         -
         -                     *
         -
         -
    0.12+
         -
         -                               *
         -
         -                                         *
    0.00+                                                  *         *         *         *
         --+---------+---------+---------+---------+---------+--C1
          0.0       1.5       3.0       4.5       6.0       7.5
```

MINITAB BOX 6.23

In the following, 100 random samples of size 5 from the Poisson distribution with a mean of 1 are selected and stored in columns c1 to c5, respectively. Column c6 will contain the means of the 100 samples of size 5.

```
MTB > random 100 c1-c5;
SUBC> poisson 1.
MTB > add c1-c5 c6
MTB > let c6=c6/5
MTB > desc c6
                  N      MEAN    MEDIAN    TRMEAN     STDEV    SEMEAN
C6              100    0.9680    1.0000    0.9644    0.4519    0.0452

                MIN       MAX        Q1        Q3
C6           0.0000    2.0000    0.6000    1.2000

MTB > hist c6

Histogram of C6   N = 100
Midpoint    Count
     0.0        2  **
     0.2        4  ****
     0.4       12  ************
     0.6        9  *********
     0.8       18  ******************
     1.0       18  ******************
     1.2       13  *************
     1.4       10  **********
     1.6        9  *********
     1.8        3  ***
     2.0        2  **
```

MINITAB BOX 6.24

The following shows the distribution of sample means of size 40.

```
MTB > random 100 c1-c40;
SUBC> pois 1.
MTB > add c1-c40 c41
MTB > let c41=c41/40
MTB > desc c41
                N      MEAN   MEDIAN   TRMEAN   STDEV  SEMEAN
C41           100    1.0007   1.0000   0.9983  0.1453  0.0145

              MIN       MAX       Q1       Q3
C41        0.7000    1.4500   0.9000   1.1000

MTB > hist c41

 Histogram of C41   N = 100
 Midpoint    Count

     0.7         3  ***
     0.8        10  **********
     0.9        17  *****************
     1.0        35  ***********************************
     1.1        15  ***************
     1.2        15  ***************
     1.3         4  ****
     1.4         0
     1.5         1  *

MTB > boxp c41

                            ----------------
             ---------------I        +     I--------------             *
                            ----------------
        -----+---------+---------+---------+---------+---------+C41
           0.75      0.90      1.05      1.20      1.35      1.50
```

Box 6.24 are used in the examination of the distribution of the 100 sample means of size 40. All three graphical methods support the idea that the sample mean could be normally distributed as n increases even though the variable itself is clearly not normally distributed.

Besides showing that the central limit theorem holds for one very nonnormal distribution, this demonstration also showed the effect of sample size on the estimate of the population mean. From the DESC command, we see that the mean of the 100 sample means from samples of size 40 is 1.0007, which is very close to the population mean of 1.0. The standard deviation of the sample means is 0.1453. This value is close to the

MINITAB BOX 6.25

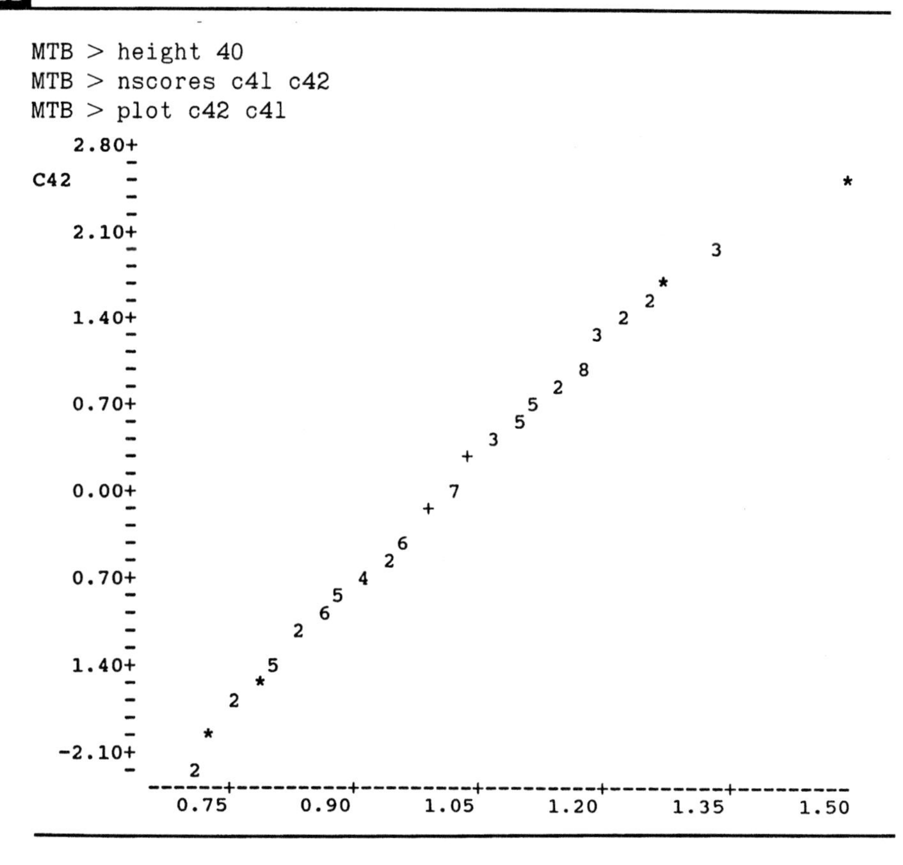

theoretical value of 0.1581 ($= 1/\sqrt{40}$) and, in addition, is much smaller than the corresponding standard deviation of the means from samples of size 5. The sample range also attests to the much smaller variation in the sample means from samples of size 40 compared with samples of size 5. The range of the 100 sample means from samples of size 40 is 0.75, with the sample means ranging from 0.70 to 1.45. The corresponding range based on samples of size 5 is 2.00, with the sample means ranging from 0.00 to 2.00. This example reinforces the idea that the mean from a very small sample may not be close to the population mean.

V. APPROXIMATIONS TO THE BINOMIAL AND POISSON DISTRIBUTIONS

As was mentioned earlier, another reason for the use of the normal distribution is that, under certain conditions, it provides a good approximation to some other distributions, in particular, the binomial and Poisson distributions. This was more important in the past when there was not such a widespread availability of computer packages for calculating binomial and Poisson probabilities for parameter values far exceeding those shown in tables in most textbooks; however, it is still important today as computer packages have limitations in their ability to calculate binomial probabilities for large sample sizes or for extremely large values of the Poisson parameter. For example, when the binomial proportion is 0.5, MINITAB is unable to calculate binomial probabilities for samples larger than 125. In the following sections, we show the use of the normal distribution as an approximation to the binomial and Poisson distributions. As there are conditions when the normal distribution does not provide a good approximation, we also show the use of the Poisson distribution to approximate the binomial.

A. Normal Approximation to the Binomial Distribution

In the plots of the binomial probability mass functions, we saw that as the binomial proportion approached 0.5, the plot began to look like the normal distribution. This was true for sample sizes even as small as 10. It is therefore not surprising that the normal distribution can *sometimes* serve as a good approximation to the binomial distribution. The following plots of the binomial probability mass function for different values of n and π demonstrate why we used the modifier sometimes in the above sentence.

In Box 6.26, $n * \pi$ is less than 5 while $n * (1 - \pi)$ is greater than 5. The plots in Box 6.27 show plots for which both $n * \pi$ and $n * (1 - \pi)$ are greater than 5. The two plots in Box 6.26 show two skewed distributions. The normal approximation would not provide good fits in either of these two cases, particularly in the second situation. The first plot in Box 6.27 is symmetric, as π is 0.5, and the normal distribution should provide a reasonable approximation here. The second plot in Box 6.27 also uses the same proportions of 0.2 and 0.8 as in the first plot in Box 6.26; however, the sample size is much larger in the latter plot than in the former plot (50 versus 20), and the latter plot is beginning to look more like a normal distribution with tails in both directions, although the distribution is still skewed.

The central limit theorem provides a rationale why the normal distribution can provide a good approximation to the binomial. In the binomial setting, there are two outcomes, for example, disease and no disease. Let us assign the numbers 1 and 0 to the outcomes of disease and no disease, respectively. The sum of these numbers over the entire sample is the number of diseased persons in the sample. The mean then is simply the number of diseased sample persons divided by the sample size. And, according to the central limit theorem, the sample mean should approximately follow a normal distribution as n increases. But, if the sum of values divided by a constant approximately follows a normal distribution, the

MINITAB BOX 6.26

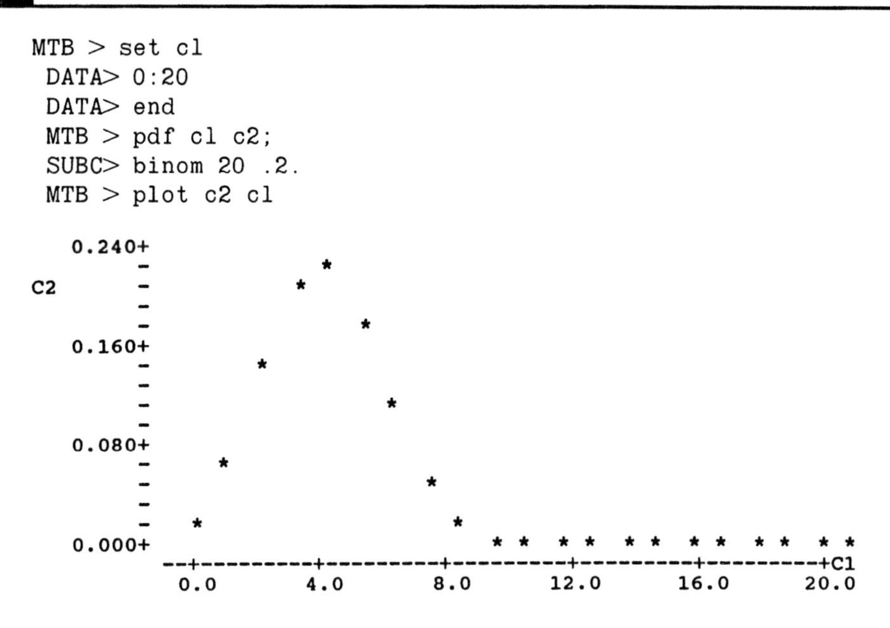

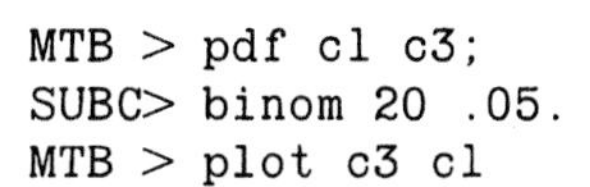

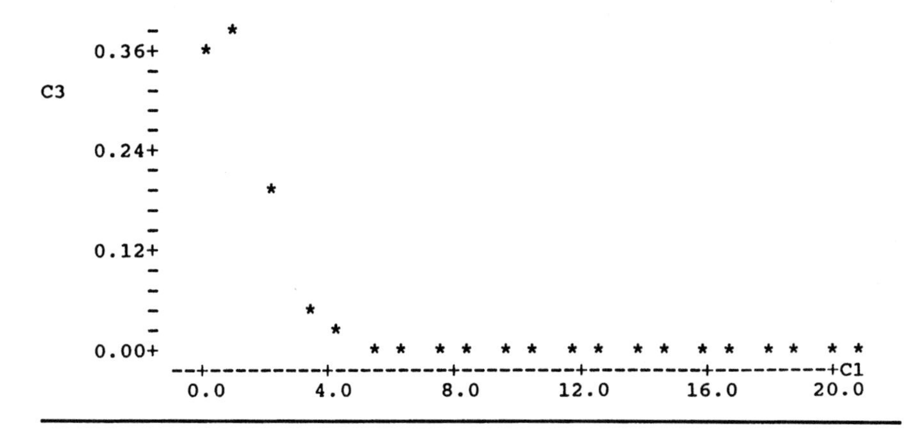

sum of the values itself also approximately follows a normal distribution. The sum of the values in this case is the binomial variable, and hence, it also approximately follows the normal distribution.

MINITAB BOX 6.27

```
MTB > pdf cl c4;
SUBC> binom 20 .5.
MTB > plot c4 cl
```

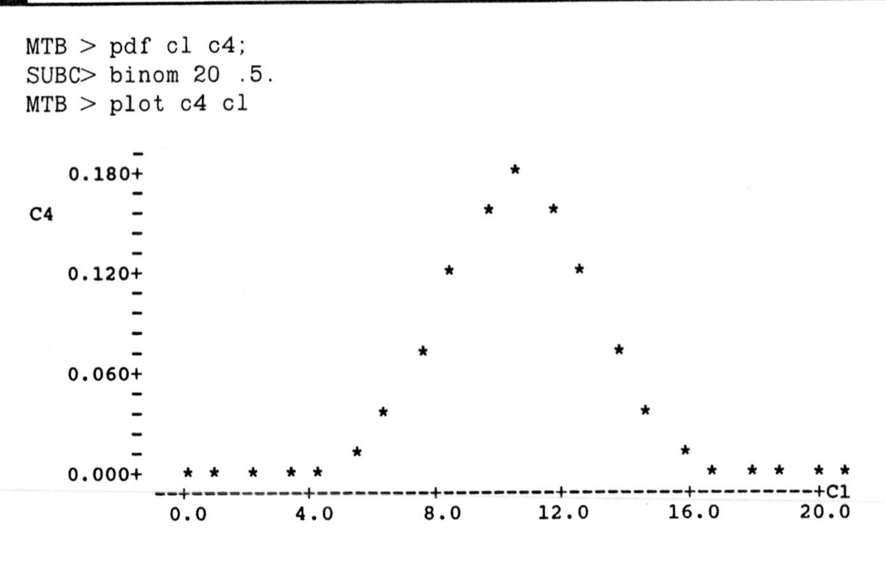


```
MTB > set c5
DATA> 0:50
DATA> end
MTB > pdf c5 c6;
SUBC> binom 50 .2.
MTB > plot c6 c5
```

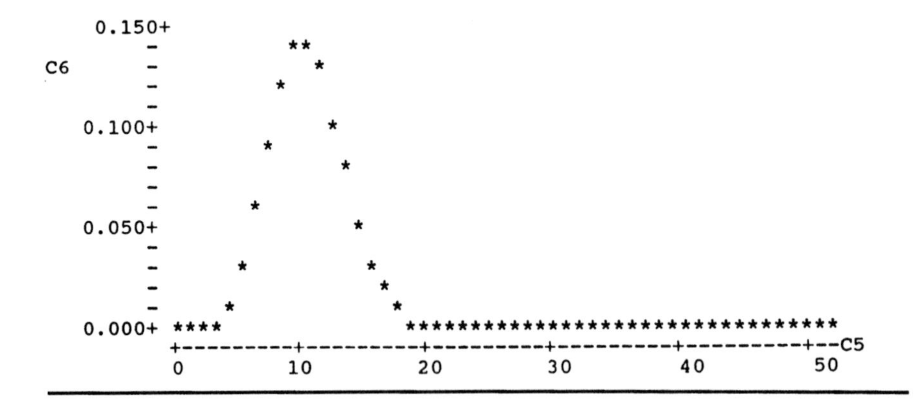

Unfortunately, there is no consensus as to when the normal approximation can be used, that is, when n is large enough for the central limit theorem to apply. This issue has been examined in a number of recent articles (6–8). Based on work by Samuels and Lu (8) and on some calculations we performed, Table 6.7 shows our recommendations for the size of the sample required, as a function of π, for the normal distribution to serve as a good approximation to the binomial distribution. Use of these sample sizes guarantees that the maximum difference between the binomial probability and its normal approximation is less than or equal to 0.0060 and that the average difference is less than 0.0017.

The mean and variance to be used in the normal approximation to the binomial are the mean and variance of the binomial, $n * \pi$ and $n * \pi * (1 - \pi)$, respectively. As we are using a continuous distribution to approximate a discrete distribution, we have to take this into account. We do this by using an interval to represent the integer. For example, the interval 5.5 to 6.5 would be used with the continuous variable in place of the discrete variable value of 6. This adjustment is called the *correction for continuity*.

As an example, we use the normal approximation to the binomial for the c-section deliveries example shown above. We wanted to find the probability of 22 or more c-section deliveries in a sample of 62 deliveries. The values of the binomial mean and variance, assuming that π is 0.235, are 14.57 (= 62 * 0.235) and 11.146 (= 62 * 0.235 * 0.765), respectively. The standard deviation of the binomial is then 3.339. Finding the probability of 22 or more c-sections for the discrete binomial variable is approximately equivalent to finding the probability that a normal variable with a mean of 14.57 and a standard deviation of 3.339 is greater than 21.5.

Before using the normal approximation, we must first check to see if the sample size of 62 is large enough. From Table 6.7, we see that because the assumed value of π is between 0.20 and 0.25, our sample size is large enough. Therefore it is okay to use the normal approximation to the binomial. Figure 6.7 shows the area under the normal curve corresponding to values greater than 21.5.

To find the probability of being greater than 21.5, we convert 21.5 to a

TABLE 6.7 Sample Size Required for the Normal Distribution to Serve as a Good Approximation to the Binomial Distribution as a Function of the Binomial Proportion π

π	.05	.10	.15	.20	.25	.30	.35	.40	.45	.50
n	440	180	100	60	43	32	23	15	11	10
Difference[a]	.0041	.0048	.0054	.0059	.0059	.0057	.0059	.0060	.0049	.0027
Mean difference[b]	.0010	.0012	.0013	.0016	.0016	.0016	.0016	.0017	.0016	.0013

[a] Maximum difference between binomial probability and normal approximation.
[b] Mean of absolute value of difference between binomial probability and normal approximation for all nonzero probabilities.

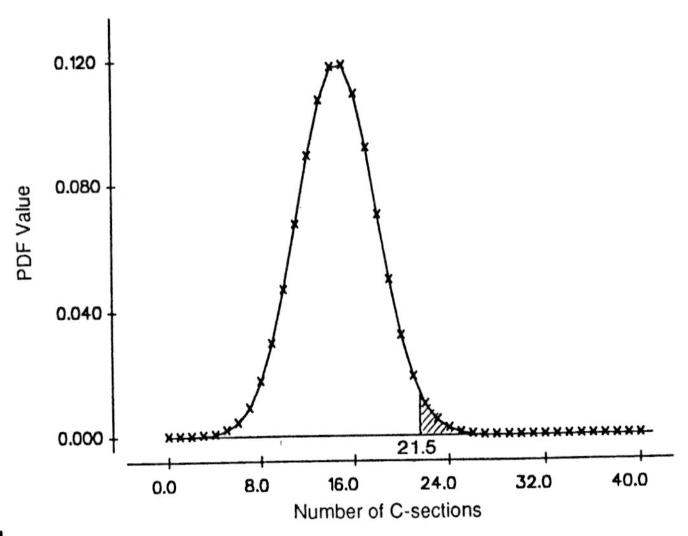

FIGURE 6.7 Area corresponding to a value of greater than 21.5.

standard normal value by subtracting the mean and dividing by the standard deviation. The corresponding z value is 2.075 [= (21.5 − 14.57)/3.339]. Looking in Table B4, we find the probability of a standard normal variable being less than 2.075 is about 0.9810. Subtracting this value from one gives the value of 0.0190, very close to the exact binomial value of 0.0224 found above.

A second example involves marijuana use among high school seniors. According to data reported in Table 65 of "Health, United States, 1991" (9), 14.0 percent of high school seniors admitted that they used marijuana during the 30 days previous to a survey conducted in 1990. If this percentage applies to all seniors in high school, what is the probability that in a survey of 140 seniors, the number reporting use of marijuana will be between 15 and 25? We want to use the normal approximation to the binomial, but we must first check our sample size with Table 6.7. As a sample of size 100 is required for a binomial proportion of 0.15, our sample of 140 for an assumed binomial proportion of 0.14 is large enough to use the normal approximation.

The mean of the binomial is 19.6 and the variance is 140 * 0.14 * 0.86, which is 16.856. Thus the standard deviation is 4.106. These values are used in converting the values of 15 and 25 to z scores. Taking the continuity correction into account means that interval is really 14.5 to 25.5. Figure 6.8 shows this interval.

We convert 14.5 and 25.5 to z scores by subtracting the mean of 19.6 and dividing by the standard deviation of 4.106. The z scores are −1 [= [= (14.5–19.6)/4.106] and 1.44 [= (25.5–19.6)/4.106]. To find the probabi of

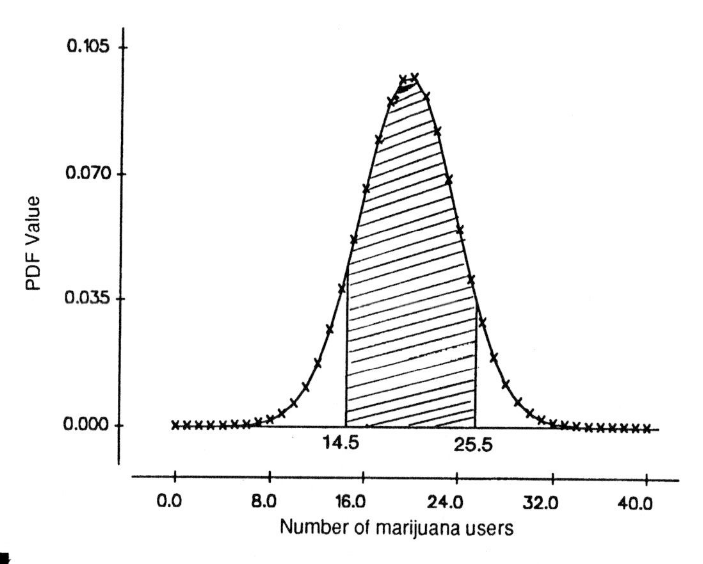

FIGURE 6.8 Area corresponding to the interval from 14.5 to 25.5.

of being between −1.24 and 1.44, we first find the probability of being less than 1.44. From that, we subtract the probability of being less than −1.24. This subtraction yields the probability of being in the interval.

These probabilities are found from Table B4 in the following manner. First, we read down the z column until we find the value of 1.44. We go across to the .00 column and read the value of 0.9251; this is the probability of a standard normal value being less than 1.44. The probability of being less than −1.24 is 0.1075. Subtracting 0.1075 from 0.9251 yields 0.8176. This is the probability that, out of a sample of 140, between 15 and 25 high school seniors would admit to using marijuana during the 30 days previous to the question being asked.

B. Poisson Approximation to the Binomial Distribution

As was pointed out above, sometimes the graphs of the binomial and Poisson distributions look similar; for example, the plot of the probability mass function for the binomial distribution with a sample size of 10 and π equal to 0.2 looks similar to the corresponding Poisson plot with a mean of 2. This is also a case in which the normal approximation should not be used because, for a π of 0.2, n should be at least 60. The similarity of the binomial and Poisson plots means that one of these distributions could be used to approximate the other under certain conditions. Because the Poisson is easier to calculate and has more expansive tables, it is used to approximate the binomial. As is shown below, the Poisson approximation complements the normal approximation. The Poisson is used for some of the situations when the requirements for the use of the normal approximation are not satisfied.

As the mean and variance of the Poisson distribution are equal, plots of the probability mass functions of the binomial and Poisson distributions will be similar when the mean and variance of the binomial are approximately equal. The mean of the binomial is $n * \pi$ and its variance is $n * \pi * (1 - \pi)$. Thus, these two values are approximately equal when $1 - \pi$ is close to 1, or when π is close to 0. When π is close to 0, n must be very large for the normal approximation to be used. Hence the requirements for use of the normal approximation to the binomial are usually not satisfied with very small values of π. It is in this situation that the Poisson approximation to the binomial can be used.

The following three plots in Boxes 6.28, 6.29, and 6.30 show the probability mass functions for the binomial and Poisson distributions for small values of π. To use the Poisson approximation, the mean of the binomial is also used as the mean of the Poisson distribution. In the three examples, the binomial means are 2, 2.5, and 1, and the corresponding variances are 1.8, 2.375, and 0.9, respectively. Hence the Poisson should provide a good approximation in these three examples.

In all three plots, there is little difference between the binomial and Poisson probability mass functions. The value 2 which is plotted indicates that both the binomial and Poisson probabilities were plotted at the same value. This high level of agreement is also shown in the printouts of the probabilities. Of the three situations, the Poisson approximation is the poorest for the smallest sample size and the largest value of the binomial proportion. For the sample size of 20, the largest difference between the binomial probability and its Poisson approximation is about 0.015. The agreement between the value of the probabilities and their approximations improves as n increases and as π decreases. The plots also show that the normal distribution would not be a good approximation to the binomial in these cases.

C. Normal Approximation to the Poisson Distribution

As the Poisson tables do not show every possible value of the parameter μ, and as the tables and computer packages do not provide probabilities for extremely large values of μ, it is useful to be able to approximate the Poisson distribution. As can be seen from the above plots, the Poisson does not look like a normal distribution for small values of μ; however, as the two plots in Box 6.31 show, the Poisson does resemble the normal distribution for large values of μ. The first plot shows the probability mass function for the Poisson with a mean of 10 and the second plot shows the probability mass function for the Poisson distribution with a mean of 20.

MINITAB BOX 6.28

In the following, we use the values from 0 to 12 in c1 because the probability of counts larger than 12 is so close to zero to be of no interest.

```
MTB > set c1
DATA> 0:12
DATA> end
MTB > pdf c1 c2;
SUBC> binom 20 .1.
MTB > pdf c1 c3;
SUBC> poiss 2.
MTB > print c2 c3
 ROW        C2         C3
   1    0.121577   0.135335
   2    0.270170   0.270671
   3    0.285180   0.270671
   4    0.190120   0.180447
   5    0.089779   0.090224
   6    0.031921   0.036089
   7    0.008867   0.012030
   8    0.001970   0.003437
   9    0.000356   0.000859
  10    0.000053   0.000191
  11    0.000006   0.000038
  12    0.000001   0.000007
  13    0.000000   0.000001

MTB > mplot c2 c1, c3 c1

0.30+
    -        2    2
    -
    -
    -
0.20+                  A
    -                  B
    -
    - B
    - A
0.10+                       B
    -                       A
    -
    -                            2
    -                                 B
0.00+                                 A    2    2    2    2    2    2
    -
     +---------+---------+---------+---------+---------+-
    0.0       2.5       5.0       7.5      10.0      12.5

         A = C2 vs. C1                  B = C3 vs. C1
```

MINITAB BOX 6.29

```
MTB > pdf cl c4;
SUBC> binom 50 .05.
MTB > pdf cl c5;
SUBC> poiss 2.5.
MTB > print c4 c5
 ROW        C4          C5
   1  0.076945    0.082085
   2  0.202487    0.205213
   3  0.261101    0.256516
   4  0.219875    0.213763
   5  0.135975    0.133602
   6  0.065841    0.066801
   7  0.025990    0.027834
   8  0.008598    0.009941
   9  0.002432    0.003106
  10  0.000597    0.000863
  11  0.000129    0.000216
  12  0.000025    0.000049
  13  0.000004    0.000010

MTB > mplot c4 cl, c5 cl

0.30+
    -
    -         2
    -
    -            2
0.20+     2
    -
    -
    -               2
    -
0.10+
    - 2
    -                  2
    -
    -                     2
0.00+                        2  2  2  2  2  2
    -
     +---------+---------+---------+---------+---------+--
    0.0       2.5       5.0       7.5      10.0      12.5

         A = C4 vs. C1              B = C5 vs. C1
```

As can be seen from these plots, the normal distribution should be a reasonable approximation to the Poisson distribution for values of μ greater than 10. As additional evidence for the use of the normal distribu-

MINITAB BOX 6.30

```
MTB > pdf c1 c6;
SUBC> binom 100 .01.
MTB > pdf c1 c7;
SUBC> poiss 1.
MTB > print c6 c7
 ROW         C6          C7
   1   0.366032    0.367879
   2   0.369730    0.367879
   3   0.184865    0.183940
   4   0.060999    0.061313
   5   0.014942    0.015328
   6   0.002898    0.003066
   7   0.000463    0.000511
   8   0.000063    0.000073
   9   0.000007    0.000009
  10   0.000001    0.000001
  11   0.000000    0.000000
  12   0.000000    0.000000
  13   0.000000    0.000000

MTB > mplot c6 c1, c7 c1

        -
  0.36+ 2   2
        -
        -
        -
        -
  0.24+
        -
        -       2
        -
        -
  0.12+
        -
        -           2
        -
        -               2
  0.00+                     2   2   2   2   2   2   2   2
        +---------+---------+---------+---------+---------+---
       0.0       2.5       5.0       7.5      10.0      12.5
            A = C6 vs. C1                B = C7 vs. C1
```

tion as an approximation, consider the plot in Box 6.32 of the normal scores for the Poisson distribution with the mean of 10. The normal scores plot appears to be a straight line, additional confirmation that the normal distribution provides a good approximation to a Poisson distribution with a mean of 10.

MINITAB BOX 6.31

```
MTB > set c1
DATA> 0:25
DATA> end
MTB > pdf c1 c2;
SUBC> poiss 10.
MTB > plot c2 c1

          -                   * *
     0.120+
          -                 *     *
C2        -
          -                         *
          -               *
     0.080+
          -                           *
          -             *
          -                             *
          -
     0.040+           *
          -                               *
          -                                 *
          -         *                         *
          -       *                             *
     0.000+ * * *                                 * * * * * * *
            +---------+---------+---------+---------+---------+-C1

           0.0       5.0      10.0      15.0      20.0      25.0
```

```
MTB > set c1
DATA> 0:40
DATA> end
MTB > pdf c1 c3;
SUBC> poiss 20.
MTB > plot c3 c1
     0.090+                          **
          -                         *  *
C3        -                       *      *
          -
          -                      *        *
     0.060+
          -                     *          *
          -
          -                                 *
          -                    *              *
     0.030+                  *
          -                                    *
          -                 *                   *
          -                *                     *
          -               *                        ***
     0.000+  *** *** ****                             * **** ***
           --+---------+---------+---------+---------+---------C1
            0.0       8.0      16.0      24.0      32.0      40.0
```

MINITAB BOX 6.32

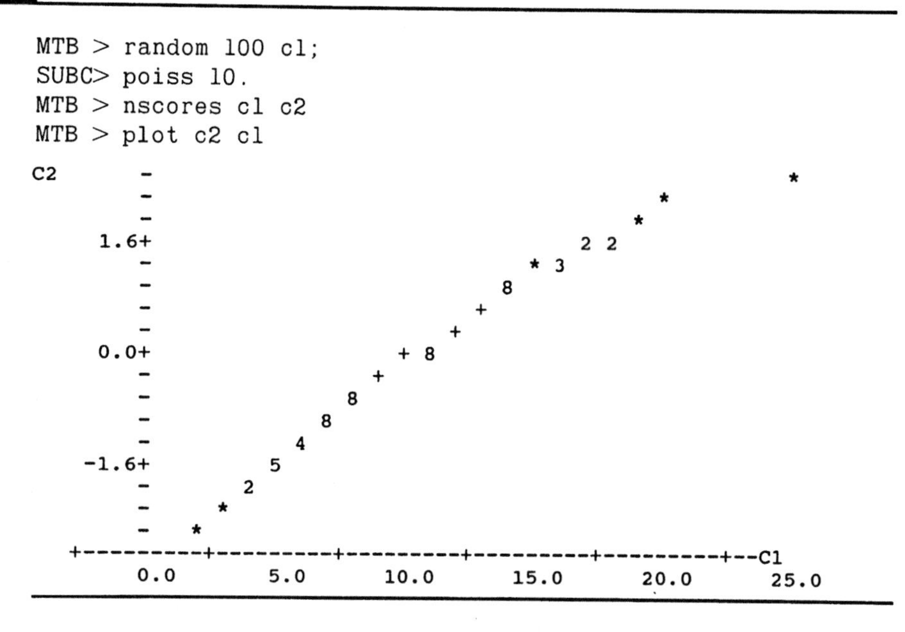

```
MTB > random 100 c1;
SUBC> poiss 10.
MTB > nscores c1 c2
MTB > plot c2 c1
```


The normal approximation to the Poisson uses the mean and variance from the Poisson distribution for the normal mean and variance. We use the pertussis example from above to demonstrate the normal approximation to the Poisson distribution. In the pertussis example, we wanted to find the probability of 18 or fewer cases of pertussis, given that the mean of the Poisson distribution was 35.31. This value, 35.31, is used for the mean of the normal and its square root, 5.942, for the standard deviation of the normal. As we are using a continuous distribution to approximate a discrete one, we must use the continuity correction. Therefore, we want to find the probability of values less than 18.5. To do this, we convert 18.5 to a z value by subtracting the mean of 35.31 and dividing by the standard deviation of 5.942. The z value is -2.829. The probability of a Z variable being less than -2.829 or -2.83 is found from Table B4 to be 0.0023, close to the exact value of 0.001 given above.

VI. CONCLUDING REMARKS

Three of the more useful probability distributions—the binomial, the Poisson, and the normal—were introduced in this chapter. Examples of their use in describing data were provided. The examples also suggested that the distributions could be used to examine whether or not the data came from population A or some other population. This use is explored in more depth in Chapter 9 on hypothesis testing and in several of the subsequent chapters. Another use of these distributions is demonstrated in the next chapter on interval estimation.

EXERCISES

6.1. Explain why the cumulative distribution function of X either stays the same or increases as X increases in value.

6.2. According to data from NHANES II (9, Table 70), 26.8 percent of persons 20 to 74 years of age had high serum cholesterol values ($\geq$240 mg/dl).

   a. In a sample of 20 persons ages 20 to 74, what is the probability that 8 or more persons had high serum cholesterol? Use Table B2 to approximate this value first and then provide a more accurate answer.

   b. How many persons out of the 20 would be required to have high cholesterol before you would think that the population from which your sample was drawn differs from the U.S. population of persons ages 20 to 74?

   c. In a sample of 200 persons ages 20 to 74, what is the probability that 80 or more persons had high serum cholesterol?

6.3. Based on reports from state health departments, there were 10.33 cases of tuberculosis per 100,000 population in the United States in 1990 (9, Table 50). What is the probability of a health department, in a county of 50,000, observing 10 or more cases in 1990 if the U.S. rate held in the county? What is the probability of fewer than 3 cases if the U.S. rate held in the county?

6.4. Create a normal probability plot or plot the normal scores for the 33 caloric intakes shown in Table 4.1. Based on the plot, do you think that the caloric intakes could be normally distributed?

6.5. Assume that systolic blood pressure for 5-year-old boys is normally distributed with a mean of 94 mm Hg and a standard deviation of 11 mm Hg. What is the probability of a 5-year-old boy having a blood pressure less than 70 mm Hg? What is the probability that the blood pressure of a 5-year-old boy will be between 80 and 100 mm Hg?

6.6. Less than 10 percent of the U.S. population is hospitalized in a typical year; however, the per capita hospital expenditure in the United States is generally large, for example, in 1990, it was approximately $975. Do you think that the expenditure for hospital care (at the person level) follows a normal distribution? Explain your answer.

6.7. In Harris County, Texas, in 1986, there were 173 cases of hepatitis A in a population of 2,942,550 (5, p. 8-2). The corresponding rate for the

United States was 10.0 per 100,000 population. What is the probability of a rate as low as or lower than the Harris County rate if the U.S. rate held in Harris County?

6.8. Approximately 6.5 percent of women ages 30 to 49 were iron deficient based on data from NHANES II (10, Table II-99). In a sample of 30 women ages 30 to 49, 6 were found to be iron deficient. Is this result so extreme that you would want to investigate why the percentage is so high?

6.9. Based on data from the Hispanic Health and Nutrition Examination Survey (HHANES) and reported in "Nutrition Monitoring in the United States" (10, Table II-40), the mean serum cholesterol for Mexican-American men ages 20 to 74 was 203 mg/dl. The standard deviation was approximately 44 mg/dl. Assume that serum cholesterol follows a normal distribution. What is the probability that a Mexican-American man 20 to 74 years old has a serum cholesterol value greater than 240 mg/dl?

6.10. In 1988, 71 percent of 15- to 44-year-old U.S. women who have ever been married have used some form of contraception (9, Table 15). What is the probability that, in a sample of 200 women in these childbearing years, fewer than 120 of them have used some form of contraception?

6.11. In ecology, the frequency distribution of the number of plants of a particular species in a square area is of interest. Skellam (11) presented data on the number of plants of *Plantago major* present in squares of 100 cm^2 laid down in grassland. There were 400 squares and the numbers of plants in the squares are as follows:

Plants per square	0	1	2	3	4	5	6	≥7
Frequency	235	81	43	18	9	6	4	4

Create a Poissonness plot to examine whether or not these data follow the Poisson distribution.

6.12. The Bruce treadmill test is used to assess exercise capacity in children and adults. Cumming *et al.* (12) studied the distribution of Bruce treadmill test endurance times in normal children. The mean endurance time for a sample of 36 girls 4 to 5 years old was 9.5 minutes, with a standard deviation of 1.86 minutes. If we assume that these are the true population mean and standard deviation, and if we also assume that the endurance times follow a normal distribution, what is the probability of observing a 4-year-old girl with an endurance time of less than 7 minutes? The 36 values shown below are based on summary statistics from the research by Cumming *et al.* (12). Do you believe that these data are normally distributed? Explain your answer.

Hypothetical Endurance Times in Minutes for 36 Girls 4 to 5 Years of Age:

5.3	6.5	7.0	7.2	7.5	8.0	8.0	8.0	8.0	8.2	8.5	8.5
8.8	8.8	8.9	9.0	9.0	9.0	9.0	9.5	9.8	9.8	10.0	10.0
10.6	10.8	11.0	11.2	11.2	11.3	11.5	11.5	12.2	12.4	12.7	13.3

6.13. Seventy-nine firefighters were exposed to burning polyvinyl chloride (PVC) in a warehouse fire in Plainfield, New Jersey, on March 20, 1985. A study was conducted in an attempt to determine whether there were short- and long-term respiratory effects of the PVC exposure (13). At the long-term follow-up visit 22 months after the exposure, 64 firefighters who had been exposed during the fire and 22 firefighters who were not exposed reported on the presence of various respiratory conditions. Eleven of the PVC-exposed firefighters had moderate to severe shortness of breath compared with only one of the nonexposed firefighters. What is the probability of finding 11 or more of the 64 exposed firefighters reporting moderate to severe shortness of breath if the rate of moderate to severe shortness of breath is one case per 22 persons? What are two possible confounding variables in this study that could affect the interpretation of the results?

REFERENCES

1. Public Citizen Health Research Group (1992). Unnecessary caesarean sections: Halting a National Epidemic. *Public Citizen Health Research Group Health Lett.* 8(6); 1–6.
2. Boyer, C. B. (1985). "A History of Mathematics," Princeton Univ. Press, Princeton, NJ.
3. Student (1907). On the error of counting with a haemacytometer. *Biometrika* **5,** 351–360.
4. Hoaglin, D. C. (1980). A Poissonness plot. *Amer. Stat.* **34,** 146–149.
5. Canfield, M., ed. (1990). "The Health Status of Harris County Residents: Births, Deaths and Selected Measures of Public Health, 1980–1986." Harris County Health Department, Houston, TX.
6. Blyth, C. R., and Still, H. A. (1983). Binomial confidence intervals. *J. Amer. Stat. Assoc.* **78,** 108–116.
7. Schader, M., and Schmid, F. (1989). Two rules of thumb for the approximation of the binomial distribution by the normal distribution. *Amer. Stat.* **43,** 23–24.
8. Samuels, M. L., and Lu, T. C. (1990). Sample size requirements for the back-of-the-envelope binomial confidence interval. *Amer. Stat.* **46,** 228–231.
9. National Center for Health Statistics (1992). "Health, United States, 1991 and Prevention Profile," DHHS Publ. No. 92-1232. Public Health Service, Hyattsville, MD.
10. Life Sciences Research Office, Federation of American Societies for Experimental Biology (1989). "Nutrition Monitoring in the United States: An Update Report on Nutrition Monitoring," DHHS Publ. No. (PHS) 89-1255. U.S. Department of Agriculture and the U.S. Department of Health and Human Services, Public Health Service, Washington, U.S. Government Printing Office.
11. Skellam, J. G. (1952). Studies in statistical ecology. *Biometrika* **39,** 346–362.
12. Cumming, G. R., Everatt, D., and Hastman, L. (1978). Bruce treadmill test in children: Normal values in a clinic population. *Am. J. Cardiol.* **41,** 69–75.
13. Markowitz, J. S. (1989). Self-reported short- and long-term respiratory effects among PVC-exposed firefighters. *Arch. Environ. Health* **44,** 30–33.

CHAPTER 7

Interval Estimation

In Chapter 6 we saw that there is variation that occurs when we use a sample instead of the entire population. For example, in the presentation of the binomial distribution, we saw that the sample estimates of the population proportion varied considerably from sample to sample. In this chapter, we present prediction, confidence, and tolerance intervals, quantities that allow us to take the variation in sample results into account in describing the data. These intervals represent specific types of *interval estimation*, the provision of limits that are likely to contain either (1) the population parameter of interest or (2) future observations of the variable. Interval estimation thus provides more information about the population parameter than the point estimation approach discussed in Chapter 4. In that chapter, we provided a single value as the estimate of the population parameter without giving any information about the sampling variability of the estimator. For example, knowledge of the value of the sample mean, a point estimate of the population mean, does not tell us anything about the variability of the sample mean. Interval estimation addresses this variability.

I. PREDICTION, CONFIDENCE, AND TOLERANCE INTERVALS

The material in this and the following section is based on material presented by Vardeman (1) and Walsh (2). To understand the difference between these three intervals (prediction, confidence, and tolerance), consider the following. Dairies add vitamin D to milk for the purpose of fortification. The recommended amount of vitamin D to be added to a quart of milk is 400 IU (10 μg). If a dairy adds too much vitamin D, perhaps more than 5000 IU, there is the possibility that a consumer will develop hypervitaminosis D, that is, vitamin D toxicity.

A *prediction interval* focuses on a single observation of the variable, for example, the amount of vitamin D in the next bottle of milk. A *confidence interval* focuses on a population parameter, for example, the mean or median amount of vitamin D per bottle in a population of bottles of milk. Thus the prediction interval is of more interest to the consumer of the next bottle of milk, whereas the confidence interval is of more interest to the dairy. A *tolerance interval* provides limits such that there is a high level of confidence that a large proportion of values of the variable will fall within them. For example, besides being interested in the mean, the dairy owner or a regulatory agency also wants to be confident that a large proportion of the bottles' vitamin D contents are within a specified tolerance of the value of 400 IU.

We begin our treatment of these intervals with distribution-free intervals.

II. DISTRIBUTION-FREE INTERVALS

When the method for forming the different intervals is independent of how the data are distributed, the resultant intervals are said to be *distribution free*. Distribution-free intervals are based on the rank order of the sample values and a notation that captures the rank order is the following. The smallest of the x values is indicated by $x_{(1)}$, the second smallest by $x_{(2)}$, and so on, to the largest value which is denoted by $x_{(n)}$. The $x_{(i)}$ are called *order statistics* as the subscripts show the order of the values.

We use hypothetical data showing the amount of vitamin D in 30 bottles of milk selected at random from one dairy. The values are shown in rank order in Table 7.1. Based on this sample, $x_{(1)}$ equals 289 IU, $x_{(2)}$ is 326 IU, and so on to $x_{(30)}$ which equals 485 IU.

A. Prediction Interval

As a consumer of milk, our major concern about vitamin D is that the milk does not contain an amount of vitamin D that is toxic to us. We are not too

TABLE 7.1 Values of Vitamin D (IU) in a Hypothetical Sample of 30 Bottles

289	355	376	392	406	433
326	363	379	395	410	434
339	364	384	396	413	456
346	370	386	398	422	471
353	373	389	403	427	485

concerned about there being too little vitamin D in the bottle. Based on the hypothetical sample of vitamin D contents in 30 bottles of milk, we can form a one-sided prediction interval—our concern focuses on the upper limit—for the amount of vitamin D in the bottle of milk that we are going to purchase.

A natural one-sided prediction interval in this case is from 0 to the maximum observed value of vitamin D (485 IU) in the sample. The level of confidence associated with this interval, from 0 to 485 IU, is 96.8 percent (= 30/31). This value can be found from the consideration of the order statistics and the real number line. For example, we have the line

```
|_1_|____2_|_______3_______|____________________|__30__|__31____
0   x(1)       x(2)              x(3)       . . .      x(30)
```

and there are 31 intervals along this line. The vertical marks indicate the location of the order statistics along the line and the numbers above the line between the vertical marks indicate the interval numbers. There are 31 intervals and the next observation can fall into any one of the intervals. Of these 31 intervals, 30 have values less than the maximum value. Hence, we are 96.8 percent confident that the vitamin D content in the next bottle will be between zero and the observed maximum value.

Note that we used the word *confidence* instead of *probability* here. We use confidence because we are using the sample data as the basis of estimating the probability distribution of the vitamin D content. If we used the probability distribution of the vitamin D content instead of using its sample estimate, the empirical distribution function, we would use the word *probability*. In repeated sampling, we expect that 96.8 percent of the prediction intervals, ranging from zero to the observed maximum in each sample of size 30, would contain the next observed vitamin D content.

The use of the second largest value, $x_{(29)}$, as the upper limit of the interval results in a prediction confidence level of 93.5 percent (= 29/31). An attraction of this interval is that it provides a slightly shorter interval with a maximum of 471 IU, but we are slightly less confident about it. Based on either of these intervals, the consumer should not be worried about purchasing a bottle that has a value of vitamin D that would cause vitamin D poisoning.

For a two-sided interval, a natural interval would be from the minimum observed value, $x_{(1)}$, to the maximum observed value, $x_{(30)}$. In this case, the two-sided interval is from 289 to 485 IU. The confidence level associated with this prediction interval is 93.5 percent (= 29/31). Of the 31 intervals shown above, there is one below the minimum value and also one above the maximum value. Hence there are 29 chances out of 31 that the next observed value will fall between the minimum and maximum values.

With a sample size of 30, it is not possible to have a distribution-free, two-sided, 95 percent prediction interval. The smallest sample size that attains the 95 percent level is 39. When n is 39, there are 40 intervals, and 2/40 equals 0.05. This calculation shows that it is easy to determine how large a sample is required to satisfy prediction interval requirements.

B. Confidence Interval

The dairy wants to know, on average, how much vitamin D is being added to the milk. If the interval estimate for the central tendency differs much from 400 IU, the dairy may have to change its process for adding vitamin D. One way of obtaining the interval estimate is to use a distribution-free confidence interval.

Distribution-free confidence intervals are used to provide information about population parameters, for example, the median and other percentiles. There are two approaches to finding confidence intervals for percentiles: (1) the use of order statistics and (2) the use of the normal approximation to the binomial distribution. The first approach is generally used for smaller samples, whereas the second approach is used for larger samples.

1. Use of Order Statistics and the Binomial Distribution

The lower and upper limits of the $(1 - \alpha) * 100$ percent confidence interval for the pth percentile of X are the order statistics $x_{(j)}$ and $x_{(k)}$, where the values of j and k, j less than k, are to be determined. The limits of the confidence interval for the pth percentile of X are the values $x_{(j)}$ and $x_{(k)}$ that satisfy the inequality

$$\Pr\{x_{(j)} < p\text{th percentile} < x_{(k)}\} \geq 1 - \alpha$$

and this is equivalently

$$\Pr\{x_{(j)} \geq p\text{th percentile}\} + \Pr\{x_{(k)} \leq p\text{th percentile}\} \leq \alpha.$$

If we require that both terms in the sum be less than or equal to $\alpha/2$, from the first term, we have

$$\Pr\{\text{at most } j - 1 \text{ observations} < p\text{th percentile}\} \leq \alpha/2.$$

This situation has two outcomes: an observation is less than the pth percentile or it is greater than or equal to the pth percentile. The probability that an observation is less than the pth percentile is p. The variable of interest is the number of observations, out of n, that are less than the pth percentile. Thus this variable follows a binomial distribution with parameters n and p. Knowing the values of n and p enables us to find the value of j because j must satisfy the inequality

$$\sum_{i=0}^{j-1} \frac{n!}{i!\,(n-i)\,!} p^i(1-p)^{n-i} \leq \alpha/2.$$

The inequality used to find the value of k is

$$\sum_{i=k}^{n} \frac{n!}{i!\,(n-i)\,!} p^i(1-p)^{n-i} \leq \alpha/2.$$

Putting these two inequalities together means that the binomial sum from j to $k - 1$ must be greater than or equal to $1 - \alpha$. Here we have dropped the requirement that the sums of the probabilities from 0 to $j - 1$ and from k to n both must be less than $\alpha/2$. The values of j and k are found from the binomial table, Table B2, or by using MINITAB.

For example, suppose we want to find a 95 percent confidence interval for the median, the 50th percentile, for the vitamin D values from the dairy used in Table 7.1. The sample estimate of the median is the average of the 15th and 16th smallest values, that is, 390.5 IU [= (389 + 392)/2].

To find the 95 percent confidence interval for the median in the population of bottles of milk from the selected dairy, we use the binomial distribution. As Table B2 does not have values for n larger than 20, we use MINITAB to find the confidence interval as shown in Box 7.1. There may be more than one pair of values of j and k that satisfy the requirement that the sum of the binomial probabilities from j to $k - 1$ is greater than or equal to $1 - \alpha$. To choose from among these pairs, we select the pair whose difference $(k - j)$ is the smallest. In the special case of the median, we require that k equals $n - j + 1$; this requirement gives the same number of observations in both tails of the distribution.

The sum of the probabilities from j to $k - 1$ must be greater than or equal to 0.95. Examination of the cumulative probabilities tells us that j is 9 and k is 21. The sum of the probabilities between 9 and 20 is 0.9572 (= 0.9786 − 0.0214). If j were 10 and k were 20, the sum of the probabilities between 10 and 19 would be 0.9012, less than the required value of 0.95. Thus the approximate 95 percent (really closer to 96 percent) confidence interval for the median is from 370 IU (= $x_{(9)}$) to 406 IU (= $x_{(21)}$). The use of distribution-free intervals does not necessarily provide intervals that are symmetric about the sample estimator. For example, the sample median value, 390.5 IU, is not in the exact middle of the confidence interval.

Note that the confidence interval for the median is much narrower than the approximate 95 percent prediction interval, from 289 to 485 IU, for

MINITAB BOX 7.1

The CDF command will be used to help us find the values of j and k.

```
MTB > cdf;
SUBC> binom 30 0.50.
    BINOMIAL WITH N =   30
             P = 0.500000
       K  P( X LESS OR = K)
       0         0.0000
       1         0.0000
       2         0.0000
       3         0.0000
       4         0.0000
       5         0.0002
       6         0.0007
       7         0.0026
       8         0.0081
       9         0.0214
      10         0.0494
      11         0.1002
      12         0.1808
      13         0.2923
      14         0.4278
      15         0.5722
      16         0.7077
      17         0.8192
      18         0.8998
      19         0.9506
      20         0.9786
      21         0.9919
      22         0.9974
      23         0.9993
      24         0.9998
      25         1.0000
```

a single observation. As we saw in Chapter 4, much less variability is associated with a mean or median than with a single observation and this is additional confirmation of that.

As we can observe from the above, the use of distribution-free intervals does not provide exactly 95 percent levels. The level of confidence associated with these intervals is a function of the sample size as well as which order statistics are used in the creation of the interval.

It is also possible to create one-sided confidence intervals for parameters. For example, if the goal were to create an upper one-sided confidence interval for the median, we would find the value of k such that

$$\sum_{i=k}^{n} \frac{n!}{i!\,(n-i)\,!} p^i (1-p)^{n-i} \leq \alpha$$

for a p having the value of 0.50. The upper one-sided confidence interval for the median is from 0 to $x_{(k)}$ where k's value is found from the above inequality.

2. Use of the Normal Approximation to the Binomial

For larger sample sizes, the normal approximation to the binomial distribution can be used to find the values of j and k. The sample size must be large enough to satisfy the requirements for the use of the normal approximation. As p is 0.50, the sample size of 30 bottles from the dairy is large enough.

As above, we want to find the value of j such that the probability of the binomial variable, Y, being less than or equal to $j - 1$ is less than or equal to $\alpha/2$, that is,

$$\Pr\{Y \leq j - 1\} \leq \alpha/2.$$

Use of the continuity correction converts this to

$$\Pr\{Y \leq j - 0.5\} \leq \alpha/2.$$

To convert Y to the standard normal variable, we must subtract $n * p$, the estimate of the mean, and divide by $\sqrt{n * p * (1 - p)}$, the estimate of the standard error. This yields

$$\Pr\left\{\frac{Y - n * p}{\sqrt{n * p * (1 - p)}} \leq \frac{j - 0.5 - n * p}{\sqrt{n * p * (1 - p)}}\right\} \leq \frac{\alpha}{2}.$$

This can be reexpressed as

$$\Pr\left\{Z \leq \frac{j - 0.5 - n * p}{\sqrt{n * p * (1 - p)}}\right\} \leq \frac{\alpha}{2}.$$

If we change this inequality to an equality, that is, the probability is equal to $\alpha/2$, we can find a unique value for j. The value of the term on the right side of the inequality inside the braces is simply $z_{\alpha/2}$ and, hence, we can find the value of j from the equation

$$j - 0.5 - n * p = z_{\alpha/2} * \sqrt{n * p * (1 - p)}$$

or

$$j = z_{\alpha/2} * \sqrt{n * p * (1 - p)} + 0.5 + n * p$$

In the above example, p was 0.50, n was 30, and α was 0.05. As the value of $z_{0.025}$ is -1.96, we have

$$j = -1.96 * \sqrt{30 * 0.50 * 0.50} + 0.5 + 30 * 0.50$$

or j is 10.13. To ensure that the level of the confidence interval is at least $(1 - \alpha) * 100$ percent, we must round down the value of j to the next smaller integer, 10, and we round up the value of k, found below, to the next larger integer.

The value of k is found from the equation

$$k = z_{1-\alpha/2} * \sqrt{n * p * (1 - p)} + 0.5 + n * p$$

which yields a k equal to 20.87, which is rounded to 21. Thus, the 95 percent confidence interval is from 373 IU ($= x_{(10)}$) to 406 IU ($= x_{(21)}$). In this case, both the binomial and the normal approximation approaches used $x_{(21)}$ as the upper limit, but the binomial approach used $x_{(9)}$ as the lower limit whereas the normal approximation used $x_{(10)}$.

C. Tolerance Interval

As mentioned above, tolerance intervals are of most interest to the dairy or to a regulatory agency. The tolerance limits are values such that we have a high level of confidence that a large proportion of the bottles have vitamin D contents located between the lower and upper tolerance limits. These upper and lower limits of the tolerance interval can be used in determining whether the process for adding vitamin D is under control. If the limits are too wide, the dairy may have to modify its process for adding vitamin D to the milk.

The dairy does not want to add too much vitamin D to the milk because of the possible problems for the consumer and the extra cost associated with using more vitamin D than required. At the same time, the dairy must add enough vitamin D to be in compliance with truth in advertising legislation.

As with the prediction interval, it is reasonable to use the smallest and largest observed values for the lower and upper limits of the tolerance interval, although other values could be used. We also have to specify the proportion of the population, p, that we want to include within the tolerance interval. Given the tolerance interval limits and the proportion of values to be included within it, we can calculate the confidence level, γ, associated with the interval.

In symbols, the tolerance interval limits are the order statistics $x_{(j)}$ and $x_{(k)}$ such that

$$\Pr\{\Pr\{X \leq x_{(k)}\} - \Pr\{X \leq x_{(j)}\} \geq p\} = \gamma$$

The quantity, $\Pr\{X \leq x_{(k)}\} - \Pr\{X \leq x_{(j)}\}$, is the proportion of the population values contained in the tolerance interval for this sample. Let us call the above quantity W_{kj}. In symbols we then have $\Pr\{W_{kj} \geq p\} = \gamma$. The variable W_{kj} is either less than p or greater than or equal to p. This is a binomial situation and, therefore, we can use the same approach as in the confidence interval section to find the value of γ. The value of γ can be expressed in terms of the binomial summation as

$$\gamma = \sum_{i=0}^{k-j-1} \frac{n!}{i!\,(n-i)\,!} p^i (1-p)^{n-i}.$$

If we use the minimum, $x_{(1)}$, and the maximum, $x_{(n)}$, for the limits, $k - j - 1$ becomes $n - 1 - 1$ which equals $n - 2$. It is therefore easy to find the value of this summation for i ranging from 0 to $n - 2$ because that sum is equal to 1 minus the binomial sum from $n - 1$ to n. In symbols, the value of γ is

$$1 - [p^n] - [np^{n-1}\,(1-p)].$$

Suppose we want our tolerance interval to contain 95 percent of the observations. Let us calculate the confidence level associated with the tolerance interval of 289 to 485 IU. In this case, n is 30 and p is 0.95. The value of γ is found by taking $1 - 0.95^{30} - 30 * (0.95)^{29} * (1 - 0.95)$, which equals 0.4465. There is not a high level of confidence associated with this tolerance interval. This confidence level is contrasted with the 0.935 level associated with the prediction interval. It is not surprising that the confidence level of the prediction interval is much higher than that of the tolerance interval because the prediction interval is based on the location of a single future value whereas the tolerance interval is based on the location of a large proportion of the population values.

The interval from 289 to 485 IU is the widest interval we can have using the sample data as these are the minimum and maximum observed values. We can increase our confidence either (1) by decreasing p, the proportion of the population to be included in the tolerance interval, or (2) by taking a larger sample.

Let us reduce p to 90 percent. The confidence level for this interval is increased to 0.8162, a much more reasonable value. Instead of reducing p, let us increase the sample size from 30 to 60. The confidence level associated with the increased sample size is 0.8084, also a much more reasonable value. Table 7.2 shows the sample size required to have 90, 95, and 99 percent confidence associated with tolerance intervals that have 80, 90, 95, and 99 percent coverage of the distribution, based on the use of $x_{(1)}$ and $x_{(n)}$.

From these calculations and the general formula for calculating γ, we can see the relationships between p, the values of k and j, n, and γ. We can investigate the values of these quantities before we have performed the study and can modify the proposed study design if we are not satisfied with the values of p and γ.

A one-sided tolerance interval is sometimes of interest. Suppose that there was interest in the upper one-sided tolerance interval. In this case, the tolerance interval ranges from 0 to $x_{(n)}$ and the confidence associated

TABLE 7.2 Sample Size Required for the Tolerance Interval to Have the Indicated Confidence Level for the Specified Coverage Proportions Based on the Use of $x_{(1)}$ and $x_{(n)}$

Coverage proportion	Confidence level		
	90%	95%	99%
0.80	18	22	31
0.90	38	46	64
0.95	77	93	130
0.99	388	473	662

with this interval is found by taking $1 - p^n$, that is, one minus the binomial term calculated for i equal to n.

III. INTERVALS BASED ON THE NORMAL DISTRIBUTION

If the data are from a known probability distribution, knowledge of this distribution allows more informative (smaller) intervals to be constructed for the parameters of interest or for future values. We begin this presentation by showing how to create confidence intervals for a variety of population parameters, assuming that the data come from a normal distribution. Following the material on confidence intervals, we show how to use the normal distribution in the creation of prediction and tolerance intervals. We begin the confidence interval presentation with the population mean and follow it with the confidence interval for the population proportion which can also be viewed as a mean.

A. Confidence Interval for the Mean

In the material above, we saw how to construct a confidence interval for the population median. That confidence interval gave information to the dairy about the average amount of vitamin D being added to the milk. As an alternative to the median, a confidence interval for the mean could have been used. To find a confidence interval for the mean, assuming that the data follow a specific distribution, we must know the sampling distribution of its estimator. We must also specify how confident we wish to be that the interval contains the population parameter. The sample mean is the estimator of the population mean, and the sampling distribution of the sample mean is easily found.

Because we are assuming the data follow a normal distribution, the sample mean, the average of the sample values, also follows a normal distribution; however, this assumption is not crucial. Even if the data are not normally distributed, the central limit theorem states that the sample mean, under appropriate conditions, will approximately follow a normal distribution. To specify the normal distribution completely, we also have to provide the mean and variance of the sample mean.

1. Known Variance

In Chapter 6, we saw that the mean of the sample mean was μ, the population mean, and its variance was σ^2/n. The standard deviation of the sample mean is thus $\sigma/\sqrt{n}$, and it is called the standard error of the sample mean ($\bar{x}$). The use of the word *error* is confusing as no mistake has been made; however, it is the traditional term used in this context. The term *standard error* is used instead of standard deviation when we are discussing the variation in a sample statistic. The term *standard deviation* is usually reserved for discussion of the variation in the sample data themselves.

We now address the issue of how confident we wish to be that the interval contains the population mean (μ). From the material on the normal distribution in Chapter 6, we know that

$$\Pr\{-1.96 < Z < 1.96\} = 0.95,$$

where Z is the standard normal variable. In terms of the sample mean, this is

$$\Pr\left\{-1.96 < \frac{\bar{x} - \mu}{(\sigma/\sqrt{n})} < 1.96\right\} \; 0.95.$$

But we want an interval for μ, not for Z. Therefore we must perform some algebraic manipulations to convert this to an interval for μ. First we multiply all three terms inside the braces by $\sigma/\sqrt{n}$. This yields

$$\Pr\left\{-1.96 * \left(\frac{\sigma}{\sqrt{n}}\right) < \bar{x} - \mu < 1.96 * \left(\frac{\sigma}{\sqrt{n}}\right)\right\} = 0.95.$$

We next subtract $\bar{x}$ from all the expressions inside the braces and this gives

$$\Pr\left\{-1.96 * \left(\frac{\sigma}{\sqrt{n}}\right) - \bar{x} < -\mu < 1.96 * \left(\frac{\sigma}{\sqrt{n}}\right) - \bar{x}\right\} = 0.95.$$

This interval is about $-\mu$; to convert it to an interval about μ, we multiply each term in the braces by -1. Before doing this, we must be aware of the effect of multiplying an inequality by a minus number. For example, we know that 3 is less than 4; however, -3 is greater than -4, so the result of multiplying both sides of an inequality by -1 changes the direction of the inequality. Therefore, we have

$$\Pr\left\{1.96 * \left(\frac{\sigma}{\sqrt{n}}\right) + \bar{x} > \mu > -1.96 * \left(\frac{\sigma}{\sqrt{n}}\right) + \bar{x}\right\} = 0.95.$$

We reorder the terms to have the smallest of the three quantities to the left, that is,

$$\Pr\left\{\bar{x} - 1.96 * \left(\frac{\sigma}{\sqrt{n}}\right) < \mu < \bar{x} + 1.96 * \left(\frac{\sigma}{\sqrt{n}}\right)\right\} = 0.95$$

or, more generally,

$$\Pr\left\{\bar{x} - z_{1-\alpha/2} * \left(\frac{\sigma}{\sqrt{n}}\right) < \mu < \bar{x} + z_{1-\alpha/2} * \left(\frac{\sigma}{\sqrt{n}}\right)\right\} = 1 - \alpha.$$

The $(1 - \alpha) * 100$ percent confidence interval limits for the population mean can be expressed as

$$\bar{x} \pm z_{1-\alpha/2} * \left(\frac{\sigma}{\sqrt{n}}\right).$$

The result of these manipulations is an interval for μ in terms of σ, n, 1.96 (or some other z value), and $\bar{x}$. The sample mean, $\bar{x}$, is the only one of these quantities that varies from sample to sample. Once we draw a sample, however, the interval is fixed as the sample mean's value, $\bar{x}$, is known. As the interval will either contain or not contain μ, we no longer talk about the probability of the interval containing μ.

Although we do not talk about the probability of an interval containing μ, we do know that in repeated sampling, intervals of the form above will contain the parameter, μ, 95 percent of the time. Thus, instead of discussing the probability of an interval containing μ, we say that we are 95 percent confident that the interval from $\bar{x} - 1.96 * (\sigma/\sqrt{n})$ to $\bar{x} + 1.96 * (\sigma/\sqrt{n})$ will contain μ. Intervals of this type are therefore called *confidence intervals*. This reason for the use of the word *confidence* is the same as that discussed in the distribution-free material above. The limits of the confidence interval usually have the form of the sample estimate plus or minus some distribution percentile—in this case, the normal distribution—times the standard error of the sample estimate.

The 95 percent confidence interval for the mean caloric intake for suburban middle school boys in the Houston area can be found based on the data shown in Table 4.1. We assume that the standard deviation for this population is 700 calories. As the sample mean, $\bar{x}$, based on a sample size of 33 observations, was found to be 2314 calories, the 95 percent confidence interval for the population mean ranges

$$\text{from} \quad 2314 - 1.96 * \left(\frac{700}{\sqrt{33}}\right) \quad \text{to} \quad 2314 + 1.96 * \left(\frac{700}{\sqrt{33}}\right),$$

that is, from 2075.2 to 2552.8 calories.

Box 7.2 illustrates the concept of confidence intervals. It shows the results of drawing 50 samples of size 60 from a normal distribution with a mean of 94 and a standard deviation of 11. These values are close to the mean and standard deviation of the systolic blood pressure variable for 5-year-old boys in the United States as reported by the NHLBI Task Force on Blood Pressure Control in Children (3).

In the following demonstration, 4 percent of the intervals did not contain the population mean and 96 percent did. If we draw many more samples, the proportion of the intervals containing the mean will be 95

MINITAB BOX 7.2

The command ZINTERVAL, shortened to ZINT, is used to create the confidence intervals. If no percentage is specified, a 95 percent confidence interval is created for the data in the listed columns. The command requires the value of σ and the columns containing the data for which the confidence intervals are to be created.

```
MTB > random 60 c1-c50;
SUBC> normal 94 11.
MTB > zint 95 11 c1-c50

 THE ASSUMED SIGMA =11.0
        N   MEAN   STDEV SE MEAN 95 PERCENT CI.

C1    60  94.75  10.25  1.42 ( 91.96,  97.54)
C2    60  94.85  10.86  1.42 ( 92.06,  97.63)
C3    60  94.71  10.09  1.42 ( 91.92,  97.50)
C4    60  94.03  12.27  1.42 ( 91.24,  96.82)
C5    60  93.77  10.05  1.42 ( 90.98,  96.56)
C6    60  92.54   9.32  1.42 ( 89.76,  95.33)
C7    60  93.40  12.07  1.42 ( 90.62,  96.19)
C8    60  93.97  11.02  1.42 ( 91.18,  96.75)
C9    60  96.33   9.26  1.42 ( 93.54,  99.12)
C10   60  93.56  12.01  1.42 ( 90.78,  96.35)
C11   60  94.94  10.81  1.42 ( 92.15,  97.73)
C12   60  94.66  12.08  1.42 ( 91.88,  97.45)
C13   60  94.21  11.02  1.42 ( 91.42,  97.00)
C14   60  94.55   9.98  1.42 ( 91.76,  97.34)
C15   60  93.57  11.50  1.42 ( 90.79,  96.36)
C16   60  95.99  12.01  1.42 ( 93.20,  98.78)
C17   60  93.86  12.53  1.42 ( 91.08,  96.65)
C18   60  92.02  13.58  1.42 ( 89.23,  94.81)
C19   60  95.16  12.03  1.42 ( 92.38,  97.95)
C20   60  94.99  12.00  1.42 ( 92.20,  97.78)
C21   60  94.65  11.18  1.42 ( 91.86,  97.43)
C22   60  92.86  12.52  1.42 ( 90.07,  95.64)
C23   60  93.99  11.76  1.42 ( 91.20,  96.78)
C24   60  91.44  10.75  1.42 ( 88.65,  94.22)
C25   60  96.07  11.89  1.42 ( 93.28,  98.86)
```

	N	MEAN	STDEV	SE MEAN	95 PERCENT CI.	
C26	60	94.61	11.49	1.42	(91.82,	97.39)
C27	60	92.79	9.36	1.42	(90.00,	95.58)
C28	60	96.00	12.19	1.42	(93.22,	98.79)
C29	60	95.99	11.36	1.42	(93.20,	98.78)
C30	60	93.98	11.74	1.42	(91.19,	96.76)
C31	60	95.36	13.08	1.42	(92.57,	98.15)
C32	60	91.10	8.69	1.42	(88.31,	93.89)*
C33	60	93.85	12.94	1.42	(91.06,	96.63)
C34	60	96.01	9.63	1.42	(93.22,	98.79)
C35	60	95.20	8.94	1.42	(92.41,	97.99)
C36	60	95.64	9.41	1.42	(92.85,	98.43)
C37	60	94.74	10.31	1.42	(91.95,	97.53)
C38	60	93.52	10.30	1.42	(90.73,	96.31)
C39	60	92.92	10.27	1.42	(90.13,	95.71)
C40	60	95.08	10.07	1.42	(92.30,	97.87)
C41	60	93.88	10.53	1.42	(91.09,	96.66)
C42	60	95.38	9.98	1.42	(92.59,	98.17)
C43	60	94.38	11.65	1.42	(91.59,	97.17)
C44	60	91.55	10.63	1.42	(88.76,	94.33)
C45	60	95.41	12.79	1.42	(92.62,	98.20)
C46	60	92.40	10.57	1.42	(89.62,	95.19)
C47	60	96.00	11.45	1.42	(93.21,	98.78)
C48	60	95.39	10.56	1.42	(92.60,	98.18)
C49	60	97.69	10.89	1.42	(94.90,	100.47)*
C50	60	95.01	10.61	1.42	(92.22,	97.79)

MEAN is the sample mean, STDEV is the sample standard deviation, and SE MEAN (standard error of the sample mean) is the population standard deviation divided by the square root of n. The lower limit of the 95 PERCENT interval is the sample mean minus 1.96 times SE MEAN and the upper limit is the sample mean plus 1.96 times SE MEAN. We have marked 2 intervals, out of the 50, that did not contain 94, the population mean.

---

percent. This is the basis for the statement that we are 95 percent confident that the confidence interval, based on our single sample, will contain the population mean.

If we use a different value for the standard normal variable, the level of confidence changes accordingly. For example, if we had started with a value of 1.645, $z_{0.95}$, instead of 1.96, $z_{0.975}$, the confidence level would be 90 percent instead of 95 percent. The $z_{0.95}$ value is used with the 90 percent level because we want 5 percent of the values to be in each tail. The lower and upper limits for the 90 percent confidence interval for the population mean for the data in c1, the first sample of 60 observations, are 92.41 ($= 94.75 - 1.645 * 1.42$) and 97.09 ($= 94.75 + 1.645 * 1.42$), respectively.

This interval is narrower than the corresponding 95 percent confidence interval of 91.96 to 97.54. This makes sense because, if we wish to be more confident that the interval contains the population mean, the interval will have to be wider. The 99 percent confidence interval uses $z_{0.995}$, which is 2.576, and the corresponding interval is 91.09 (= 94.75 −2.576 * 1.42) to 98.41 (= 94.75 + 2.576 * 1.42).

The 50 samples shown above had sample means, based on 60 observations, ranging from a low of 91.1 to a high of 97.7. This is the amount of variation in sample means expected if the data came from the same normal population with a mean of 94 and a standard deviation of 11. The Second National Task Force on Blood Pressure Control in Children had study means ranging from 85.6 mm Hg (based on 181 values) to 103.5 mm Hg (based on 61 values) (3), far outside the range shown above. These extreme values suggest that these data do not come from the same population, and this then calls into question the Task Force's combination of the data from these diverse studies.

The size of the confidence interval is also affected by the sample size which appears in the $\sigma/\sqrt{n}$ term. As n is in the denominator, increasing n decreases the size of the confidence interval. For example, if we doubled the sample size from 60 to 120 in the above example, the SE MEAN term changes from 1.42 (= $11/\sqrt{60}$) to 1.004 (= $11/\sqrt{60 * 2}$). Doubling the sample size reduces the confidence interval to about 71 percent (= $1/\sqrt{2}$) of its former width. Thus we know more about the location of the population mean, because the confidence interval is shorter, as the sample size increases.

The size of the confidence interval is also a function of the value of σ, but to change σ means that we are considering a different population. If, however, we are willing to consider homogeneous subgroups of the population, the value of the standard deviation for a subgroup should be less than that for the entire population. For example, instead of considering the blood pressure of 5-year-old boys, we consider the blood pressure of 5-year-old boys grouped according to height intervals. The standard deviation of systolic blood pressure in the different height subgroups should be much less than the overall standard deviation.

Another factor affecting the size of the confidence interval is whether it is a one-sided or two-sided interval. If we are concerned only about higher blood pressure values, we could use an upper one-sided confidence interval. The lower limit would be zero, or $-\infty$ for a variable that had positive and negative values, and the upper limit would be

$$\bar{x} + z_{1-\alpha} * \left(\frac{\sigma}{\sqrt{n}}\right).$$

This is similar to the two-sided upper limit except for the use of $z_{1-\alpha}$ instead of $z_{1-\alpha/2}$.

a. *Sample Size*

One important point about the confidence interval is that its width can be calculated before the sample is selected. The width of the 95 percent confidence interval is the upper limit minus the lower limit, that is,

$$\left[\bar{x} + 1.96 * \left(\frac{\sigma}{\sqrt{n}}\right)\right] - \left[\bar{x} - 1.96 * \left(\frac{\alpha}{\sqrt{n}}\right)\right]$$

which simplifies to

$$2 * 1.96 * \left(\frac{\sigma}{\sqrt{n}}\right).$$

As σ and n are known, the width can be calculated. If the interval is viewed as being too wide to be informative, we can change one of the values used, the z value, the sample size, or σ, in calculating the width to see if we can reduce it to an acceptable value. The two most common ways of reducing its size are by decreasing our level of confidence and by increasing the sample size; however, there are limits for both of these choices. Most researchers prefer to use at least the 95 percent level for the confidence interval although the use of the 90 percent level is not uncommon. To drop below the 90 percent level is usually unacceptable. Researchers may be able to increase the sample size somewhat, but the increase requires additional resources which are limited.

Suppose that we wish to estimate the mean systolic blood pressure of girls who are 120 to 130 cm (approximately 4 ft to 4 ft 3 in.) tall. We assume that the standard deviation of the systolic blood pressure variable for girls in this height group is 7 mm Hg. Given this information, how large a sample is required so that the 99 percent confidence interval is no more than 6 mm Hg wide? From above, we saw that the width of the confidence interval is

$$2 * (\text{a selected z value}) * \left(\frac{\sigma}{\sqrt{n}}\right).$$

Because we are using the 99 percent level, $1 - \alpha$ is 0.99 or α is 0.01. Then $z_{1-\alpha/2}$ is $z_{1-0.005}$ or $z_{0.995}$, which is 2.576. Thus, we have

$$2 * 2.576 * \left(\frac{7}{\sqrt{n}}\right) = 6$$

and we must solve this equation for n. Multiplying both sides by $\sqrt{n}$ gives

$$2 * 2.576 * 7 = 6 * \sqrt{n} \quad \text{or} \quad \sqrt{n} = \frac{(2 * 2.576 * 7)}{6}$$

and squaring both sides gives

$$n = \left(\frac{2 * 2.576 * 7}{6}\right)^2 = 36.13.$$

As n must be an integer, the next highest integer value, 37, is taken to be the value of n.

The formula for n, given a specified width, d, for the $(1 - \alpha) * 100$ percent confidence interval is

$$n = \left(\frac{2 * z_{1-\alpha/2} * \sigma}{d}\right)^2.$$

So far, we have been assuming that σ is known; however, in practice, we seldom know the population standard deviation. Sometimes the literature or a pilot study provides an estimate of its value which we may use for σ. In cases when we have no information about σ, the method shown in the following section is used to find the confidence interval for the mean.

2. Unknown Variance

When the population variance, σ^2, is unknown, it is reasonable to substitute its sample estimator, s^2, in the confidence interval calculation. There is a problem in doing this though. Although $(\bar{x} - \mu)/(\sigma/\sqrt{n})$ follows the standard normal distribution, $(\bar{x} - \mu)/(s/\sqrt{n})$ does not. In the first expression, there is only one random variable, $\bar{x}$, whereas the second expression involves the ratio of two random variables, $\bar{x}$ and s. We need to know the probability distribution for this ratio of random variables.

Fortunately, Gosset, who we encountered in Chapter 6, already discovered the distribution of $(\bar{x} - \mu)/(s/\sqrt{n})$. The distribution is called Student's t, crediting Student, the pseudonym used by Gosset, or more simply, the t distribution. For large values of n, sample values of s are very close to σ and, hence, the t distribution looks very much like the standard normal. For small values of n, however, the sample values of s vary considerably and the t and standard normal distributions have different appearances. Thus the t distribution has one parameter, the number of independent observations used in the calculation of s. In Chapter 4, we saw that this value was $n - 1$, and we called this value the degrees of freedom. Hence the parameter of the t distribution is the degrees of freedom associated with the calculation of the standard error. The degrees of freedom is shown as a subscript, that is, as t_{df}. For example, a t with 5 degrees of freedom is written as t_5.

The MINITAB commands in Box 7.3 allow us to compare the appearance of different t distributions with the standard normal distribution over the range -3.8 to 3.8. As we can see from these plots, the t distribution with one degree of freedom, the lowest curve, is considerably flatter, that is, there is more variability, than the standard normal distribution, the top curve in the figure. This is to be expected, as the sample mean divided by the sample standard deviation is more variable than the sample mean alone. As the degrees of freedom increases, the t distributions become

MINITAB BOX 7.3

```
MTB > set cl
DATA> -3.8:3.8/.1
DATA> end
MTB > pdf cl c2 (standard normal probability density function)
MTB > pdf cl c3;
SUBC> t 1.        (t distribution with df=1)
MTB > pdf cl c4;
SUBC> t 5.        (t distribution with df=5)
MTB > gplot;
SUBC> lines c2 cl;
SUBC> lines c3 cl;
SUBC> lines c4 cl.
```


closer and closer to the standard normal in appearance. The tendency for the t to approach the standard normal distribution as the number of degrees of freedom increases can also be seen in Table 7.3, which shows selected percentiles for several t distributions and the standard normal distribution. A more complete t table is found in Appendix Table B5.

Now that we know the distribution of $(\bar{x} - \mu)/(s/\sqrt{n})$, we can form confidence intervals for the mean even when the population variance is unknown. The form for the confidence interval is similar to that above for the mean with known variance except that s replaces σ and the t distribution is used instead of the standard normal distribution. Therefore, the lower and upper limits for the $(1 - \alpha) * 100$ percent confidence interval for the mean when the variance is unknown are $\bar{x} - t_{n-1,1-\alpha/2} * (s/\sqrt{n})$ and $\bar{x} + t_{n-1,1-\alpha/2} * (s/\sqrt{n})$, respectively.

Let us calculate the 90 percent confidence interval for the population mean of the systolic blood pressure for 5-year-old boys based on the sam-

TABLE 7.3 Selected Percentiles for Several t Distributions and the Standard Normal Distribution

Distribution	Percentile 0.80	0.90	0.95	0.99
t_1	1.376	3.078	6.314	31.821
t_5	0.920	1.476	2.015	3.365
t_{10}	0.879	1.372	1.813	2.764
t_{30}	0.854	1.310	1.697	2.457
t_{60}	0.848	1.296	1.671	2.390
t_{120}	0.845	1.289	1.658	2.358
Standard normal	0.842	1.282	1.645	2.326

ple data in column c1 above. A 90 percent [$= (1 - \alpha) * 100$ percent] confidence interval means that α is 0.10. Based on a sample of 60 observations, the sample mean was 94.75 and the sample standard deviation was 10.25 mm Hg. Thus we need the 95 th ($= 1 - \alpha/2$) percentile of a t distribution with 59 degrees of freedom; however, neither Table 7.3 nor Table B5 show the percentiles for a t distribution with 59 degrees of freedom. Based on the small changes in the t distribution for larger degrees of freedom, there should be little error if we use the 95th percentile for a t_{60} distribution. Therefore, the lower and upper limits are

$$94.75 - 1.671 * \left(\frac{10.25}{\sqrt{60}}\right) \quad \text{and} \quad 94.75 + 1.671 * \left(\frac{10.25}{\sqrt{60}}\right)$$

which are 92.54 and 96.96 mm Hg, respectively.

If we use MINITAB to find the 95th percentile value for a t_{59} distribution, we find its value is 1.6711. Hence, there is little error introduced in this example by using the percentiles from a t_{60} instead of a t_{59} distribution.

Corresponding to the ZINTERVAL (ZINT) command in MINITAB is the TINTERVAL (TINT) command for finding a confidence interval for the mean when the population variance is unknown. The command has the same form as the ZINT command; that is, you specify the level of confidence desired and the columns containing the data of interest. For example, suppose you wanted a 90 percent confidence interval for the population mean based on the sample data in column c1. The command is TINT 90 c1.

B. Confidence Interval for a Proportion

We are frequently exposed to the confidence interval for a proportion. Most surveys about opinions or voting intentions today report the margin of error. This quantity is simply one-half the width of the 95 percent confidence interval for the proportion. Finding the confidence interval for a proportion, π, can be based on either the binomial or normal distribution. The binomial distribution is generally used for smaller samples and it provides an exact interval, whereas the normal distribution is used with larger samples and provides an approximate interval.

1. Use of the Binomial Distribution

Suppose we wish to find a confidence interval for the proportion of restaurants that are in violation of local health ordinances. A simple random sample of 20 restaurants are selected, and of those, 4 are found to have violations. The sample proportion, p, which is equal to 0.20 (= 4/20), is the point estimate of π, the population proportion. How can we use this sample information to create the $(1 - \alpha) * 100$ percent confidence interval for the population proportion?

This is a binomial situation as there are only two outcomes for a restaurant: a restaurant either does or does not have a violation. The binomial variable is the number of restaurants with a violation and we have observed its value to be 4 in this sample.

The limits of the confidence interval for the proportion are those values that make this outcome appear to be unusual. Another way of stating this is that the lower limit is the proportion for which the probability of 4 or more restaurants is equal to $\alpha/2$. Correspondingly, the upper limit is the proportion for which the probability of 4 or fewer restaurants is equal to $\alpha/2$. Box 7.4 shows an attempt to find these values by trial and error.

Table B6 provides two charts that can be used to find the 95 and 99 percent confidence intervals. The charts eliminate the need for the calculations shown in Box 7.4. The detail provided by these charts is less than that shown above, but the accuracy from the charts should be sufficient for most applications.

Suppose that instead of the 90 percent confidence interval for the proportion of restaurants with violations of the health code, we wanted the 95 percent interval. We use Table B6a and, because the sample proportion is less than 0.50, we read across the bottom until we find the sample proportion value of 0.20. We then move up along the line corresponding to 0.20 until it intersects the first curve for a sample size of 20. As p is less than 0.50, we read the value of the lower limit from the left vertical axis; it is slightly less than 0.06. To find the upper limit, we continue up the vertical line corresponding to 0.20 until we reach the second curve for a sample size of 20. We read the upper limit from the left vertical axis and its value is slightly less than 0.44. The values from MINITAB are 0.0574 and 0.4364.

2. Use of the Normal Approximation to the Binomial

The sample proportion, p, is the binomial variable, x, divided by a constant, the sample size. As the normal distribution was shown in Chapter 6

MINITAB BOX 7.4

Suppose that we wish to find the 90 percent confidence interval. This means that α is 0.10 and $\alpha/2$ is 0.05. We wish to find the probability of being less than or equal to 4 and being greater than or equal to 4 for different binomial proportions. We start out with the upper limit. As the sample estimate's value is 0.20, we know the upper limit must be greater than this, and thus we try 0.35.

```
MTB > set cl;
DATA> 4
DATA> end
MTB > cdf cl;
SUBC> binom 20 .35.
        K   P( X LESS OR = K)
     4.00              0.1182
```

Because the value of 0.1182 is greater than 0.05, we try a larger value for the proportion.

```
MTB > cdf cl;
SUBC> binom 20 .40.
        K   P( X LESS OR = K)
     4.00              0.0510
```

This is very close to the value of 0.05, but we can examine a few more values for the proportion in an attempt to get closer to 0.05.

```
MTB > cdf cl;
SUBC> binom 20 .41.
        K   P( X LESS OR = K)
     4.00              0.0423

MTB > cdf cl;
SUBC> binom 20 .401.
        K   P( X LESS OR = K)
     4.00              0.0500
```

The value of the upper limit of the confidence interval is thus 0.401. For the lower limit, we want the probability of 4 or more restaurants to be equal to 0.05 or, equivalently, the probability of less than or equal to 3 to be 0.95. Therefore, we store the value of 3 in column c2.

```
MTB > set c2
DATA> 3
DATA> end
MTB > cdf c2;
SUBC> binom 20 .05.
        K   P( X LESS OR = K)
     3.00              0.9841
```

Using a proportion of 0.05 gives a probability that is too large. Therefore we increase the trial proportion.

```
MTB > cdf c2;
SUBC> binom 20 .07.
        K   P( X LESS OR = K)
     3.00              0.9529

MTB > cdf c2;
SUBC> binom 20 .071.
        K   P( X LESS OR = K)
     3.00              0.9508
```

The value of the lower limit is 0.071 to three decimal places. Thus, based on the point estimate of 0.20, the 90 percent confidence interval for the proportion of restaurants with violations of the health code is 0.071 to 0.401. Note that this interval is not symmetric about the point estimate.

```
MTB > cdf c2;
SUBC> binom 20 .0713.
        K   P( X LESS OR = K)
     3.00              0.9501
```

to be a good approximation to the distribution of x when the sample size is large enough, it also serves as a good approximation to the distribution of p. The variance of p is expressed in terms of the population proportion, π, and it is $\pi * (1 - \pi)/n$. Because π is unknown, we estimate the variance by substituting p for π in the formula.

The sample proportion can also be viewed as a mean as was discussed in Chapter 6. Therefore, the confidence interval for a proportion has the same form as that of the mean, and the limits of the interval are

$$p - z_{1-\alpha/2} * \sqrt{\frac{p * (1 - p)}{n}} + \frac{1}{2n} \quad \text{and} \quad p + z_{1-\alpha/2} * \sqrt{\frac{p * (1 - p)}{n}} + \frac{1}{2n}.$$

The $1/2n$ is the continuity correction term required because a continuous distribution is used to approximate a discrete distribution. For large values of n, the term has little effect and many authors drop it from the presentation of the confidence interval.

The local health department is concerned about the protection of children against diphtheria, pertussis, and tetanus (DPT). To determine if there is a problem in the level of DPT immunization, the health department decides to estimate the proportion immunized by drawing a simple random sample of 150 children who are 5 years old. If the proportion of children in the community who are immunized against DPT is clearly less than 75 percent, the health department will mount a campaign to increase the immunization level. If the proportion is clearly greater than 75 percent, the health department will shift some resources from immunization to prenatal care. The department decides to use a 99 percent confidence interval for the proportion to help it reach its decision.

Based on the sample, 86 families claimed that their child was immunized, and 54 said their child was not immunized. There were 10 children for whom immunization status could not be determined. As was mentioned in Chapter 3, there are several approaches to dealing with the unknowns. As there are only 10 unknowns, we shall ignore them in the calculations. Thus, the value of p is 0.614 (= 86/140) which is much lower than the target value of 0.75. If all 10 of the children with unknown status had been immunized, then p would have been 0.640, not much different from the value of 0.614, and still much less than the target value of 0.75.

The 99 percent confidence interval ranges from

$$0.614 - 2.576 * \sqrt{\frac{0.614 * 0.386}{140}} + \frac{1}{2 * 140}$$

to

$$0.614 + 2.576 * \sqrt{\frac{0.614 * 0.386}{140}} + \frac{1}{2 * 140}$$

or from 0.504 to 0.724. Because the upper limit of the 99 percent confidence interval is less than 0.75, the health department decides that it is highly unlikely that the proportion of 5-year-old children who are immunized is as large as 0.75. Therefore the health department will mount a campaign to increase the level of DPT immunization in the community.

If the issue facing the health department was whether or not to add resources to the immunization program, not to shift any resources away from the program, a one-sided interval could have been used. The 99 percent upper one-sided interval uses $z_{0.99}$ instead of $z_{0.995}$ in its calculation and it ranges from 0 to

$$0.614 + 2.326 * \sqrt{\frac{0.614 * 0.386}{140}} + \frac{1}{2 * 140} = 0.713.$$

This interval also does not contain 0.75. Therefore resources should be added to the immunization program.

The next section shows how to construct confidence intervals for crude and adjusted rates, parameters that are very similar to proportions.

C. Confidence Intervals for Crude and Adjusted Rates

In Chapter 4, we presented crude, specific, and direct and indirect adjusted rates; however, we did not present any estimate for the variance or standard deviation of a rate which is necessary for the calculation of the confidence interval. Therefore we begin a discussion of this material with a section on how to estimate the variance of a rate.

Rates are usually based on the entire population. If this is the case, there is really no need to calculate their variances or confidence intervals for them. However, we often view a population rate in some year as a sample in location or time. From this perspective, there is justification for calculating variances and confidence intervals. If the value of the rate is estimated from a sample, as is often done in epidemiology, then it is important to estimate the variance and the corresponding confidence interval for the rate. If the rate is based on the occurrence of a very small number of events, for example, deaths, the rate may be unstable and should not be used in this case. We shall say more about this later.

1. Variances of Crude and Adjusted Rates

The crude rate is calculated as the number of events in the population during the year divided by the midyear population. This is not really a proportion, but it is very similar to a proportion and we shall treat it as if it were a proportion. The variance of a sample proportion, p, is $\pi * (1 - \pi)/n$. Thus the *variance of a crude rate* is approximated by the product of the rate (converted to a decimal value) and one minus the rate divided by the population total.

From the data on rates in Chapter 4, we saw that the crude death rate for Harris County, Texas, in 1986 was 529.6 per 100,000. The corresponding estimated 1986 Harris County population was 2,294,550. Thus the estimated standard error, the square root of the variance estimate, for this crude death rate is

$$\sqrt{\frac{0.005296 * (1 - 0.005296)}{2{,}294{,}550}} = 0.0000479$$

or 4.8 deaths per 100,000 population.

The direct method's age-adjusted rate is a sum of the age-specific rates, sr_i's, in the population under study weighted by the age distribution, w_i's, in the standard population. In symbols, this is $\Sigma\ (w_i * sr_i)$, where w_i is the proportion of the standard population and sr_i is the age-specific rate in the ith age category. The age-specific rate is calculated as the number of events in the age category divided by the midyear population in that age category. Again, this is not a proportion, but it is very similar to a proportion. We approximate the variance of the age-specific rates by treating them as if they were proportions. As the w_i's are from the standard population which is usually very large and stable, we treat the w_i's as constants as far as the variance calculation is concerned. Because the age-specific rates are independent of one another, the *variance of the direct method's adjusted rate*, that is, the variance of this sum, is simply the sum of the individual variances

$$\text{var}(\Sigma\ w_i * \text{sr}_i) = \Sigma\ \text{var}(w_i * \text{sr}_i) = \Sigma\ w_i^2 * \text{var}(\text{sr}_i) = \Sigma\ w_i^2 * \left(\frac{\text{sr}_i * (1 - \text{sr}_i)}{n_i}\right)$$

where n_i is the number of persons in the ith age subgroup in the population under study.

Considering the Harris County mortality data as a sample in time, we can calculate the approximate variance of the direct method's age-adjusted death rate. The data to be used are the Harris County age-specific death rates along with the Harris County population totals and the U.S. population proportions by age from Table 4.10. Table 7.4 repeats the relevant data and shows the calculations. The entries in the last column are all quite small, less than 0.00000001; therefore, only their sum is shown. The standard error of the direct method's age-adjusted mortality rate is 0.00007 (= square root of variance). The direct method's age-adjusted rate was 860.9 deaths per 100,000 population, and the standard error of the rate is 7 deaths per 100,000. The magnitude of the standard error here is not unusual, and it shows why the sampling variation of the adjusted rate is often ignored in studies involving large samples.

TABLE 7.4 Calculation of the Approximate Variance for the Age-Adjusted Death Rate by the Direct Method for Harris County in 1986

Age i	Harris county age-specific rate sr_i	Harris county population n_i	U.S. population proportion w_i	$\frac{w_i^2 * sr_i * (1 - sr_i)}{n_i}$
0–4	0.002502	253,776	0.0753	
5–14	0.000196	469,446	0.1404	
15–24	0.000998	489,053	0.1618	
25–34	0.001468	640,813	0.1774	
35–44	0.002185	444,366	0.1372	
45–54	0.004647	275,007	0.0946	
55–64	0.013202	190,352	0.0922	
65–74	0.028328	111,870	0.0719	
≥75	0.081011	67,867	0.0491	
Total		2,942,550	0.9999	4.9×10^{-9}

For the indirect method, the adjusted rate can be viewed as the observed crude rate in the population under study multiplied by a ratio. The ratio is the standard population's crude rate divided by the rate obtained by weighting the standard population's age-specific rates by the age distribution from the study population. This ratio is viewed as a constant in terms of approximating the variance. Hence the *approximation of the variance of the indirect method's adjusted rate* is simply the square of the ratio multiplied by the variance of the study population's crude rate.

Using the data from Chapter 4, the standard population's (the U.S.) crude rate was 873.2 deaths per 100,000 population. Combination of the standard population's age-specific rates with the study population's (Harris County) age distribution yielded 534.6 deaths per 100,000 population. The crude rate in Harris County was 529.6 deaths per 100,000 population. Thus the approximate variance of the indirect method's age-adjusted mortality rate is

$$\left(\frac{0.008732}{0.005346}\right)^2 * \left[\frac{0.005296 * (1 - 0.005296)}{2{,}942{,}550}\right] = 0.0000000047.$$

The standard error of the indirect method's age-adjusted death rate is the square root of the variance, and it is also 0.00007.

2. Formation of the Confidence Interval

To form the confidence interval for a rate, we require knowledge of its sampling distribution. As we are treating crude and specific rates as if they are proportions, the confidence intervals for these rates will be based on the normal approximation as shown above for the proportion. Therefore, the confidence interval for the population crude rate (θ) is

$$\mathrm{cr} - z_{1-\alpha/2} * \sqrt{\frac{\mathrm{cr} * (1 - \mathrm{cr})}{n}} < \theta < \mathrm{cr} + z_{1-\alpha/2} * \sqrt{\frac{\mathrm{cr} * (1 - \mathrm{cr})}{n}}$$

where cr is the value of the crude rate based on the observed sample.

For example, the 95 percent confidence interval for the 1986 Harris County crude death rate is

$$0.005296 - 1.96 * 0.0000479 < \theta < 0.005296 + 1.96 * 0.0000479$$

or from 0.005202 to 0.005390. Thus the confidence interval for the crude death rate is 520.2 to 539.0 deaths per 100,000 population.

The confidence intervals for the rates from the direct and indirect methods of adjustment have the same form as that of the crude rate. For example, the 95 percent confidence interval for the direct method's 1986 age-adjusted mortality rate for Harris County is found by taking

$$860.9 \pm 1.96 * 7.0 = 860.9 \pm 13.7$$

and thus the limits are 847.2 to 874.6 deaths per 100,000 population.

3. Minimum Number of Events Required for a Stable Rate

As we mentioned above, rates based on a small number of occurrences of the event of interest may be unstable. To deal with this instability, a health agency for a small area often will combine its mortality data over several years. By using the estimated coefficient of variation, the estimated standard error of the estimate divided by the estimate and multiplied by 100 percent, we can determine when there are too few events for the crude rate to be stable.

Recall that in Chapter 4, we said that if the coefficient of variation was large, the data had too much variability for the measure of central tendency to be very informative. Values of the coefficient of variation greater than 30 percent—others might use slightly larger or smaller values—are often considered to be large. We use this idea with the crude rate to determine how many events are required so that the rate is stable.

For example, the coefficient of variation for the 1986 crude mortality rate of Harris County is 0.904 percent [= (0.0000479/0.005296) * 100 percent]. This rate, less than 1 percent, is very reliable from the coefficient of variation perspective. It turns out that the coefficient of variation of the crude rate can be approximated by $(1/\sqrt{d}) * 100$ percent where d is the number of events. For example, the total number of deaths for Harris County in 1986 was 12,152 and $(1/\sqrt{12152}) * 100$ percent is 0.907 percent, essentially the same result as above.

Thus we can use the approximation $(1/\sqrt{d}) * 100$ percent for the coefficient of variation. Setting this value equal to large values of the coefficient of variation, say 20, 30, and 40 percent, yields 25, 12, and 7 events, respectively. If the crude rate is based on fewer than 7 events, it certainly should not be reported. If we require that the coefficient of variation be less than 20 percent, there must be at least 25 occurrences of the event for the crude rate to be reported.

Besides forming confidence intervals for measures of central tendency or location, there is also interest in constructing confidence intervals for other population parameters. The following sections show the creation of confidence intervals for the population variance and the correlation coefficient.

D. Confidence Interval for the Variance

Besides being useful in describing the data, the variance is also frequently used in quality control situations. It is one way of stating how reliable the process under study is. For example, in Chapter 2 we presented data on the measurement of blood lead levels by different laboratories. We saw from that example that great variability in the measurements made by laboratories exists, and the variance is one way to characterize that variability. Variability within laboratories can be due to different technicians, failure to calibrate the equipment, and so on. It is critically important that measurements of the same sample within a laboratory have variability less than or equal to a prespecified small amount. Thus, based on the sample variance for a laboratory for measuring blood lead, we wish to determine whether or not the laboratory's variance is in compliance with the standards. The confidence interval for the population variance provides one method of doing this.

To construct the confidence interval for the population variance, we need to know the sampling distribution of its estimator, the sample variance, s^2. We can use MINITAB to examine the sampling distribution of s^2 for a few different situations. A reason for using MINITAB here is that it has the capability of storing a set of commands and then executing this set a number of times. The stored set of commands is called a *macro*. A macro is particularly useful when studying the sampling distribution of a statistic as shown in Box 7.5.

Box 7.6 shows the execution of the macro shown in Box 7.5

All 200 sample standard deviations are printed, but most have not been shown because they themselves are of little interest. The mean of the sample variance from the 200 observations is 25.50, very close to the population value of 25.00. There is tremendous variability in the sample variances as they range from 0.01 to 130.28 in value. This large variation is

MINITAB BOX 7.5

We use the STORE command to store in a file a set of commands that (1) draws a sample from a specified distribution, (2) calculates the sample standard deviation and variance, and (3) places the standard deviation and variance values in columns c1 and c2. The EXECUTE 'filename' command then causes the set of commands in the file to be executed.

```
MTB > store 'samdist'
STORE> noecho
STORE> random k1 c3;
STORE> normal 0 k2.
STORE> stdev c3 k3
STORE> let k4=k3*k3
STORE> let k5=k5+1
STORE> let c1(k5)=k3
STORE> let c2(k5)=k4
STORE> end
MTB > let k1=3
MTB > let k2=5
MTB > let k5=0
```

The NOECHO statement tells MINITAB not to print each command that it encounters. The constant k1 is the sample size to be drawn; here we have initially specified a very small sample of size 3. The constant k2 is the value of the standard deviation to be used; in this case we are drawing a sample of size 3 from a normal distribution with a mean of 0 and a standard deviation of 5. The constants k3 and k4 are the sample standard deviation and variance, respectively. The constant k5 is a counter which is initialized to be 0 and increases by 1 every time the 'samdist' set of commands is executed.

---

expected as each sample variance was based on only three observations. As was pointed out in Chapter 4, the average or expected value of the sample standard deviation slightly underestimates the population value and this is demonstrated here. The sample estimate is 4.42, slightly less than the population value of 5.

The histogram shows that most of the sample variance values are in the ranges 0 to 5 and 5 to 15, underestimates of the population value. Large proportions of the values are also in the ranges 15 to 25 and 25 to 35. The distribution is very asymmetric with a long tail to the right, and it does not look like either a normal or a *t* distribution.

In Box 7.7 we examine drawing a sample of 21 observations from the same normal distribution as above. The reduced variation in the sample variance and standard deviation reflect the increase in the sample size from 3 to 21 observations. The mean of the sample standard deviations is also closer to the population value with the increase in the sample size. This

MINITAB BOX 7.6

The EXECUTE command is used to run the macro that draws 200 samples of size 3.

```
MTB > execute 'samdist' 200
   ST.DEV. =        10.107
   ST.DEV. =        5.9390
       .               .
       .               .
   ST.DEV. =        5.0300

MTB > desc cl c2
                 N      MEAN    MEDIAN    TRMEAN    STDEV    SEMEAN
C1             200     4.421     4.474     4.328    2.447     0.173
C2             200     25.50     20.02     22.72    25.57      1.81

               MIN       MAX        Q1        Q3
C1           0.095    11.414     2.631     5.870
C2            0.01    130.28      6.92     34.45

MTB > hist c2

Histogram of C2   N = 200

Midpoint   Count
       0      44  ********************************************
      10      43  *******************************************
      20      32  ********************************
      30      32  ********************************
      40      16  ****************
      50      11  ***********
      60       5  *****
      70       3  ***
      80       5  *****
      90       4  ****
     100       2  **
     110       0
     120       2  **
     130       1  *
```

agrees with our expectations from Chapter 4. The sample variance now ranges from 6.9 to 52.4, a much smaller range than from 0.01 to 130.28. The histogram no longer shows so many small sample variances and the categories with the greatest frequencies are 17.5 to 22.5, 22.5 to 27.5, and 27.5 to 32.5. The distribution is not so asymmetric and the tail to the right is much shorter than in the first histogram.

MINITAB BOX 7.7

```
MTB > let k1=21
MTB > let k5=0
MTB > execute 'samdist' 200
      (We have not shown the standard deviations.)

MTB > desc c1 c2
                 N      MEAN    MEDIAN    TRMEAN     STDEV    SEMEAN
C1             200    5.0269    4.9667    5.0085    0.8618    0.0609
C2             200    26.009    24.668    25.543     8.938     0.632

               MIN       MAX        Q1        Q3
C1          2.6297    7.2377    4.4713    5.5296
C2           6.915    52.385    19.992    30.577

MTB > hist c2

Histogram of C2   N = 200
Midpoint   Count
       5       1  *
      10       6  ******
      15      25  *************************
      20      49  *************************************************
      25      39  ***************************************
      30      38  **************************************
      35      21  *********************
      40      10  **********
      45       5  *****
      50       6  ******
```

In Box 7.8 we increase the sample size to 61. The sample statistics show much less variability in the sample variance and standard deviation, reflecting the increase in the sample size from 21 to 61. For example, the interquartile range containing the middle 50 percent of the values of the sample variances goes from 21.15 to 26.68 for the sample size of 61, compared with 19.99 to 30.58 for an n of 21 and 6.92 to 34.45 for an n of 3. The histogram reflects this reduction in variability as well. We can see that the sampling distributions for the three sample sizes are very different; that is, they depend on the sample size. The distributions, particularly for the smaller sample sizes, also are very nonnormal.

It appears that the distribution of the sample variance does not match any of the probability distributions we have encountered so far. Fortunately, when the data come from a normal distribution, the distribution of the sample variance is known. The sample variance, multiplied by

MINITAB BOX 7.8

```
MTB > let k1=61
MTB > let k5=0
MTB > execute 'samdist' 200 (Again the sample standard deviations are not
      shown.)

MTB > desc c1 c2
                 N      MEAN    MEDIAN    TRMEAN     STDEV    SEMEAN
C1             200    4.9116    4.9376    4.9087    0.4527    0.0320
C2             200    24.328    24.380    24.214     4.487     0.317

               MIN       MAX        Q1        Q3
C1          3.6778    6.3761    4.5987    5.1656
C2          13.526    40.655    21.148    26.684

MTB > hist c2

Histogram of C2   N = 200
Midpoint   Count
      14       1  *
      16       7  *******
      18      14  **************
      20      24  ************************
      22      35  ***********************************
      24      32  ********************************
      26      40  ****************************************
      28      21  *********************
      30      13  *************
      32       7  *******
      34       3  ***
      36       0
      38       1  *
      40       2  **
```

$(n - 1)/\sigma^2$, follows a chi-square (χ^2) distribution. Two eminent 19th-century French mathematicians, Laplace and Bienaymé, played important roles in the development of the chi-square distribution. Karl Pearson, an important British statistician previously encountered in connection with the correlation coefficient, popularized the use of the chi-square distribution in the early 20th century. As we saw above, the distribution of the sample variance depends on the sample size, actually on the number of independent observations (degrees of freedom) used to calculate s^2. Therefore Table B7 shows percentiles of the chi-square distribution for different values of the degrees of freedom parameter.

To create a confidence interval for the population variance, we begin with the probability statement

$$\Pr\left\{\chi^2_{n-1,\alpha/2} < \frac{(n-1)s^2}{\sigma^2} < \chi^2_{n-1,1-\alpha/2}\right\} = 1 - \alpha.$$

This statement indicates that the confidence interval will be symmetric in the sense that the probability of being less than the lower limit is the same as that of being greater than the upper limit; however, the confidence limit will not be symmetric about s^2. This probability statement is in terms of s^2 however, and we want a statement about σ^2. To convert it to a statement about σ^2, we first divide all three terms in the bracket by $(n-1)s^2$. This yields

$$\Pr\left\{\frac{\chi^2_{n-1,\alpha/2}}{(n-1)s^2} < \frac{1}{\sigma^2} < \frac{\chi^2_{n-1,1-\alpha/2}}{(n-1)s^2}\right\} = 1 - \alpha.$$

The interval is now about $1/\sigma^2$, not σ^2. Therefore, we next take the reciprocal of all three terms which changes the direction of the inequalities. For example, we know that 3 is greater than 2, but the reciprocal of 3, which is 1/3 or 0.333, is less than the reciprocal of 2, which is 1/2 or 0.500. Thus we have

$$\Pr\left\{\frac{(n-1)s^2}{\chi^2_{n-1,\alpha/2}} > \sigma^2 > \frac{(n-1)s^2}{\chi^2_{n-1,1-\alpha/2}}\right\} = 1 - \alpha.$$

and reversing the directions of the inequalities to have the smallest term on the left yields

$$\Pr\left\{\frac{(n-1)s^2}{\chi^2_{n-1,1-\alpha/2}} < \sigma^2 < \frac{(n-1)s^2}{\chi^2_{n-1,\alpha/2}}\right\} = 1 - \alpha.$$

It is also possible to create one-sided confidence intervals for the population variance. For example, the lower one-sided confidence interval for the population variance is

$$\frac{(n-1)s^2}{\chi^2_{n-1,1-\alpha}} < \sigma^2 < \infty.$$

Let us apply this formula to an example. From 1988 to 1991, eight persons in Massachusetts were identified as having vitamin D intoxication due to receiving large doses of vitamin D_3 in fortified milk (4). The problem was traced to a local dairy which had tremendous variability in the amount of vitamin D added to individual bottles of milk. Homogenized whole milk showed the greatest variability based on measurements made in April and June 1991, with a low value of less than 40 IU and a high of 232,565 IU of vitamin D_3 per quart. These values are contrasted with the requirement for at least 400 IU (10 μg) to no more than 500 IU of vitamin D per quart of milk in Massachusetts.

The Food and Drug Administration (FDA) found poor compliance with the requirement for 400 IU of vitamin D per quart of vitamin D-fortified milk in a 1988 survey (5). Based on this poor compliance, the FDA urged that the problem be corrected; otherwise it would institute a regulatory program. Suppose that compliance is defined in terms of the mean and standard error of the mean vitamin D concentration in milk. The mean concentration should be 400 IU with a variance of less than 1600 IU. To determine if a milk producer is in compliance, a simple random sample of milk cartons from the producer is selected and the amount of vitamin D in the milk is ascertained. It is decided that if the 90 percent lower one-sided confidence interval for the variance contains 1600 IU, the process used by the producer to add vitamin D is within the acceptable limits for variability. This is an approach for determining compliance that greatly favors the producer.

A random sample of 30 cartons is selected and the sample variance for the vitamin D in the milk is found to be 1700 IU. The 90 percent confidence interval uses $\chi^2_{29,0.90}$, where the first subscript is the degrees of freedom parameter and the second subscript is the percentile value. The value from Table B7 is 39.09. The lower limit is found from (29 * 1700)/39.09, which gives the value of 1261.3. As the 90 percent confidence interval does contain 1600 IU, the producer is said be in compliance with the variability requirement. To find that a producer is not in compliance requires that the sample variance be at least 2156.5.

A key assumption in calculating the confidence interval for the population variance is that the data come from a normal distribution. If the data are from a very nonnormal distribution, the use of the above formula for calculating the confidence interval can be very misleading.

To find the confidence interval for the population standard deviation, we take the square root of the variance's confidence interval limits. Thus the lower limit of the confidence interval for σ in the above example is 35.5 IU.

E. Confidence Interval for the Pearson Correlation Coefficient

In Chapter 4, we presented ρ, the Pearson correlation coefficient, which is used in assessing the strength of the linear relationship between two jointly normally distributed variables. We presented a formula for finding r, the sample Pearson correlation coefficient. We also found the correlation between protein and total fat, based on the 33 observations in Table 4.1, to be 0.648, suggestive of a strong positive relation. Although this point estimate of ρ is informative, more information is provided by the interval estimate. For example, if the sampling variation of r is so large that the 95 percent confidence interval for ρ contains zero, we would not be impressed by the strength of the relationship between total fat and protein.

It turns out that the sampling distribution of r is not easily characterized; however, the father of modern statistics, Ronald Fisher, showed that a transformation of r approximately follows a normal distribution. This transformation is

$$z' = 0.5 * [\log_e(1 + r) - \log_e(1 - r)],$$

and it provides the basis for the confidence interval for ρ. The mean of z' is $[\log_e(1 + \rho) - \log_e(1 - \rho)]$ and its standard deviation, $\sigma_{z'}$, is $1/\sqrt{(n - 3)}$. Note that for convenience, log_e is often written as ln and we do that below. Thus we can employ the procedures we have used above for finding the confidence interval for the transformed value of ρ, that is,

$$z' - z_{1-\alpha/2} * \sigma_{z'} < 0.5 * [\ln (1 + \rho) - \ln (1 - \rho)] < z' - z_{1-\alpha/2} * \sigma_{z'}.$$

There is one simplification we can make that allows us to have to take only one natural logarithm in the calculation instead of finding two natural logarithms. In the presentation of the geometric mean in Chapter 4, we saw that the sum of logarithms of two terms is the logarithm of the product of the terms, that is,

$$\ln x_1 + \ln x_2 = \ln (x_1 * x_2).$$

In the same way, the difference of logarithms of two terms is the logarithm of the quotient of the terms, that is,

$$\ln x_1 - \ln x_2 = \ln \left(\frac{x_1}{x_2}\right).$$

Thus we have the relationship

$$z' = 0.5 * [\ln (1 + r) - \ln (1 - r)] = 0.5 * \ln \left[\frac{1 + r}{1 - r}\right].$$

Let us apply these formulas for finding the 95 percent confidence interval for the correlation between total fat and protein. As r is 0.648, z' is

$$0.5 * \ln \left(\frac{1 + 0.648}{1 - 0.648}\right) = 0.5 * \ln \left(\frac{1.648}{0.352}\right) = 0.5 * \ln 4.682$$
$$= 0.5 * 1.5437 = 0.77185.$$

The standard deviation of z' is $1/\sqrt{30}$, which is 0.18257. Thus the interval for $0.5 * \ln [(1 + \rho)/(1 - \rho)]$ is $0.77185 - 1.96 * 0.18257$ to $0.77185 + 1.96 * 0.18257$, or 0.4140 to 1.1297.

These calculations are easily performed with MINITAB as shown in Box 7.9.

MINITAB BOX 7.9

```
 MTB > let k1=0.648
 MTB > let k2=(1+k1)/(1-k1)
 MTB > let k3=0.5*loge(k2)
 MTB > print k3
K3          0.77185

 MTB > let k4=1/sqrt(30)
 MTB > let k5=1.96*k4
 MTB > let k6=k3-k5
 MTB > let k7=k3+k5
 MTB > print k6 k7

K6          0.413998
K7          1.12969
```

To find the confidence interval for ρ, we first perform the inverse transformation on twice the lower and upper limits of the interval just calculated. The inverse transformation of the natural logarithm, ln, is the exponential transformation. This means that

$$\exp(\ln x) = x.$$

After obtaining the exponential of twice a limit, call it a, further manipulation leads to the following equation:

$$\text{limit for } \rho = \frac{a - 1}{a + 1}.$$

The exponential of twice the lower limit, that is, two times 0.4140, is the exponential of 0.8280, which is 2.28874, and this is the value used for a for the lower limit. The lower limit for ρ is then

$$\frac{2.28874 - 1}{2.28874 + 1} = 0.392.$$

The exponential of twice the upper limit, that is, two times 1.1297, is the exponential of 2.2594, which is 9.57734, and this is the value used for a for the upper limit. The upper limit for ρ is then

$$\frac{9.57734 - 1}{9.57734 + 1} = 0.811.$$

Therefore, the 95 percent confidence for the Pearson correlation coefficient between total fat and protein in the population is 0.392 to 0.811. Thus it is reasonable to conclude that there is a strong positive association between

MINITAB BOX 7.10

```
MTB > let k8=2*k6
MTB > let k9=2*k7
MTB > let k10=exp(k8)
MTB > let k11=exp(k9)
```

(The constants k10 and k11 are the exponentials of twice the lower and upper limits, respectively.)

```
MTB > let k12=(k10-1)/(k10+1)
MTB > let k13=(k11-1)/(k11+1)
MTB > print k12 k13
K12        0.391862
K13        0.810913
```

total fat and protein in the diet of suburban middle school boys in the Houston area.

These calculations are again easily performed in MINITAB as shown in Box 7.10.

This material also applies to the Spearman correlation coefficient for sample sizes greater than or equal to 10.

So far, all the confidence intervals presented have been for a single parameter. The following sections address confidence intervals for the comparison of parameters from two populations.

F. Confidence Interval for the Difference of Two Means

1. Independent Means

We often wish to compare the mean from one population with that of another population. Examples include the following. Is the mean change in blood pressure for men with mild to moderate hypertension the same for men taking different doses of an angiotensin-converting enzyme inhibitor? Is the mean length of stay in a psychiatric hospital equal for patients with the same diagnosis but under the care of two different psychiatrists? Given the following, there is an interest in the mean change in air pollution, specifically, in carbon monoxide, from 1991 to 1992 for neighboring states A and B. There was no change in gasoline formulation in State A, whereas on January 1, 1992, State B required that gasoline consist of 10 percent ethanol during the November to March period.

One reason for interest in the confidence interval for the difference of two means is that it can be used to address the question of the equality of the two means. If there is no difference in the two population means, the confidence interval for their difference is likely to include zero.

a. ***Known Variances***

The confidence interval for the difference of two means has the same form as that for a single mean, that is, it is the difference of the sample means plus or minus some distribution percentile times the standard error of the difference of the sample means. Let us convert these words to symbols. Suppose that we draw samples of sizes n_1 and n_2 from two independent populations. All the observations are assumed to be independent of one another; that is, the value of one observation does not affect the value of any other observation. The unknown population means are μ_1 and μ_2, the sample means are $\bar{x}_1$ and $\bar{x}_2$, and the known population variances are $\sigma_1{}^2$ and $\sigma_2{}^2$, respectively. The variances of the sample means are $\sigma_1{}^2/n_1$ and $\sigma_2{}^2/n_2$, respectively. As the means are from two independent populations, the standard error of the difference of the sample means is the square root of the sum of the variances of the two sample means:

$$\sqrt{\frac{\sigma_1^2}{n_1} + \frac{\sigma_2^2}{n_2}}.$$

The central limit theorem implies that the difference of the sample means will approximately follow the normal distribution for reasonable sample sizes. This can be expressed as

$$Z = \frac{(\bar{x}_1 - \bar{x}_2) - (\mu_1 - \mu_2)}{\sqrt{\sigma_1^2/n_1 + \sigma_2^2/n_2}}.$$

Therefore, the $(1 - \alpha) * 100$ percent confidence interval for the difference of population means, $\mu_1 - \mu_2$, ranges

$$\text{from } \left((\bar{x}_1 - \bar{x}_2) - z_{1-\alpha/2} * \sqrt{\frac{\sigma_1^2}{n_1} + \frac{\sigma_2^2}{n_2}}\right) \text{ to } \left((\bar{x}_1 - \bar{x}_2) - z_{1-\alpha/2} * \sqrt{\frac{\sigma_1^2}{n_1} + \frac{\sigma_2^2}{n_2}}\right).$$

Suppose we wish to construct a 95 percent confidence interval for the effect of different doses of ramipril, an angiotensin-converting enzyme inhibitor, used in treating high blood pressure. A study reported changes in diastolic blood pressure using the values at the end of a 4-week run-in period as the baseline and measured blood pressure after 2, 4, and 6 weeks of treatment (6). We shall form a confidence interval for the difference in mean decreases from baseline to 2 weeks after treatment was begun between doses of 1.25 and 5 mg of ramipril. The sample mean decreases are 10.6 ($\bar{x}_1$) and 14.9 mm Hg ($\bar{x}_2$) for the 1.25- and 5-mg doses, respectively, and n_1 and n_2 are both equal to 53. Both σ_1 and σ_2 are assumed to be 9 mm Hg. The 95 percent confidence interval for $\mu_1 - \mu_2$ is calculated as ranging

$$\text{from } \left((10.6 - 14.9) - 1.96 * \sqrt{\frac{81}{53} + \frac{81}{53}}\right)$$

$$\text{to } \left((10.6 - 14.9) + 1.96 * \sqrt{\frac{81}{53} + \frac{81}{53}}\right),$$

or from −7.98 to −0.62. The value of 0 is not contained in this interval. As the difference in mean decreases is negative, it appears that the 5-mg dose of ramipril is associated with a greater decrease in diastolic blood pressure during the first 2 weeks of treatment when considering only these two doses.

b. *Unknown but Equal Population Variances*

If the variances are unknown but assumed to be equal, data from both samples can be combined to form an estimate of the common population variance. Use of the sample estimator of the variance calls for use of the t, instead of the normal, distribution in the formation of the confidence interval. The pooled estimator of the common variance, s_p^2, is defined as

$$s_p^2 = \frac{\sum_{i=1}^{n_1} (x_{1i} - \bar{x}_1)^2 + \sum_{i=1}^{n_2} (x_{2i} - \bar{x}_2)^2}{n_1 + n_2 - 2}$$

and this can be rewritten as

$$s_p^2 = \frac{(n_1 - 1)s_1^2 + (n_2 - 1)s_2^2}{(n_1 - 1) + (n_2 - 1)} = \frac{(n_1 - 1)s_1^2 + (n_2 - 1)s_2^2}{n_1 + n_2 - 2}.$$

The pooled estimator is a weighted average of the two sample variances, weighted by the respective degrees of freedom associated with the individual sample variances and divided by sum of the degrees of freedom associated with each of the two sample variances.

Now that we have an estimator of σ^2, we can use it in estimating the standard error of the difference of the sample means, $\bar{x}_1$ and $\bar{x}_2$. As we are assuming that the population variances for the two groups are the same, the standard error of the difference of the sample means is

$$\sqrt{\frac{\sigma^2}{n_1} + \frac{\sigma^2}{n_2}} = \sigma \sqrt{\frac{1}{n_1} + \frac{1}{n_2}}$$

and its estimator is

$$s_p \sqrt{\frac{1}{n_1} + \frac{1}{n_2}}.$$

The corresponding t statistic is

$$t = \frac{(\bar{x}_1 - \bar{x}_2) - (\mu_1 - \mu_2)}{s_p \sqrt{1/n_1 + 1/n^2}}$$

and the $(1 - \alpha) * 100$ percent confidence interval for $\mu_1 - \mu_2$ ranges from

$$\left((\bar{x}_1 - \bar{x}_2) - t_{n-2,1-\alpha/2}\, s_p \sqrt{\frac{1}{n_1} + \frac{1}{n_2}}\right) \text{ to } \left((\bar{x}_1 - \bar{x}_2) + t_{n-2,1-\alpha/2}\, s_p \sqrt{\frac{1}{n_1} + \frac{1}{n_2}}\right)$$

where n is the sum of n_1 and n_2.

Suppose that we wish to calculate the 95 percent confidence interval for the difference in the proportion of caloric intake that comes from fat for fifth and sixth grade boys compared with seventh and eighth grade boys in suburban Houston. The sample data shown in Table 4.1 will be used in the calculation. The proportion of caloric intake that comes from fat is found by converting the grams of fat to calories by multiplying by 9 (9 calories result from 1 g of fat) and then dividing by the number of calories consumed. Table 7.5 shows these variables.

The sample mean for the 14 fifth and sixth grade boys is 0.329, compared with 0.353 for the 19 seventh and eighth grade boys. These values of percent of intake from fat are slightly above the recommended value of 30 percent (7, p. 51). The corresponding standard deviations are 0.0895 and 0.0974, which support the assumption of equal variances.

TABLE 7.5 Total Fat,[a] Calories, and the Proportion of Calories from Total Fat for the 33 Boys in Table 4.1

Grades 7 and 8			Grades 5 and 6		
Total fat	**Calories**	**Proportion from fat**	**Total fat**	**Calories**	**Proportion from fat**
567	1823	0.311	1197	3277	0.365
558	2007	0.278	891	2039	0.437
297	1053	0.282	495	2000	0.248
1818	4322	0.421	756	1781	0.424
747	1753	0.426	1107	2748	0.403
927	2685	0.345	792	2348	0.337
657	2340	0.281	819	2773	0.295
2043	3532	0.578	738	2310	0.319
1089	2842	0.383	738	2594	0.285
621	2074	0.299	882	1898	0.465
225	1505	0.150	612	2400	0.255
783	2330	0.336	252	2011	0.125
1035	2436	0.425	702	1645	0.427
1089	3076	0.354	387	1723	0.225
621	1843	0.337			
666	2301	0.289			
1116	2546	0.438			
531	1292	0.411			
1089	3049	0.357			

[a]Total fat has been converted to calories by multiplying the number of grams by 9.

The estimate of the pooled standard deviation is therefore

$$s_p = \sqrt{\frac{13 * 0.0895^2 + 18 * 0.0974^2}{14 + 19 - 2}} = 0.094.$$

The estimate of the standard error of the difference of the sample means is

$$0.094 * \sqrt{\frac{1}{14} + \frac{1}{19}} = 0.033.$$

To find the confidence interval, we require $t_{31,0.975}$. This value is not shown in Table B5, but based on the values for 29 and 30 degrees of freedom, an approximate value for it is 2.04. Therefore, the lower and upper limits are

$$(0.329 - 0.353) - 2.04 * 0.033 \quad \text{and} \quad (0.329 - 0.353) + 2.04 * 0.033$$

which are -0.092 and 0.044. As zero is contained in the 95 percent confidence interval, there does not appear to be a difference in the mean proportions of calories that come from fat for fifth and sixth grade boys compared with seventh and eighth grade boys in suburban Houston.

These calculations are easily carried out with MINITAB as shown in Box 7.11.

MINITAB BOX 7.11

Recall that column c2 contains the caloric intake and c3 contains the total fat values. These data are arranged such that the values for the seventh and eighth graders are followed by the values for the fifth and sixth graders. Column c7 is the proportion of total calories that come from fat, and c8 is an indicator column that identifies the seventh and eighth graders (c8 = 0) and the fifth and sixth graders (c8 = 1). The COPY command used here has two columns, for example, c7 and c9. Some or all of the data from c7 are copied into c9. If the USE subcommand is specified, only a subset of the data in c7 are copied into c9. The subset includes only the values in c7 for which the corresponding values in c8 are 0. Thus columns c9 and c10 contain the proportions of total calories that come from fat for these two groups of boys. The DESCRIBE command is then used to obtain the sample means and standard deviations required for the calculations, and the INVCDF command is used to obtain the value of $t_{31,0.975}$.

```
MTB > let c7=9*c3/c2
MTB > set c8
DATA> 19(0) 14(1)
DATA> end
MTB > copy c7 c9;
SUB > use c8=0.
MTB > copy c7 c10;
SUB > use c8=1.
MTB > desc c9 c10
MTB > invcdf 0.975;
SUBC> t 31.
```

c. Unknown and Unequal Population Variances

If the population variances are different, this poses a problem. There is a procedure for obtaining an exact confidence interval for the difference in the means when the population variances are unequal, but it is much more complex than the other methods in this book (8, pp. 141–146). Because of this complexity, most researchers use an approximate approach to the problem. The following shows one of the approximate approaches.

As the population variances are unknown, we again use a *t*-like statistic. This statistic is

$$t' = \frac{(\bar{x}_1 - \bar{x}_2) - (\mu_1 - \mu_2)}{\sqrt{s_1^2/n_1 + s_2^2/n_2}}.$$

The *t* distribution with the degrees of freedom shown next can be used to obtain the percentiles of the t' statistic. The degrees of freedom value, df, is

$$\text{df} = \frac{(s_1^2/n_1 + s_2^2/n_2)^2}{(s_1^2/n_1)^2/(n_1 - 1) + (s_2^2/n_2)^2/(n_2 - 1)}.$$

This value for the degrees of freedom was suggested by Satterthwaite (9). It is unlikely to be an integer and it should be rounded to the nearest integer.

The approximate $(1 - \alpha) * 100$ percent confidence interval for the difference of two independent means when the population variances are unknown and unequal is

$$(\bar{x}_1 - \bar{x}_2) - t_{\text{df},1-\alpha/2}\, s_{\bar{x}_1-\bar{x}_2} < (\mu_1 - \mu_2) < (\bar{x}_1 - \bar{x}_2) + t_{\text{df},1-\alpha/2}\, s_{\bar{x}_1-\bar{x}_2}$$

where the estimate of the standard error of the difference of the two sample means is

$$s_{\bar{x}_1-\bar{x}_2} = \sqrt{\frac{s_1^2}{n_1} + \frac{s_2^2}{n_2}}.$$

In Exercise 4.5, we presented survival times from Exercise Table 3.3 in Lee (10) on 71 patients who had a diagnosis of either acute myeloblastic leukemia (AML) or acute lymphoblastic leukemia (ALL). In one part of the exercise, we asked for additional variables that should be considered before comparing the survival times of these two diagnostic groups of patients. One such variable is age. Let us examine these two groups to determine if there appears to be an age difference. If there is a difference, it must be taken into account in the interpretation of the data. To examine if there is a difference, we find the 99 percent confidence interval for the difference of the mean ages of the AML and ALL patients. As we have no knowledge about the variation in the ages, we assume that the variances will be different. Table 7.6 shows the ages and survival times for these 71 patients.

The sample mean age for the AML patients, $\bar{x}_1$, is 49.86, and s_1 is 16.51 based on the sample size, n_1, of 51 patients. The sample mean, $\bar{x}_2$, for the 20 ALL patients is 36.65 years, and s_2 is 17.85. This is the information needed

TABLE 7.6 Ages and Survival Times of the AML and ALL Patients[a]

AML patients:													
Age	20	25	26	26	27	27	28	28	31	33	33	33	34
	36	37	40	40	43	45	45	45	45	47	48	50	50
	51	52	53	53	56	57	59	59	60	60	61	61	61
	62	63	65	71	71	73	73	74	74	75	77	80	
Survival time (months)	18	31	31	31	36	01	09	39	20	04	45	36	12
	08	01	15	24	02	33	29	07	00	01	02	12	09
	01	01	09	05	27	01	13	01	05	01	03	04	01
	18	01	02	01	08	03	04	14	03	13	13	01	
ALL patients:													
Age	18	19	21	22	26	27	28	28	28	28	34	36	37
	47	55	56	59	62	83	19						
Survival time (months)	16	25	01	22	12	12	74	01	16	09	21	09	64
	35	01	07	03	01	01	22						

[a]Age and survival times are in the same order.

to calculate the confidence interval. Let us first calculate the sample estimate of the standard error of the difference of the means:

$$s_{\bar{x}_1-\bar{x}_2} = \sqrt{\frac{16.51^2}{51} + \frac{17.85^2}{20}} = 4.61.$$

We next calculate the degrees of freedom, df, to be used and we find it from

$$\text{df} = \frac{(16.51^2/51 + 17.85^2/20)^2}{\left(\frac{(16.51^2/51)^2}{51-1} + \frac{(17.85^2/20)^2}{20-1}\right)}$$

which equals 32.501, and this is rounded to 33. The 99.5 percentile of the t distribution with 33 degrees of freedom is about midway between the values of 2.750 (30 degrees of freedom) and 2.724 (35 degrees of freedom) in Appendix Table B5. We interpolate and use a value of 2.7344 for the 99.5 percentile of the t distribution with 33 degrees of freedom. Therefore, the 99 percent confidence interval for the difference of the mean ages is

$$(49.86 - 36.65) - 2.7344 * 4.61 < \mu_1 - \mu_2 < (49.86 - 36.65) + 2.7344 * 4.61$$

or

$$0.60 < \mu_1 - \mu_2 < 25.82.$$

As zero is not contained in this confidence interval, there is an indication of a difference in the mean ages. If the survival patterns differ between patients with these two diagnoses, it may be due to a difference in the age of the patients.

How large would the confidence interval have been if we had assumed that the unknown population variances were equal? Using the approach in Section F.1.b., the pooled estimate of variance, s_p^2, is

$$\frac{(51 - 1) * 16.51^2 + (20 - 1) * 17.85^2}{51 + 20 - 2} = 285.26$$

The pooled estimate of the standard deviation is thus 16.89, and this leads to an estimate of the standard error of the difference of the two means of

$$16.89 * \sqrt{\frac{1}{51} + \frac{1}{20}} = 4.456.$$

Thus the confidence interval, using an approximation of 2.65 to the 99.5 percentile of the t distribution with 69 degrees of freedom, is

$$(49.86 - 36.65) - 2.65 * 4.456 < \mu_1 - \mu_2 < (49.86 - 36.65) + 2.65 * 4.456$$

or

$$1.20 < \mu_1 - \mu_2 < 25.02.$$

This interval is slightly narrower than the confidence interval found above; however, both intervals lead to the same conclusion about the ages in the two diagnosis groups.

In practice, we usually know little about the magnitude of the population variances. This makes it difficult to decide which approach, equal or unequal variances, should be used. We recommend that the unequal variances approach be used in those situations when we have no knowledge about the variances and no reason to believe that they are equal. Fortunately, as we saw above, often there is little difference in the results of the two approaches. Some textbooks and computer packages recommend that we first test to see if the two population variances are equal and then decide which procedure to use. Several studies have been conducted recently and conclude that this should not be done (11–13).

These sections have focused on the situation in which two population means are independent of one another, for example, men who have received different doses of medication, boys in different classes, and patients with different diagnoses. The next section deals with the creation of a confidence interval for two dependent means.

2. Confidence Interval for the Difference of Two Dependent Means

Dependent means occur in a variety of situations. One example of interest comprises a preintervention measurement, some intervention, and a postintervention measurement. Another dependent mean situation occurs when there is a matching or pairing of subjects with similar characteristics.

One subject in the pair receives one type of treatment and the other member in the pair receives another type of treatment. Measurements on the variable of interest are made on both members of the pair. In both of these situations, there is some relationship between the values of the observations in a pair. For example, the preintervention measurement for a subject is likely to be correlated with the postintervention measurement on the same subject. If there is a nonzero correlation, this violates the assumption of independence of the observations. To deal with this relationship (dependency), we form a new variable which is the difference of the observations in the pair. We then analyze the new variable, the difference of the paired observations.

Consider the blood pressure example presented earlier. Suppose that we focus on the 1.25-mg dose of ramipril. We have a value of the subject's blood pressure at the end of a 4-week run-in period and the corresponding value after 2 weeks of treatment for 53 subjects. There are 106 measurements, but only 53 pairs of observations and only 53 differences for analysis. The mean decrease in diastolic blood pressure after 2 weeks of treatment for the 53 subjects is 10.6 mm Hg, and the sample standard deviation of the difference is 8.5 mm Hg. The confidence interval for this difference has the form of the confidence interval for the mean from a single population. If the population variance is known, we use the normal distribution; otherwise we use the t distribution. We assumed that the population standard deviation was 9 mm Hg above and we use that value here. Thus the confidence interval will use the normal distribution, that is,

$$\bar{x}_d - z_{1-\alpha/2} * \left(\frac{\sigma_d}{\sqrt{n}}\right) < \mu_d < \bar{x}_d + z_{1-\alpha/2} * \left(\frac{\sigma_d}{\sqrt{n}}\right)$$

where the subscript d denotes difference.

Let us calculate the 90 percent confidence interval for the mean decrease in diastolic blood pressure. Table B4 shows that the 95th percentile of the standard normal is 1.645. Thus the confidence interval is

$$10.6 - 1.645 * \frac{9}{\sqrt{53}} < \mu_d < 10.6 + 1.645 * \frac{9}{\sqrt{53}}$$

which gives an interval ranging from 8.57 to 12.63 mm Hg. As zero is not contained in the interval, it appears that there is a decrease from the end of the run-in period to the end of the first 2 weeks of treatment.

If we had ignored the relationship between the pre- and postintervention values and used the approach for independent means, how would that have changed things? The mean difference between the pre and post values does not change, but the standard error of the mean difference does change. We assume that the population variances are known and that σ_1, for the preintervention value, is 7 mm Hg and σ_2 is 8 mm Hg. The standard error of the differences, wrongly ignoring the correlation between the pre and post measures, is then

$$\sqrt{\frac{7^2}{53} + \frac{8^2}{53}} = 1.46.$$

This is larger than the value of $9/\sqrt{53}$ (= 1.236) found above when taking the correlation into account. This larger value for the standard error of the difference (1.46 versus 1.236) makes the confidence interval larger than it would be had the correct method been used.

This experiment was done to examine the dose–response relationship of ramipril. It consisted of a comparison of the changes in the pre- and postintervention blood pressure values for three different doses of ramipril. If the purpose had been different, for example, to determine whether or not the 1.25-mg dose of ramipril has an effect, this type of design may not have been the most appropriate. One problem with this type of design—measurement, treatment, measurement—when used to establish the existence of an effect is that we have to assume that nothing else relevant to the subjects' blood pressure values occurred during the treatment period. If this assumption is reasonable, then we can attribute the decrease to the treatment. If this assumption is questionable, however, then it is problematic to attribute the change to the treatment. In this case, the patients received a placebo—here, a capsule that looked and tasted liked the medication to be taken later—during the 4-week run-in period. There was little evidence of a placebo effect, a change that occurs because the subject believes that something has been done. A placebo effect, when it occurs, is real and may reflect the power of the mind to affect disease conditions. This lack of a placebo effect here lends credibility to attributing the decrease to the medication, but it is no guarantee. More will be said about experimental designs in the next chapter.

G. Confidence Interval for the Difference of Two Proportions

In this section, we want to find the $(1 - \alpha) * 100$ percent confidence interval for the difference of two independent proportions, that is, $\pi_1 - \pi_2$. We assume that the sample sizes are large enough so that it is appropriate to use the normal distribution as an approximation to the distribution of $p_1 - p_2$. In this case, the confidence interval for the difference of the two proportions is approximate. Its form is very similar to that for the difference of two independent means when the variances are not equal.

The variance of the difference of the two independent proportions is

$$\frac{\pi_1 * (1 - \pi_1)}{n_1} + \frac{\pi_2 * (1 - \pi_2)}{n_2}.$$

As the population proportions are unknown, we substitute the sample proportions, p_1 and p_2, for them in the variance formula. The $(1 - \alpha) * 100$ percent confidence interval for $\pi_1 - \pi_2$ then is

$$(p_1 - p_2) - z_{1-\alpha/2}\sqrt{\frac{p_1(1-p_1)}{n_1} + \frac{p_2(1-p_2)}{n_2}} < \pi_1 - \pi_2 < (p_1 - p_2) + z_{1-\alpha/2}\sqrt{\frac{p_1(1-p_1)}{n_1} + \frac{p_2(1-p_2)}{n_2}}.$$

Because we are considering the difference of two proportions, the continuity correction terms cancel out in taking the difference.

Holick *et al.* (5) conducted a study of 13 milk processors in five Eastern states. They found that only 12 of 42 randomly selected samples of milk they collected contained 80 to 120 percent of the amount of vitamin D stated on the label. Suppose that 10 milk processors in the Southwest are also studied and that 21 of 50 randomly selected samples of milk contain 80 to 120 percent of the amount of vitamin D stated on the label. Construct a 99 percent confidence interval for the difference of proportions of milk that contain 80 to 120 percent of the amount of vitamin D stated on the label between these Eastern and Southwestern producers.

As the sample sizes and the proportions are relatively large, the normal approximation can be used. The estimate of the standard error of the sample difference is

$$\sqrt{\frac{(12/42)(1 - 12/42)}{42} + \frac{(21/50)(1 - 21/50)}{50}}$$

which is 0.0987. The value of $z_{0.995}$ is found from Table B4 to be 2.576. Therefore the 99 percent confidence interval is

$$(0.286 - 0.420) - 2.576 * 0.0987 < \pi_1 - \pi_2 < (0.286 - 0.420) + 2.576 * 0.0987$$

which is

$$-0.388 < \pi_1 - \pi_2 < 0.120.$$

As zero is contained in the confidence interval, there is little indication of a difference in the proportion of milk samples with a vitamin D content within the 80 to 120 percent range of the amount stated on the label between these Eastern and Southwestern milk producers.

H. Prediction and Tolerance Intervals Based on the Normal Distribution

As we have seen, knowledge that the data follow a specific distribution can be used effectively in the creation of confidence intervals. This knowledge can also be used in the formation of prediction and tolerance intervals, and this use is shown below.

1. Prediction Interval

The distribution-free method for forming intervals used specific observed values of the variable under study. In contrast, the formation of intervals based on the normal distribution uses the sample estimates of its parameters, the mean and standard deviation. Assuming that the data follow the normal distribution, the prediction interval is formed by taking the sample mean plus or minus some value. This form is the same as that used in the construction of the confidence interval for the population mean; however, we know that the prediction interval will be much wider than the confidence interval because the prediction interval focuses on a single future observation.

The confidence interval for the mean, when the population variance is unknown, is

$$\bar{x} \pm t_{n-1,1-\alpha/2}\left(\frac{s}{\sqrt{n}}\right).$$

The estimated standard error of the sample mean, $s/\sqrt{n}$, can also be expressed as $\sqrt{[s^2 * (1/n)]}$. The variance of a future observation is the sum of the variance of an observation about the sample mean and the variance of the sample mean itself, that is, $\sigma^2 + \sigma^2/n$. Thus the estimated standard error of a future observation is $\sqrt{[s^2 * (1 + 1/n)]}$ and the corresponding prediction interval is

$$\bar{x} \pm t_{n-1,1-\alpha/2}s\sqrt{1 + \frac{1}{n}}.$$

Let us calculate the prediction interval for the systolic blood pressure data used above in the calculation of the 90 percent confidence interval for the mean. The sample mean was 94.75 mm Hg and the sample standard deviation was 10.25 mm Hg based on a sample size of 60. The value of $t_{59,0.95}$ used in the 90 percent confidence interval was 1.671. The value of $s * \sqrt{(1 + 1/n)}$ is 10.335 [= $10.25 * \sqrt{(1 + 1/60)}$]. Therefore the prediction interval is

$$94.75 \pm 1.671 * 10.335$$

and the lower and upper limits are 77.48 and 112.02 mm Hg, respectively. These values are contrasted with 92.54 and 96.96 mm Hg, limits of the confidence interval for the mean. Thus, as expected, the 90 percent prediction interval for a single future observation is much wider than the corresponding 90 percent confidence interval for the mean.

2. Tolerance Interval

The tolerance interval is also formed by taking the sample mean plus or minus some quantity, k, multiplied by the estimate of the standard deviation. As the derivation of k is beyond the level of this book, we simply use its value found in Table B8. In symbols, the $(1 - \alpha) * 100$ percent tolerance interval containing p percent of the population based on a sample of size n is

$$\bar{x} \pm k_{n,p,1-\alpha} * s.$$

Let us use Table B8 to find the 90 percent tolerance interval containing 95 percent of the systolic blood pressure values in the population based on the first sample of 60 observations from above. From Table B8 we find that the value of $k_{60,0.95,0.90}$ is 2.248. Therefore, the tolerance interval is

$$94.75 \pm 2.248 * 10.25$$

which gives limits of 71.71 and 117.79.

One-sided prediction and tolerance intervals based on the normal distribution are also easy to construct.

IV. CONCLUDING REMARKS

In this chapter, the concept of interval estimation was introduced. We presented prediction, confidence, and tolerance intervals and explained their applications. We showed how distribution-free intervals and intervals based on the normal distribution were calculated. The idea and use of confidence intervals discussed in this chapter are explored further to introduce methods of testing statistical hypotheses in Chapter 13. Parenthetically, it is worth pointing out that the idea of confidence interval is often expressed as a margin of error in journalistic reporting, which refers to one-half of the width of a two-sided confidence interval.

We also pointed out that characteristics, for example, size, of the intervals could be examined before actually conducting the experiment. If the characteristics of the interval are satisfactory, the investigator uses the proposed sample size. If the characteristics are unsatisfactory, the design of the experiment, the topic of the next chapter, needs to be modified.

EXERCISES

7.1. Assume that the AML patients shown in Exercise 4.7 can be considered a simple random sample of all AML patients.

a. Calculate the 95 percent confidence interval for the population mean survival time after diagnosis for AML patients.
b. Interpret this confidence interval so that someone who knows no statistics can understand it.
c. Calculate the approximate 95 percent confidence interval for the median survival time. Compare the intervals for the population mean and median.
d. There are two methods for forming the tolerance interval. Use both methods to form the approximate 95 percent tolerance interval containing 90 percent of the survival times for the population of AML patients. Which method do you think is the more appropriate one to use here? Provide your rationale.

7.2. Calculate a 90 percent confidence interval for the population median length of stay based on the data from the patient sample shown in Exercise 4.10. Is it appropriate to calculate a confidence interval for the population mean based on these data? Support your answer.

7.3. Find a study from the health literature that uses confidence intervals for one of the statistics covered in this chapter. Provide a reference for the study and briefly explain how confidence intervals were used.

7.4. The following table shows the average annual fatality rate per 100,000 workers based on the 1980–1988 period by state along with

State	*Fatality rate*[a]	*NSWI score*[b]	*State*	*Fatality rate*	*NSWI score*	*State*	*Fatality rate*	*NSWI score*
CT	1.9	65	SC	6.7	26	LA	11.2	31
MA	2.4	73	VT	6.8	38	NE	11.3	27
NY	2.5	76	IL	6.9	76	NV	11.5	30
RI	3.3	59	NC	7.2	47	TX	11.7	72
NJ	3.4	80	WA	7.7	55	KY	11.9	32
AZ	4.1	40	IN	7.8	47	NM	12.0	14
MN	4.3	64	ME	7.8	67	AR	12.5	11
NH	4.5	56	TN	8.1	24	UT	13.5	26
OH	4.8	55	OK	8.7	53	ND	13.8	21
MI	5.3	63	AL	9.0	25	MS	14.6	25
MO	5.3	42	KS	9.1	15	SD	14.7	25
MD	5.7	46	IA	9.2	54	WV	16.2	47
DE	5.8	40	CO	9.3	52	ID	17.2	22
HI	6.0	25	FL	9.3	48	MT	21.6	28
PA	6.1	55	VA	9.9	60	WY	29.5	12
WI	6.3	58	GA	10.3	36	AK	33.1	59
CA	6.5	81	OR	11.0	63			

[a]Average annual fatality rate per 100,000 workers based on 1980–1988 data.
[b]National Safe Workplace Institute Score (116 is the maximum and a higher score is better).

the state's composite score on a scale created by the National Safe Workplace Institute (NSWI). The scale takes into account prevention and enforcement activities and compensation paid to the victim. The data are taken from the Public Citizen Health Research Group (14). During the 1980–1988 period, the National Institute of Occupational Safety and Health reported that there were 56,768 deaths in the workplace. The above rates are based on that number. The National Safety Council reported 105,500 deaths for the same period. Do you think that there should be any relationship between the fatality rates and the NSWI scores? If you think that there is a nonzero correlation, will it be positive or negative? Explain your reasoning. Calculate the Pearson correlation coefficient for these data. Is there any reason to calculate a confidence interval based on the correlation value you calculated? Why or why not?

7.5. There is some concern today about excessive intakes of vitamins and minerals, possibly leading to nutrient toxicity. For example, many persons take vitamin and mineral supplements. It is estimated that 35 percent of the adult U.S. population consumes vitamin C in the form of supplements (7, p. 62). Based on survey results, among users of vitamin C supplements, the median intake was 333 percent of the recommended daily allowance. Suppose that you take a tablet that claims to contain 500 mg vitamin C. Which type of interval—prediction, confidence, or tolerance—about the vitamin C content of the tablets is of most interest to you? Explain your reasoning.

7.6. In a test of a laboratory's measurement of serum cholesterol, 15 samples containing the same known amount (190 mg/dl) of serum cholesterol are submitted for measurement as part of a larger batch of samples, one sample each day over a 3-week period. Suppose that the following daily values in mg/dl for serum cholesterol for these 15 samples were reported from the laboratory:

180 190 197 199 210 187 192 199 214 237 188 197 208 220 239.

Assume that the variance for the measurement of serum cholesterol is supposed to be no larger than 100 mg/dl. Construct the 95 percent confidence interval for this laboratory's variance. Does 100 mg/dl fall within the confidence interval? What might be an explanation for the pattern shown in the reported values?

7.7. The proportion of persons in the United States without health insurance in 1991 was 14.1 percent, or approximately 35.5 million persons. The following data show the percentages of persons without health insurance in 1991 by state (15) along with the 1990 population of the state (16). The District of Columbia is treated as a state in this presentation. Calculate the sample Pearson correlation coefficient between the state population total and its percent without health insurance.

How can these counts be viewed as a sample? Calculate a 95 percent confidence interval for the Pearson correlation coefficient in the population. Does there appear to be a strong linear relationship between these two variables? Provide at least one additional variable that may be related to the proportion without health insurance in each state and provide a rationale for your choice.

State	*Population*[a]	*Percent without health insurance*
New England		
ME	1.23	11.1
NH	1.11	10.1
VT	0.56	12.7
MA	6.02	10.9
RI	1.00	10.2
CT	3.29	7.5
Mid-Atlantic		
NY	17.99	12.3
NJ	7.73	10.8
PA	11.88	7.8
East North Central		
OH	10.85	10.3
IN	5.54	13.0
IL	11.43	11.5
MI	9.30	9.0
WI	4.89	8.0
West North Central		
ND	0.64	7.6
SD	0.70	9.9
NE	1.58	8.3
KS	2.48	11.4
MN	4.38	9.3
IA	2.78	8.8
MO	5.12	12.2
South Atlantic		
DE	0.67	13.2
MD	4.78	13.1
VA	6.19	16.3
WV	1.79	15.7
FL	12.94	18.6
NC	6.63	14.9
SC	3.49	13.2
GA	6.48	14.1
DC	0.61	25.7
East South Central		
KY	3.69	13.1
TN	4.88	13.4
AL	4.04	17.9
MS	2.57	18.9
West South Central		
AR	2.35	15.7
LA	4.22	20.7
OK	3.15	18.2
TX	16.99	22.1
Mountain		
MT	0.80	12.7
ID	1.01	17.8
WY	0.45	11.3
CO	3.29	10.1
NM	1.52	21.5
AZ	3.67	16.9
UT	1.72	13.8
NV	1.20	18.7
Pacific		
WA	4.87	10.4
OR	2.84	14.2
CA	29.76	18.7
AK	0.55	13.2
HI	1.11	7.0

[a]Population is expressed in millions.

7.8. Calculate the mean state proportion of those without health insurance from data in Exercise 7.7. Is this number the same as the overall

U.S. percentage? Explain how the state information can be used to obtain the overall U.S. percentage of 14.1.

7.9. Suppose you are planning a simple random sample survey to estimate the mean family out-of-pocket expenditures for health care in your community during the last year. In 1990, the approximate per capita (not per family) out-of-pocket expenditure was $525 (17, Table 121). From previous studies in the literature, you think that the population standard deviation for family out-of-pocket expenditures is $500. You want the 90 percent confidence interval for the community mean family out-of-pocket expenditures to be no wider than $100.

a. How many families do you require in the sample to satisfy your requirement for the width of the confidence interval for the mean?
b. Do you believe that family out-of-pocket expenditures follow the normal distribution? Support your answer.
c. Regardless of your answer, assume that you said that the family out-of-pocket expenditures do not follow a normal distribution. Discuss why it is still appropriate to use the material based on the normal distribution in finding the confidence interval for the population mean.
d. In the conduct of the survey, how would you overcome reliance on a person's memory for out-of-pocket expenditures for health care for the past year?

7.10. In 1979, the Surgeon General's Report on Health Promotion and Disease Prevention and its follow-up in 1980 established health objectives for 1990. One of the objectives was that the proportion of 12- to 18-year-old adolescents who smoked should be reduced to below 6 percent (17, p. 85). Suppose that you have monitored progress in your community toward this objective. In a survey conducted in 1983, you found that seventeen of ninety 12- to 18-year-old adolescents admitted that they were smokers. In your 1990 simple random sample survey, you found eleven of eighty-five 12- to 18-year-old adolescents who admitted that they smoked.

a. Construct a 95 percent confidence interval for the proportion of smokers among 12- to 18-year-old adolescents in your community. Is 6 percent contained in the confidence interval?
b. Construct a 99 percent confidence interval for the difference in the proportion of smokers among 12- to 18-year-old adolescents from 1983 to 1990. Do you believe that there is a difference in the proportion of smokers among the 12- to 18-year-old adolescents between 1983 and 1990? Explain your answer.
c. Briefly describe how you would conduct a simple random sample of 12- to 18-year-old adolescents in your community. Do you have confidence in the response to the question about smoking? Pro-

vide the rationale for your answer. What is a method that might improve the accuracy of the response to the smoking question?

7.11. Construct the 95 percent confidence interval for the difference in the population mean survival times between the AML and ALL patients shown in Table 7.6. As there appears to be a difference in mean ages between the AML and ALL patients, perhaps we should adjust for age. One way to do this is to calculate age-specific confidence intervals. For example, calculate the confidence interval for the difference in population mean survival times for AML and ALL patients who are 40 years old or younger. Is the confidence interval for those 40 years of age or younger consistent with the confidence interval that has ignored the ages? How else might we adjust for the age variable in the comparison of the AML and ALL patients?

7.12. Suppose that we wish to investigate the claims of a weight loss clinic. We randomly select 20 individuals who have just entered the program and we follow them for 6 weeks. The clinic claims that its members will lose on the average 10 pounds during the first 6 weeks of membership. The beginning weights and the weights after 6 weeks are shown below. Based on this sample of 20 individuals, is the clinic's claim plausible?

Person	*Beginning weight*	*Weight at 6 wk*
1	147	143
2	163	151
3	198	184
4	261	245
5	233	229
6	227	220
7	158	161
8	154	147
9	162	155
10	249	254
11	246	239
12	218	222
13	143	135
14	129	124
15	154	136
16	166	159
17	278	263
18	228	205
19	173	164
20	135	122

7.13. In a study of aplastic anemia patients, 16 of 41 patients on one treatment achieved complete or partial remission after 3 months of treatment, compared with 28 of 43 patients on another treatment (18). Construct a 99 percent confidence interval on the difference in proportions that achieved complete or partial remission. Does there appear to be a difference in the population proportions of the patients who would achieve complete or partial remission on these two treatments?

7.14. In 1970, Japanese-American women had a fertility rate (number of live births per 1000 women ages 15–44) of 51.2, considerably lower than the rate of 87.9 for all U.S. women in this age group. Use the following data to calculate an age-adjusted fertility rate for Japanese-American women and approximate the standard deviation of the age-adjusted rate.

Age	*U.S. age-specific fertility rate*	*Number of Japanese-American women*
15–19	69.6	24,964
20–24	167.8	23,435
25–29	145.1	22,093
30–34	73.3	23,055
35–39	31.7	32,935
40–44	8.6	34,044

Source: U.S. Population Census, 1970, P(2)-1G, and U.S. Vital Statistics, 1970 (19)

REFERENCES

1. Vardeman, S. B. (1992). What about the other intervals? *Am. Stat.* **46,** 193–197.
2. Walsh, J. E. (1962). Nonparametric confidence intervals and tolerance regions. *In* "Contributions to Order Statistics" (A. E. Sarhan and B. G. Greenberg, eds.). Wiley, New York.
3. NHLBI Task Force on Blood Pressure Control in Children (1987). The report of the second task force on blood pressure control in children, 1987. *Pediatrics* **79,** 1–25.
4. Jacobus, C. H., Holick, M. F., Shao, Q., Chen, T. C., Holm, I. A., Kolodny, J. M., Fuleihan, G. E., and Seely, E. W. (1992). Hypervitaminosis D associated with drinking milk. *N. Engl. J. Med.* **326,** 1173–1177.
5. Holick, M. F., Shao, Q., Liu, W. W., and Chen, T. C. (1992). The Vitamin D content of fortified milk and infant formula. *N. Engl. J. Med.* **326,** 1178–1181.
6. Walter, U., Forthofer, R., and Witte, P. U. (1987). Dose-response relation of angiotensin converting enzyme inhibitor ramipril in mild to moderate essential hypertension. *Am. J. Cardiol.* **59,** 125D–132D.
7. Life Sciences Research Office, Federation of American Societies for Experimental Biology (1989). "Nutrition Monitoring in the United States: An Update Report on Nutrition Monitoring," U.S. Department of Agriculture and the U.S. Department of Health and Human Services, Public Health Service, Washington, U.S. Government Printing Office, DHHS Publ. No. (PHS) 89-1255.

8. Kendall, M. G., and Stuart, A. (1967). "The Advanced Theory of Statistics," 2nd ed., Vol. 2. Hafner, New York.
9. Satterthwaite, F. E. (1946). An approximate distribution of estimates of variance components. *Biometrics Bull.* **2,** 110–114.
10. Lee, E. T. (1980). "Statistical Methods for Survival Data Analysis." Wadsworth, Belmont, CA.
11. Markowski, C. A., and Markowski, E. P. Conditions for the effectiveness of a preliminary test of variance. *Am. Stat.* **44,** 322–326.
12. Gans, D. J. (1991). Letter to the Editor. Preliminary test on variances. *Am. Stat.* **45,** 258.
13. Moser, B. K., and Stevens, G. R. (1992). Homogeneity of variance in the two-sample means test. *Am. Stat.* **46,** 19–21.
14. Public Citizen Health Research Group (1992). Work-related injuries reached record level last year. *Public Citizen Health Res. Group Health Lett.* **8**(12), 1–3, 9.
15. Public Citizen Health Research Group (1993). The growing epidemic of uninsurance. *Public Citizen Health Res. Group Health Lett.* **9**(1), 1–2.
16. U.S. Bureau of the Census (1991). 1990 Census of Population and Housing, Summary Tape File 1A. on CD-ROM Technical Documentation/prepared by the Bureau of the Census. Washington: The Bureau, 1991.
17. National Center for Health Statistics (1992). "Health, United States, 1991 and Prevention Profile," Public Health Service, DHHS Publ. No. 92-1232. Hyattsville, MD.
18. Frickhofen, N., Kaltwasser, J. P., Schrezenmeier, H., Raghavachar, A., Vogt, H. G., Herrmann, F., Freund, M., Meusers, P., Salama, A., and Heimpel, H. (1991). Treatment of aplastic anemia with antilymphocyte globulin and methylprednisolone with or without cyclosporine. *N. Engl. J. Med.* **324,** 1297–1304.
19. U.S. Bureau of the Census (1973). 1970 Census of Population, Subject Reports: Japanese, Chinese, and Filipinos in the United States, P(2)-1G. U.S. Govt. Printing Office, Washington, DC; National Center for Health Statistics (1975). "U.S. Vital Statistics," 1970, Vol. I. U.S. Govt. Printing Office, Washington, D.C.

CHAPTER 8

Designed Experiments

This chapter introduces the designed experiment, one of the two methods used in statistics for producing data. We previously met the other method, the sample survey, in Chapter 3. Designed experiments have been used in biostatistics in the evaluation of: (1) the efficacy and safety of drugs or medical procedures; (2) the effectiveness and cost of different health care delivery systems; and (3) the effect of exposure to possible carcinogens. In the following, we present the principles underlying such experiments. Limitations of experiments and ethical issues related to experiments, especially when applied to humans, are also raised.

I. SAMPLE SURVEYS AND EXPERIMENTS

There are many similarities among as well as some differences between sample surveys and experiments. From sample surveys, we learn the characteristics of some population. The sample survey design focuses on the sampling of individuals from the population. From experiments, we discover the effect of applying a stimulus to subjects. The experimental design focuses on the formation of comparison groups that allow conclusions about the effect of the stimulus to be drawn.

As emphasized in Chapter 3, a good survey begins with a carefully drawn blueprint or design and the same holds true for an experiment. The blueprint or design of an experiment is based on both statistical and substantive considerations. Chapter 7 provided one example of statistical considerations that should be part of the study design. We saw how the analysis of the relationship among sample size, size of the interval, and level of confidence associated with the study can be used in the creation of the study design before any data are collected.

An experiment is different from a sample survey in that the experimenter actively intervenes with the experimental subjects through the assignment of the subjects to groups, whereas the survey researcher passively observes or records responses of the survey subjects. Experiments and surveys often have different goals as well.

The goal in an experiment is to determine whether or not there is an association between the independent or predictor variables and the dependent or response variable. The different groups to which the subjects are assigned usually represent the levels of the independent variable. Independent and dependent were chosen as names for the variable types because it was thought that the response variable depended on the levels of the predictor variables. To determine whether or not there is an association, the experimenter assigns subjects to different levels of one variable, for example, to different doses of some medication. The effects of the different levels—the different doses—are found by measuring the values of an outcome variable, for example, blood pressure. An association exists if there is a relationship between the blood pressure values and the dosage levels.

In a survey, the primary goal is to describe the population and a secondary goal is to investigate the association between variables. In a survey, variables are usually not referred to as independent or dependent because all the variables can be viewed as being response variables. The survey researcher usually has not manipulated the levels of any of the variables as the experimenter does.

Let us consider an example to illustrate the essential points in the experimental design.

II. EXAMPLE OF AN EXPERIMENT

The Hypertension Detection and Follow-up Program (HDFP) was a community-based, clinical trial conducted in the early 1970s by the National Heart, Lung and Blood Institute (NHLBI) with the cooperation of 14 clinical centers and other supporting groups (1). The purpose of the trial was to assess the effectiveness of treating hypertension, a major risk factor for several different forms of heart disease. For this trial, it was decided that the major outcome variable would be total mortality.

At the time of designing the HDFP trial, results of a Veterans Administration (VA) Cooperative Study were known. This study had already demonstrated the effectiveness of antihypertensive drugs in reducing morbidity and mortality due to hypertension among middle-aged men with sustained elevated blood pressure; however, the VA study included only a subset of the entire community. Applicability of its findings to those with undetected hypertension in the community, to women, and to minority persons was uncertain. It was therefore decided to perform a study, the HDFP study, in the general community. Instead of including only people who knew they had high blood pressure, subjects were recruited by screening people in the community.

In this clinical trial, antihypertensive therapy was the independent or predictor variable and mortality rate was the dependent or response variable. To determine the effectiveness of the antihypertensive therapy, a comparison group was required. Thus the study was intended to have a treatment group, those who received the therapy, and a control group, those who did not receive the therapy. This classic experimental design could not be used, however. As the antihypertensive therapy was already known to be effective, it could not ethically be withheld from the control group. Recognizing this, the HDFP investigators decided to compare a systematic antihypertensive therapy given to those in the treatment group (Stepped Care) with the therapy received from their usual sources of care for those in the control group (Regular Care). As a result, no one was denied treatment.

III. COMPARISON GROUPS AND RANDOMIZATION

A simple experiment may be conducted without any comparison group. For example, a newly developed AIDS education course was taught to a class of ninth graders in a high school for a semester. The level of knowledge regarding AIDS was tested before and after the course to assess the effect of the course on students' knowledge. The difference in test scores between the pre- and posttests would be taken as the effect of the instructional program. It may, however, be inappropriate to attribute the change in scores to the instructional program. The change may be entirely or partially due to some influence outside the AIDS course, for example, mass media coverage of AIDS-related information. Therefore, we have to realize that when this simple experimental design is used, the outside influence, if any, is mixed with the effect of the course and it is not possible to separate them.

Thus in studying the effect of an independent variable on a dependent variable, we have to be aware of the possible influence of an extraneous variable(s) on the dependent variable. When the effects of the independent variable and the extraneous variable cannot be separated, the variables are said to be *confounded*. In observational studies such as sample surveys, all variables are confounded with one another and the analytical task is to untangle the comingled influence of many variables that are measured at the same time. In experimental studies, the effects of extraneous variables are separated from the effect of the independent variable by adopting an appropriate design.

The basic tool for separating the influence of extraneous variables from that of the independent variable is the use of comparison groups. For example, giving the treatment to one of two equivalent groups of subjects and withholding it from the other group means that the observed difference in the outcome variable between the two groups can be attributed to the effect of the treatment. In this design, any extraneous variables would presumably influence both groups equally and, thus, the difference between the two groups would not be influenced by the extraneous variables. The key to the successful use of this design is that the groups being compared are really equivalent before the experiment begins.

Matching is one method that is used in an attempt to make groups equivalent. For example, subjects are often matched on age, gender, race, and other characteristics and then one member of each matched pair receives the treatment and the other does not. It is difficult, however, to match subjects on many variables; in addition, the researcher may not know all the important variables that should be used in the matching process. A method for dealing with these difficulties with matching is the use of randomization.

Randomization is the random assignment of subjects to groups. By using randomization, the researcher is attempting to: (1) eliminate intentional or nonintentional selection bias, for example, the assignment of healthier subjects to the treatment group and sicker subjects to the control group; (2) remove the effect of any extraneous variables. With large samples, the random assignment of subjects to groups should cause the distributions of the extraneous variables to be equivalent in each group, thus removing their effects.

IV. RANDOM ASSIGNMENT AND SAMPLE SIZE

One way of randomly assigning subjects to groups is the use of the random sampling without replacement procedure. For example, the demonstration in Box 8.1 randomly assigns 50 sequentially numbered subjects to two

MINITAB BOX 8.1

```
MTB > set c1
DATA> 1:50
DATA> end
MTB > sample 25 c1 c2
```

Twenty-five observations from c1 are randomly selected and placed in c2.

```
MTB > sort c2 c2
MTB > print c2
C2
     2     4     5     6    11    12    16
    17    18    20    21    25    26    27
    30    31    32    33    35    36    40
    41    44    47    48
```

groups. The subjects whose numbers are shown in Box 8.1 are assigned to the treatment group and the remaining 25 subjects form the control group. In many randomized experiments, subjects are assigned to the groups sequentially as soon as subjects are identified, as in the HDFP trial. In that case, the above results can be put into the following sequence of letters T (treatment group) and C (control group) that can be used to show the assignment:

```
C T C T T T C C C C
T T C C C T T T C T
T C C C T T T C C T
T T T C T T C C C T
T C C T C C T T C C
```

If one were to assign 60 subjects to three groups, the first random sample of 20 will be assigned to the first group, the second random sample of 20 to the second group, and the remaining 20 subjects to the third group.

The random assignment of subjects to groups does not guarantee the equivalence of the distributions of the extraneous variables in the groups. There must be a sufficiently large number of subjects in each group for randomization to have a high probability of causing the distributions of the extraneous variables to be similar across groups. As discussed in Chapters 3 and 7, use of larger random samples decreases the sample-to-sample variability and increases our confidence that the sample estimates are closer to the population parameters. In the same way, a greater number of subjects in the treatment and control groups increases our confidence that the two groups are equivalent with respect to all extraneous factors.

To make this point clearer, consider the following example. A sample of 10 adults is taken from the Second National Health and Nutrition Examination Survey (NHANES II) data file and 5 of the 10 persons are randomly assigned to the treatment group and the other 5 are assigned to the control group. The two groups are compared with respect to five characteristics. The same procedure is repeated for sample sizes of 40, 60, and 100 and the results are shown in Table 8.1. The treatment and control groups are not very similar when n is 10. As the sample size increases, the treatment and control groups become more similar. When n is 100, the two groups are very similar. It appears that at least 30 to 50 persons are needed in each of the treatment and control groups for them to be reasonably similar.

The confidence interval for the difference between two means (proportions) that was discussed in Chapter 7 could also be used to address this issue. For example, we can determine the sample size required for the width of the confidence interval for the difference between proportions of two equal-size groups of subjects to be a small value. Suppose that we decide to use a 90 percent confidence level ($z = 1.645$) and we require the width of the confidence interval to be 0.20. We shall assume that the sample size will be large enough that the normal distribution approximation to the binomial can be used. We shall also assume that the proportion is 0.50 in the control group and 0.60 in the treatment group. Using these assumptions, the value of the sample size is found by

$$n = 2\left(\frac{2 * 1.645 * \sqrt{0.5 * 0.5 + 0.6 * 0.4}}{0.20}\right)^2$$

which yields a sample size of 266, allocating 133 in each group. This calcu-

TABLE 8.1 Comparison of Treatment and Control Groups for Different Group Sizes

Characteristics[a]	Treatment	Control	Treatment	Control
	($n_1 = 5$)	($n_2 = 5$)	($n_1 = 20$)	($n_2 = 20$)
Percent male	60	20	60	35
Percent black	0	20	5	20
Mean years of education	12.6	11.2	12.9	13.0
Mean age	38.8	41.6	40.7	34.0
Percent smokers	60	40	27	23
	($n_1 = 30$)	($n_2 = 30$)	($n_1 = 50$)	($n_2 = 50$)
Percent male	43	50	42	44
Percent black	17	10	16	16
Mean years of education	12.7	12.9	11.7	12.5
Mean age	39.7	40.2	42.1	42.5
Percent smokers	32	35	34	34

[a]Observations are weighted using the NHANES II sampling weights.

lation supports the need for large sample sizes in the groups to have confidence that randomization will provide equivalent groups.

In the HDFP clinical trial, more than 10,000 hypertensive persons were screened through community surveys and included in the study. These subjects were randomly assigned to either the Stepped Care or Regular Care group. Because of this random assignment and the large number of subjects included in the trial, the Stepped Care and Regular Care groups were very similar with respect to many important characteristics at the beginning of the trial. Table 8.2 is a demonstration of the similarities. The randomization and the sufficiently large sample size also give us confidence that these two groups were equivalent with respect to other characteristics that are not listed in Table 8.2.

The sample size required for an experiment depends on three factors: (1) the amount of variation among the experimental subjects; (2) the magnitude of the effect to be detected; and (3) the level of confidence associated with the study. When the experimental subjects are similar, a smaller sample size can be used than when the subjects differ. For example, a laboratory experiment using genetically engineered mice does not require as large a sample size as the same experiment using mice trapped in the wild. There is less likelihood of extraneous variables existing in the study using the genetically engineered mice. Hence a smaller sample should be acceptable as there is less need to control for extraneous variables. The fact

TABLE 8.2 Comparison of Stepped Care and Regular Care Participants by Selected Characteristics at Entry to the Hypertension Detection and Follow-up Program

Characteristics	Stepped Care	Regular Care
Number of participants	5485	5455
Mean age	50.8	50.8
Percent black men	19.4	19.9
Percent black women	24.5	24.8
Mean systolic blood pressure (mm Hg)	159.0	158.5
Mean diastolic blood pressure (mm Hg)	101.1	101.1
Mean pulse (beats/minute)	81.7	82.2
Mean serum cholesterol (mg/dl)	235.0	235.4
Mean plasma glucose (mg/dl)	178.5	178.9
Percent smoking >10 cigarettes/day	25.6	26.2
Percent with history of stroke	2.5	2.5
Percent with history of myocardial infarction	5.1	5.2
Percent with history diabetes	6.6	7.5
Percent taking antihypertensive medication	26.3	25.7

Source: Hypertension Detection and Follow-up Program Cooperative Group (1).

that the sample size for the experiment depends on the size of the effect to be detected is not surprising. Because it should be more difficult to detect a small effect of the independent variable than a large effect, the sample size must reflect this. This is one of the reasons that the HDFP trial used a large sample size. The benefit of being in the Stepped Care group compared with the Regular Care group was expected to be small. Thus a large sample size was required to detect this small benefit. A smaller sample size would have been sufficient had the HDFP trial compared a group receiving medication with a group that did not receive hypertensive medication. In this case, a larger effect would have been expected. The last point concerns the relationship between the sample size and the confidence associated with the study. As was demonstrated in Chapter 7, the confidence level associated with a study increases as the sample size increases.

V. SINGLE- AND DOUBLE-BLIND EXPERIMENTS

So far we have been concerned with the statistical aspects of the design of an experiment. This means the use of comparison groups, the random assignment of subjects to the groups, and the need for an adequate number of subjects in the groups. An additional concern is the possible bias that can be introduced in an experiment. Let us consider some possible sources of bias and possible ways to avoid them.

In drug trials, particularly in those involving a placebo, the subjects are often blinded; that is, they are not informed whether they have received the active medication or a placebo. This is done because knowledge of which treatment has been provided may affect the subject's response. For example, those assigned to the control group may lose interest, whereas those receiving the active medication, because of expectations of a positive result, may react more positively. Studies in which the treatment providers know, but the subjects are unaware of, the group assignment are called *single-blind experiments*.

In most drug trials, both the subjects and the treatment providers are unaware of the group assignment. The treatment providers are blinded because they also have expectations about the reaction to the treatment. These expectations may affect how the experimenter measures or interprets the results of the experiment. An experiment in which both the subjects and the experimenters are uninformed of the group assignment is called a *double-blind experiment*.

Let us examine one double-blind, randomized experiment conducted by a VA research team (2). They used the experimental design in Figure 8.1 to determine whether antiplatelet therapies improve saphenous vein graft patency after coronary artery bypass grafting. In this design, there are four treatment groups (four regimens of drug therapy) and a control group

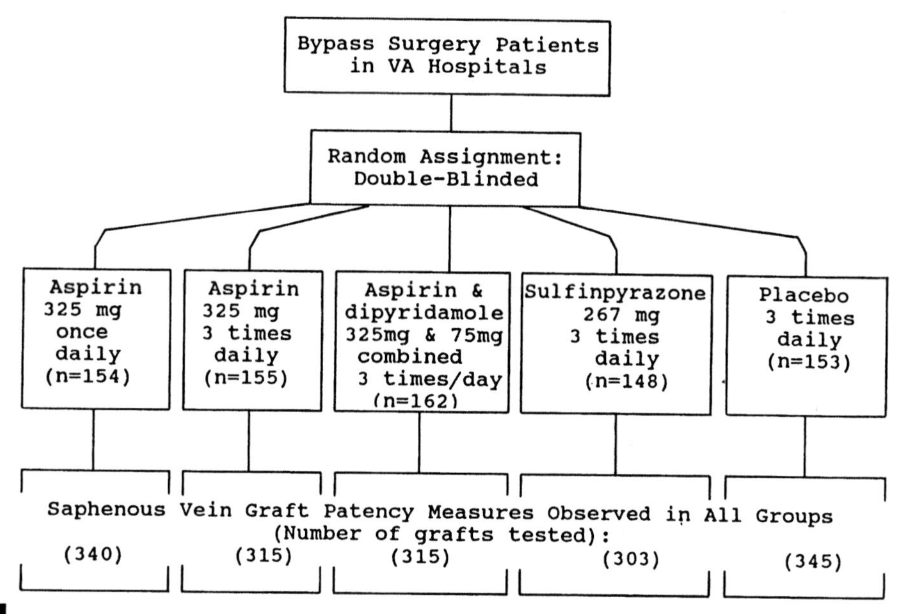

FIGURE 8.1 Experimental design for Veterans Administration Cooperative Study on Effect of Antiplatelet Therapy.

(placebo). Both the patients and the doctors were blinded, and only the designers of the trial, who were not directly involved in patient treatment, knew the group assignment. A total of 772 consenting patients were randomized and postoperative treatment was started 6 hours after surgery and continued for 1 year.

As was to be expected, this experiment encountered problems in retaining subjects during the course of the experiment. The final analysis was based on 502 patients who underwent the late catheterization. These patients had a total of 1618 grafts. Of the 270 patients not included in the final analysis, 154 refused to undergo catheterization, 32 were lost to follow-up, 31 died during treatment, 42 had medical complications, and data on 11 patients were not available in the central laboratory (3). Although we may expect that these problems are fairly evenly distributed among the groups because of the random assignment of subjects, the sample size was reduced considerably. This suggests that we need to increase the initial sample size in anticipation of the loss of some subjects during the experiment.

Other types of precautions must be taken to avoid potential biases. In addition to statistical aspects, the experiment designer must provide detailed procedures for handling experimental subjects, monitoring compliance of all participants, and collecting data. For this purpose a study protocol must be developed, and the experimenter is responsible for adherence to the protocol by all participants. Similar to the problem of nonresponse in sample surveys, the integrity of experiments is often threatened by unexpected happenings such as the loss of subjects during the experiment and changes in the experimental environment. Steps must be taken to minimize such threats.

VI. BLOCKING AND EXTRANEOUS VARIABLES

Thus far we have considered the simplest randomization, the random assignment of subjects to groups without any restriction. This design is known as a completely randomized design. The role of this design in experimental design is the same as that of the simple random sample design in survey sampling. As was mentioned earlier, in completely randomized designs, we attempt to remove the effects of extraneous variables by randomization; however, a reasonably large sample size is required before we can have confidence in the randomization process.

Another experimental design for eliminating the effects of extraneous variables known or thought to be related to the dependent variable uses blocking. *Blocking* means directly taking these extraneous variables into account in the design. For example, in a study of the effects of different diets on weight loss, subjects are often blocked or grouped into different initial weight categories. Within each block, the subjects are then randomly assigned to the different diets. The reason for the blocks based on initial weight is that it is thought that weight loss may be related to the initial weight. Designs using blocking do not rely entirely on randomization to remove the effects of these important extraneous variables. Blocking guarantees that each diet has subjects with the same distribution of initial weights; randomization cannot guarantee this. Blocking in experiments is similar to stratification in sample surveys. The experimental design that uses blocks to control the effect of one extraneous variable is called a *randomized block design*. This name indicates that randomization is performed separately within each block.

Blocking is also used for administrative convenience. The VA Cooperative Study discussed in the previous section had 11 participating hospitals located throughout the United States. As the subjects were randomized separately at each site, each participating hospital was a block. In this case, the blocking was done for administrative convenience while also controlling for the variation among hospitals.

In Chapter 3, we saw that the simple random sample design can be modified and extended as required to meet the demands of a wide variety of sampling situations. The completely randomized experimental design can similarly be expanded to accommodate many different needs in experimentation. The randomized block design is one of the many ways the basic design can be extended. More complex designs deal with more than one independent variable or block on more than one extraneous variable.

VII. LIMITATIONS OF EXPERIMENTS

The results of an experiment apply to the population from which the experimental subjects were selected. Sometimes this population may be very limited; for example, patients may be selected from only one hospital or from one clinic within the hospital. In such situations, does this mean that we must perform similar experiments in many more hospitals to determine if the results can be generalized to a larger population, for example, to all patients with the condition being studied? From a statistical perspective, the answer is yes; however, if on the basis of substantive reasons, we can argue that there is nothing unique about this hospital or clinic that should affect the experiment, then it may be possible to generalize the results to the larger population of all patients with the condition. This generalization is based on substantive reasoning, not on statistical principles.

For example, the results of the VA Cooperative Study may be valid only for male veterans. It certainly would be difficult to generalize the results to females without more information. It may be possible to generalize the results to all males who are known to have hypertension, but this requires careful scrutiny. We must know whether or not the VA medical treatment of hypertension is comparable to that received by males in the general population. Does the fact that the men served in the military cause any difference, compared with those who were not in the military, in the effect of the medical intervention? If differences are suspected, then we should not generalize beyond the VA system.

On the other hand, the results of the HDFP should apply more widely, as the subjects were screened from random samples of residents in 14 different communities and then randomly assigned to the comparison groups. This use of accepted statistical principles of random sampling from the target population and randomization of these subjects to comparison groups make it reasonable to generalize the results.

Another limitation of an experiment stems from its dependency on the experimental conditions (4). Often experiments take place in a highly controlled, artificial environment, and the observed results may be confounded with these factors. Dr. Lewis Thomas' experience (5, Chapter 9) is a case in point. While he was waiting to return home from Guam at the end of World War II, he conducted an experiment on several dozen rabbits left in a medical science animal house. He tested a mixed vaccine consisting of heat-killed streptococci and a homogenate of normal rabbit heart tissue and the test produced spectacular and unequivocal results. All the rabbits receiving the mixture of streptococci and heart tissue became ill and died within 2 weeks. The histologic sections of their hearts showed the most violent and diffuse myocarditis he had ever seen. The control rabbits injected with streptococci alone or with heart tissue alone remained healthy and showed no cardiac lesions. On returning to the Rockefeller Institute, he replicated the experiment using the Rockefeller stock of rabbits. He repeated the experiment over and over, but he never saw a single sick rabbit. One explanation for the spectacular results of the Guam experiment is that there may have been some type of a latent virus in the Guam rabbit colony. As Dr. Thomas said, "I had all the controls I needed; I wasn't bright enough to realize that Guam itself might be a control."

As Dr. Thomas's experience shows, we have to be careful not to deceive ourselves and extrapolate beyond our data. The experimental data consist of not only the observed difference between the treatment and control groups, but also the conditions and circumstances under which the experiment was conducted. These include the method of investigation, the time and place, the duration of the test, and other conditional factors. For example, in interpreting the results of drug trials, there is no statistical method by which to extrapolate the safety record of a drug beyond the period of the experiment, nor to a higher level of dosage, nor to other types of patients. The toxic effect of the medication may manifest itself only after a longer exposure, at higher levels of dosage, or for other types of patients. Therefore extrapolation of experimental results must be done with great care, if at all. Better than extrapolation is a replication of the study for different types of subjects under different conditions.

Implicit in the naming of experimental variables as being dependent and independent is the idea of cause and effect; that is, changes in the levels of the independent variables cause corresponding changes in the dependent variable. It is, however, difficult to demonstrate a cause-and-effect relationship. It is sometimes possible to demonstrate this in very carefully designed experiments; however, in most situations in which statistics are used, positive results do not mean a cause-and-effect relationship, but only the existence of an association between the dependent and independent variables.

Finally, statistical principles of experimentation can sometimes be in conflict with our cultural values and ethical standards. Experimenting, especially on human beings, can lead to many problems. If the experiment can potentially harm the subjects or impinge on their privacy or individual rights, then serious ethical questions arise. The harm can be direct physical, psychological, or mental damage, or it may be the withholding of potential benefits. As was seen in the HDFP study, to avoid withholding the benefits of antihypertensive therapy, the study designers used the Regular Care group instead of a placebo group as the control group. When the potential direct harm is obvious, we cannot subject human beings to an experiment.

To protect human subjects from potential harm or from an invasion of privacy, an informed consent is required for experiments and even for interviews in sample surveys. This consent has to be voluntary; however, it is not difficult to recognize the possibility for pressuring patients to participate in a clinical trial. To prevent undue pressure from being applied to patients or other potential study participants, all organizations receiving funds from the federal government are required to have an institutional review committee (6). It is this committee's task to evaluate the study protocol to see if it provides adequate safeguards for the rights of the study participants.

VIII. CONCLUDING REMARKS

In this chapter we have studied the basic principles of the designed experiment. A requirement of a good experimental design is that it prevents extraneous variables from being confounded with the experimental variables. Randomization and blocking are basic tools for preventing this confounding. When these tools are used appropriately, it is possible to analyze the data to determine whether or not it is likely that an association exists between the dependent variable and the independent variables. The test of hypothesis, the topic of the next chapter, is the statistical method for determining this. Even after performing the experiment appropriately, care must be used in interpreting the experimental results. We must not unduly extrapolate the findings from our experiment, but recognize that replication may be necessary for the appropriate generalization to the target population.

EXERCISES

8.1. Choose the most appropriate response from the choices listed under each question:

a. Which of the following is not required in an experiment?
 __ designation of independent and dependent variables
 __ random selection of the subjects from the population
 __ use of a control group
 __ random assignment of the subjects to groups

b. The main purpose of randomization is to balance between experimental groups the effects of extraneous variables that are
 __ known to the researchers
 __ not known to the researcher
 __ both known and unknown to the researcher

c. The experimental groups obtained by randomization may fail to be equivalent to each other, especially when
   __ the sample size is very small
   __ blocking is not used
   __ matching is not used

d. Which, if any, of the following is an inappropriate analogy between random sampling and randomized experiments?
   __ simple random sampling–completely randomized experiment
   __ stratified random sampling–randomized complete block design
   __ random selection–random assignment

e. A randomized experiment is intended to eliminate the effect of the
   __ independent variable
   __ confounded extraneous variables
   __ dependent variable

f. If the number of subjects randomly assigned to experimental groups increases, then the treatment and control groups are likely to be
   __ more similar to each other
   __ less similar to each other
   __ neither of the above

8.2. A middle school principal wants to implement a newly developed health education curriculum for 30 classes of seventh graders that are taught by six teachers. The available budget for teacher training and resource material is, however, sufficient for implementing the new course in only half of the classes. A teacher suggests that an experiment can be designed to compare the effectiveness of the new and old curricula.

a. Design an experiment to make this comparison, explaining how you would carry out the random assignment of classes and what precautions you would take to minimize hidden biases.

b. How would you select teachers for the new curriculum?

8.3. To examine the effect of the seat belt laws on traffic accident casualties, the National Highway Traffic Safety Administration compared fatalities among those jurisdictions that were covered by seat belt laws (Covered Group) with those jurisdictions that were not covered by seat belt laws (Other Group). They found that among the Covered Group, 24 belt law jurisdictions, fatalities were 6.6 percent lower than the number forecasted from past trends. In the Other Group, observed fatalities were 2 percent above the forecasted level (7).

a. Explain whether or not you attribute the difference between these two groups to seat belt laws.

b. Provide some possible extraneous variables that might have influenced the effect difference and explain why these variables may have had an effect.

8.4. A large-scale experiment was carried out in 1954 to test the effectiveness of the Salk poliomyelitis vaccine (8). After considerable debate, the randomized placebo (double-blind) design was used in approximately half of the participating areas and the "observed control" design was used in the remaining areas. In the latter areas, children in the second grade were vaccinated and children in the first and third grades were considered as controls (no random assignment was used). In both areas, volunteers participated in the study, but polio cases were monitored among all children in participating areas. The following results were announced on April 12, 1955, at the University of Michigan:

Study type and group	*Study subjects*	*Polio case rate[a] (per 100,000)*			
		Total	*Paralytic*	*Nonparalytic*	*Fatal*
Placebo control areas					
Vaccinated	200,745	28	16	12	0
Placebo	201,229	71	57	13	2
Not inoculated[b]	338,778	46	36	11	0
Observed control areas					
Vaccinated	221,998	25	17	8	0
Controls	725,173	54	46	8	2
Not inoculated[c]	123,605	44	35	9	0

[a]Based on confirmed cases
[b]Nonvolunteers in the participating areas
[c]Second graders not inoculated
Source: Francis *et al.* (8, Tables 2 and 3).

a. Why was it necessary to use so many subjects in this trial?
b. What extraneous variables could have been confounded with the vaccination in the observed control areas?

8.5. To test whether or not oat-bran cereal diet lowers serum lipid concentrations (as compared with a corn flakes diet), an experiment was conducted (9). In this experiment 12 men with undesirably high serum total cholesterol concentrations were randomly assigned to one of the two diets on admission to the metabolic ward. After completing the first diet for 2 weeks, the subjects were switched to the other diet for another 2 weeks. This is a crossover design in which each subject received both diets in sequence. Eight subjects were hospitalized in the metabolic ward for a continuous 4-week period and the remaining subjects were allowed a short leave of absence, ranging from 3 to 14 days, between diet regiments for family emergencies or holidays. The results indicated that compared with the corn flakes diet, the oat-bran cereal diet lowered serum total cholesterol and serum LDL cholesterol concentrations significantly by 5.4 and 8.5 percent, respectively.

a. Discuss how this crossover design is different from the two-group comparison design studied in this chapter. What are the advantages of a crossover design?
b. The nutritional effects of the first diet may persist during the administration of the second diet. Is the carryover effect effectively controlled in this experiment?
c. Discuss any other factors that may have been confounded with the type of cereal.

8.6. To determine the efficacy of six different antihypertensive drugs in lowering blood pressure, a large experiment was conducted at 15 clinics (10). After a washout phase lasting 4 to 8 weeks (using a placebo without informing the subjects), a total of 1292 male veterans whose diastolic blood pressure was between 95 and 109 mm Hg were randomly assigned in a double-blind manner to one of the six drugs or a placebo. Each medication was prepared in three dose levels (low, medium, and high). Average age of the subjects was 59, 48 percent were black, and 71 percent were already on antihypertensive treatment at screening. All medications were started at the lowest dose and the dose was increased every 2 weeks, as required, until a diastolic blood pressure of less than 90 mm Hg was reached without intolerance to the drug on two consecutive visits or until the maximal drug dose was reached. The blood pressure measurement during treatment was taken as the mean of the blood pressures recorded during the last two visits. The following table shows the number of subjects assigned, the number that withdrew during the treatment, and the results on reduction in diastolic blood pressure:

			Reduction in diastolic BP		
Experimental group	*Number assigned*	*Number withdrawn*	*Mean*	*Std*	*Percent success*[a]
1. Hydrochlorothiazide	188	15	10	6	57
2. Atenolol	178	16	12	6	65
3. Captopril	188	23	10	7	54
4. Clonidine	178	13	12	6	65
5. Diltiazem	185	12	14	5	75
6. Prazosin	188	29	11	7	56
7. Placebo	187	29	5	7	33
Total	1292	137			

[a]Proportion of patients reaching the target blood pressure (diastolic blood pressure < 90 mm Hg).
Source: Materson *et al.* (10, Tables 2 and 3, Figure 1).

a. Discuss why the patients were not informed about the use of a placebo during the initial washout period.
b. More than 10 percent of the subjects withdrew from the study during the treatment and there were more withdrawals in some groups than in other groups. Discuss how the withdrawals may affect the experimental results.
c. Discuss how widely you can generalize the results of this experiment.

8.7. A randomized trial was conducted to test the effects of an educational program to reduce the use of psychoactive drugs in nursing homes. Six matched pairs of nursing homes were selected for this trial. The matching was based on the size of nursing home, type of ownership, and level of drug use. Professional staff and aides participated in an educational program at one randomly selected nursing home in each pair. At baseline, drug use status was determined for all residents of the nursing homes ($n = 823$), and a blinded observer performed standardized clinical assessments of the residents who were taking psychoactive medications. After the 5-month program, drug use and patient clinical status were reassessed and the educational program was found to have reduced the use of psychoactive drugs in the nursing homes (11).
a. How would you characterize the experimental design used in this study?
b. If the effectiveness of the educational program is related to the organizational and leadership types of the nursing home staff, is the effect of this confounder effectively controlled in this study? If not, how would you modify the experimental design?
c. Obviously not all the nursing homes that could be matched were included in this study. How might this limitation affect the study findings?
d. Discuss to what extent the study findings can be extrapolated to nursing homes in other states.

REFERENCES

1. Hypertension Detection and Follow-up Program Cooperative Group (1979). Five-year findings of the hypertension detection and follow-up program. *JAMA, J. Am. Med. Assoc.* **242,** 2562–2571.
2. Goldman, S., Copeland, J., Moritz, T., Henderson, W., Zadina, K., Ovitt, T., Doherty, J., Read, R., Chesler, E., Sako, Y., Lancaster, L., Emery, R., Sharma, G. V. R. K., Josa, M., Pacold, I., Montoya, A., Parikh, D., Sethi, G., Holt, J., Kirklin, J., Shabetai, R., Moores, W., Aldridge, J., Masud, Z., DeMots, H., Ploten, S., Haakenson, C., and Harker, L. A. (1988). Improvement in early saphenous vein graft patency after coronary artery bypass surgery with antiplatelet therapy: Results of a Veterans Administration cooperative study. *Circulation* **77,** 1324–1332.

3. Goldman, S., Copeland, J., Moritz, T., Henderson, W., Zadina, K., Ovitt, T., Doherty, J., Read, R., Chesler, E., Sako, Y., Lancaster, L., Emery, R., Sharma, G. V. R. K., Josa, M., Pacold, I., Montoya, A., Parikh, D., Sethi, G., Holt, J., Kirklin, J., Shabetai, R., Moores, W., Aldridge, J., Masud, Z., DeMots, H., Floten, S., Haakenson, C., and Harker, L. A. (1989). Saphenous vein graft patency 1 year after coronary artery bypass surgery and effects of antiplatelet therapy: Results of a Veterans Administration cooperative study. *Circulation* **80,** 1190–1197.
4. Deming, W. E. (1975). On probability as a basis for action. *Am. Stat.* **29,** 146–152.
5. Thomas, L. (1984). "The Youngest Science: Notes of a Medicine Watcher." Bantam Books, New York.
6. Office of Science and Technology Policy (1991). Federal policy for the protection of human subjects: Notices and rules. *Fed. Regist.* **56,** 28003–28032.
7. Campbell, B. J., and Campbell, F. A. (1988). Injury reduction and belt use associated with occupant restraint laws. *In* "Preventing Automobile Injury: New Findings from Evaluation Research" (J. D. Graham, ed.), Chapter 2. Auburn House Publ. Co., Dover, MA.
8. Francis, T., Jr., Korns, R. F., Voight, R. B., Boisen, M., Hemphill, F. M., Napier, J. A., and Tolchinsky, E. (1955). An evaluation of the 1954 poliomyelitis vaccine trials: Summary report. *Am. J. Public Health* **45,** Suppl., 1–63.
9. Anderson, J. W., Spencer, D. B., Hamilton, C. C., Smith, S. F., Tietyen, J., Bryant, C. A., and Oeltgen, P. (1990). Oat-bran cereal lowers serum total and LDL cholesterol in hypercholesterolemic men. *Am. J. Clin. Nutr.* **52,** 495–499.
10. Materson, B. J., Reda, D. J., Cushman, W. C., Massie, B. M., Freis, E. D., Kochar, M. S., Hamburger, R. J., Fye, C., Lakshman, R., Gottdiener, J., Ramirez, E. A., and Henderson, W. G. (1993). Single-drug therapy for hypertension in men: A comparison of six antihypertensive agents with placebo. *N. Eng. J. Med.* **328,** 914–921.
11. Avorn, J., Soumerai, S. B., Everitt, D. E., Ross-Degnan, D., Beers, M. H., Sherman, D., Salem-Schatz, S. R., and Fields, D. (1992). A randomized trial of a program to reduce the use of psychoactive drugs in nursing homes. *N. Eng. J. Med.* **327,** 168–173.

CHAPTER 9

Tests of Hypotheses

In this chapter, we formally introduce the testing of hypotheses and define key terms to help us succinctly communicate the ideas of hypothesis testing. Hypothesis testing is a way of organizing and presenting evidence that helps us reach a decision. For example, the decision may be to proceed with the marketing of a new drug for reducing cholesterol. This decision was reached because it is unlikely that the greater mean reduction of serum cholesterol observed in a sample of patients receiving a new drug, when compared with the reduction achieved for a sample of patients who received the standard treatment, was due to chance. Or, the decision may be for the local health department to allocate more resources to an immunization campaign for childhood diseases. This decision was reached because, based on the sample proportion immunized, it is unlikely that the proportion of 5-year-old children in the community that have the required immunizations equals the targeted value of 95 percent.

There are negative outcomes associated with making a wrong decision and these have to be weighed carefully. If the decision to market the new drug is wrong, that is, it is not an improvement over the standard treatment, patients may pay more money for no additional benefit or for a treatment that does not work. If, however, the decision were not to market and the drug was better, patients would lose by not having access to a better treatment and the company would lose because it did not realize the profit from this drug. If the health department's decision to conduct an immunization campaign is wrong, that is, the proportion of 5-year-old children immunized in the community is at least 95 percent, scarce resources would be misdirected. Other needy programs would not receive additional resources. If, however, the decision were not to conduct the campaign when it was needed, there would be increased risk of unnecessary disease in preschool children.

We use the following example to clarify these notions and to lead into the definitions used in hypothesis testing.

I. DEFINITIONS OF TERMS USED IN HYPOTHESIS TESTING

Suppose two diets are proposed for losing weight. We have 12 pairs of individuals, matched on age (±5 years), sex, initial weight (±10 pounds), and level of exercise. One member of the pair is assigned at random to diet 1 and the other member is assigned to diet 2. Individuals remain on their diets for 6 weeks and are then reweighed. We wish to determine whether or not the diets are equivalent from a weight loss perspective. Table 9.1 shows how the data, the weight losses for those on diets 1 and 2 and the within-pair difference, may be presented.

There are several ways of analyzing these data. In this chapter, we demonstrate a very simple approach and, in later chapters, show other approaches. We examine the proportion of pairs in which the person on diet 1 had the greater weight loss. If the diets do not differ with respect to weight loss, assuming there are no ties in weight loss, the proportion should be 0.50. Deviations from 0.50 suggest that there is a difference in the diets in terms of weight loss. If there are ties in the weight losses, the hypothesis being tested is that the proportion of pairs in which the person

TABLE 9.1 Weight Losses (Pounds) by Diet for 12 Pairs of Individuals

	Pair											
Diet	1	2	3	4	5	6	7	8	9	10	11	12
1	x_1	x_2	x_3	x_4	x_5	x_6	x_7	x_8	x_9	x_{10}	x_{11}	x_{12}
2	y_1	y_2	y_3	y_4	y_5	y_6	y_7	y_8	y_9	y_{10}	y_{11}	y_{12}
Difference	d_1	d_2	d_3	d_4	d_5	d_6	d_7	d_8	d_9	d_{10}	d_{11}	d_{12}

on diet 1 had the greater weight loss is the same as the proportion of pairs in which the person on diet 2 had the greater weight loss. Note that we have converted the hypothesis in words into something that we can deal with analytically.

A. Null and Alternative Hypotheses

The hypothesis being tested is called the null hypothesis and is denoted by H_0. The null hypothesis is that π, the proportion of pairs in the population for which persons on diet 1 would show the greater weight loss, is 0.50. The alternative hypothesis, denoted by H_a or H_1, to the null hypothesis is that π is not equal to 0.50. In symbols, these hypotheses are

$$H_0: \pi = 0.50$$

$$H_a: \pi \neq 0.50.$$

We either reject or fail to reject the null hypothesis. If we reject the null hypothesis, we are expressing a belief that the alternative hypothesis is true. If there are ties in the weight losses, the alternative hypothesis is that the proportion of pairs in which the person on diet 1 had the greater weight loss is not equal to the proportion of pairs in which the person on diet 2 had the greater weight loss.

B. Type I and Type II Errors

If we reject the null hypothesis in favor of the alternative hypothesis, there are two possible outcomes. Either we have correctly rejected the null hypothesis or we have falsely rejected it. Falsely rejecting the null hypothesis is called a type I error. In this example, the type I error is claiming that the proportion of pairs for which diet 1 showed the greater weight loss is not equal to 0.50 when, in fact, it is 0.50.

If we fail to reject the null hypothesis, again there are two possible outcomes. Either we have failed to reject the null hypothesis when it should have been rejected or we have correctly failed to reject the null hypothesis. Failing to reject the null hypothesis when it should have been rejected is called a type II error. The type II error in this example is claiming that the proportion of pairs for which diet 1 showed the greater weight loss is 0.50 when, in fact, the proportion is different from 0.50.

Figure 9.1 shows these four possibilities. The probability of a type I error is usually labeled α and the probability of a type II error is usually labeled β. Ideally we would like to keep both of these probabilities as small as possible, although we usually focus more on the type I error and its probability.

Our Decision about the Null Hypothesis	Reality: Null Hypothesis is True	Reality: Null Hypothesis is False
True	Good	Type II Error
False	Type I Error	Good

FIGURE 9.1 Possibilities associated with a test of hypothesis.

C. The Test Statistic

The test statistic, the basis for the test of hypothesis, is the number of pairs out of the 12 sample pairs for which those on diet 1 achieved the greater weight loss. Equivalently, the observed sample proportion of pairs for which those on diet 1 achieved the greater weight loss, p, could be used. The test is based on the sign of the difference and, therefore, this particular test is called the sign test. Now that we know what hypothesis is to be tested and what test statistic is to be used, we must specify the decision rule to be used.

II. DETERMINATION OF THE DECISION RULE

The decision rule specifies which values of the test statistic (or some function of it) will cause us to reject the null hypothesis in favor of the alternative hypothesis.

The decision rule is based on the probabilities of the type I and II errors. The probabilities of type I and type II errors are found from consideration of the distribution of the test statistic. In this example, the test statistic follows the binomial distribution. The binomial is used because there are only two outcomes: diet 1 is better or diet 2 is better (again ignoring the possibility of a tie in weight loss). We begin by assuming that the null hypothesis is true, that is, π is 0.50. Because we know that n is 12, we know both parameters of the binomial distribution. The probability distribution of the possible outcomes is shown in the following table and in Figure 9.2.

Number of times diet 1 is better	*Probability*	*Number of times diet 1 is better*	*Probability*
0	0.0002	7	0.1934
1	0.0030	8	0.1208
2	0.0161	9	0.0537
3	0.0537	10	0.0161
4	0.1208	11	0.0030
5	0.1934	12	0.0002
6	0.2256		

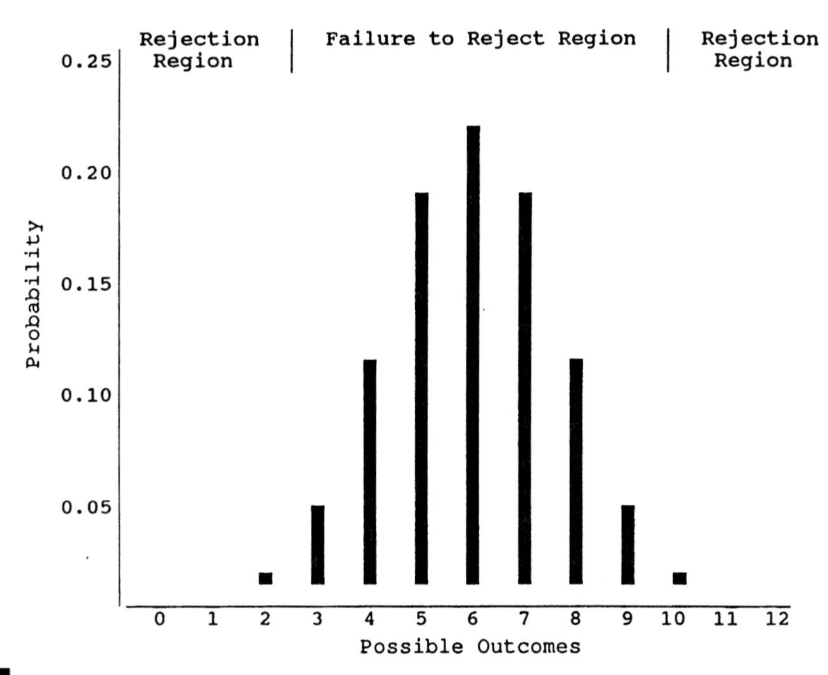

FIGURE 9.2 Bar chart showing the binomial probability distribution for $n = 12$ and $\pi = 0.50$.

What values of the test statistic would cause us to reject the null hypothesis that π is 0.50 in favor of the alternative hypothesis? Large deviations from six pairs for which diet 1 was better, that is, large deviations from a π of 0.50, are suggestive that the diets have different effects. Thus either very large or very small values of the test statistic would cause us to question the null hypothesis. As we can see from Figure 9.2, it is highly unlikely to observe either very large or very small values of the test statistic if π is really 0.50.

A. One- and Two-Sided Tests

The test we are considering is called a *two-sided test* because either large or small values of the test statistic cause us to question the truth of the null hypothesis. A *one-sided test* occurs when only values in one direction cause us to question the null hypothesis. For example, if we were the developers of diet 1, we might only be interested in whether diet 1 was better than diet 2, not whether it was worse than diet 2. If this were the situation, the null hypothesis remains that π is equal to 0.50, but the alternative hypothesis becomes that π is greater than 0.50. In symbols, this is

$$H_0: \pi = 0.50 \qquad \text{versus} \qquad H_a: \pi > 0.50.$$

In this case, only large values of the test statistic would cause us to reject the null hypothesis in favor of the alternative hypothesis.

Use of a one-sided test makes it easier to detect departures from the null hypothesis in the indicated direction, that is, π greater than 0.50; however, use of a one-sided test means that if the departure is in the other direction, that is, π is less than 0.50, it will be missed.

B. Calculation of the Probabilities of Type I and Type II Errors

Suppose that we decide to reject the null hypothesis whenever we observe a test statistic of 0 or 12 pairs; that is, the values of 0 and 12 form the rejection or critical region. The values from 1 to 11 then form the failure to reject or acceptance region. The probability of a type I error, α, is thus the probability of observing 0 or 12 pairs in which diet 1 had the greater weight loss when π is actually 0.50. From the above probability mass function, we see that α is 0.0004. That's great! There is almost no chance of making this error and this is almost as small as we can make it. Of course, we could decide never to reject the null hypothesis and, then, there would be zero probability of a type I error. That is unrealistic, however.

We are pleased with this decision rule because it has an extremely small probability of a type I error. However, what is the value of β, the probability of a type II error, associated with this decision rule? To be able to calculate β, we have to be more specific about the alternative hypothesis. The above alternative hypothesis is quite general in that it only says π is not equal to 0.50; however, just as we used a specific value, the value 0.50, for π in calculating the probability of a type I error, we must specify a value of π other than 0.50 to be used in calculating the probability of a type II error. We must move from the general alternative to a specific alternative hypothesis to be able to calculate a value for β. This means that there is not merely one β associated with the decision rule; rather, there is a value of β corresponding to each alternative hypothesis.

What specific value of π should be used in the alternative hypothesis? We should have little interest in the alternative that π is 0.51 instead of the null hypothesis value of 0.50. The difference between 0.51 and 0.50 is of little practical interest. For all practical intent, if π is really 0.51, there is little difference in the diets. As the value of π departs more and more from 0.50, the ability to detect these departures becomes more important. We may not all agree at which point π differs enough from 0.50 to be important. Some may say 0.60 is different enough, whereas others may say that π must be at least 0.70 for the difference to be important. Most would certainly agree that we should reject the equality of the diets if diet 1 provides for greater weight loss in 80 percent of the pairs.

Let us assume that π is really 0.80, not 0.50, and find the value for β. The binomial distribution for an n of 12 and a proportion of 0.80 is shown next.

Number of times diet 1 is better	*Probability*	*Number of times diet 1 is better*	*Probability*
0	0.0000	7	0.0532
1	0.0000	8	0.1328
2	0.0000	9	0.2363
3	0.0001	10	0.2834
4	0.0005	11	0.2062
5	0.0033	12	0.0687
6	0.0155		

Type II error is failing to reject the null hypothesis when it should be rejected. As our decision rule is to reject only when we observe a test statistic of 0 or 12, we will fail to reject for the values of 1 through 11. The probability of 1 through 11 when π is actually 0.80 is 0.9313 (= 1 − 0.0000 − 0.0687). Therefore, use of this decision rule yields an α of 0.0004 and a β of 0.9313. The probability of the type I error is very small, but the probability of the type II error, corresponding to the value of 0.80 for π, is quite large.

III. RELATIONSHIP OF THE DECISION RULE, α AND β

If we change our decision rule to reject the null hypothesis more often, we will increase α but decrease β, that is, there is an inverse relationship between α and β. For example, if we increase the rejection region by including values 1 and 11 in addition to 0 and 12, the value of α becomes 0.0064 (= 0.0002 + 0.0030 + 0.0030 + 0.0002). These probabilities are found from the probability distribution based on the value for π of 0.50. The new value for β, based on this expansion of the rejection region, and using 0.80 for π, is 0.7251 (= 1 − 0.2062 − 0.0687). The probability of a type I error remains quite small but the probability of a type II error is still large.

If the decision rule is to reject for values of the test statistic of 0 to 2 and 10 to 12, then the value of α is increased to 0.0386 [= 2 * (0.0161 + 0.0030 + 0.0002)] and the value of β is reduced to 0.4417 (= 1 − 0.0687 − 0.2062 − 0.2834). The probability of a type I error is still reasonable, whereas although the probability of type II error is much smaller than above, it is still quite large. A further change in the decision rule to include test statistic values of 3 and 9, however, increases the value of α to 0.1460 (= 0.0386 + 2 * 0.0537), which is now becoming large.

A. What Are Reasonable Values for α and β?

There are no absolute values that indicate that the probability of error is too large. It is a matter of personal choice, although convention suggests that an α greater than 0.10 is unacceptable. Most investigators set α to 0.05, and some set it to 0.01. There is less guidance for the choice of β. It again is a matter of personal choice; however, the implications of the type II error play a role in how large a β can be tolerated. A value of 0.20 for β is used frequently in the literature. Investigators often ignore the type II error because (1) the hypothesis has been framed such that the type I error is of much greater interest than the type II error, or (2) it is often difficult to find the value of β.

B. Ways to Decrease β without Increasing α

We were in a bind when we left the example above. The value of β was too large, and if we tried to reduce it by further enlargement of the rejection region, we made α too large. One way of decreasing β without increasing α is to change the alternative hypothesis.

1. Changing the Alternative Hypothesis

The specific alternative hypothesis that we had used above in calculating β was that π was equal to 0.80. We selected the value of 0.80 because, if diet 1 performed better for 80 percent of the pairs, we thought that this indicated a really important difference between the diets. If we are willing to change what we consider to be a really important difference, we can reduce β. For example, by increasing the value of π in the alternative hypothesis from 0.80 to 0.90, β will decrease. This means that we no longer consider it to be important to detect that π was really 80 percent instead of the hypothesized 50 percent. We focus on the test's ability to detect a really large difference, that is, the difference between 0.90 and 0.50, and not worry that the test has a small chance of detecting smaller differences.

The following table shows the probability mass function for the binomial with a sample size of 12 and a proportion of 0.90.

Number of times diet 1 is better	*Probability*	*Number of times diet 1 is better*	*Probability*
0	0.0000	7	0.0038
1	0.0000	8	0.0213
2	0.0000	9	0.0853
3	0.0000	10	0.2301
4	0.0000	11	0.3766
5	0.0001	12	0.2824
6	0.0004		

If we again use the rejection region of 0 to 2 and 10 to 12, the probability of a type I error is still 0.0386, as that was calculated based on π being 0.50; however, β is the probability of not rejecting that π is 0.50 when it is actually 0.90. This probability is the sum of the probabilities of the outcomes 3 through 9 in the above distribution, and that sum is 0.1109. Now the values of both α and β are reasonable.

By changing the alternative hypothesis, we have not changed the value of β for the alternative of π being 0.80. The β, corresponding to a π of 0.80 and a rejection region of 0 to 2 and 10 to 12, remains 0.4417. What has changed is what we consider to be an important difference. If a lesser difference is considered to be important, the probability of the type II error for that value of π can be calculated. Table 9.2 shows the values of the type II errors for several values of the alternative hypothesis based on a rejection region of 0 to 2 and 10 to 12.

The probability of a type II error decreases as the value of π used in the alternative hypothesis moves farther away from its value in the null hypothesis. This makes sense because it should be easier to detect greater differences than smaller ones. As this table shows, there is a very high chance of failing to detect departures from 0.50 less than 0.30 to 0.35 in magnitude. The last column in the table is power, the probability of rejecting the null hypothesis when it should be rejected, that is, when the alternative hypothesis is true. From the table we can see that power is 1 minus β. Power is often used in the literature when discussing the properties of a test statistic instead of using the probability of a type II error. From the values in Table 9.2, it is possible to create a *power curve*, that is, to graph the values of power versus the values of π used in the alternative hypothesis. Figure 9.3 shows a portion of the power curve for values of π greater than 0.50. Statisticians use power curves to compare different test statistics.

The above trade-off as a way of reducing β may not be very satisfactory. We still may feel that 80 percent is very different from 50 percent. As an alternative, we could increase the sample size instead of changing the alternative hypothesis.

2. Increasing the Sample Size

None of the calculations shown so far have required the observed sample data. All these calculations are preliminary to the actual collection of the

TABLE 9.2 Probability of Type II Error and Power Values for Specific Alternative Hypotheses Based on a Rejection Region of 0 to 2 and 10 to 12

Alternative hypothesis	Probability of type II error	Power
π = 0.55	0.9507	0.0493
π = 0.60	0.9137	0.0863
π = 0.65	0.8478	0.1522
π = 0.70	0.7470	0.2530
π = 0.75	0.5778	0.4222
π = 0.80	0.4416	0.5584
π = 0.85	0.2642	0.7358
π = 0.90	0.1109	0.8891
π = 0.95	0.0195	0.9805

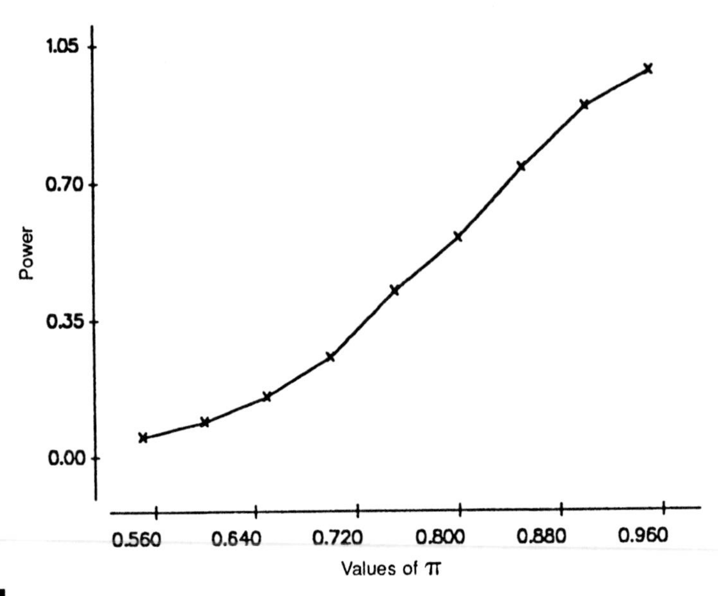

FIGURE 9.3 Portion of the power curve for π values greater than 0.5.

data. Therefore, if the probabilities of errors are too large, we can still change the experiment. As mentioned above, increasing the sample size is one way of decreasing the error probabilities, but doing this increases the resources required to perform the experiment. There is a trade-off between the sample size and the error probabilities.

Suppose we can afford to find and follow 15 pairs instead of the 12 pairs we initially intended to use. We still use the binomial distribution in the calculation of the error probabilities where π remains 0.50, but now n is equal to 15. The binomial probability mass function with these parameters is shown next.

Number of times diet 1 is better	*Probability*	*Number of times diet 1 is better*	*Probability*
0	0.0000	8	0.1964
1	0.0005	9	0.1527
2	0.0032	10	0.0917
3	0.0139	11	0.0416
4	0.0416	12	0.0139
5	0.0917	13	0.0032
6	0.1527	14	0.0005
7	0.1964	15	0.0000

Let us use a rejection region of 0 to 3 and 12 to 15. If we do this, the probability of a type I error is 0.0352 [$= 2 * (0.0005 + 0.0032 + 0.0139)$]. The probability of a type II error, based on the alternative that π is 0.80, uses the binomial distribution with parameters 15 and 0.80 and this probability mass function is now shown.

Number of times diet 1 is better	*Probability*	*Number of times diet 1 is better*	*Probability*
0	0.0000	8	0.0139
1	0.0000	9	0.0430
2	0.0000	10	0.1031
3	0.0000	11	0.1876
4	0.0000	12	0.2502
5	0.0001	13	0.2309
6	0.0007	14	0.1319
7	0.0034	15	0.0352

The probability of failing to reject the null hypothesis when it should be rejected, that is, of being in the acceptance region, values 4 to 11, when π is 0.80, is 0.3518 (= 0.0001 + 0.0007 + 0.0034 + 0.0139 + 0.0430 + 0.1031 + 0.1876). The probability of a type I error, 0.0352, is similar to its value above, 0.0386, when we considered this same alternative hypothesis. The probability of a type II error has decreased from 0.4417 above when n was 12 to 0.3518 now for an n of 15. A further increase in the sample size can reduce β to a more acceptable level. For example, when n is 20, use of values 0 to 5 and 15 to 20 for the rejection region leads to an α of 0.0414 and a β of 0.1958.

IV. CONDUCTING THE TEST

The procedure used in conducting a test begins with a specification of the null and alternative hypotheses. In this example, they are

$$H_0\colon \pi = 0.50 \qquad \text{versus} \qquad H_a\colon \pi \neq 0.50.$$

We must decide on the significance level to be used in conducting the test. The *significance level* is the probability of a type I error we are willing to accept. We use the conventional significance level of 0.05 in this example.

Based on the calculations above, we have decided to increase the sample size to 20. We will reject the null hypothesis if the test statistic's value is 0 to 5 or 15 to 20. Use of this sample size and decision rule keeps the probability of a type I error less than 0.05 and also keeps β reasonably small when considering large departures from the null hypothesis. With discrete data, the probability of a type I error usually does not equal the significance level exactly. The decision rule used with discrete data is chosen so that it results in the probability of a type I error being as close as possible to and less than the desired significance level. The data are collected and shown in Table 9.3.

TABLE 9.3 Weight Losses (Pounds) by Diet for 20 Pairs of Individuals

Diet	Pair 1	2	3	4	5	6	7	8	9	10
1	9	4	11	7	−4	13	6	3	8	10
2	7	6	9	12	3	8	5	−1	14	8
Difference	2	−2	2	−5	−7	5	1	4	−6	2

Diet	Pair 11	12	13	14	15	16	17	18	19	20
1	8	9	14	11	5	−3	6	7	13	9
2	6	8	15	7	7	4	−2	4	10	5
Difference	2	1	−1	4	−2	−7	8	3	3	4

There are 13 pairs for which persons on diet 1 had the greater weight loss. As 13 does not fall into the rejection region of 0 to 5 or 15 to 20, we fail to reject the null hypothesis in favor of the alternative hypothesis at the 0.05 significance level. The observed result is not statistically significant.

A. The *p* Value

Another statistic often reported is the p value of the test, the probability of a type I error associated with the smallest rejection region which includes the observed value of the test statistic. Another way of stating this is that the p value is the level at which the observed result would just be statistically significant. In this example, because we are conducting a two-sided test, the smallest rejection region including the observed result of 13 is the region 0 to 7 and 13 to 20. Examination of Table B2 for an n of 20 and a π of 0.50 yields a probability of being in this region of 0.2632 [$= 2 * (0.0370 + 0.0739) + 0.0414$]. The value of 0.0414 is the value associated with the region 0 to 5 and 15 to 20 and, to that, we have added the probabilities associated with the outcomes 6, 7, 13, and 14. The p value is thus 0.2632.

Some statisticians do not believe in the decision rule approach to testing hypotheses. They believe that the p value provides information regardless of whether the hypothesis is rejected. The p value tells how likely the observed result is assuming that the null hypothesis is true. For example, these statisticians see little difference in p values of 0.05001 and 0.04999, although in the first case we would fail to reject the null hypothesis at the 0.05 significance level, whereas in the second case, we would reject the null hypothesis. For these statisticians, the key information to be obtained from the study is that there is roughly 1 chance in 20 that we would have obtained the observed result if the null hypothesis were true. Using the p value in this way is very reasonable.

B. Statistical and Practical Significance

We must not confuse statistical significance with practical significance. For example, in the diet study, if we had a large enough sample, a π value of 0.51 could be significantly different statistically from the null hypothesis value of 0.50; however, this finding would be of little use practically. For a result to be important, it should be both statistically and practically significant. The test determines statistical significance, but the investigator must determine whether or not the observed difference is large enough to be practically significant.

V. TESTS OF HYPOTHESES AND CONFIDENCE INTERVALS

When reporting the results of a study, many researchers have simply indicated whether or not the result was statistically significant and/or given only the p value associated with the test statistic. This is useful information, but it is more informative to include the confidence interval for the parameter as well. For example, in the weight loss experiment, the parameter is the proportion of pairs in the population that would show a greater weight loss for diet 1. Its sample estimate was 0.65 (= 13/20). Because we performed the test of hypothesis at the 0.05 significance level, we shall form the 95 percent confidence interval for the population proportion.

We can use either MINITAB or Table B6a to find the confidence interval. Use of MINITAB gives a 95 percent confidence interval with limits of 0.408 and 0.846. Based on the confidence interval containing the value of 0.50, we would say that there is no difference in the likelihood of either diet providing the greater weight loss. This result agrees with the result of the test of hypothesis. For the parameters that we shall study, the results of the confidence interval and the test of the hypothesis agree when the test statistic is a continuous variable.[1]

[1] When the test statistic is a discrete variable, as in the above example, the results usually agree. The results may disagree because, although the test was performed at the 0.05 level, the actual probability of a type I error may be slightly less than 0.05. In the above example, the actual probability of a type I error that we finally used was 0.0414. Hence, use of the 95.86 percent confidence interval is more appropriate than the 95 percent interval. Use of confidence intervals based on the actual probability of a type I error will cause the hypothesis test and confidence intervals to agree in all cases we consider in this book.

VI. CONCLUDING REMARKS

In this chapter we have introduced hypothesis testing and the associated terminology. A key point is that the probabilities of errors should be calculated before the study is performed. By doing this, we can determine whether or not the study, as designed, can deliver answers to the question of interest with reasonable error levels.

In the following chapters, we present different tests of hypotheses and the appropriate test statistics for performing the tests. In Chapter 10, we begin the presentation with tests that are called distribution-free or nonparametric tests. These names were given because these tests do not require that the data follow any particular distribution and hence there are no parameters that have to be estimated.

EXERCISES

9.1. Provide at least one way that the weight loss study in the chapter could be compromised. What would you do in an attempt to deal with this potential problem?

9.2. In the diet study with a sample size of 20 pairs, suppose that we used a rejection region of 0 to 4 and 16 to 20. The null and alternative hypotheses are the same as in the chapter, and we are still interested in the specific alternative that π is 0.80. What are the values of α and β based on this decision rule? What is the power of the test for this specific alternative? We again observed 13 pairs favoring diet 1. What is the p value for this result?

9.3. Suppose that the null and alternative hypotheses in the diet study were

$$H_0: \pi = 0.50 \qquad \text{versus} \qquad H_a: \pi > 0.50$$

Conduct the test at the 0.05 significance level. What is the decision rule that you will use? What are the probabilities of type I and type II errors for a sample size of 20 pairs and the specific alternative that π is 0.80?

9.4. What specific alternative value for π do you think indicates an important difference in the diet study? Provide an example of another study for which the binomial distribution could be used. What value would you use for the specific alternative for π in your study? What is the rationale for your choice for π in this new study?

9.5. Complete Table 9.2 by providing the values of power for π ranging from 0.05 to 0.50 in increments of 0.05. Graph the values of the power function versus the values of π for π ranging from 0.05 to 0.95. This graph is the power curve of the binomial test using the critical region of 0 to 2 and 10 to 12. What is the value of power when π is 0.50? Is there a specific name for this value? Describe the shape of the power curve. Discuss why the power curve, when the null hypothesis is π is equal to 0.50, must have this shape.

9.6. Construct the 95.86 percent confidence interval for the population proportion of pairs that show the greater weight loss for diet 1 based on the sample size of 20 and the sample estimate of 0.65.

9.7. Frickhofen *et al.* (1) performed a study on the effect of using cyclosporine in addition to antilymphocyte globulin and methylprednisolone in the treatment of aplastic anemia patients. A sample of 43 patients received the cyclosporine in addition to the other treatment. Assume that the use of antilymphocyte globulin and methylprednisolone without cyclosporine results in complete or partial remission in 40 percent of aplastic anemia patients at the end of 3 months of treatment. We wish to determine if the use of cyclosporine can increase significantly the percentage of patients with complete or partial remission. What are the appropriate null and alternative hypotheses? Assume that the test is to be performed at the 0.01 significance level. What is the decision rule to be used? What is the probability of a type II error based on the sample size of 43 and your decision rule? Twenty-eight patients achieved complete or partial remission at the end of 3 months. Is this a statistically significant result at the 0.01 level? What is the p value of the test?

REFERENCE

1. Frickhofen, N., Kaltwasser, J. P., Schrezenmeier, H., Raghavachar, A., Vogt, H. G., Herrmann, F., Freund, M., Meusers, P., Salama, A., and Heimpel, H. (1991). Treatment of aplastic anemia with antilymphocyte globulin and methylprednisolone with or without cyclosporine. *N. Eng. J. Med.* **324,** 1297–1304.

CHAPTER 10

Nonparametric Tests

In this chapter we present several statistics for testing whether or not probability distributions have the same medians. The use of these statistics does not require that the sample data follow any particular probability distribution and, thus, there are no distributional parameters to be estimated. Because of these features, these tests are called distribution-free or nonparametric tests. The only requirement of the data is that they come from continuous distributions. We begin with the sign test, a test we met in the last chapter.

I. THE SIGN TEST

The sign test is one of the oldest tests used in statistics. For example, in 1710 John Arbuthnot, a British physician and collaborator of Jonathan Swift, performed what was in effect a sign test on the sex ratio of births over an 82-year period (1, pp. 225–226).

As we saw in the last chapter, the sign test can be used to compare different interventions for matched pairs. Individuals were assigned to a pair based on age, sex, weight, and exercise level; then one member within the pair was randomly assigned to diet 1 and the other member assigned to diet 2. The sign test was used to determine which of the two diets was more likely to be associated with the greater weight loss for each pair. Another way of stating this null hypothesis is that each difference of weight losses has a median of zero. The sign test can also be used with a single population, for example, in the comparison of multiple measurements made on the same individual as is shown next.

One problem often encountered in research designs involving pre- and posttest measurements is the reversion or regression toward the mean effect (2). Briefly, persons scoring high on one test tend not to score as high on a subsequent test, and low scorers on the first test tend to score higher on the next test; that is, the test scores tend to revert toward the mean score. We wish to see if there is reversion toward the mean in dietary data. We consider the caloric intake for the 33 boys we first encountered in Chapter 4.

Table 10.1 shows the caloric intake for the boys for the first two of three randomly selected days during a 2-week period. The more extreme—the seven highest and seven lowest—day 1 values are shown in bold type.

A. Descriptive Statistics

We can examine whether or not there is reversion toward the mean, as shown in Box 10.1, for those with the seven highest and those with the seven lowest day 1 intakes separately. Based on the descriptive statistics in

TABLE 10.1 Two Days of Caloric Intake for 33 Boys Enrolled in Two Middle Schools outside of Houston

ID	Day 1	Day 2	ID	Day 1	Day 2	ID	Day 1	Day 2
10	1823	1623	39	2330	2339	118	**1781**	1844
11	2007	1748	40	2436	2189	120	2748	2104
13	**1053**	2484	41	**3076**	2431	127	2348	2122
14	**4322**	2926	44	1843	2907	130	**2773**	3236
16	**1753**	1054	46	2301	4120	137	2310	1569
17	2685	2304	47	2546	1732	139	2594	2867
26	2340	3182	50	**1292**	810	141	1898	1236
27	**3532**	3289	51	**3049**	2573	145	2400	2554
30	**2842**	2849	101	**3277**	2185	148	2011	1566
32	2074	3312	105	2039	1905	149	**1645**	2269
33	**1505**	1925	107	2000	1797	150	**1723**	3163

MINITAB BOX 10.1

We use the DESCRIBE command. Column c1 contains the 14 extreme day 1 intakes and c2 contains the corresponding day 2 intakes. Column c3 contains an indicator that tells whether the day 1 value is a low value (c3 = 1) or a high value (c3 = 0). Columns c4 and c5 contain the values for the boys with the seven lowest intakes on day 1, and c6 and c7 contain the day 1 and day 2 values for the boys with the highest day 1 intakes.

```
MTB > set cl
DATA> 1053 4322 1753 3532 2842 1505 3076 1292 3049 3277 1781
DATA> 2773 1645 1723
DATA> set c2
DATA> 2484 2926 1054 3289 2849 1925 2431  810 2573 2185 1844
DATA> 3236 2269 3163
DATA> set c3
DATA> 1 0 1 0 0 1 0 1 0 0 1 0 1 1
DATA> end
MTB > copy cl c2, c4 c5;
SUBC> use c3=1.
MTB > copy cl c2, c6 c7;
SUBC> use c3=0.
MTB > desc c4-c7
                 N      MEAN    MEDIAN    TRMEAN     STDEV    SEMEAN

C4               7      1536      1645      1536       273       103
C5               7      1936      1925      1936       814       307
C6               7      3267      3076      3276       531       201
C7               7      2784      2849      2784       411       155
```

(We have not shown the last part of the MINITAB output.)

Box 10.1, it appears that there could be a reversion toward the mean effect here. The seven lowest values had a mean of 1536 calories on the first day of intake compared with a mean of 1936 calories on the second day. The seven highest values had a mean of 3267 calories on the first day compared with a mean of 2784 calories on the second day. However, we wish to go beyond a descriptive presentation of the sample in our consideration of the question. We wish to test a hypothesis about the population values and we show two ways of doing this.

B. Hypothesis of Interest

If there is no reversion toward the mean effect here, of the boys with extreme day 1 values, the proportion of those whose day 2 values move in the direction of the mean should be equal to 0.50 (ignoring the possibility that the day 1 and day 2 values are the same). If there is reversion toward the mean, the proportion should be greater than 0.50. The null and alternative hypotheses are therefore

$$H_0\text{: } \pi = 0.50 \qquad \text{versus} \qquad H_a\text{: } \pi > 0.50$$

If there are few ties (a subject has same values for day 1 and day 2) in the data, convention is that these observation pairs are dropped from the data. For example, if one of the 14 boys had the same day 1 and day 2 values, the sample size for the binomial would then be 13 instead of 14, reflecting the deletion of the tied pair. When there are many ties, indicating no difference in the day 1 and day 2 values, there is little reason to perform the test for the remaining untied pairs.

C. Population

As we mentioned in Chapter 4, the population from which this sample is drawn consists of middle schools in a northern suburb of Houston. Although the population is limited, perhaps the results from this population can be generalized to boys in suburban middle schools throughout the United States, not just to those in one suburb of Houston. As was mentioned in Chapter 8, this generalization does not flow from statistical properties because we did not sample from this larger population, but is based on substantive considerations. If there are differences in dietary practices between the one Houston suburb and others, this generalization to the larger population is then questionable.

D. Conducting the Test

We conduct the test of hypothesis at the 0.05 level. The test statistic is the number of boys with an extreme day 1 value whose day 2 value moves toward the mean, which is found to be 10 from the data. The critical region for the test can be found from the binomial distribution. For larger sample sizes, the normal approximation to the binomial can be used. We could use Table B2 or MINITAB, as shown in Box 10.2, to find the probabilities for a binomial distribution with parameters 14 and 0.50. We are interested only in the upper tail of the binomial distribution; therefore, c1 includes only values above the expected value of 7.

Because we wish to perform the test at the 0.05 level, the rejection region consists of the values of 11 to 14. If 10 were included in the rejection region, the probability of a type I error would exceed the significance level of 0.05. Ten of the fourteen boys with an extreme day 1 value had day 2 values that moved in the direction of the mean. As 10 is not included in the rejection region, we fail to reject the null hypothesis in favor of the alternative at the 0.05 significance level. Although we fail to reject the null hypothesis, the p value of this result is 0.0898.

MINITAB BOX 10.2

```
MTB > set cl
DATA> 8:14
DATA> end
MTB > pdf cl;
SUBC> binom 14 .50.
        K              P( X = K)
     8.00                0.1833
     9.00                0.1222
    10.00                0.0611
    ------------------------------
    11.00                0.0222 Rejection
    12.00                0.0056
    13.00                0.0009     Region
    14.00                0.0001
```

E. Power of the Test

What is the power of the test; that is, What is the probability of rejecting the null hypothesis when it should be rejected? As we saw in Chapter 9, to find a value for power, we must provide a specific alternative. Let us work with the alternative that π is 0.70. Then the power is easily found from MINITAB as shown in Box 10.3. The power is 0.3552 (= 0.1943 + 0.1134 + 0.0407 + 0.0068), not a large value.

F. The STEST Command

It is even easier to perform the sign test using MINITAB as shown in Box 10.4. Of the 14 pairs, 10 have a positive sign and 4 have a negative sign. The test is not statistically significant at the 0.05 level as the p value is

MINITAB BOX 10.3

```
MTB > pdf cl;
SUBC> binom 14 .70.
        K              P( X = K)
     8.00                0.1262
     9.00                0.1963
    10.00                0.2290
    ------------------------------
    11.00                0.1943 Rejection
    12.00                0.1134
    13.00                0.0407     Region
    14.00                0.0068
```

MINITAB BOX 10.4

We use the command STEST. We store the changes toward (+) and away from (−) the mean in column c1 for the 14 boys with the seven highest and seven lowest calorie intakes on day 1 and then use the STEST command. The subcommand alt 1 tells MINITAB that this is a one-sided test and that the alternative is greater than. The subcommand alt −1 also indicates a one-sided test, but the alternative is now less than. When there is no subcommand, MINITAB assumes that the test is two-sided.

```
MTB > set c1
DATA> 1431 1396 -699 243 -7 420 645 -482 476 1092 63 -463 624 1440
DATA> end
MTB > stest c1;
SUBC> alt 1.

SIGN TEST OF MEDIAN = 0.00000 VERSUS  G.T.  0.00000
                 N  BELOW  EQUAL  ABOVE   P-VALUE     MEDIAN
 C1             14      4      0     10    0.0898      448.0
```

greater than 0.05. MINITAB uses the binomial distribution to calculate the *p* value for sample sizes of 50 or less. For larger sample sizes, it uses the normal approximation with the continuity correction in the calculation of the *p* value. The median value of the 14 numbers is 448. In this case the median is of little interest.

The sign test is easy to perform as the test statistic is simply a count of the occurrences of some event, for example, a move toward the mean or a positive difference. The test can also be used with nonnumerical data, for example, in situations in which the outcome is that the subject does or does not feel better. The simplicity of the test is attractive but, with numeric data, in ignoring the magnitude of the values, the sign test does not use all the information in the data. The other tests in this chapter use more of the available information in the data.

II. WILCOXON SIGNED RANK TEST

Another much more recently developed test that can be used to examine whether or not there is reversion toward the mean in the above data is the Wilcoxon signed rank (WSR) test. An American statistician who worked in the chemical industry, Frank Wilcoxon developed this test in 1945. Unlike the sign test which can be used with nonnumeric data, the WSR test requires that the differences in the paired data come from a continuous distribution.

The data for the 14 boys with an extreme day 1 value are shown in Table 10.2. In this table, the differences between the day 1 and day 2 values

TABLE 10.2 Day 1 and 2 Caloric Intakes for the 14 Boys with the More Extreme Caloric Intakes on Day 1

ID	Day 1	Day 2	Change (+) toward the mean	Change (−) away from the mean	Rank +	Rank −
13	1053	2484	1431		13	
14	4322	2926	1396		12	
16	1753	1054		699		10
27	3532	3289	243		3	
30	2842	2849		7		1
33	1505	1925	420		4	
41	3076	2431	645		9	
50	1292	810		482		7
51	3049	2573	476		6	
101	3277	2185	1092		11	
118	1781	1844	63		2	
130	2773	3236		463		5
149	1645	2269	624		8	
150	1723	3163	1440		14	
				Sum of ranks:	82	23

are shown as either a change in the direction of the mean (+) or away from the mean (−). The differences are ranked from smallest to largest, and the ranks are summed separately for those changes in the direction of the mean and for those changes away from the mean. We use R_{WSR} to represent the signed rank sum statistic for the positive differences, in this case, those changes toward the mean.

A. Expected Value of the Signed Rank Sum under the Null Hypothesis

When there are n pairs of data, the sum of the ranks is the sum of the integers from 1 to n and that sum is $n * (n + 1)/2$. The average rank for an observation is therefore $(n + 1)/2$.

The null hypothesis is that the changes have a median of zero, and the alternative hypothesis in this example is that the median is greater than zero.[1] If changes toward the mean and away from the mean are equally likely, that is, the changes have a median of zero, there should be $n/2$ changes in the direction of the mean and $n/2$ away from the mean. Therefore, if there is no reversion toward the mean, the sum of the ranks for the changes toward the mean, R_{WSR}, should be $(n/2) * (n + 1)/2$. If there is regression toward the mean, R_{WSR} should be greater than $n * (n + 1)/4$.

[1]If H_0 is true, the distribution of the differences will be symmetric.

B. Determination of the Critical Region

The test statistic is the sum of the ranks of the changes toward the mean. Table B9 provides boundaries for the critical region for the sum of the ranks of the positive changes, in this case, changes toward the mean. To give an idea how these boundaries were determined, let us consider 5 pairs of observations instead of the 14 pairs above. The boundaries result from the enumeration of possible outcomes as shown in Table 10.3.

In Table 10.3, there is no need to show the sum of ranks for 3, 4, and 5 positive ranks because their values are already shown under the sum of the negative rank column. For example, when there are 0 positive ranks, there are 5 negative ranks with a sum of 15. But the sum of 5 positive ranks must also be 15. When there is 1 positive rank, there are 4 negative ranks with the indicated sums. But these are also the sums for the possibilities with 4 positive ranks. The same reasoning applies for 2 and 3 positive ranks.

Based on Table 10.3, we can form Table 10.4 which shows all the possible values of the sum and their relative frequency of occurrence. Using Table 10.4, we see that the smallest rejection region for a two-sided test is 0 or 15 and this gives the probability of a type I error of 0.062. Thus, in Table B9, there is no rejection region shown for a sample size of 5 and a significance level of 0.05. If the test of interest were a one-sided test, then it would be possible to have a type I error probability less than 0.05.

TABLE 10.3 Positive Ranks for a Sample of Size 5 for 0, 1, and 2 Positive Ranks

Number of positive ranks	Possible ranks	Sum of positive ranks	Sum of negative ranks
0		0	15
1	1	1	14
	2	2	13
	3	3	12
	4	4	11
	5	5	10
2	1,2	3	12
	1,3	4	11
	1,4	5	10
	1,5	6	9
	2,3	5	10
	2,4	6	9
	2,5	7	8
	3,4	7	8
	3,5	8	7
	4,5	9	6

TABLE 10.4 All Possible Sums and Their Relative Frequency

Sum	Frequency	Relative frequency
0 or 15	1	0.031
1 or 14	1	0.031
2 or 13	1	0.031
3 or 12	2	0.063
4 or 11	2	0.063
5 or 10	3	0.094
6 or 9	3	0.094
7 or 8	3	0.094

C. Conducting the Test

Let us return to the 14 pairs now. We perform the test at the 0.05 significance level, the same level used in the sign test. As this is a one-sided test and Table B9 shows the critical values for a two-sided test, we use the boundary shown for an $\alpha \leq 0.10$. The critical region consists of values greater than or equal to 80. Because our test statistic is 82, greater than 80, we reject the null hypothesis of no regression toward the mean in favor of the alternative that there is regression toward the mean.

D. The WTEST Command

Box 10.5 shows how to conduct the Wilcoxon signed rank test using MINITAB. These are the same results we obtained above except that we have an approximation to the p value instead of simply knowing that the result was statistically significant. The p value is obtained from the normal approximation to the test which is discussed in Section II.G.

MINITAB BOX 10.5

We use the WTEST command in MINITAB to perform this test. The WTEST command uses the same form of the ALT subcommand as was used with STEST.

```
MTB > wtest c1;
SUBC> alt 1.
TEST OF MEDIAN = 0.000000 VERSUS MEDIAN G.T. 0.000000
                    N FOR   WILCOXON              ESTIMATED
              N      TEST   STATISTIC   P-VALUE      MEDIAN
C1           14        14        82.0     0.034       433.5
```

E. The Sign Test and the WSR Test

The WSR result is inconsistent with the result of the sign test and reflects the greater power of the WSR test. This greater power is due to the use of more of the information in the data by the WSR test compared with the sign test. The WSR test incorporates the fact that the average rank for the four changes away from the mean is 5.75 [= (1 + 5 + 7 + 10)/4], less than the average rank of 7.50. This lower average rank of these four changes, along with the fact that there were only four changes away from the mean, caused the WSR test to be significant. The sign test used only the number of changes toward the mean, not the ranks of these changes, and was not significant. Although the sign test failed to reject the null hypothesis, its p value of 0.0898 was not that different from 0.05.

F. Ties in the Data

Two types of ties can occur in the data. One type is that the day 1 and day 2 values for a boy are the same. If this type of tie occurs in an observational pair, that pair is deleted from the data set and the sample size is reduced by one for every pair deleted. Again, this procedure is appropriate when there are only a few ties in the data. If there are many ties, there is little reason to perform the test.

The other type of tie occurs when two or more differences have exactly the same nonzero value. This has an impact on the ranking of the differences. In this case, convention is that the differences are assigned the same rank. For example, if two differences were tied as the smallest value, each would receive the rank of 1.5, the average of ranks 1 and 2. If three differences were tied as the smallest value, each would receive the rank of 2, the average of ranks 1, 2, and 3. If there are few ties in the differences, the rank sum can still be used as the test statistic; however, the results of the test are now approximate. If there are many ties, an adjustment for the ties (3, p. 28) must be made or one of the methods in the next chapter should be used.

G. The Normal Approximation

If at least 16 pairs of observations are used in the calculations, R_{WSR} will approximately follow a normal distribution. As we saw above, the expected value of R_{WSR}, under the assumption that the null hypothesis is true, is $n * (n + 1)/4$, and its variance can be shown to be $n * (n + 1) * (2n + 1)/24$. Therefore, the statistic

$$\frac{|R_{WSR} - [n(n + 1)/4]| - 0.5}{\sqrt{n(n + 1)(2n + 1)/24}}$$

approximately follows the standard normal distribution. The two vertical lines in the numerator indicate the absolute value of the difference; that is, regardless of the sign of the difference, it is now a positive value. The 0.5 term is the continuity correction term, required because the signed rank sum statistic is not a continuous variable.

Let us calculate the normal approximation to see if it agrees with the p value reported by MINITAB. The expected value of R_{WSR} is 52.5 (= 14 * 15/4) and the standard error is 15.93 (= $\sqrt{14 * 15 * 29/24}$). Therefore, the statistic's value is

$$\frac{|82 - 52.5| - 0.5}{15.93} = 1.82$$

What is the probability that Z is greater than 1.82? This probability is found from Table B4 to be 0.0344, and this is the value MINITAB reported. This also agrees very closely with the exact p value of 0.0338. The exact p value is based on 554 of the 16,384 possible signed rank sums having a value of 82 or greater. Thus, even though n is less than 16, the normal approximation worked quite well in this case.

The sign and Wilcoxon signed rank tests are both used most frequently in the comparison of paired data, although they can be used with a single population to test that the median has a specified value. In the use of these tests with pre- and postintervention measurement designs, care must be taken to ensure that no extraneous factors could have an impact during the study. Otherwise, the possibility of confounding of the extraneous factor with the intervention variable is raised. In addition, the research designer must consider whether or not reversion to the mean is a possibility. If extraneous factors or reversion to the mean cannot be ruled out, the research design should be augmented to include a control group to help account for the effect of these possibilities.

III. WILCOXON RANK SUM TEST

Another test developed by Wilcoxon, the Wilcoxon rank sum (WRS) test, is used to determine, at some significance level, whether or not the probability that a randomly selected observation from one population being greater than a randomly selected observation from another population is equal to 0.5. This test is sometimes referred to as the Mann–Whitney test after Mann and Whitney, who later independently developed a similar test procedure for unequal sample sizes. The WRS test also requires that the data come from independent continuous distributions. We demonstrate this test using the proportion of calories that comes from fat data shown in Chapter 7.

TABLE 10.5 Proportion of Calories from Fat for Boys in Grades 5 and 6 and Grades 7 and 8

Grades 5 and 6		Grades 7 and 8	
Proportion from fat	Rank	Proportion from fat	Rank
0.365	21	0.311	13
0.437	30	0.278	6
0.248	4	0.282	8
0.424	26	0.421	25
0.403	23	0.426	28
0.337[a]	16.5	0.345	18
0.295	11	0.281	7
0.319	14	0.578	33
0.285	9	0.383	22
0.465	32	0.299	12
0.255	5	0.150	2
0.125	1	0.336	15
0.427	29	0.425	27
0.225	3	0.354	19
		0.337[b]	16.5
		0.289	10
		0.438	31
		0.411	24
		0.357	20
Sum of ranks	224.5		336.5

[a]To four decimals, the value is 0.3373.
[b]To four decimals, the value is 0.3370.

The proportion of calories from fat for boys in grades 5 and 6 and grades 7 and 8 are shown in Table 10.5. The ranks of these values are also shown, from smallest to largest. We have rounded the proportions to three decimal places, and as a result, there is one tie in the data. The tied values were the 16th and 17th smallest observations and hence were assigned the rank of 16.5, the average of 16 and 17. We could have used the fourth decimal place to break the tie, but we chose not to because we wanted to demonstrate how to calculate the ranks when there was a tie.

A. Expected Value of the Rank Sum under the Null Hypothesis

The test statistic, R_{WRS}, is the sum of the ranks for the first sample, in this case, for the 14 fifth and sixth grade boys. If there were no difference in the magnitudes of the proportion of calories from fat variables in the two groups, the rank sum for the smaller sample would be the product of the number of boys in the smaller sample and the average rank, that is,

$14 * 34/2$, which is 238. Values of R_{WRS} that deviate greatly from 238 suggest that the null hypothesis of no difference in magnitudes—the probability of the fifth and sixth grade boys having a higher proportion of calories coming from fat than the seventh and eighth grade boys is equal to 0.5—should be rejected in favor of the alternative hypothesis that one group has larger values than the other.

B. Determination of the Critical Region

The critical region, shown in Table B10,[2] is determined in a manner similar to that for the Wilcoxon signed rank statistic. All possible arrangements of size 14 of the 33 ranks are listed and the sum of the 14 ranks in each arrangement is found. The p value of the R_{WRS} is then determined. For a two-sided test, if R_{WRS} is less than the expected sum, the p value is twice the proportion of the rank sums that are less than or equal to the test statistic. If R_{WRS} is greater than the expected sum, the p value is twice the proportion of the rank sums that are greater than or equal to R_{WRS}. For a lower-tail, one-sided test, the p value is the proportion of the rank sums that are less than or equal to R_{WRS}. For an upper-tail, one-sided test, the p value is the proportion of the rank sums that are greater than or equal to R_{WRS}.

As an example of determining the rejection region, consider a situation with four observations in each of two samples. The possible ranks are 1 through 8. Table 10.6 shows all possible arrangements of size 4 of these ranks and Table 10.7 shows the relative frequency of the rank sums.

For a two-sided test that is to be performed at the 0.05 significance level, the rejection region consists of rank sums of 10 and 26. The probability of these two values is 0.0286, which is less than the 0.05 level. Including 11 and 25 in the rejection region increases the probability of the rejection region to 0.0571, which is greater than the 0.05 value. For a lower-tail, one-sided test to be performed at the 0.05 level, the rejection region is 10 and 11. It is not possible to perform the test at the 0.01 level because the probability of each rank sum in the Table 10.7 is greater than 0.01. The rejection region we have found here agrees with that shown in Table B10 ($2\alpha = 0.05$, $N_1 = 4$, $N_2 = 4$), the critical region for the WRS test at the 0.05 significance level.

Now let us return to the example dealing with the proportion of calories coming from fat.

[2]In Table B10, the value 2α refers to the two-sided significance level and N_1 and N_2, respectively, refer to the number of observations in the first and second groups. For a one-sided test at $\alpha = 0.05$, the page with $2\alpha = 0.10$ is used.

TABLE 10.6 Listing of Sets of Size 4 from the Ranks 1 to 8

Set	Sum of ranks	Set	Sum of ranks
1,2,3,4	10	2,3,4,5	14
1,2,3,5	11	2,3,4,6	15
1,2,3,6	12	2,3,4,7	16
1,2,3,7	13	2,3,4,8	17
1,2,3,8	14	2,3,5,6	16
1,2,4,5	12	2,3,5,7	17
1,2,4,6	13	2,3,5,8	18
1,2,4,7	14	2,3,6,7	18
1,2,4,8	15	2,3,6,8	19
1,2,5,6	14	2,3,7,8	20
1,2,5,7	15	2,4,5,6	17
1,2,5,8	16	2,4,5,7	18
1,2,6,7	16	2,4,5,8	19
1,2,6,8	17	2,4,6,7	19
1,2,7,8	18	2,4,6,8	20
1,3,4,5	13	2,4,7,8	21
1,3,4,6	14	2,5,6,7	20
1,3,4,7	15	2,5,6,8	21
1,3,4,8	16	2,5,7,8	22
1,3,5,6	15	2,6,7,8	23
1,3,5,7	16	3,4,5,6	18
1,3,5,8	17	3,4,5,7	19
1,3,6,7	17	3,4,5,8	20
1,3,6,8	18	3,4,6,7	20
1,3,7,8	19	3,4,6,8	21
1,4,5,6	16	3,4,7,8	22
1,4,5,7	17	3,5,6,7	21
1,4,5,8	18	3,5,6,8	22
1,4,6,7	18	3,5,7,8	23
1,4,6,8	19	3,6,7,8	24
1,4,7,8	20	4,5,6,7	22
1,5,6,7	19	4,5,6,8	23
1,5,6,8	20	4,5,7,8	24
1,5,7,8	21	4,6,7,8	25
1,6,7,8	22	5,6,7,8	26

TABLE 10.7 Frequency and Relative Frequency of the Rank Sums for Two Samples of Four Observations Each

Rank sum	Frequency	Relative frequency
10 or 26	1	0.0143
11 or 25	1	0.0143
12 or 24	2	0.0286
13 or 23	3	0.0429
14 or 22	5	0.0714
15 or 21	5	0.0714
16 or 20	7	0.1000
17 or 19	7	0.1000
18	8	0.1143

C. Conducting the Test

Let us perform the test of the null hypothesis of no difference in the magnitudes of the variable in the two independent populations at the 0.01 significance level. The alternative hypothesis is that there is a difference in the magnitudes. As this is a two-sided test, extremely large or small values of the test statistic will cause us to reject the null hypothesis. The test statistic is the rank sum of the smaller sample. Because the test is being performed at the 0.01 significance level, we use Table B10 ($2\alpha = 0.01$) with sample sizes of 14 and 19. If R_{WRS} is less than or equal to 168 or greater than or equal to 308, we reject the null hypothesis in favor of the alternative hypothesis. As R_{WRS} is 224.5, a value not in the rejection region, we fail to reject the null hypothesis. Based on this test, there is no evidence that fifth and sixth grade boys differ from seventh and eighth grade boys in terms of the proportion of calories coming from fat. As discussed above, this conclusion applies to a population of boys in a northern suburb of Houston. The generalization of this result to a larger population of boys may be appropriate; however, the extension to a larger population does not follow from statistics, but must come from substantive considerations.

D. Mann–Whitney Command

Box 10.6 shows how to conduct the WRS test using MINITAB.

MINITAB BOX 10.6

The MINITAB command for performing the WRS test is Mann–Whitney. Data from the smaller sample are in column c1, and c2 contains the data from the larger sample.

```
MTB > set c1
DATA> .365 .437 .248 .424 .403 .337 .295 .319 .285 .465
DATA> .255 .125 .427 .225
MTB > set c2
DATA> .311 .278 .282 .421 .426 .345 .281 .578 .383 .299
DATA> .150 .336 .425 .354 .337 .289 .438 .411 .357
DATA> end
MTB > mann-whit c1 c2

Mann-Whitney Confidence Interval and Test
C1          N =  14     Median =       0.3280
C2          N =  19     Median =       0.3450

Point estimate for ETA1-ETA2 is      -0.0175
95.3 pct c.i. for ETA1-ETA2 is (-0.0890,0.0490)
W = 224.5
Test of ETA1= ETA2  vs. ETA1 n.e. ETA2 is significant at 0.6358
The test is significant at 0.6358 (adjusted for ties)
 Cannot reject at alpha = 0.05
```

The medians, estimates of the magnitudes of the variable, in each sample are shown followed by an estimate of their difference. An approximate 95 percent confidence interval for the difference is also provided. The test statistic, W, is shown next followed by its approximate p value. The W statistic is the same as our R_{WRS}. As no ALT subcommand was used, MINITAB assumed that this was a two-sided test in the calculation of the p value. The calculation of the approximate p value is based on the normal distribution approximation, shown below, to the distribution of R_{WRS}. As the p value is much larger than the significance level, we fail to reject the null hypothesis. If a confidence interval other than 95 percent were desired, the numerical value would be entered before the columns containing the data are specified. The following MINITAB command shows the use of an interval other than 95 percent.

```
MTB > mann-whit 0.90 c1 c2
```

E. Normal Approximation

Once we exceed the sample sizes shown in Table B10, or for n_1 and n_2 both greater than or equal to 8, we can use a normal distribution as an approximation for the distribution of the R_{WRS} statistic. As we saw above, the expected value of R_{WRS} is expressed in terms of the sample sizes. Let n_1 be the sample size of the smaller sample, n_2 be the sample size of the other sample, and n be their sum. The mean and variance of R_{WRS}, assuming that the null hypothesis is true, are $n_1 * (n + 1)/2$ and $n_1 * n_2 * (n + 1)/12$, respectively. Therefore the statistic

$$\frac{|R_{WSR} - n_1(n + 1)/2| - 0.5}{\sqrt{n_1 n_2 (n + 1)/12}}$$

approximately follows the standard normal distribution. The 0.5 term is the continuity correction term, required because the rank sum statistic is not a continuous variable.

Let us calculate the normal approximation for these data. We already calculated the expected value of R_{WRS} above and its value is 238. The standard error is 27.453 $(= \sqrt{14 * 19 * 34/12})$. Therefore the statistic's value is

$$\frac{|224.5 - 238| - 0.5}{27.453} = 0.4735.$$

Because this is a two-sided test, the p value is twice the probability that a standard normal variable is greater than 0.4735. Using linear interpolation in Table B4, we find that

$$\Pr\{Z > 0.4735\} = 0.3179$$

and hence the p value is twice that, or 0.6358. This is the value reported by MINITAB.

If there are many ties between the data in the two groups, an adjustment for the ties should be made (3, p. 69) or a procedure in the next chapter should be used in the analysis of the data.

IV. KRUSKAL–WALLIS TEST

The above procedures are limited to the consideration of two populations. In this section, a method for the comparison of the locations (medians) from two or more populations is presented. This method, the Kruskal–Wallis (KW) test, a generalization of the Wilcoxon test, is named after the two prominent American statisticians who developed it in 1952. The KW test also requires that the data come from continuous probability distributions. The hypothesis being tested by the KW statistic is that all the medians are equal to one another; the alternative hypothesis is that the medians are not all equal.

We demonstrate this test in a study of the effect of weight loss without salt restriction on blood pressure in overweight hypertensive patients (4). Patients in the study all weighed at least 10 percent above their ideal weight and all were hypertensive. The patients either were not taking any medication or were on medication that had not reduced their blood pressure below 140 mm Hg systolic or 90 mm Hg diastolic. Three groups of patients were formed. Group I consisted of patients who were not taking any antihypertensive medication and who were placed on a weight reduction program; group II patients were placed on a weight reduction program in addition to continuing their antihypertensive medication; and group III patients simply continued with their antihypertensive medication. Patients already receiving medication were randomly assigned to group II or III. Patients were followed initially for 2 months, and the baseline value was the blood pressure reading at the end of the 2-month period. Patients were then followed for 4 additional months. Changes in weight and blood pressure between Month 2 and Month 6 were measured.

Table 10.8 contains simulated values that are consistent with those reported in the study by Reisin *et al.* (4). Besides the simulated values, the only data shown are from the female patients. We wish to determine whether or not there are differences in the median reductions in diastolic blood pressure in the populations of females from which these samples were drawn.

TABLE 10.8 Simulated Reductions (mm Hg) in Diastolic Blood Pressure for Females from Month 2 to Month 6 of Follow-up in Each of the Three Treatment Groups

Only weight reduction ($n_1 = 8$)				**Medication and weight reduction ($n_2 = 15$)**				**Only medication ($n_3 = 16$)**			
38	10	10	28	19	36	16	36	12	16	0	−12
6	8	33	8	38	28	36	22	14	16	−10	4
				42	24	40	34	−20	−6	18	16
				6	16	30		−14	6	−16	6

A. The Test Statistic and Its Distribution

It is possible, although not feasible for any reasonable sample sizes, to use the sums of the ranks as we had done in the Wilcoxon tests. As it is not feasible to determine the distribution of the rank sums, we use an alternative approach suggested by Kruskal and Wallis. They suggested that H, a statistic defined in terms of n_i and R_i, the sample size and rank sum for the ith group, be used as the test statistic. The definition of H is

$$H = \frac{12}{n(n+1)} \sum_{i=1}^{k} \frac{R_i^2}{n_i} - 3(n+1)$$

where n is the sum of the group sample sizes and k is the number of groups. This statistic follows the chi-square distribution with $k - 1$ degrees of freedom when the null hypothesis is true.[3] Thus we reject the null hypothesis when the observed value of H is greater than $\chi^2_{k-1,1-\alpha}$ and we fail to reject the null hypothesis otherwise.

B. Conducting the Test

We have to find the rank sums for the three groups to calculate the value of H and we must also choose the significance level for the test. Let us perform the test at the 0.10 significance level. Table 10.9 shows the ranks of the values in Table 10.8. The smallest observation has the rank of 1 and the largest has the rank of 39, as there are 39 observations. Observations with the same value receive the same average rank as above.

We now have the information required to calculate H. The observed value of H is found from

$$H = \frac{12}{39 * 40}\left(\frac{164.5^2}{8} + \frac{436.5^2}{15} + \frac{179^2}{16}\right) - 3 * 40$$

and this is 19.133. If the null hypothesis, equality of medians, is true, H follows the chi-square distribution with 2 degrees of freedom. As 19.133 is greater than 4.61 ($= \chi^2_{2,0.90}$), we reject the null hypothesis in favor of the alternative. From Table B7, we see that the p value of H is less than 0.005. There appears to be a difference in the effects of the different interventions on diastolic blood pressure. Weight reduction can play an important role in blood pressure reduction for overweight patients.

C. Kruskal–Wallis Command

Box 10.7 shows how to conduct the Kruskal–Wallis test using MINITAB.

[3]The statistic H follows the chi-square distribution because H can be shown to be proportional to the sample variance of the rank sums which follows a chi-square distribution.

TABLE 10.9 Ranks of the Simulated Values in Table 10.8

Only weight reduction ($n_1 = 8$)				Medication and weight reduction ($n_2 = 15$)				Only medication ($n_3 = 16$)			
36.5	15.5	15.5	28.5	25	34	21	34	17	21	7	4
10.5	13.5	31	13.5	36.5	28.5	34	26	18	21	5	8
				39	27	38	32	1	6	24	21
				10.5	21	30		3	10.5	2	10.5
$R_1 = 164.5$				$R_2 = 436.5$				$R_3 = 179$			

MINITAB BOX 10.7

The command for this test is Kruskal–Wallis. The data for this command are entered in two columns. One column contains the actual data values and the other column contains an indicator variable that tells MINITAB to which group the data value belongs. c1 contains the 39 data values and c2 contains the corresponding group identification. All the values from group I are entered first, followed by the group II values and then the group III data.

```
MTB > set cl
DATA> 38 10 10 28 6 8 33 8
DATA> 19 36 16 36 38 28 36 22 42 24 40 34 6 16 30
DATA> 12 16 0 -12 14 16 -10 4 -20 -6 18 16 -14 6 -16 6
DATA> set c2
DATA> 8(1) 15(2) 16(3)
DATA> end
MTB > kruskal-wallis cl c2

LEVEL     NOBS     MEDIAN   AVE. RANK   Z VALUE
    1        8     10.000        20.6      0.16
    2       15     30.000        29.1      3.94
    3       16      5.000        11.2     -4.03
OVERALL     39                   20.0

H = 19.13  d.f. = 2  p = 0.000
H = 19.21  d.f. = 2  p = 0.000 (adj. for ties)
```

The median for each group (level) is presented along with the average rank of observations in each group. A z value is also shown for each group. This value is found by subtracting the overall average rank from the group's average rank and dividing by the standard error of the group's average rank. If the null hypothesis were true, the z values would follow a standard normal distribution. As it is extremely unlikely to observe values of 3.94 and −4.03 from a standard normal distribution, these values suggest that the null hypothesis should be rejected. The Kruskal–Wallis statistic, *H*, is shown last and its value is the same as we calculated above.

V. CONCLUDING REMARKS

In this chapter, we have introduced the reversion toward the mean idea as well as several of the more frequently used nonparametric tests for continuous data. Reversion toward the mean is important because of its possible effect on test results (5,6). The nonparametric tests are attractive because they do not require an assumption of the normal distribution or the equality of variances. Even when the data do come from normal distributions, these nonparametric tests do not sacrifice much power in comparison to tests based on the normality assumption. Although these tests were designed to be used with continuous data, they are often used with ordered data as well. Their use with ordered data can create problems as there are likely to be more ties for ordered data than for continuous data. In the next chapter, we introduce methods for testing hypotheses about ordered or nominal data as well as about continuous data that are grouped into categories.

EXERCISES

10.1. The table below shows the annual average fatality rate per 100,000 workers for each state, data originally introduced in Exercise 7.2. A state is placed into one of three groups according to the National Safe Workplace Institute (NSWI) score. Group 1 consists of states whose NSWI score was above 55, group 2 consists of states with scores of 31 to 55, and group 3 consists of states with scores less than or equal to 30. In Exercise 7.2, we examined the correlation between the fatality rates and the NSWI scores. Here we wish to determine whether or not we believe that the median fatality rates for the three groups of states are the same.

State Fatality Rates per 100,000 Workers by the National Safe Workplace Institute Scores

NSWI group								
Low (≤30)			*Middle (31–55)*			*High (>55)*		
State	*Rate*	*Rank*	*State*	*Rate*	*Rank*	*State*	*Rate*	*Rank*
AR	12.5	41	LA	11.2	35	NH	4.5	8
WY	29.5	49	KY	11.9	39	WI	6.3	16
NM	12.0	40	GA	10.3	33	RI	3.3	4

					NSWI group			
Low (≤30)			*Middle* (31–55)			*High* (>55)		
State	*Rate*	*Rank*	*State*	*Rate*	*Rank*	*State*	*Rate*	*Rank*
KS	9.1	28	VT	6.8	19	AK	33.1	50
ND	13.8	43	AZ	4.1	6	VA	9.9	32
ID	17.1	47	DE	5.8	13	MI	5.3	10.5
TN	8.1	25	MO	5.3	10.5	OR	11.0	34
HI	6.0	14	MD	5.7	12	MN	4.3	7
AL	9.0	27	NC	7.2	21	CT	1.9	1
MS	14.6	44	IN	7.8	23.5	ME	7.8	23.5
SD	14.7	45	WV	16.2	46	TX	11.7	38
SC	6.7	18	FL	9.3	30.5	MA	2.4	2
UT	13.5	42	CO	9.3	30.5	NY	2.5	3
NE	11.3	36	OK	8.7	26	IL	6.9	20
MT	21.6	48	IA	9.2	29	NJ	3.4	5
NV	11.5	37	OH	4.8	9	CA	6.5	17
			PA	6.1	15			
			WA	7.7	22			
Sum of ranks		584			420			271

Is there any need to use a statistical test of hypothesis to determine whether or not the median fatality rates of these three groups of states are the same? If there is, what test would you use?

10.2. A study was conducted to determine the effect of short-term, low-level exposure of demolition workers to asbestos fibers and silica-containing dusts (7). Twenty-three demolition workers were exposed for 26 consecutive days during the destruction of a three-story building. The dependent variable is the percent reduction in the baseline value of the ratio of forced expiratory volume in the first second to forced vital capacity (FEV_1/FVC) compared with the same ratio at the end of the demolition project. None of the exposures to asbestos or silica were above the permissible values. The following table shows the data for the 23 workers, grouped according to level of exposure to asbestos and silica.

Percent Reduction in Pre- and Postproject FEV_1/FVC Values by Level of Exposure to Asbestos and Silica-Containing Dusts

Higher exposure (*n* = 10)					*Lower exposure* (*n* = 13)				
0.73	0.72	0.70	0.33	0.54	0.42	0.70	0.65	0.62	0.81
0.75	0.67	0.73	0.69	0.59	0.64	0.63	0.60	0.66	0.61
					0.68	0.76	0.65		

Test the hypothesis that there is no difference in the median percent reduction between those with the higher level of exposure and those with the lower level of exposure. Use a 5 percent significance level.

10.3. A study was conducted to compare the effectiveness of the applied relaxation method and the applied relaxation method with biofeedback in patients with chronic low back pain (8). Twenty female patients were randomly assigned to each treatment group and the treatments were then provided. One of the dependent variables studied was the change in the pain rating index, based on the McGill Pain Questionnaire, between pre- and posttreatment. Patients were also followed for a longer period, but those results are not used in this exercise. The actual change data were not shown in the article, but the following table contains hypothetical changes for the two groups.

Hypothetical Data Showing the Changes in Pre- and Posttreatment Values of the McGill Pain Questionnaire for 40 Women Randomly Assigned to the Different Treatments

Relaxation only										*Relaxation with biofeedback*									
10	11	21	18	16	16	15	9	2	19	9	12	7	14	4	2	11	8	9	11
5	18	16	14	12	13	11	13	14	20	6	10	9	7	8	10	6	13	7	8

Use the appropriate one- or two-sided test for the null hypothesis of no difference in the median changes in pain rating between the two groups at the 0.10 significance level. Provide the rationale for your choice of either the one-sided or two-sided test.

10.4. The following data are from the 1971 census for Hull, England (9). The data show by ward, roughly equivalent to a census tract, the number of households per 1000 without a toilet and the corresponding incidence of infectious jaundice per 100,000 population reported

Ward	*No toilet*	*Jaundice*	*Ward*	*No toilet*	*Jaundice*
1	222	139	12	1	128
2	258	479	13	276	263
3	39	88	14	466	469
4	389	589	15	443	339
5	46	198	16	186	189
6	385	400	17	54	198
7	241	80	18	749	401
8	629	286	19	133	317
9	24	108	20	25	201
10	5	389	21	36	419
11	61	252			

between 1968 and 1973. Group the wards into three groups based on the rate of households without a toilet. Use the Kruskal–Wallis test to determine whether or not there is a difference in the median incidence of jaundice for the three groups at the 0.05 significance level.

10.5. Exercise 10.4 provides an example of ecological data, data aggregated for a group of subjects. Care must be taken in the use of this type of data (10). For example, suppose in Exercise 10.4 there was a statistically significant difference in the median incidence of jaundice for the three groups of wards. Is it appropriate to conclude that there is an association between the presence or absence of a toilet in a household and the occurrence of jaundice? Provide the rationale for your answer.

10.6. In the study on ramipril, introduced in Chapter 7, there was a 4-week baseline period during which patients took placebo tablets (11). Of the 160 patients involved in the study, 24 had previously taken medication for high blood pressure, but it had been greater than 7 days since they had last taken their medication. These 24 patients had some expectation that medication works. We will examine hypothetical data based on the summary statistics reported to determine whether or not there is a placebo effect: a reduction in blood pressure values associated with taking the placebo. The hypothetical systolic blood pressure (SBP in mm Hg) values are the following:

Patient number	*Week 0 SBP*	*Week 4 SBP*	*Patient number*	*Week 0 SBP*	*Week 4 SBP*
1	171	182	13	148	178
2	172	167	14	182	166
3	166	186	15	210	183
4	181	175	16	171	164
5	194	177	17	165	163
6	200	200	18	201	175
7	200	168	19	189	165
8	181	178	20	197	174
9	173	189	21	187	167
10	178	189	22	174	180
11	206	167	23	197	185
12	199	185	24	169	149

Use the sign test to test the hypothesis that the proportion of decreases in SBP between Week 0 and Week 4 is equal to 0.50 versus the alternative that the proportion of decreases in SBP is greater than 0.50. Use the 0.05 significance level. If there were reversion or regression to the mean here, would that affect our conclusion about the placebo effect? Test the hypothesis of no reversion to the mean at the 0.05 level.

10.7. Use the Wilcoxon signed rank test to test the hypothesis that the median change in SBP in Exercise 10.6 is zero versus the alternative hypothesis that the median change is greater than zero. Perform the test at the 0.05 level. Compare your results with those of the sign test. Do you think that there is a placebo effect here?

REFERENCES

1. Stigler, S. M. (1986). "The History of Statistics: The Measurement of Uncertainty before 1900." Harvard Univ. Press (Belknap), Cambridge, MA.
2. Samuels, M. L. (1991). Statistical reversion toward the mean: More universal than regression toward the mean. *Am. Stat.* **45,** 344–346.
3. Hollander, M., and Wolfe, D. (1973). "Nonparametric Statistical Methods." Wiley, New York.
4. Reisin, E., Abel, R., Modan, M., Silverberg, D. S., Eliahou, H. E., and Modan, B. (1978). Effect of weight loss without salt restriction on the reduction of blood pressure in overweight hypertensive patients. *N. Engl. J. Med.* **298,** 1–6.

4a. Kruskal, W. H., and Wallis, W. A. (1952). "Use of ranks in one-criterion variance analysis." *J. Amer. Statist. Assoc.* **47,** 583–621.

5. Davis, C. E. (1976). The effect of regression to the mean in epidemiologic and clinical studies. *Am. J. Epidemiol.* **104,** 493–498.
6. Nesselroade, J. R., Stigler, S. M., and Baltes, P. B. (1980). Regression toward the mean and the study of change. *Psychol. Bull.* **88,** 622–637.
7. Kam, J. K. (1989). Demolition worker hazard: The effect of short-term, low-level combined exposures. *J. Environ. Health* **52,** 162–163.
8. Strong, J., Cramond, T., and Mass, F. (1989). The effectiveness of relaxation techniques with patients who have chronic low back pain. *Occup. Ther. J. Res.* **9,** 184–192.
9. Goldstein, M. (1982). Preliminary inspection of multivariate data. *Am. Stat.* **36,** 358–362.
10. Piantadosi, S., Byar, D. P., and Green, S. B. (1988). The ecological fallacy. *Am. J. Epidemiol.* **127,** 893–904.
11. Walter, U., Forthofer, R., and Witte, R. U. (1987). Dose-response relation of angiotensin converting enzyme inhibitor ramipril in mild to moderate essential hypertension. *Am. J. Cardiol.* **59,** 125D–132D.

CHAPTER 11

Analysis of Categorical Data

In this chapter, we present some additional nonparametric tests that are used with nominal, ordinal, and continuous data that have been grouped into categories. The data in this chapter are presented in the form of frequency or contingency tables. In Chapter 4, we demonstrated how one- and two-way frequency tables could be used in data description. In this chapter, we show how frequency or contingency tables can be used to test whether the distribution of the variable of interest agrees with some hypothesized distribution or whether there is an association among two or more variables. For example, in the material on the normal distribution in Chapter 6, we examined the distribution of blood pressure. In this chapter, we show how to test the null hypothesis that the data follow a particular distribution. In Chapter 5, we considered the association between birth weight and the timing of the initiation of prenatal care. In this chapter, we show how to test the null hypothesis that an association exists between two discrete variables versus the alternative hypothesis that there is no association. A goodness-of-fit statistic is used to test these hypotheses and it follows a chi-square distribution if the null hypothesis is true.

I. GOODNESS-OF-FIT TEST

The *goodness-of-fit test* can be used to examine the fit of a one-way frequency distribution for X, the variable of interest, to the distribution expected under the null hypothesis. This test, developed in 1900, is another contribution of Karl Pearson, also known for the Pearson correlation coefficient. The X variable is usually a discrete variable, but it could also be a continuous variable that has been grouped into categories.

To facilitate the presentation, we use the following symbols. Let O_i represent the number of sample observations at level i of X, and E_i represent the expected number of observations at level i assuming that the null hypothesis is true. The E_i are found by multiplying the population probability of level i, π_i, by n, the total number of observations. As the sum of the π_i is one, the sum of the E_i is n.

A natural statistic for this comparison would seem to be the sum of the differences of O_i and E_i, that is, $\Sigma(O_i - E_i)$; however, as both the O_i and the E_i sum to n, the sum of their differences is always zero. Thus this statistic is not very useful; however, the sum of the squares of the differences, $\Sigma(O_i - E_i)^2$, will be different from zero except when there is a perfect fit. Squaring the differences is the same strategy used in defining the variance in Chapter 4.

One problem remains with $\Sigma(O_i - E_i)^2$. If the sample size is large, even very small differences in the observed and expected proportions at each level of X become large in terms of the O_i and E_i. Therefore, we must take the magnitude of the O_i and E_i into account. Pearson suggested dividing each squared difference by the expected number for that category and using the result, $\Sigma(O_i - E_i)^2/E_i$, as the test statistic.

It turns out that this statistic, for reasonably large values of E_i, follows the chi-square distribution. In the early 1920s, Sir Ronald A. Fisher showed that this statistic has $k - 1 - m$ degrees of freedom, where k is the number of levels of the X variable and m is the number of estimated parameters. For the chi-square distribution to apply, no cell should have an expected count that is less than five times the proportion of cells with E_i that are less than 5 (1). For example, if k is 10 and two cells have expected counts less than 5, then no expected cell count should be less than one [$=5 * (2/10)$]. If some of the E_i are less than this minimum value, categories with small expected values may be combined with adjacent categories. The combinations of categories must make sense substantively; otherwise the categories should not be combined.

Note that the goodness-of-fit test is a one-sided test. Only large values of the chi-square test statistic will cause us to reject the null hypothesis of good agreement between the observed and expected counts in favor of the alternative hypothesis that the observed counts do not provide a good fit to the expected counts. Small values of the test statistic support the null hypothesis.

A. No Parameter Estimation Required

The study of genetics has led to the discovery and understanding of the role of heredity in many diseases, for example, hemophilia, colorblindness, Tay-Sach's disease, phenylketonuria, and diabetes insipidus (2). The father of genetics, Abbe Gregor Mendel, presented his research on garden peas in 1865, but the importance of his results was not appreciated until 1900. One of Mendel's discoveries was the 1 : 2 : 1 ratio for the number of dominant, heterozygous, and recessive offspring from hybrid parents, that is, from parents with one dominant and one recessive gene.

Although doubts have been raised about Mendel's data, we use data from one of his many experiments. Table 11.1 (3, p. 328) shows the number of offspring by type from the crossbreeding of smooth seeds (*A*), the dominant type, with wrinkled seeds (*a*), the recessive type. Dominant means that when both a smooth gene and a wrinkled gene are present, the pea will be smooth. The pea will be wrinkled only when both genes are wrinkled.

The question of interest is whether this experiment supports Mendel's theoretical ratio of 1 : 2 : 1. The null hypothesis is that the observed data are consistent with Mendel's theory. The alternative hypothesis is that the data are not consistent with his theory. Let us test this hypothesis at the 0.10 significance level.

We must first calculate the expected cell counts for this one-way contingency table. As the expected counts are based on the theoretical 1 : 2 : 1 ratio, the ratio tells us that we expect one-fourth of the observations to be *AA*, two-fourths (one-half) to be *Aa* or *aA*, and one-fourth to be *aa*. One-fourth of 529 is 132.25 and one-half of 529 is 264.5; therefore the expected counts are 132.25, 264.5, and 132.25, respectively. The test statistic is

$$X^2 = \frac{(138 - 132.25)^2}{132.25} + \frac{(265 - 264.5)^2}{264.5} + \frac{(126 - 132.25)^2}{132.25} = 0.546.$$

This statistic follows the chi-square distribution if the null hypothesis is true. The number of degrees of freedom is $k - 1 - m$. In this example, the value of k is 3, for the three types of possible offspring. As we did not estimate any parameters, m is 0. Therefore there are 2 degrees of freedom. The critical value, $\chi^2_{2,0.90}$, is 4.61. Because 0.546 is less than 4.61, we fail to reject the null hypothesis. It appears that these data support Mendel's theoretical results.

TABLE 11.1 Mendel's Data on Garden Peas Number of Smooth and Wrinkled Offspring from Hybrid Parents

AA	*Aa*	*aa*	Total
138	265	126	529

In the next section, we consider an example in which we estimate two parameters.

B. Parameter Estimation Required

The goodness-of-fit chi-square statistic can also be used to test the hypothesis that the data follow a particular probability distribution. Thus it can be used to complement the graphic approaches, for example, the Poissonness and normal probability plots, presented in Chapter 6. The test of hypothesis provides a number, the p value, that can be used alone or together with the graphic approach to help us decide whether we will reject or fail to reject the null hypothesis.

Let us test the hypothesis, at the 0.01 significance level, that the systolic blood pressure values, shown in Table 4.4, come from a normally distributed population. In testing the hypothesis that data are from a normally distributed population, we must specify the particular normal distribution. This specification means that the values of the population mean and standard deviation are required. Because we do not know these values for the systolic blood pressure variable for 12-year-old boys in the United States, we estimate their values. Table 11.2 shows the systolic blood pressure values along with the sample estimates of the mean and standard deviation.

The goodness-of-fit test is based on the variable of interest being discrete or being grouped into k categories. We must therefore group the systolic blood pressures into categories. We use the same categories portrayed in the histogram shown in Figure 4.8, and these categories are shown in Table 11.3.

TABLE 11.2 Systolic Blood Pressure Values (mm Hg) and Their Sample Mean and Standard Deviation for the Data in Table 4.4

Value	Frequency	Value	Frequency	Value	Frequency
80	3	102	5	118	1
84	1	104	6	120	5
88	1	105	4	122	1
90	5	106	3	124	1
92	1	108	6	125	2
94	2	110	11	126	1
95	4	112	5	128	2
96	1	114	2	130	4
98	2	115	1	134	1
100	13	116	4	140	2

Sample mean = 107.41
Sample standard deviation = 12.66

TABLE 11.3 Number of Boys Observed and Expected[a] in the Systolic Blood Pressure Categories

Systolic blood pressure (mm Hg)	Number	
	Observed	Expected
≤80.5	3	1.66
80.51— 87.5	1	4.16
87.51— 94.5	9	9.57
94.51—101.5	20	16.53
101.51—108.5	24	21.67
108.51—115.5	19	20.30
115.51—122.5	11	14.41
122.51—129.5	6	7.61
129.51—136.5	5	3.02
≥136.5	2	1.07
Total	100	100.00

[a]Expected calculated assuming the data follow the $N(107.41, 12.66)$ distribution.

The expected values are found by converting the category boundaries to standard normal values and then finding the probability associated with each category. For example, the probability associated with the first category, a systolic blood pressure less than 80.5 mm Hg, is found in the following manner. First 80.5 is converted to a standard normal value by subtracting the mean and dividing by the standard deviation. Thus 80.5 is converted to -2.13 [= (80.5 − 107.41)/12.66]. The probability that a standard normal is less than -2.13 is 0.0166. The expected number of observations is found by taking the product of n, 100, and the probability of being in the category. Thus the expected number of observations is 1.66.

The expected number of observations in the second category is found in the following manner. The upper boundary of the second category, 87.5, is converted to the standard normal value of -1.57 [= (87.5 − 107.41)/12.66]. The probability that a standard normal value is less than -1.57 is 0.0582. The probability of being in the second category is then 0.0416 (= 0.0582 − 0.0166). Multiplying this probability by 100 yields the expected count of 4.16 for the second category. The other expected cell counts are calculated in this same way. If the sum of the expected counts does not equal the number of observations, with allowance for rounding, an error has been made.

The calculation of the chi-square goodness-of-fit statistic is

$$X^2 = \frac{(3 - 1.66)^2}{1.66} + \frac{(1 - 4.16)^2}{4.16} + \cdots + \frac{(2 - 1.07)^2}{1.07} = 7.832.$$

TABLE 11.4 Guideline for the Number of Intervals to Be Used with a Continuous Variable

Sample size	200	400	600	800	1000	1500	2000
Number of intervals	15	20	24	27	30	35	39

Source: Cochran (4).

The value of k, the number of categories, is 10, and that of m, the number of parameters estimated, is 2. Therefore, there are 7 degrees of freedom $(= 10 - 1 - 2)$. The value of this test statistic is compared with 18.48 $(= \chi^2_{7,0.99})$. As 7.832 is less than 18.48, we fail to reject the null hypothesis. Based on this sample, there is no evidence to suggest that the systolic blood pressures of 12-year-old boys are not normally distributed.

In dealing with continuous variables, for example, the blood pressure variable, we have to decide how many intervals and what interval boundaries should be used. In the above example, we used the same intervals that were shown in Figure 4.8. Some research has been conducted on the relationship between power considerations and the number and size of intervals, and as we might expect, the number of intervals depends on the sample size. Table 11.4, based on a review by Cochran (4), shows the suggested number of intervals to be used with a continuous variable. The size of the intervals may also vary. The intervals are chosen so that the expected number of observations in each interval are approximately equal. Thus some intervals may be much narrower than other intervals. For ease of computation, it is reasonable to choose the intervals so that the observed numbers of observations in each interval are approximately equal. These suggestions for the choice of the number and size of intervals differ from those used in Chapter 4, but the goals of the analyses are also different. In Chapter 4, the guidelines used were for the presentation of data in histograms. In this chapter, we wish to determine whether or not it appears that the data follow a particular distribution.

In the following sections, we extend the use of the chi-square goodness-of-fit test statistic to two-way contingency tables. This extension allows a determination of whether or not there is an association between two variables.

II. 2 BY 2 CONTINGENCY TABLE

We begin the study of the association of two discrete random variables with the simplest two-way table, the 2 by 2 contingency table. The following quote by M. H. Doolittle in 1888 (5, p. 131) states the purpose of our analysis.

> The general problem may be stated as follows: Having given the number of instances respectively *in which things are both thus and so*, in which they are *thus but not so*, in which they are *so but not thus*, and in which they are *neither thus nor so*, it is required to eliminate the general quantitative relativity inhering in the mere thingness of the things, and to determine the special quantitative relativity subsisting between the *thusness and the soness* of the things. (emphasis added)

A restatement of the purpose is that we wish to determine, at some significance level, whether there is an association between the variables.

For example, is there is an association between the occurrence of iron deficiency in women and their level of education? If we use two levels of education, for example, less than 12 years and greater than or equal to 12 years, the 2 by 2 table to use in this investigation would look like Table 11.5. The entries in the table, the n_{ij}, are the observed number of women in the ith row (level of education) and jth column (iron status) in the sample. The symbol $n_{i\cdot}$ represents the sum of the frequencies in the ith row, $n_{\cdot j}$ is the sum of the frequencies in the jth column, and n, the sample size, is the sum of the frequencies in the entire table. There are several ways of answering the question about whether an association exists between these two variables. We begin with the approach from Chapter 7.

A. Comparing Two Independent Binomial Proportions

The 2 by 2 table is one way of presenting the data used in the calculation of two independent binomial proportions. If there is no association between iron status and education, then the probability of iron deficiency for women with less than 12 years of education, π_1, should equal the corresponding probability, π_2, for women with 12 or more years of education. We can construct a confidence interval for the difference of π_1 and π_2. If the interval contains zero, there is no evidence of an association between iron status and education. The confidence interval is based on the sample estimates of π_1 and π_2 and these are $n_{11}/n_{1\cdot}$ and $n_{21}/n_{2\cdot}$, respectively.

B. Expected Cell Counts Assuming No Association

We use the symbol m_{ij} to represent the expected number of women in the ith row and jth column assuming that the null hypothesis is true. In the

TABLE 11.5 Iron Status by Level of Education

	Iron status		
Education	Deficient	Acceptable	Total
<12 years	n_{11}	n_{12}	$n_{1\cdot}$
≥12 years	n_{21}	n_{22}	$n_{2\cdot}$
Total	$n_{\cdot 1}$	$n_{\cdot 2}$	n

material on two-way tables, we are using n and m to represent the observed and expected cell counts instead of the O and E used in the previous section. For the null hypothesis of no association between iron status and education, the expected proportion of women with low iron status at each level of education, $m_{i1}/n_{i\cdot}$, equals the overall proportion of iron-deficient women, $n_{\cdot 1}/n$. This is equivalent to saying that the proportion of women with low iron status is the same for those with less than 12 years of education as for those who have at least 12 years of education. Thus when there is no association, the expected number of iron-deficient women at the ith level of education can be found from the relationship

$$\frac{m_{i1}}{n_{i\cdot}} = \frac{n_{\cdot 1}}{n}$$

which yields

$$m_{i1} = \frac{n_{i\cdot} * n_{\cdot 1}}{n}.$$

The same type of relationship holds true for women with acceptable levels of iron. Therefore the general formula for the expected cell count, assuming no association, is

$$m_{ij} = \frac{n_{i\cdot} * n_{\cdot i}}{n}.$$

We can use these observed and expected values to calculate the chi-square goodness-of-fit statistic to test the hypothesis of no association between the two variables.

C. The Odds Ratio: A Measure of Association

A useful statistic for measuring the level of association in contingency tables is the odds ratio, θ. For example, in Table 11.5, an estimator of the odds that a woman with less than a high school education is iron deficient is n_{11}/n_{12}. The corresponding estimator of the odds that a woman with at least a high school education is iron deficient is n_{21}/n_{22}. If there is no association between education and iron status, these two odds should be equal. If the odds are equal, their ratio equals one. A sample estimator of the odds ratio OR is

$$\hat{\theta} = \text{OR} = \frac{n_{11}/n_{12}}{n_{21}/n_{22}} = \frac{n_{11} * n_{22}}{n_{21} * n_{12}}.$$

Thus, an OR far from one calls into question the assumption (hypothesis) of no association. If the estimated odds ratio is much less than one, this means that the denominator is much larger than the numerator; that is, the product of the off-diagonal cells in the 2 by 2 table is larger than the product of the diagonal cells. For Table 11.5, an odds ratio of less than one indicates that the proportion of women with 12 or more years of education who are iron deficient is greater than the corresponding proportion for women with fewer than 12 years of education. An odds ratio greater than one indicates that women with fewer than 12 years of education have the greater proportion of iron deficiency.

A problem with the estimated odds ratio occurs if any of the cell frequencies are zero. The estimated odds ratio is zero if n_{11} or n_{22} are zero, and it is undefined if n_{12} or n_{21} are zero. To avoid this problem, some statisticians base the calculation of the estimated odds ratio on $n_{ij} + 0.5$ instead of the n_{ij}.

D. Data Collection

The two choices for data collection used most often in practice are a SRS of n women and (2) stratified samples of $n_{1\cdot}$ and $n_{2\cdot}$ women. In the SRS case, the test for no association is a test of the independence of the row and column variables. In the stratified sampling case, the test for no association is a test of the homogeneity of the proportions in row i with those in row j. Regardless of which of these two sample selection processes is used, the expected cell counts for the hypothesis of no association are calculated as shown above.

Suppose that we select a SRS of 100 women 20 to 44 years old and we obtain information on their educational level and iron status. The hypothetical data, based on Figure 6.13 in "Nutrition Monitoring in the United States" (6), are shown in Table 11.6.

E. Calculation of the Confidence Interval for the Difference in Proportions

The estimated conditional probability of a woman being iron deficient given that she has less than 12 years of education is 0.133 (= 4/30). This is contrasted with the estimated probability of 0.057 (= 4/70) for a woman with 12 or more years of education. The 95 percent confidence interval for

TABLE 11.6 Hypothetical Frequency Data for Iron Status by Education

	Iron status		
Education	Deficient	Acceptable	Total
<12 years	4	26	30
≥12 years	4	66	70
Total	8	92	100

the difference of π_1 and π_2 is found from

$$(0.133 - 0.057) \pm z_{0.975}\sqrt{\frac{0.133 * 0.867}{30} + \frac{0.057 * 0.943}{70}}$$

which yields an interval from −0.057 to 0.209. As zero is contained in the interval for the difference, there is no evidence of an association between iron status and education based on this sample.

F. The Test Statistic and Its Calculation

On the basis of these data, the expected values, assuming the independence of the row and column variables, are

$$\begin{aligned} m_{11} &= 30 * 8/100 = 2.4 \\ m_{12} &= 30 * 92/100 = 27.6 \\ m_{21} &= 70 * 8/100 = 5.6 \\ m_{22} &= 70 * 92/100 = 64.4 \\ \text{Total} &= 100.0 \end{aligned}$$

The sum of the expected values in the first row is 30, the first row total. The sum of the expected values in the first column is 8, the first column total. Hence, once we calculate m_{11}, we can determine the value of m_{12} by subtracting m_{11} from 30. In the same way, we can determine the value of m_{21} by subtracting m_{11} from 8. Because we now know the value of m_{12}, we can also determine the value of m_{22} by subtracting m_{12} from 92. Hence, once we calculate any cell's expected value, the expected values of the other three cells are determined. This means that only one degree of freedom is associated with the test of no association for a 2 by 2 contingency table.

The expected cell frequency for the cell in the intersection of the first row and first column is 2.4. This is the only expected frequency less than 5, and according to the guideline given above, the minimum acceptable value for an expected cell frequency is 1.25 [= 5 * (1/4)]. As none of the expected frequencies are less than 1.25, we can use the chi-square test statistic.

Now that we have both the observed and expected cell counts, we can test the hypothesis of no association (independence) of iron status and education. We perform the test at the 0.05 significance level. The test uses a modified version of the chi-square goodness-of-fit statistic. The modified form, called the Yates' corrected chi-square after the British statistician Frank Yates who suggested it, is

$$X^2_{YC} = \sum_i \sum_j \frac{(|n_{ij} - m_{ij}| - 0.5)^2}{m_{ij}}.$$

The modification consists of subtracting 0.5 from the absolute value of the difference of the observed and expected cell counts (7). The p value associated with the Yates' corrected chi-square statistic agrees more closely with the p value of the exact test statistic developed by Ronald Fisher (8).[1] The calculation of Fisher's exact test statistic for 2 by 2 tables, or for its extension to r by c tables (9), is quite involved and, therefore, we have chosen not to present it in this text.

The calculated X^2_{YC} is compared with 3.84 (= $\chi^2_{1,0.95}$). If X^2_{YC} is greater than 3.84, we reject the hypothesis of independence in favor of the alternative that there is some association between iron status and education. If X^2_{YC} is less than 3.84, we fail to reject the null hypothesis. The test statistic is

$$X^2_{YC} = \frac{(|4 - 2.4| - 0.5)^2}{2.4} + \frac{(|26 - 27.6| - 0.5)^2}{27.6} + \frac{(|4 - 5.6| - 0.5)^2}{5.6} + \frac{(|66 - 64.4| - 0.5)^2}{64.4}$$

or 0.783. As X^2_{YC} is less than 3.84, we fail to reject the null hypothesis. Based on this sample, there does not appear to be any association between iron status and education. Note that the uncorrected X^2 value is 1.656.

MINITAB can be used to analyze these data, as shown in Box 11.1, but MINITAB calculates X^2, the uncorrected value of the test statistic. As MINITAB does not calculate X^2_{YC}, we shall present a formula that is easier to use in place of the defining formula. The easier-to-use formula is

$$X^2_{YC} = \frac{n(|n_{11}n_{22} - n_{12}n_{21}| - n/2)^2}{n_{1\cdot}\, n_{2\cdot}\, n_{\cdot 1}\, n_{\cdot 2}}.$$

The calculation of this statistic yields

$$X^2_{YC} = \frac{100(|4 * 66 - 4 * 26| - 100/2)^2}{30 * 70 * 8 * 92} = 0.783,$$

the same value as just described.

G. Confidence Interval for the Odds Ratio

The sample odds ratio for these data is 2.538 [= (4 * 66)/(4 * 26)]. This value seems to be different from one and, therefore, it suggests that there is an association. However, there is sampling variation associated with the sample estimate of θ and this must be taken into account.

[1]Some statisticians question the use of Fisher's exact test in 2 by 2 tables when the data arise from either of the two sampling methods discussed above. They question the application because Fisher's test was developed based on both the row and column margins being fixed in advance, a different sampling scheme than used in the other two methods. Hence they do not recommend the use of Yates' correction, but we believe that Yates' correction is appropriate (7).

MINITAB BOX 11.1

The CHISQUARE (abbreviated to CHIS) command is used to analyze the data in Table 11.6.

```
MTB > set cl
DATA> 4 4
DATA> set c2
DATA> 26 66
DATA> end
MTB > chis cl c2
Expected counts are printed below observed counts
              Cl        C2     Total
    1          4        26        30
            2.40     27.60

    2          4        66        70
            5.60     64.40

Total          8        92       100

ChiSq = 1.067 + 0.093 + 0.457 + 0.040 = 1.656
df = 1
1 cells with expected counts less than 5.0
```

Because the distribution of the natural logarithm of θ, $\ln \theta$, converges to the normal distribution for smaller sample sizes than the distribution of θ, we focus on the confidence interval for $\ln \theta$. After finding the confidence interval for $\ln \theta$, we can transform it to a confidence interval for θ. The estimated standard error for the sample estimate of $\ln \theta$ (10, pp. 54–55) is

$$\hat{\sigma}_{\ln(\mathrm{OR})} = \sqrt{\frac{1}{n_{11}} + \frac{1}{n_{12}} + \frac{1}{n_{21}} + \frac{1}{n_{22}}}.$$

The $(1 - \alpha) * 100$ percent confidence interval for $\ln \theta$ is

$$\ln \mathrm{OR} \pm z_{1-\alpha/2} * \hat{\sigma}_{\ln \mathrm{OR}}.$$

The estimated standard error for the sample estimate of $\ln \theta$ is 0.7441, which is obtained from $\sqrt{1/4 + 1/26 + 1/4 + 1/66}$. The value of the natural logarithm of the sample odds ratio, ln 2.538, is 0.9314. Therefore, the 95 percent confidence interval for $\ln \theta$ is $0.9314 \pm 1.96 * 0.7441$, which ranges from -0.5270 to 2.3897. Taking the exponential of these limits provides the 95 percent confidence interval for θ and its limits are 0.5904 and 10.9104. The confidence interval for the odds ratio is quite large and does include the value of one. Hence there is no evidence that the null hypothesis should be rejected.

All three approaches agree that there is no evidence of an association between iron status and education based on this hypothetical sample. These approaches will almost always agree about whether an association exists between two variables. The confidence interval for the difference of the probabilities and the uncorrected chi-square statistic will always agree in their conclusions.

Other analyses of the 2 by 2 table could be presented, but we next consider the extension of the 2 by 2 table to a table with r rows and c columns. We focus on the use of the goodness-of-fit test approach in the r by c table.

III. *r* BY *c* CONTINGENCY TABLE

The data in Table 11.7 are from a study in Los Angeles conducted to determine the knowledge and opinion of women about mammography. The study was a response to concern raised in the media about the potential radiation hazards of the long-term use of mammography (11). Two issues the study addressed were: (1) whether these articles had caused women to refuse the use of mammography screening for breast cancer; and (2) variables related to women's opinion about mammography. A telephone interview was conducted with a sample of women and approximately 60 percent of the women had heard or read something about mammography. Table 11.7 shows the opinion about mammography of those women who had heard or read about it.

A. Hypothesis of No Association

This is a 2 (the number of rows is always given first) by 3 (number of columns) table. The question of interest for this table is whether there is an association between the woman's opinion about mammography screening and the variable knowing someone with breast cancer. We test this hypothesis at the 0.01 significance level.

TABLE 11.7 Frequency of Women by Opinion about Mammography and Whether They Know Someone with Breast Cancer

Know someone with breast cancer	Opinion about mammography			
	Positive	Neutral	Negative	Total
Yes	120	45	28	193
No	77	15	8	100
Total	197	60	36	293

The same ideas used in the 2 by 2 table still apply here. If there is no association between the knowledge of someone with breast cancer variable and the opinion about mammography, the ratio of the expected cell frequency in the ith row and jth column, m_{ij}, to the ith row total, $n_{i\cdot}$, should equal the ratio of the jth column total, $n_{\cdot j}$, to the overall total. Thus, m_{ij} is still found from

$$\frac{m_{ij}}{n_{i\cdot}} = \frac{n_{\cdot j}}{n}$$

which yields

$$m_{ij} = \frac{n_{i\cdot} * n_{\cdot j}}{n}.$$

The values of m_{ij} are shown in Table 11.8.

There are 2 degrees of freedom for this table because once we know the frequencies of any two cells, we can find the values of the other frequencies by subtraction from the row and column totals. For example, knowing the frequencies for the 1,1 and 1,2 cells allows us to find the value of the 1,3 cell by subtraction of the sum of the 1,1 and 1,2 frequencies from the total of the first row. Knowledge of the frequencies in the first row then allows us to find the cell frequencies in the second row by subtraction from the column totals. For example, the frequency of the 2,1 cell is found by subtracting the frequency of the 1,1 cell from the total of the first column. There is a formula for the degrees of freedom that simplifies its determination. The formula for the degrees of freedom for the hypothesis of no association in an r by c contingency table is $(r - 1) * (c - 1)$.

The hypothesis of no association between the row and column variables is tested using the chi-square goodness-of-fit statistic. Most statisticians perform no adjustment to the test statistic when used with tables other than the 2 by 2 table. If the test statistic is greater than the value of 9.21 ($= \chi^2_{2,0.99}$), we reject the hypothesis of no association in favor of the alternative that the row and column variables are related. If the test statistic

TABLE 11.8 Expected Frequency of Women by Opinion about Mammography and Whether They Know Someone with Breast Cancer

Know someone with breast cancer	Opinion			
	Positive	Neutral	Negative	Total
Yes	129.76	39.52	23.71	192.99
No	67.24	20.48	12.29	100.01
Total	197.00	60.00	36.00	293.00

is less than 9.21, we fail to reject the null hypothesis. The value of the test statistic is found from

$$X^2 = \frac{(120 - 129.76)^2}{129.76} + \frac{(45 - 39.52)^2}{39.52} + \cdots + \frac{(8 - 12.29)^2}{12.29}$$

which equals 6.648. As 6.648 is less than 9.21, we fail to reject the null hypothesis. There does not appear to be a statistically significant association, at the 0.01 level, between opinion about mammography and whether someone with breast cancer was known.

We can use MINITAB to perform the test as shown in Box 11.2. The p value is 1 minus 0.9640, or 0.036. Although there is some suggestion (a p value of 0.036) of an association between the row and column variables, the association is not statistically significant at the 0.01 level.

MINITAB BOX 11.2

```
MTB > set c1
DATA> 120 77
MTB > set c2
DATA> 45 15
MTB > set c3
DATA> 28 8
DATA> end
MTB > chis c1-c3
Expected counts are printed below observed counts

             C1        C2        C3     Total
    1       120        45        28       193
         129.76     39.52     23.71

    2        77        15         8       100
          67.24     20.48     12.29

Total       197        60        36       293

ChiSq =   0.735 +   0.759 +   0.775 +
          1.418 +   1.465 +   1.496 = 6.648
df = 2
```

We can also find the p value of the test by using the CDF command.

```
MTB > cdf 6.648;
SUBC> chis 2.
    6.6480     0.9640
```

B. Hypothesis of No Trend

The hypothesis of no association is very general, and it is a reasonable hypothesis to test with nominal variables; however, when a variable conveys more information than the category name, it is possible to test a more specific hypothesis that uses more of the information contained in the variable. For example, in this example, opinion is an ordinal variable that ranges from positive to neutral to negative, and the test for no association ignores this ordering. In 2 by *c* contingency tables, there is a test, a test for trends, that takes the ordering of the column variable into account.[2]

In the test for no association, we examined the unconditional cell probabilities. We also could have focused on the conditional probabilities, for example, the probability of women who knew someone with breast cancer conditional on their opinion of mammography. In calculating the conditional probabilities in this fashion, we are not implying that the probability of women who knew someone with breast cancer depends on their opinion of mammography. We are calculating the conditional probabilities in this fashion simply to see if there is a trend in the probabilities of women who knew someone with breast cancer by opinion category. The sample estimates of these conditional probabilities are easily found. For the women who are positive about mammography, the estimated probability of a woman knowing someone with breast cancer is 0.609 (= 120/197). The corresponding values for the women with neutral and negative opinions are 0.750 and 0.778, respectively. If the estimates of these probabilities are related to the opinion category, this suggests that an association exists between the row and column variables.

We now consider the hypothesis of no linear trend. By no linear trend, we mean that the proportion of women who knew someone with breast cancer does not increase (decrease) consistently with the changes in opinion from positive to neutral to negative. To perform a test of this hypothesis, we assign a numerical score to the categories of the opinion variable. For example, it seems reasonable to assign scores of $+1$ to the positive category, 0 to the neutral level and -1 to the negative opinion category. This assignment of scores assumes that the distance from positive to neutral is the same as the distance from neutral to negative. The assignment of scores is subjective, and in unusual situations, the scoring system used can have an impact on the test of hypothesis. In most cases, however, different reasonable scoring systems will lead to the same conclusion about the test of hypothesis.

The hypothesis of no linear trend is basically a test of no correlation between the assigned scores and the conditional probabilities. Thus the

[2]There is also a method that can be used for *r* by *c* contingency tables. See the article by Semenya *et al.* (12) for details.

test statistic should look something like a correlation coefficient. The following notation is used in the representation of the test statistic. Let p_j be the sample estimate of the conditional probabilities of women who knew someone with breast cancer and S_j be the score assigned to the jth opinion category, where j equals 1, 2, and 3 for positive, neutral, and negative. The unconditional sample estimate of the women who knew someone with breast cancer is $\bar{p}$, and $\bar{q}$ is 1 minus $\bar{p}$. Let $\bar{S}$ be the sample mean score.

The test statistic is

$$X^2 = \frac{\left(\sum_{j=1}^{c} n_{.j}(p_j - \bar{p})(S_j - \bar{S})\right)^2}{\bar{p}\bar{q} \sum_{j=1}^{c} n_{.j}(S_j - \bar{S})^2}.$$

The numerator of this statistic is the square of the numerator of the correlation coefficient between the conditional proportion and the assigned score. Hence we can see that this statistic is a measure of the linear trend between these two variables. For sufficiently large sample sizes, this statistic can be shown to follow the chi-square distribution with one degree of freedom if there is no linear trend. The sample size is sufficiently large if, given the value of $\bar{p}$, it is larger than that shown in Table 6.7. Large values of X^2 cause us to reject the null hypothesis of no linear trend in favor of the alternative hypothesis of a linear trend.

Let us test the null hypothesis of no linear trend in the opinion about mammography data at the 0.01 significance level. The overall proportion of women who knew someone with breast cancer, $\bar{p}$, is 0.659 (= 193/293). Hence $\bar{q}$ is 0.341. As n is 293, much larger than the values in Table 6.7 for proportions of 0.30 and 0.35, we can use the test statistic shown above. The p_j are 0.609, 0.750, and 0.778 for j values of 1, 2, and 3. S_1 is +1, S_2 is 0, and S_3 is −1, and the values of the column totals, $n_{.j}$, are 197, 60, and 36, respectively. The mean of the scores, $\bar{S}$, is found from

$$\frac{(197 * 1) + (60 * 0) + (36 * (-1))}{293} = 0.5495$$

The test statistic is

$$X^2 = \frac{[197(-.050)(.4505) + 60(.091)(-.5495) + 36(.119)(-1.5495)]^2}{0.659 * 0.341 * [197(.4505)^2 + 60(-.5495)^2 + 36(-1.5495)^2]}$$

and this simplifies to

$$\frac{(-14.076)^2}{32.479} = 6.100.$$

This statistic is compared with 6.63 (= $\chi^2_{1,0.99}$). As 6.100 is less than 6.63, we fail to reject the null hypothesis of no linear trend. The p value of this test statistic is found from MINITAB to be 0.0135. Although there is not a significant linear trend in these data at the 0.01 significance level, there is a strong inverse relationship between the conditional proportion of women who knew someone with breast cancer and their opinion about mammography. We know the relationship is inverse because the sign of the numerator, before squaring, is negative. The opinion about mammography is more likely to be neutral or negative as the proportion of women who knew someone with breast cancer increases.

This test for trends is equivalent to creating a confidence interval for the difference in means from two independent populations. In this example, the two independent populations are the women who did not know someone with breast cancer and those who did know someone with breast cancer.

The test for trends is particularly appropriate for 2 by c contingency tables when there is an ordering among the column categories. If a linear trend exists, it may be missed by the general test for association, whereas the trend test has a greater chance of detecting it. The general test for association could cause us to say that there is no relationship between the rows and columns when there actually is a linear trend.

IV. MULTIPLE 2 BY 2 CONTINGENCY TABLES

Most studies involve the analysis of more than two variables at one time. Often we are interested in the relationship between an independent variable and the dependent variable, but there is an extraneous variable that must also be considered. For example, consider a study to determine if there is any association between the occurrence of upper respiratory infections (URIs) of young children and outdoor air pollution. Several variables could affect the relationship between the occurrence of infections and outdoor air pollution. One variable is the quality of the indoor air. One easily obtained variable that partially addresses the indoor air quality is whether someone smokes in the home. This variable is likely to be related to the dependent variable, the occurrence of URIs, and it may also be related to the independent variable. Hypothetical data for this situation are based on an article by Jaakkola (13) and are shown in Table 11.9.

A. Analyzing the Tables Separately

If we ignore the passive smoke variable, the X^2_{YC} for the combined table is 6.387, its p value is 0.0115 and the estimate of the odds ratio is 1.524. There is a statistically significant relationship between the outdoor pollution variable and the occurrence of URIs. The estimated odds ratio of 1.524 means that the odds of URI during the previous 12 months is about $1\frac{1}{2}$ times greater in a city with high pollution than in a city with low pollution.

TABLE 11.9 Number of 6-Year-Old Finnish Children by Respiratory Status and Pollution Level with a Control for Passive Smoke in the Home[a]

Passive smoke in home	City polluted	Upper respiratory infection during previous 12 months: Some	None	Total
Yes	High	100	20	120
	Low	124	40	164
	Total	224	60	284
No	High	128	62	190
	Low	166	119	285
	Total	294	181	475

[a]The entries in the table are based on an article by Jaakkola *et al.* (13), but the data are hypothetical.

However, this analysis has excluded the passive smoke variable, a variable that we wish to take into account.

One way of taking the passive smoke variable into account is to analyze each 2 by 2 table separately. In this example, the X^2_{YC} statistic is 2.039 and its p value is 0.1533 for homes in wihch someone smoked. The X^2_{YC} value is 3.645 and its p value is 0.0562 for those without passive smoke in the home.

The corresponding estimates of the odds ratios for these two tables are 1.613 and 1.480. The 95 percent confidence intervals for the two odds ratios are 0.887 to 2.933 and 1.007 to 2.171, respectively. The first confidence interval, which is much wider than the second interval, includes the value of one, suggesting there is no relationship between the two variables. The second interval barely misses including one. The second interval's smaller size reflects the larger sample size associated with the homes in which there was no passive smoke. Neither of these tables has a statistically significant association between outdoor air pollution and occurrence of URI at the 0.05 level based on the test statistics. The conclusion from the analyses of the separate tables is different from that from analysis of the combined table.

A problem with the use of separate tables is that the analyses are based on smaller sample sizes associated with subtables, not on the sample size of the combined table. This makes it difficult to find the presence of small but consistent trends across tables. A method for eliminating this problem is discussed in the next section. Before presenting the method, however, we should consider a problem that can occur when subtables are combined.

Besides ignoring the passive smoke variable, a potential problem in using the combined table is that it can be misleading. For example, if the data are selected from a population that does not represent the target population, strange things can occur. Suppose that we want our results to apply to all children in Finland, but that the children used in this study were sampled from those who had been hospitalized during the previous 12 months. If this were done, the population used in the study would not match the target population. Is that a problem? As has been mentioned before, the decision on the generalizability of the results to the target population depends on substantive considerations, not on statistical ideas. Let us assume that the sample data are those in Table 11.10.

In both of the subtables, the city with the lesser pollution had the greater proportion of children with no URI during the past 12 months. If we ignore the passive smoke variable, the combined table is Table 11.11. In the combined table, the city with the greater outdoor pollution now has the greater proportion of children with no URI during the past 12 months: 0.624 compared with 0.595 for the city with lesser pollution. This example points out that care must be exercised in combining tables when the population from which the sample is drawn is not representative of the target population. This was clearly pointed out by Berkson in an article in 1946 (14).

B. The Cochran–Mantel–Haenszel Test

Two biostatisticians, Nathan Mantel and William Haenszel, developed a method in 1959 for examining the relationship between two categorical variables while controlling for another categorical variable (15). This method, similar to a method published by William Cochran in 1954 (16), uses all the data in the combined table and produces one overall test statistic. The test is designed to detect the consistent effect of the indepen-

TABLE 11.10 Number of 6-Year-Old Finnish Children by Respiratory Status and Pollution Level with a Control for Passive Smoke in the Home Based on Taking Samples from a List of Hospitalized Children[a]

Passive smoke in home	City polluted	Upper respiratory infection during previous 12 months: Some	None	Total
Yes	High	35	40	75
	Low	60	80	140
	Total	95	120	215
No	High	170	300	470
	Low	15	30	45
	Total	185	330	515

[a]Hypothetical data.

TABLE 11.11 Number of Children with Occurrence of Upper Respiratory Infection by Pollution Status of City Ignoring the Passive Smoke Variable

City polluted	Upper respiratory infection during previous 12 months		
	Some	None	Total
High	205	340	545
Low	75	110	185
Total	280	450	730

dent variable on the dependent variable across the levels of the extraneous variable. Thus this method should be used only when the estimated odds ratios in the subtables are similar to one another. One very attractive feature of this test is that it can be used with extremely small sample sizes. This test has also been generalized for application to three-way tables of size other than 2 by 2 by k (17).

To facilitate the presentation of the test statistic, we use the following notation for the ith 2 by 2 contingency table, where i ranges from 1 to k, the number of levels of the extraneous variable. In our example, k is 2 as there are only two levels, presence and absence, of the passive smoke variable. The ith 2 by 2 table is shown next.

Polluted city	*Upper respiratory infection*		
	Some	*None*	*Total*
High	a_i	b_i	$a_i + b_i$
Low	c_i	d_i	$c_i + d_i$
Total	$a_i + c_i$	$b_i + d_i$	n_i

The test statistic is based on an overall comparison of the observed and expected in the (1, 1) cell in each of the k subtables. As we saw earlier in this chapter, under the hypothesis of no association between the row and column variables, only one degree of freedom is associated with the table. Hence we key on only one cell in the table and the choice of which cell is arbitrary. A statistic that could be used to examine whether there is an association is

$$Z^* = \frac{\sum_{i=1}^{k} (O_i - E_i)}{\text{s.e.}\left[\sum_{i=1}^{k} (O_i - E_i)\right]}$$

where O_i and E_i are the observed and expected values in the (1, 1) cell in the ith subtable. This statistic is very similar to a standard normal variable where E_i is analogous to the hypothesized mean in the standard normal variable.

In terms of the entries in the ith table, E_i is defined as

$$E_i = \frac{(a_i + b_i)(a_i + c_i)}{n_i},$$

the product of the row total and the column total divided by the table's sample size. The observed (1, 1) cell frequency, O_i, is a_i. V_i, the variance of O_i minus E_i, can be shown to be

$$V_i = \frac{(a_i + b_i)(c_i + d_i)(a_i + c_i)(b_i + d_i)}{n_i^2(n_i - 1)}.$$

Because we are dealing with discrete variables, we should use the continuity correction with Z^*; however, instead of using the continuity-corrected Z^* statistic, we would prefer to use a chi-square statistic as all the other tests associated with contingency tables use a chi-square statistic. This poses no problem because the square of a standard normal variable follows a chi-square distribution with 1 degree of freedom. Thus the statistic to be used to test the hypothesis of no association between air pollution and the occurrence of upper respiratory problems is the Cochran–Mantel–Haenszel chi-square statistic. Also called the Mantel–Haenszel statistic, it is defined by

$$X^2_{\text{CMH}} = \frac{(|O - E| - 0.5)^2}{V}$$

where O, E, and V are defined as the sums of the O_i, the E_i, and the V_i over the k subtables. If X^2_{CMH} is greater than $\chi^2_{1,1-\alpha}$, we reject the hypothesis of no association between air pollution and the occurrence of upper respiratory infections. Otherwise we fail to reject the null hypothesis.

As the odds ratios in the two separate subtables were similar, 1.613 in homes with passive smoke and 1.480 in the other homes, we can use the X^2_{CMH} statistic. If the odds ratios had not been similar, the effect of the independent variable on the dependent variable would not be consistent across the levels of the extraneous variable. Hence it would not make sense to use the CMH statistic to test for a consistent effect of the independent variable when we already know that such an effect does not exist. Because the values of our odds ratios are similar, we can test the hypothesis of no association (no consistent effect) between air pollution and the occurrence of URI while controlling for passive smoke and we perform the test at the 0.05 significance level. From Table 11.9 we see that O_1 is 100 and O_2 is 128 and their sum is 228. The expected values are found from

$$E_1 = \frac{120 * 224}{284} = 94.65 \quad \text{and} \quad E_2 = \frac{190 * 294}{475} = 117.60$$

and their sum is 212.25. The variances are found from

$$V_1 = \frac{120 * 164 * 60 * 224}{284^2 * 283} = 11.59 \quad \text{and} \quad V_2 = \frac{190 * 285 * 181 * 294}{475^2 * 474} = 26.94$$

and their sum is 38.53. Thus we have the pieces needed to calculate X^2_{CMH}:

$$X^2_{CMH} = \frac{(|228 - 212.25| - 0.5)^2}{38.53} = 6.036.$$

As 6.036 is greater than 3.84 ($= \chi^2_{1,0.95}$), we reject the hypothesis of no association. At the 0.05 level, we conclude that there is an association between air pollution and URI even after controlling for passive smoke in the home.

C. The Mantel–Haenszel Common Odds Ratio

Mantel and Haenszel also showed how to combine the data from the separate subtables to form a common odds ratio for the data. Again, this should be done only when the estimated odds ratios in the subtables are similar. If the estimated odds ratios for the subtables are not similar, for example, some are less than one and some are greater than one, the common odds ratio would not be very useful. The relationship between the independent and dependent variables would depend on the level of the extraneous variable, and the use of a common odds ratio would mask this. The Mantel–Haenszel estimator of the common odds ratio, θ, is

$$\mathrm{OR}_{MH} = \frac{\sum_{i=1}^{k} (a_i * d_i/n_i)}{\sum_{i=1}^{k} (b_i * c_i/n_i)}.$$

For the air pollution data, the Mantel–Haenszel estimate is found from

$$\mathrm{OR} = \frac{(100 * 40)/284 + (128 * 119)/475}{(20 * 124)/284 + (62 * 166)/475} = 1.517.$$

This value is similar to the individual odds ratios of 1.613 and 1.480 and also close to the value, 1.524, found from the overall table. The similarity of the values supports the finding that the passive smoke variable had little effect on the relationship between air pollution and URI. There are several approaches to finding an estimate of the variance of the Mantel–Haenszel estimator of the common odds ratio (18, 19), but they are quite involved and are not presented here.

V. CONCLUDING REMARKS

In this chapter, we introduced another nonparametric test, the chi-square goodness-of-fit test, and showed its use with one- and two-way contingency tables. We also showed two related methods, comparison of two binomial proportions and the calculation of the odds ratio, for determining, at some significance level, whether a relationship exists between two discrete variables with two levels each. The odds ratio is of particular interest as it is used extensively in epidemiological research. We also presented the extension of the goodness-of-fit test for no interaction to r by c contingency tables. Another test shown was the trend test, and it is of interest because it has a greater chance of detecting a linear relationship between a nominal variable and an ordinal variable than does the general chi-square test for no interaction. The Cochran–Mantel–Haenszel test and estimate of the common odds ratio were introduced for multiple 2 by 2 contingency tables. These procedures are also used extensively by epidemiologists. In the next chapter, we conclude the material on nonparametric procedures with the presentation of several nonparametric methods for the analysis of survival data.

EXERCISES

11.1. The following data are from one of the hospitals that participated in a study performed by the Veterans Administration Cooperative Duodenal Ulcer Study Group (20). The data from 148 men show the severity of an undesirable side effect, the dumping syndrome, of surgery for duodenal ulcer for four surgical procedures: (A) drainage and vagotomy, (B) 25 percent resection (antrectomy) and vagotomy, (C) 50 percent resection (hemigastrectomy) and vagotomy, (D) 75 percent resection.

	Severity of dumping syndrome			
Surgery	*None*	*Slight*	*Moderate*	*Total*
A	23	7	2	32
B	23	10	5	38
C	20	13	5	38
D	24	10	6	40
Total	90	40	18	148

Was the design used in this hospital a completely randomized design or a randomized block design? Explain your answer. Test the hypothesis of no association between the type of surgery and the severity of the side effect at the 0.05 significance level. Assuming that the procedures are equally effective, would you recommend any of the procedures over the others?

11.2. Test the hypothesis that the data from Gosset, shown in Table 6.4 and repeated here, come from a Poisson distribution at the 0.01 significance level.

Observed Frequency of Yeast Cells in 400 Squares

X	0	1	2	3	4	5	6
Frequency	103	143	98	42	8	4	2

11.3. The following data, from an article by Cochran (16), show the clinical change by degree of infiltration, a measure of a type of skin damage, present at the beginning of the study for 196 leprosy patients who received 48 weeks of treatment.

Degree of infiltration	*Improvement*					
	Worse	*Same*	*Slight*	*Moderate*	*Marked*	*Total*
0–7	11	27	42	53	11	144
8–15	7	15	16	13	1	52
Total	18	42	58	66	12	196

Test the hypothesis of no association between the degree of infiltration and the clinical change at the 0.05 significance level. Is this a test of independence or homogeneity? Explain your answer. Now assign scores from -1 to $+3$ for the clinical change categories worse to marked improvement and test the hypothesis of no linear trend at the 0.05 significance level. Is there any difference in the results of the tests? Select another reasonable set of scores and perform the trend test again using the second set of scores. Is the result consistent with the result from the first set of scores?

11.4. Mantel (21) provided data from a study to determine whether or not there is any difference in the effectiveness of immediately injecting or waiting 90 minutes before injecting penicillin into rabbits who have been given a lethal injection. An extraneous variable is the level of penicillin. The data are shown in the following table.

Penicillin level	*Delay*	*Response*		
		Cured	*Died*	*Total*
1/8	None	0	6	6
	90 min	0	5	5
1/4	None	3	3	6
	90 min	0	6	6
1/2	None	6	0	6
	90 min	2	4	6
1	None	5	1	6
	90 min	6	0	6
4	None	2	0	2
	90 min	5	0	5

Is it appropriate to use the CMH statistic here to test the hypothesis of no association between the delay and response variables while controlling for the penicillin level? Explain your answer. If you feel that it is appropriate to use the CMH statistic here, test, at the 0.01 significance level, the hypothesis of no association between the delay and response variables while controlling for the penicillin level.

11.5. Your local health department conducts a course on food handling. To evaluate this course, you select a SRS of restaurants from the list of licensed restaurants. For these restaurants in your sample, you then search the list of violations found by the health department during the last 2 years as well as the list of restaurants with employees who have attended the course during the last 2 years. For the 86 sampled restaurants, the data can be presented as follows.

Attended course	*Violation*		
	Yes	*No*	*Total*
Yes	9	28	37
No	36	13	49
Total	45	41	86

Use an appropriate procedure to test the hypothesis of no association between course attendance and the occurrence of a violation at the 0.10 significance level. On the basis of these data, discuss whether or not course attendance had any effect on the finding of a restaurant's violation of the health code.

11.6. Cochran (16) presented data on erythroblastosis foetalis, a sometimes fatal disease of newborn infants. The disease is caused by the presence of an anti-Rh antibody in the blood of an Rh+ baby. One treatment used for this disease is the transfusion of blood that is free of the anti-Rh antibody. In 179 cases in which this treatment was used in a Boston hospital, there were no infant deaths out of 42 cases when the donor was female compared with 27 deaths when the donor was male. One possible explanation for this surprising finding was that the male donors were used in the more severe cases. Therefore, the disease severity was taken into account and the data are shown in the following table.

Disease severity	Donor's sex	Survival status: Dead	Alive	Total
None	M	2	21	23
	F	0	10	10
Mild	M	2	40	42
	F	0	18	18
Moderate	M	6	33	39
	F	0	10	10
Severe	M	17	16	33
	F	0	4	4
Total		27	152	179

Use the CMH statistic to test the hypothesis of no association between donor's sex and the survival status of the infant at the 0.05 significance level.

11.7. Group the blood pressure values shown in Table 11.2 into categories of <80, 80–89, 90–99, . . . , ≥130 mm Hg. Based on this grouping, test the hypothesis that the systolic blood pressure of 12-year-old boys follows a normal distribution using the 0.05 significance level. Compare your results with those based on the grouping shown in Table 11.3.

11.8. The following data show the relationship between two types of media exposure and a person's knowledge of cancer (22, p. 36).

Media exposure: Newspapers	Radio	Knowledge of cancer: Good	Poor
Read	Listen	168	138
	Do not listen	310	357
Do not read	Listen	34	72
	Do not listen	156	494
Total		668	1061

On the basis of these data, test the hypothesis of no association between newspapers and knowledge of cancer, ignoring the radio variable. Next test the hypothesis of no association between radio and knowledge of cancer, ignoring the newspaper variable. Which variable has the stronger association with the knowledge of cancer variable? On the basis of these data, would you feel comfortable recommending one of these media over the other for the purpose of increasing the public's knowledge of cancer? If your answer is yes, what assumptions are you making about the data? If your answer is no, provide your rationale for your answer.

REFERENCES

1. Yarnold, J. K. (1970). The minimum expectation in X^2 goodness of fit tests and the accuracy of approximations for the null distribution. *J. Am. Stat. Assoc.* **65,** 864–886.
2. Snyder, L. H. (1970). Heredity. *Collier's Encycl.* **12,** 68–76.
3. Bishop, Y. M. M., Fienberg, S. E., and Holland, P. W. (1975). "Discrete Multivariate Analysis: Theory and Practice." MIT Press, Cambridge, MA.
4. Cochran, W. G. (1952). The χ^2 test of goodness of fit. *Ann. Math. Stat.* **23,** 315–345.
5. Goodman, L. A., and Kruskal, W. H. (1959). Measures of association for cross-classifications. II. Further discussion and references. *J. Am. Stat. Assoc.* **54,** 123–163.
6. Life Sciences Research Office, Federation of American Societies for Experimental Biology (1989). "Nutrition Monitoring in the United States: An Update Report on Nutrition Monitoring," DHHS Publ. No. (PHS) 89-1255. U.S. Department of Agriculture and the U.S. Department of Health and Human Services, Public Health Service, Washington, U.S. Government Printing Office.
7. Yates, F. (1984). Tests of significance for 2 × 2 contingency tables (with discussion). *J. R. Stat. Soc., Ser. A* **147,** 426–463.
8. Fisher, R. A. (1935). The logic of inductive inference (with discussion). *J. R. Stat. Soc.* **98,** 39–82.
9. Mehta, C. R., and Patel, N. R. (1983). A network algorithm for performing Fisher's exact test in r × c contingency tables. *J. Am. Stat. Assoc.* **78,** 427–434.
10. Agresti, A. (1990). "Categorical Data Analysis." Wiley, New York.
11. Berkanovic, E., and Reeder, S. J. (1979): Awareness, opinion and behavioral intention of urban women regarding mammography. *Am. J. Public Health* **69,** 1172–1174.
12. Semenya, K. A., Koch, G. G., Stokes, M. E., Forthofer, R. N. (1983). Linear models methods for some rank function analyses of ordinal categorical data. *Commun. Stat.* **12,** 1277–1298.
13. Jaakkola, J. J. K., Paunio, M., Virtanen, M., Heinonen, O. P. (1991). Low-level air pollution and upper respiratory infections in children. *Am. J. Public Health* **81,** 1060–1063.
14. Berkson, J. (1946). Limitations of the application of fourfold table analysis to hospital data. *Biometrics Bull.* **2,** 47–53.
15. Mantel, N., and Haenszel, W. (1959). Statistical aspects of the analysis of data from retrospective studies of disease. *J. Nat. Cancer Inst. (U.S.)* **22,** 719–748.
16. Cochran, W. G. (1954). Some methods for strengthening the common χ^2 tests. *Biometrics* **10,** 417–451.
17. Landis, J. R., Heyman, E. R., and Koch, G. G. (1978). Average partial association in three-way contingency tables: A review and discussion of alternative tests. *Int. Stat. Rev.* **46,** 237–254.
18. Mehta, C. R., and Walsh, S. J. (1992). Comparison of exact, mid-p, and Mantel–Haenszel confidence intervals for the common odds ratio across several 2 × 2 contingency tables. *Am. Stat.* **46,** 146–150.
19. Letters to the Editor, letters by Mantel, by Sato and Takagi, and by Mehta and Walsh relating to (18). (1993). *Am. Stat.* **47,** 86–87.
20. Grizzle, J. E., Starmer, C. F., and Koch, G. G. (1969). Analysis of categorical data for linear models. *Biometrics* **25,** 489–504.
21. Mantel, N. (1963). Chi-square tests with one degree of freedom; extensions of the Mantel–Haenszel procedure. *J. Am. Stat. Assoc.* **58,** 690–700.
22. Forthofer, R. N., and Lehnen, R. G. (1981). "Public Program Analysis: A New Categorical Data Approach." Lifetime Learning Publications, Belmont, CA.

CHAPTER 12

Analysis of Survival Data

This chapter introduces methods for analyzing data collected from a longitudinal study in which a group of subjects are followed for a defined period or until some specified event occurs. We frequently encounter such data in the health field. For example, newly diagnosed cancer patients in a registry were followed annually until they died. Another example consists of smokers who completed a smoking cessation program and were then contacted every 3 months to find out whether they had relapsed. The focus in these studies is the length of time from a meaningful starting point until the time at which either some well-defined event happens, such as death or relapse to a certain condition, or the study ends. The data from such studies are called *survival data*. We have previously encountered survival data in our consideration of the life table in Chapter 5. In this chapter, we consider a special type of life table, the follow-up life table.

We first discuss the collection and organization of the data. This discussion is followed by the presentation of two related methods for analyzing survival data. The life-table method is used for larger data sets and the product-limit method is generally used for smaller data sets. We also show how the CMH test statistic from Chapter 11 can be used for comparing two survival distributions.

I. DATA COLLECTION IN FOLLOW-UP STUDIES

Perhaps an example best illustrates the nature of the data required for a survival analysis. The California Tumor Registry identified a total of 2711 females with ovarian cancer initially diagnosed between 1942 and 1956 in 37 hospitals in California (1). The follow-up system of the Central Registry was designed to identify deaths through the statewide vital registration system and to facilitate the follow-up activities of the participating hospital registries. The Central Registry received yearly follow-up information on each case. The registry program served not only to furnish the information essential for statistical study of cancer cases, but also to stimulate periodic medical checkups of the cancer patients. Based on the data accumulated in the Central Registry up to 1957, the researchers were able to analyze ovarian cancer patients who had been followed for up to 17 years.

In this data set, patients were observed for different lengths of time and not all of the patients had died by 1957. In addition, others could not be contacted, that is, were lost to follow-up. Despite the different lengths of observation and the incomplete observations, it is possible to analyze the survival experience of these patients. An appropriate survival analysis is not restricted to those who died, but incorporates all the patients who entered the study. It is essential to include all those who entered the study because the exclusion of any patient from the analysis could introduce a selection bias as well as reduce the sample size.

The survival time cannot be calculated for those patients who were still alive at the closing date of the study or for those patients whose survival status was unknown. For these incomplete observations, the survival time is said to be *censored*. Those patients who were still alive at the closing date are known as *withdrawn alive* and those patients whose status could not be assessed (because, for example, they moved away or refused to participate) are known as *lost-to-follow-up*.

To include the censored observations in the analysis, we calculate a censored survival time from the date of diagnosis to (1) the closing date of the study for those withdrawn alive, and (2) the last known date of observation for the lost-to-follow-up. This allows the number of years from the date of diagnosis to the date of death or to the termination date to be calculated for each patient in the study.

By tabulating the uncensored and censored survival times of all 2711 female ovarian cancer patients by 1-year intervals, we obtain the data

TABLE 12.1 Survival Times for Ovarian Cancer Patients Initially Diagnosed 1942–1956, Followed to 1957

Years after diagnosis	Death d_i	Censored: Lost l_i	Censored: Withdrawn w_i	Censored: Total	Number entering interval n_i
0–1	1421	68	0	1489	2711
1–2	335	19	37	391	1222
2–3	132	17	84	233	831
3–4	64	10	47	121	598
4–5	44	12	48	104	477
5–6	20	12	39	71	373
6–7	19	10	35	64	302
7–8	14	14	19	47	238
8–9	7	10	25	42	191
9–10	7	9	19	35	149
10–11	5	4	14	23	114
11–12	5	4	17	26	91
12–13	1	4	11	16	65
13–14	3	1	15	19	49
14–15	1	0	13	14	30
15–16	0	0	7	7	16
16–17	0	0	9	9	9

Source: California Tumor Registry (1, pp. 258–259).

shown in Table 12.1. Within the first year of diagnosis, 1421 of 2711 patients had died and 68 were lost to follow-up. There were no patients in the category withdrawn alive, as every patient was followed for at least 1 year. The last column of the table can be created by adding the total column entries from the bottom. This reverse cumulative total indicates the number of patients alive at the beginning of each interval. The entry in the first row of this column is the total number of patients in the study. The other entries in this last column can also be found by subtracting the sum of the number of deaths, lost to follow-up, and withdrawn alive from the number of persons who started the previous interval. For example, the second entry in this column is 1222. It can be found by subtracting the sum of 1421, 68, and 0 from 2711, the number of subjects who began the previous interval.

The essential data items required for a survival analysis include d_i, the number of deaths, l_i, the number of patients lost to follow-up, w_i, the number of patients withdrawn alive, and n_i, the number of patients alive at the beginning of the ith interval. These data, presented in Table 12.1, are analyzed by the life-table method in the next section.

II. THE LIFE-TABLE METHOD

In Chapter 5, the population life table was introduced to illustrate the idea of probability and its connection to life expectancy. The estimated life expectancy is generally used as a descriptive statistic. To use the life-table technique as an analytical tool, we shall combine ideas from Chapter 6 on probability distributions with the life-table analysis framework.

In survival analysis, our focus is on the length of survival. Let X be a continuous random variable representing survival time. Consider a new function, the *survival function*, defined in symbols as

$$S(x) = \Pr\{X > x\}$$

This function is the probability that a subject survives beyond time x. As $F(x)$, the cdf, is defined as

$$F(x) = \Pr\{X \leq x\},$$

the survival function is one minus the cdf, that is,

$$S(x) = 1 - F(x).$$

It is more convenient to work with $S(x)$ rather than $F(x)$ because we usually talk about survival being greater than some value rather than being less than a value.

The idea of a survival function is contained in the population life table presented in Chapter 5. It is represented by the l_x column, the number of survivors at the beginning of each age interval. Specifically, $S(x)$ in the population life table is l_x/l_0. Recall that the l_x column starts with l_0, usually set at 100,000, and all subsequent l_x values are derived by multiplying the conditional probability of surviving in an age interval by the number of those who have survived all previous age intervals.

To analyze the data in Table 12.1 by the life-table method, we estimate the survival distribution in the same manner. The results of these calculations are shown in Table 12.2. The first two columns (time interval and number of deaths) are transferred from Table 12.1. The other columns show the results of the life-table analysis.

The first task is to estimate the conditional probability of dying for each interval of observation. When there is no censoring in an interval, the estimate of the probability of dying in the interval is simply the ratio of the number who died during the interval to the number alive at the beginning of the interval; however, it is not appropriate to use this ratio as the estimator of the probability of dying if censoring occurred in the interval. The use of this denominator, the number alive at the beginning of the interval, means that those who were lost to follow-up or withdrawn alive during the interval are treated as if they survived the entire interval. Thus, using this

TABLE 12.2 Estimates of Probabilities and Standard Errors for Ovarian Cancer Patients

(1) Years after diagnosis	(2) Deaths d_i	(3) Exposed to risk n_i'	Conditional probability (4) Dying q_i	(5) Surviving $(1 - q_i)$	(6) Cumulative probability surviving P_i	(7) Standard error $SE(P_i)$
0–1	1421	2677.0	0.531	0.469	1.000	0.0000
1–2	335	1194.0	0.281	0.719	0.469	0.0096
2–3	132	780.5	0.169	0.831	0.338	0.0092
3–4	64	569.5	0.112	0.888	0.280	0.0089
4–5	44	447.0	0.098	0.902	0.249	0.0087
5–6	20	347.5	0.058	0.942	0.224	0.0086
6–7	19	279.5	0.068	0.932	0.212	0.0086
7–8	14	221.5	0.063	0.937	0.197	0.0086
8–9	7	173.5	0.040	0.960	0.185	0.0087
9–10	7	135.0	0.052	0.948	0.177	0.0088
10–11	5	105.0	0.048	0.952	0.168	0.0090
11–12	5	80.5	0.062	0.938	0.160	0.0093
12–13	1	57.5	0.017	0.983	0.150	0.0097
13–14	3	41.0	0.073	0.927	0.147	0.0099
14–15	1	23.5	0.043	0.957	0.137	0.0109
15–16	0	12.5	0.000	1.000	0.131	0.0119
16–17	0	4.5	0.000	1.000	0.131	0.0119
17					0.131	0.0119

ratio when there is censoring likely results in an underestimate of the probability of dying in the interval.

The problem with the censored individuals is that we do not know their actual length of survival during the interval. We do know that it is extremely unlikely that they all survived the entire interval. The assumption used most often in practice, although there are other, perhaps, more reasonable assumptions, is that the censored individuals survived to the midpoint of the interval. Under this assumption, we can calculate q_i, an estimator of the conditional probability of dying during the ith interval,

$$q_i = \frac{d_i}{n_i - (\frac{1}{2}) * (l_i + w_i)} = \frac{d_i}{n_i'}.$$

The denominator in this equation is the effective number of subjects exposed to the risk of dying during the interval, denoted by n_i'. Table 12.2 shows the estimated effective number of patients exposed to the risk of dying in column 3 and the estimate of the conditional probability of dying in column 4.

By a simple manipulation of the definition of q_i given above, we obtain the alternative expression

$$q_i = \frac{d_i + (q_i/2) * (l_i + w_i)}{n_i}.$$

The use of n_i' above implies that those patients who were lost or withdrawn were subjected to one-half the risk of dying during the interval.

The estimator of the conditional probability of survival in the ith interval is one minus the estimator of the probability of dying, that is, $1 - q_i$. The result of this subtraction is shown in column 5.

Next, we calculate P_i, the sample estimator of the probability of surviving until the beginning of the ith interval. The set of the P_i are used to estimate the survival distribution $S(x)$. By definition, $P_1 = 1$, and the estimators of the other survival probabilities are calculated in the following manner:

$$P_2 = (1 - q_1), \qquad P_3 = (1 - q_2) * (1 - q_1),$$

and in general

$$P_i = (1 - q_{i-1}) * (1 - q_{i-2}) * \ldots * (1 - q_1) = (1 - q_{i-1}) * P_{i-1}$$

The results of these products are shown in column 6 of Table 12.2. From column 6, we see that the estimate of the 1-year survival probability for ovarian cancer patients in California who were diagnosed during the period 1942–1956 was 0.47, and the estimate of the 5-year survival probability was 0.22. More recent statistics estimate the 5-year survival probability for ovarian cancer to be 0.39 for white females and 0.38 for black females in the period 1981–1986 (2), suggesting some improvement in cancer treatment. This improvement, however, may be due more to the early detection of ovarian cancer in recent years. Cancer-related statistics, including estimates of survival rates, are routinely provided by the National Cancer Institute's Surveillance, Epidemiology and End Results (SEER) program, which includes many population-based cancer registries throughout the United States.

Besides knowing the point estimate of a population survival probability, we also wish to have a confidence interval for the survival probability. We assume that, in large samples, an estimated cumulative survival probability approximately follows a normal distribution. The variance of the estimated cumulative survival probability is estimated by

$$\widehat{\text{Var}}(P_i) = P_i^2 \sum_{j=1}^{i-1} \frac{q_j}{n_j' * (1 - q_j)}.$$

The estimated standard errors (the square root of the estimated variance) of the P_i are shown in column 7 of Table 12.2.

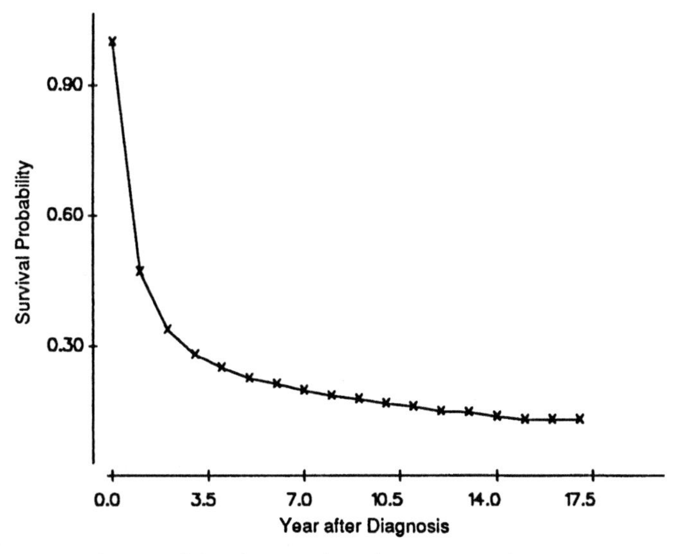

FIGURE 12.1 Estimated survival distribution of ovarian cancer patients.

Given these estimated standard errors plus the assumption of the approximate normality of the estimated survival probabilities, we can calculate confidence intervals for the survival probabilities. The approximate $(1 - \alpha) * 100$ percent confidence interval for a survival probability is given by

$$P_i \pm [z_{1-\alpha/2} * \text{s.e.}(P_i)]^1.$$

For example, an approximate 95 percent confidence interval for the 5-year survival probability is

$$0.224 - (1.96 * 0.0086) \quad \text{to} \quad 0.224 + (1.96 * 0.0086)$$

or 0.207 to 0.241.

It is also possible to calculate the confidence interval for the difference between two survival probabilities from different study groups, for example, the 5-year survival probability of ovarian cancer for white females and black females, by using the procedure discussed in Chapter 7.

Let us further explore the estimated survival distribution by creating Figure 12.1, the plot of the cumulative survival probabilities against the years after diagnosis. Although we have values of P_i for only the integer values of t, we have connected the points to show the shape of the survival

[1]Thomas and Grunkemeier (3) have shown that there are other more complicated approaches to constructing a confidence interval for P_i that cause the actual confidence level to agree more closely with the nominal confidence level, especially for small samples.

distribution. It starts with a survival probability of 1 at time 0 and drops quickly as time progresses, indicating a very high early mortality for ovarian cancer patients. Note that the survival curve does not descend all the way to zero. This is due to some women surviving more than 17 years.

The rapid decrease in the estimated survival curve suggests that the mean and the median survival times will be short. To verify this, let us estimate the mean and the median survival times from the estimated survival distribution. As some of the women survive longer than the 17 years of the study, this complicates the estimation of the population mean survival time. Instead of estimating the population mean, we therefore estimate the mean restricted to the time frame of 17 years, the length of the study. This restricted value will thus underestimate the true unrestricted mean. If no patient survived longer than the time frame of the study, the procedure shown below provides an estimate of the unrestricted mean.

The sample mean, restricted to the 17-year time frame, is found by summing the number of years (or other unit of time) survived during each time interval and dividing this sum by the sample size; however, the process of determining the number of years survived in an interval is complicated by the deaths, losses, and withdrawals that occurred during the interval. Instead of directly attempting to calculate the years survived, we use the following method to deal with this complication.

We calculate the sample mean by forming a weighted average of the years provided by each interval. The weight used with each interval is the cumulative survival probability associated with the interval. This approach deals with the complications mentioned above as the probability takes the deaths, losses, and withdrawals into account. Because there are two cumulative survival probabilities associated with each interval, the probability at the beginning, P_i, and the probability at the end, P_{i+1}, we use their average. Thus the formula for the restricted sample mean is

$$\bar{x}_r = \sum_{i=1}^{k} a_i * \left(\frac{P_i + P_{i+1}}{2}\right)$$

where k is the number of intervals and a_i is the width of the ith interval.

This formula has an interesting geometric interpretation: it provides an approximation to the area under the estimated survival curve. For example, consider a curve with three intervals (Figure 12.2). We are using rectangles to estimate the area under the curve. As we can see, some of the area under the curve is not included in the rectangles; however, this area is approximately offset by the areas included in the rectangles that are not under the curve. The formula for the area of a rectangle is the height multiplied by the width. In this case, the width is one unit, or in general, a_i units, and the height is taken to be the average of the points at the beginning and end of the interval, that is, $(P_i + P_{i+1})/2$. Hence another way of

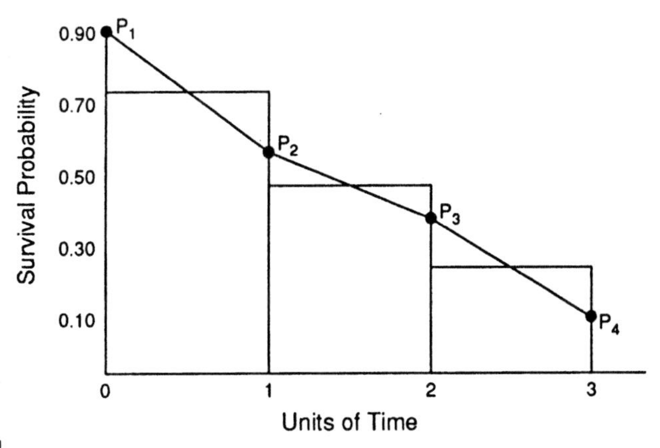

FIGURE 12.2 Survival curve with rectangles superimposed.

interpreting the mean is that it is the area under the survival curve. We approximate this area by calculating the area of the rectangles that can be superimposed on the survival curve.

When the intervals are all of the same width, for example, a, then the formula can be simplified to

$$\bar{x}_r = a * \left(\sum_{i=1}^{k+1} P_i - \frac{P_1 + P_{k+1}}{2} \right).$$

Because the intervals are all of width one in this example, the sample mean is simply the sum of the entries in column 6 of Table 12.2 minus one-half of the first and last entries in the column. This is

$$(1.000 + 0.469 + 0.338 + \ldots + 0.131) - 0.5(1.000 + 0.131)$$

which equals 3.92 years. This restricted mean survival time appears to be larger than what the first-year survival probability might suggest. As we saw in Chapter 4, the mean can be affected by a few large observations, and that is the case here. The sample mean reflects the presence of a few long-term survivors. Let us now calculate the median length of survival.

The median survival time is estimated in the following manner. First we read down the list of estimated cumulative survival probabilities, column 6 in Table 12.2, until we find the interval for which P_i is greater than or equal to 0.5 and P_{i+1} is less than 0.5. In Table 12.2, this is the first interval, as P_1 is greater than 0.5 and P_2 is less than 0.5. Thus we know that the estimated median survival time is between 0 and 1 years. As 47 percent of the patients survived the first year, we suspect that the estimated median survival time is much closer to one year than to zero years. To find a more precise value, we shall use linear interpolation.

In using linear interpolation, we are assuming that the deaths occurred at a constant rate throughout the interval. This is the same assumption we made when we connected the survival probabilities in Figure 12.1. In using linear interpolation, we know that to reach the median, we require only a portion of the interval, not the entire interval. The portion that we need is simply the ratio of the difference of P_i and 0.5 to the length of the interval. In symbols, this is

$$(P_i - 0.5)/(P_i - P_{i+1}).$$

We multiply this ratio by the width of the interval and add that to the survival time at the beginning of the interval. Replacing these words by symbols, the formula is

$$\text{sample median} = x_i + a_i * \frac{P_i - 0.5}{P_i - P_{i+1}}$$

where x_i is the survival time at the beginning of the interval and a_i is the width of the interval. In this example, the sample median survival time is

$$0 + 1 * \left(\frac{1 - 0.5}{1 - 0.469}\right) = 0.94 \text{ year.}$$

The sample median survival time of about 1 year is much shorter than the estimated restricted mean survival time. As we mentioned above, the mean survival time is affected by a small number of long-term survivors. This is why the median is more often used with survival data.

The median can also be obtained from the plot of the estimated survival curve shown in Figure 12.1. We move up the vertical axis until we reach the survival probability value of 0.5. We then draw a line parallel to the time axis and mark where it intersects the survival curve. We next draw a line, parallel to vertical axis, from the intersection point to the time line. The sample median survival time is the value where the line intersects the time axis. Figure 12.3 shows the estimated survival curve plot with these lines used to find the sample median drawn in the plot as well. The accuracy of the estimate of the median is limited by the scales used in plotting the survival curve. In Figure 12.3, the precision of the estimate is likely not to be high because of the scales used. It appears that the sample estimate of the median is approximately 1 year.

Another statistic often used in survival analysis is the *hazard rate*, which is also known as the life-table mortality rate, force of mortality, and instantaneous failure rate. It is used to measure the proneness to failure during a very short time interval. It is analogous to an age-specific death rate or interval-specific failure rate. It is the proportion of subjects dying or failing in an interval per unit of time. The hazard rate is usually estimated by the

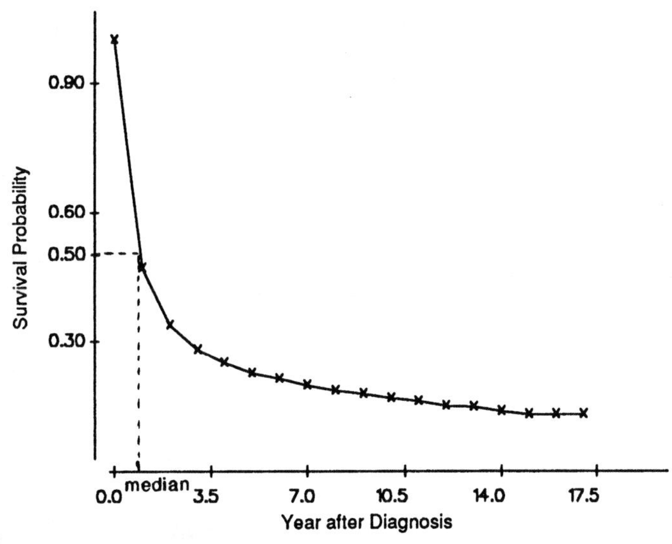

FIGURE 12.3 Using the estimated survival curve to find the median.

formula

$$h_i = \frac{d_i}{a_i * (n_i' - d_i/2)} = \frac{2 * q_i}{a_i * (2 - q_i)}.$$

The denominator of this formula uses the number of survivors, again assuming that death is occurring at a constant rate throughout the interval, at the midpoint of the interval. When the interval is very short, it makes little difference whether the number of survivors at the beginning or at the midpoint of the interval is used in the denominator. The sample hazard rates are calculated and shown in Table 12.3 for the first 10 years of follow-up. The estimate of the first-year hazard or mortality rate is quite high with 723 deaths per 1000 patients. The hazard is concentrated in the first 5 years after diagnosis and stabilizes at a low level after 5 years of survival. The variance of the sample hazard rate is estimated by

$$\widehat{\text{Var}}(h_i) = h_i^2 \left(\frac{1 - (h_i * a_i/2)^2}{n_i' * q_i} \right).$$

The estimated standard errors (the square root of the estimated variance) of the sample hazard rates are calculated and shown in Table 12.3. If we assume that the sample hazard rates are asymptotically normally distributed, these sample standard errors can be used to calculate confidence intervals for the population hazard rates. For example, the 95 percent

TABLE 12.3 Estimates of Hazard Rates and Their Standard Errors

Year	Hazard rate	Standard error	Year	Hazard rate	Standard error
0–1	0.723	0.0179	5–6	0.059	0.0132
1–2	0.326	0.0176	6–7	0.070	0.0161
2–3	0.185	0.0160	7–8	0.065	0.0174
3–4	0.119	0.0149	8–9	0.041	0.0156
4–5	0.104	0.0156	9–10	0.053	0.0201

confidence interval for the first-year hazard or mortality rate ranges from

$$0.723 - 1.96(0.0179) \quad \text{to} \quad 0.723 + 1.96(0.0179)$$

or from 0.688 to 0.758.

These life-table calculations can be performed by MINITAB as shown in Box 12.1; the results are printed in Box 12.2.

III. PRODUCT-LIMIT METHOD

When we analyze a smaller data set, for example, a sample size less than 100, the life-table method may not work very well because the grouping of survival times becomes problematic. Instead we use a method that is based on the actual survival time for each subject rather than grouping the subjects into intervals. The product-limit method, also known as the Kaplan–Meier method (4), is used to estimate the cumulative survival probability from a small data set, without relying on groupings of survival times. As can be seen below, the basic principles and computational procedures involved in the product-limit method are similar to those for the life-table method.

We start with an example. Suppose that 14 alcohol-dependent patients went through an intensive detoxification treatment for 4 years during the 1990–1993 period at a small clinic. There was a follow-up contact every month to check on their drinking status. The data shown in Table 12.4 were abstracted from the clinic patient records. The date of discharge and the date of termination are shown in year and month (9001 indicates 1990 January). The follow-up status is coded 2 if censored (withdrawn or lost-to-follow-up) and 1 if relapsed to drinking. Gender is coded 1 for females and 2 for males. The purpose of our study is to analyze the length of alcohol-free time among these 14 patients.

The first step of analysis is to calculate the survival time, x, in months for all subjects, censored and uncensored, and arrange them in order from the smallest to the largest with the censoring status indicated. If an uncen-

MINITAB BOX 12.1

```
(1)
MTB > read c1-c5
DATA>  0   1421 68  0 2711
DATA>  1    335 19 37 1222
DATA>  2    132 17 84  831
        .         .
        .         .
DATA> 16      0  0  9    9
DATA> end
(2)
MTB > let c6=c5-.5*(c3+c4)
MTB > let c7=c2/c6
MTB > let c8=1-c7
(3)
MTB > let c9(1)=1
MTB > let k1=2
MTB > store 'cumpro'
STOR> noecho
STOR> let c9(k1)=c9(k1-1)*c8(k1-1)
STOR> let k1=k1+1
STOR> end
MTB > execute 'cumpro' 17


(4)
MTB > let c1(18)=17
MTB > plot c9 c1
```

(The plot is not shown here.)

```
(5)
MTB > sum c9 k2
    SUM = 4.4861
MTB > let k3=k2-0.5*(c9(1)+c9(18))
MTB > print k3
 K3       3.9206
(6)
MTB > let c10(1)=0
MTB > let c14(1)=0
MTB > let k1=2
MTB > let c13=c7/(c6*c8)
MTB > store 'error'
STOR> noecho
STOR> let c14(k1)=c14(k1-1)+c13(k1-1)
STOR> let c10(k1)=c9(k1)*sqrt(c14(k1))
STOR> let k1=k1+1
STOR> end
```

```
MTB > execute 'error' 17
(7)
MTB > let c11=c2/(c6-0.5*c2)
MTB > let c12=(1-(.5*c11)**2)/(c6*c7)
MTB > let c12=c11*sqrt(c12)
```

(1) The data in Table 12.1 are entered: the starting points of the time intervals into c1; the number died during the interval (d_i) into c2; the number lost (l_i) into c3; the number withdrawn (w_i) into c4; and the number alive at the beginning of the interval (n_i) into c5.

(2) The effective number, stored in c6; the conditional probability of dying, stored in c7; and the conditional probability of surviving, stored in c8.

(3) The cumulative survival probability is calculated by using Macro commands and the results are stored in column c9. First, c9(1) is set to 1, the value of the survival probability at the beginning of the first interval. The constant k1, the indicator of which interval is being considered, is set to 2 because we are ready to calculate P_2. The STORE command creates a Macro file under the name of 'cumpro' which contains four statements. After the NOECHO statement, which tells MINITAB not to show the Macro commands each time they are executed, the next statement calculates the cumulative survival probability for the interval k1. The following statement increases k1 by one so that the next survival probability that is calculated is stored in the proper row of c9. "EXECUTE 'cumpro' 17" instructs the computer to use these formulas 17 times.

(4) The survival probabilities in c9 are plotted against the years after diagnosis in c1.

(5) The mean survival time is calculated in k3.

(6) The standard error for P_1 is set to 0. The standard errors of P_i, calculated by a MINITAB macro, are stored in c10. As the calculation is difficult to do by MINITAB in a single step, the intermediate computational steps are stored in c13 and c14. The constant k2 is an indicator of which interval is being considered. It is initially set to 2 because the estimated standard error of P_1 was set to zero. Column c13 contains the conditional probability of dying divided by the product of the effective sample size and the conditional probability of surviving the ith interval.

(7) The hazard rates are calculated in c11. The standard errors of the hazard rates can be calculated in a single step and the results are stored in c12.

---

sored subject and a censored subject have survival times of the same length, the uncensored one precedes the corresponding censored observation. For the data shown in Table 12.4, the ordered list of alcohol-free times in months, with the censored observations marked by asterisks, is:

4, 6, 6, 9*, 10, 14*, 16, 17*, 19, 20, 28, 31, 34*, 47*.

The second step is to create a worksheet like that shown in Table 12.5. In Table 12.5, the column headings refer to death and survival. For this problem, death is equated with relapse and survival is remaining alcohol

MINITAB BOX 12.2

```
MTB > print c1 c9-c12

ROW C1       C9        C10        C11        C12

  1  0  1.00000  0.0000000  0.722604  0.0178743
  2  1  0.46918  0.0096454  0.326352  0.0175915
  3  2  0.33754  0.0092394  0.184745  0.0160112
  4  3  0.28046  0.0089133  0.119070  0.0148573
  5  4  0.24894  0.0087390  0.103529  0.0155867
  6  5  0.22444  0.0086243  0.059259  0.0132450
  7  6  0.21152  0.0085980  0.070370  0.0161341
  8  7  0.19714  0.0086231  0.065268  0.0174343
  9  8  0.18468  0.0086974  0.041176  0.0155599
 10  9  0.17723  0.0087906  0.053232  0.0201127
 11 10  0.16804  0.0089949  0.048780  0.0218088
 12 11  0.16004  0.0092511  0.064103  0.0286528
 13 12  0.15010  0.0096858  0.017544  0.0175432
 14 13  0.14749  0.0098628  0.075949  0.0438178
 15 14  0.13669  0.0109335  0.043478  0.0434680
 16 15  0.13088  0.0119155  0.000000          *
 17 16  0.13088  0.0119155  0.000000          *
 18 17  0.13088  0.0119155
```

TABLE 12.4 Status of 14 Alcohol-Dependent Patients

Patient number	Date of discharge	Date of termination	Follow-up status	Gender
1	9001	9312	2 Still sober (withdrawn)	1 Female
2	9003	9009	1 Relapsed	1 Female
3	9005	9209	1 Relapsed	2 Male
4	9009	9111	2 Lost to follow-up	2 Male
5	9011	9306	1 Relapsed	1 Female
6	9102	9312	2 Still sober (withdrawn)	1 Female
7	9104	9211	1 Relapsed	1 Female
8	9108	9304	1 Relapsed	1 Female
9	9110	9202	1 Relapsed	2 Male
10	9203	9308	2 Lost to follow-up	2 Male
11	9207	9311	1 Relapsed	2 Male
12	9212	9310	1 Relapsed	1 Female
13	9303	9312	2 Still sober (withdrawn)	2 Male
14	9304	9310	1 Relapsed	2 Male

TABLE 12.5 Kaplan–Meier Estimates of Survival Probabilities

(1) Survival time x_i	(2) Number of deaths d_x	(3) Number at risk n_x	(4) Conditional probability of survival $(1 - q_x)$	(5) Cumulative probability of survival P_x	(6) Standard error $SE(P_x)$
0	0	14	1.000	1.000	—
4	1	14	0.929	0.929	0.066
6	2	13	0.846	0.786	0.101
10	1	10	0.900	0.707	0.121
16	1	8	0.875	0.619	0.124
19	1	6	0.833	0.516	0.146
20	1	5	0.800	0.412	0.141
28	1	4	0.750	0.309	0.128
31	1	3	0.667	0.206	0.106

free. The first three columns in the worksheet are created according to the following procedures.

1. List the uncensored alcohol-free times in order. These are 4, 6, 10, 16, 19, 20, 28, and 31. We refer to these times as $x_1, x_2, \ldots, x_8$, respectively.
2. Count the number of relapses at each of the x_i. There is one relapse at each time unless there are ties. The numbers are 1, 2, 1, 1, 1, 1, 1, and 1.
3. Count the number of subjects who are at risk of relapse at the x_i. For example, when the survival time is 10 months, three people have already relapsed and one person was withdrawn. Thus, only 10 persons are at risk of relapse at 10 months. These numbers are 14, 13, 10, 8, 6, 5, 4, and 3.

The fourth and fifth columns, estimates of the conditional probability of survival $(1 - q_x)$ and the cumulative probability of survival (P_x), are calculated next followed by the calculation of estimated variance of P_x, shown in column 6. The estimator of the conditional probability of relapse is the number of relapses divided by the number at risk, that is, $q_x = d_x/n_x$. The estimator of the conditional probability of survival is

$$1 - q_x = 1 - \frac{d_x}{n_x} = \frac{n_x - d_x}{n_x}.$$

The estimator of the cumulative probability of survival is found from the estimators of the conditional probabilities of survival in the same way as in the life-table method, that is,

$$P_x = \prod_{t \le x} (1 - q_t) = \prod_{t \le x} \frac{n_t - d_t}{n_t}.$$

The product symbol, Π, means that we multiply each term in the expression by one another for the indicated values of t. For example,

$$\prod_{t \leq 10} (1 - q_t) = (1 - q_4) * (1 - q_6) * (1 - q_{10}).$$

We could have included $1 - q_0$ in the product, but as q_0 is defined to be zero, its inclusion would not have changed the product.

As we have seen above, the censored observations have not been excluded from the analysis. They played a role in the determination of the number at risk at each time of relapse. If the censored observations were totally excluded from the analysis, the estimate of the conditional survival probabilities for the uncensored observations would be different.

The variance of P_x is estimated by

$$\widehat{\text{Var}}(P_x) = P_x^2 \left(\sum_{t \leq x} \frac{1}{(n_t - d_t)n_t} \right) \doteq P_x^2 \left(\frac{q_x}{n_x} \right).$$

The approximation shown in the above equation is much simpler to calculate and it works reasonably well in most situations (5). Taking the square root of the variance, calculated by the use of the approximation, we obtain the estimated standard errors of the P_x that are shown in column 6. As a demonstration of the accuracy of the approximation, the approximate estimate of the standard error of P_4 is 0.066, compared with the value of 0.069 obtained from the use of the first expression for the sample variance.

Figure 12.4 graphically displays the estimated survival distribution shown in the fifth column of Table 12.5. The plot includes a survival proba-

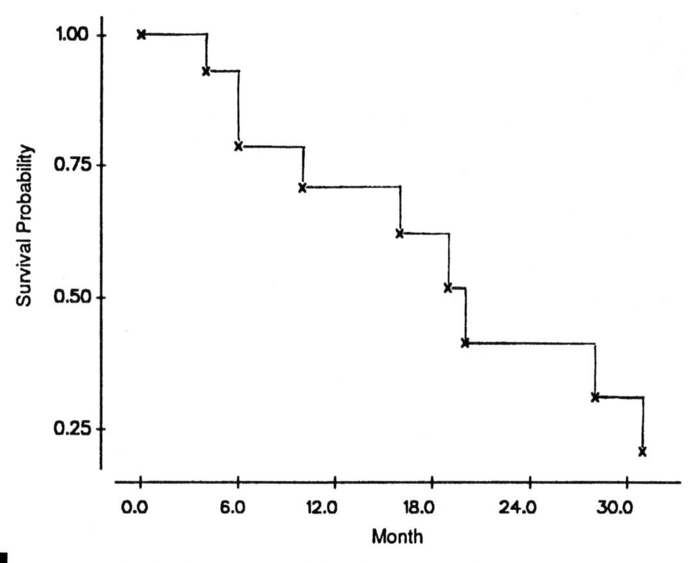

FIGURE 12.4 Survival distribution estimated by the product-limit method.

bility of 1 at time 0. The plot of the survival probabilities is referred to as a step function because it looks like a stairstep. It has this appearance because the probability of survival stays the same over a time period—this causes the horizontal lines—and then drops whenever there is another relapse—the vertical lines. However, long horizontal lines, showing no change in survival probability for a long period, should not be interpreted as a period with no risk, for these may occur because only small numbers of subjects are under observation during those periods.

We can estimate the mean survival time from the survival distribution. Again, just as in the life-table method, if the largest survival time is a censored time, we are really estimating a restricted mean. If the largest survival time is uncensored, then the survival probability will decrease to zero, and we will be estimating the unrestricted mean. As in the life table, the mean survival time is the area under the curve. We again use rectangles to approximate this area. Because of the step nature of the survival curve here, the rectangles are already formed for us. Unlike the life-table method, the widths of the intervals here are usually different. The following formula shows the area of each rectangle being calculated as the product of the height of the rectangle, the estimated cumulative survival probability associated with x_i, by the width, $x_{i+1} - x_i$. In symbols, this is

$$\bar{x}_r = \sum_{i=0}^{k-1} P_{x_i} * (x_{i+1} - x_i)$$

where k is the number of distinct time points when someone relapsed, x_0 is defined to be zero, and P_0 is defined to be one.

For these data, the estimate of the restricted mean alcohol-free time, restricted to a 31-month window, is given by

$$\begin{aligned}\bar{x}_r &= 1 * (4 - 0) + 0.929 * (6 - 4) + 0.786 * (10 - 6) + 0.707 * (16 - 10) \\ &\quad + 0.619 * (19 - 16) + 0.516 * (20 - 19) + 0.412 * (28 - 20) \\ &\quad + 0.309 * (31 - 28) = 18.4 \text{ months.}\end{aligned}$$

This is an underestimate of the true mean alcohol-free time because we are restricted to the study time frame and there were still people free of alcohol at the end of the study.

From Table 12.5, we can see that the median survival time, the point at which the cumulative survival probability is 0.5, occurs between the 19th and 20th months and is closer to month 19. We interpolate to find the sample median in the same way as in the life-table method. From our data, the sample median survival time is found as follows:

$$19 + (20 - 19) * \left(\frac{0.516 - 0.5}{0.516 - 0.412}\right) = 19.2 \text{ months.}$$

We should not use interpolation to find the median if there is a large gap in time between the two survival times in which we will be using the interpolation.

MINITAB can be used to calculate the entries in Table 12.5 as well as the sample mean and medians. The commands, not shown here, are a little more complicated than those in Box 12.1, but they follow the same ideas.

Because the product-limit method is based on the ranking of individual survival times, it is cumbersome to apply with a large data set. We would not consider using it with the ovarian cancer data from the California Tumor Registry which comprised more than 2000 observations. For a large data set, the life-table method simplifies the calculation and gives results similar to those of the product-limit method.

So far we have focused on describing the survival experience of a single population; however, we are often interested in comparing the survival experiences of two or more groups of subjects who differ on some account, for example, patients who have received different therapies for cancer or patients who belong to different age or sex groups. The comparison of two survival distributions is the topic of the following section.

IV. COMPARISON OF TWO SURVIVAL DISTRIBUTIONS

When comparing the survival experience of two or more groups, the description of the differences in the estimated survival distributions and the plot of the survival curves are only the beginning of the analysis. In addition to these descriptive techniques, researchers require a statistical test to determine whether the observed differences are statistically significant or due to chance variation.

In the analysis of survival data, we generally do not assume that the data follow any particular probability distribution. In the analysis, we also use the median survival time, rather than the mean, to summarize the survival experience. Because of these features, it seems as if a nonparametric test should be used when comparing survival distributions.[2]

For small data sets with no censored observations, the Wilcoxon rank sum test (Mann–Whitney test) can be used to test the null hypothesis of no difference in survival distributions for two independent samples. However, as survival time data usually contain censored observations, the Wilcoxon test cannot be directly applied. In this section, we show how the Cochran–Mantel–Haenszel (CMH) test statistic, described in Chapter 11, can be used in testing the hypothesis of no difference between two survival distributions (7). There are a number of other tests, extensions of the Wilcoxon and other rank tests, that could be used as well, but the CMH test seems to perform as well, if not better, than these other tests. Hence, the CMH test is the only one presented here.

[2]If we know that the survival data follow a particular distribution, we should take advantage of that knowledge. Parametric tests are available that can be used when we know the probability distribution of the survival data (6).

The key to the use of the CMH method with survival data is to realize that the data in each time interval can be formulated as a 2 by 2 table. The number of deaths and the number of survivors (the number exposed minus the number of deaths) for the two groups can be put in a 2 by 2 table for each time interval as shown next.

	Number of deaths	*Number of survivors*	*Total*
Group 1	d_{1i}	$(n'_{1i} - d_{1i})$	n'_{1i}
Group 2	d_{2i}	$(n'_{2i} - d_{2i})$	n'_{2i}
Total	$d_{.i}$	$(n'_{.i} - d_{.i})$	$n'_{.i}$

It can be shown that the time intervals are uncorrelated with one another which allows us to use the CMH statistic here.

Let us consider an example. The Hypertension Detection and Follow-up Program examined the effect of serum creatinine on 8-year mortality among hypertensive persons under care (8). We are interested in testing whether the survival experience of persons with a serum creatinine concentration less than 1.7 mg/dl at the time of screening is more favorable than those with a serum creatinine concentration greater than or equal to 1.7 mg/dl. The data for testing this hypothesis are shown in Table 12.6.

First, we use these data to estimate the cumulative survival probabilities for the two groups, applying the methods discussed earlier. The results are shown graphically in Figure 12.5. The survival distribution appears to be more favorable for the hypertensive persons with a serum creatinine concentration less than 1.7 mg/dl than those with a serum creati-

TABLE 12.6 Sample Sizes and Numbers of Deaths by Year and Level of Serum Creatinine Concentration in the HDFP Study

	Serum creatinine (mg/dl)					
Year under care	**<1.7**		**≥1.7**		**Total**	
	d_{1i}	n'_{1i}	d_{2i}	n'_{2i}	$d_{.i}$	$n'_{.i}$
0–1	93	10469.5	21	297.0	114	10766.5
1–2	115	10374.5	16	276.0	131	10650.5
2–3	125	10254.0	13	260.0	138	10514.0
3–4	181	10121.5	14	246.5	195	10368.0
4–5	160	9930.5	17	232.0	177	10162.5
5–6	212	9763.0	10	215.0	222	9978.0
6–7	191	9551.0	14	205.0	205	9756.0
7–8	203	9147.5	8	186.5	211	9334.0
Total	1280		113		1393	

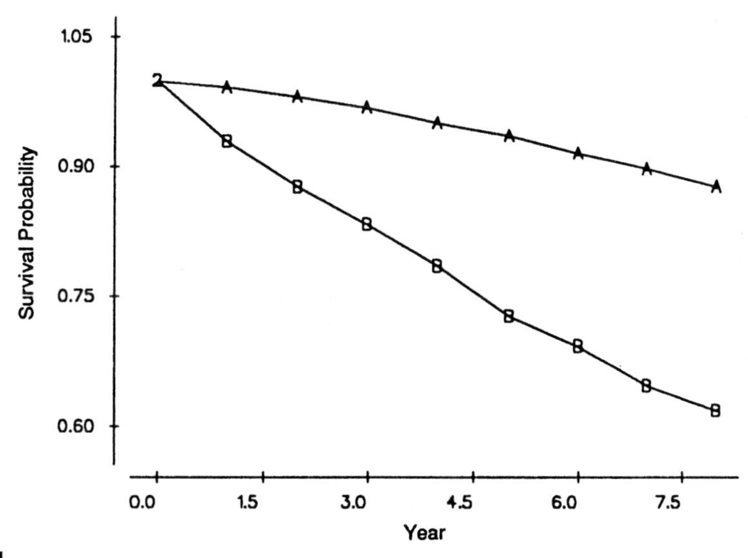

FIGURE 12.5 Estimated survival distributions by level of serum creatinine concentration. A, serum creatinine < 1.7; B, serum creatinine $\geqslant 1.7$.

nine concentration greater than or equal to 1.7 mg/dl. The two survival curves are consistently diverging, suggesting that the odds ratios in each time interval are similar to one another. Therefore, we do not have any problem using the Cochran–Mantel–Haenszel test to compare the two survival distributions.

To apply this test to the data in Table 12.6, we need to find the expected number of deaths and the variance for the (1,1) cell in each of the eight 2 by 2 tables. For example, the 2 by 2 table for the year 0–1 is shown next.

Creatinine level	*Number of deaths*	*Number of survivors*	*Total*
<1.7 mg/dl	93	10,376.5	10,469.5
≥1.7 mg/dl	21	276.0	297.0
Total	114	10,652.5	10,766.5

The expected number of deaths in the (1,1) cell is the product of the total of the first row and the first column divided by the table total. Thus the expected value is

$$10{,}469.5 * 114/10{,}766.5 = 110.86.$$

The variance of the (1,1) cell is the product of the four marginal totals divided by the square of the table total times the table total minus one.

Thus the sample variance is

$$\frac{10{,}469.5 * 297 * 114 * 10{,}652.5}{(10{,}766.5)^2 * 10{,}765.5} = 3.03.$$

Table 12.7 shows the expected number of deaths and the variances for the eight (1,1) cells based on the data in Table 12.6. The observed number of deaths in group 1 (creatinine < 1.7 mg/dl) is 1280 and the expected number of deaths is 1361, suggesting that group 1 has a favorable survival experience. We shall test the hypothesis of no difference in the survival distributions of the two groups at the 0.01 significance level. The test statistic, X^2_{CMH}, is calculated based on the data in Tables 12.6 and 12.7 as

$$X^2_{CMH} = \frac{(|O - E| - 0.5)^2}{V} = \frac{(|1280 - 1361.07| - 0.5)^2}{30.65} = 211.80$$

As the test statistic is greater than 6.63 (= $\chi^2_{1,0.99}$), we reject the null hypothesis and conclude that persons with a serum creatinine concentration less than 1.7 mg/dl had a more favorable survival distribution than those with a higher creatinine value at the time of screening.

The Cochran–Mantel–Haenszel test can also be used with a smaller data set along with the product-limit method. Let us apply this method to the data in Table 12.4 in comparing the survival distributions of male and female patients at the 0.05 significance level. The data and the calculation of the test statistic for making this comparison are shown in Table 12.8.

The first column of the table shows the observed alcohol-free times (x_i) with the censoring status and gender indicated. The second column is the total number of subjects under observation at time x. The third and fourth columns show, respectively, the number of females (group 1) and number of males (group 2) under observation at time x. The fifth column shows the observed number of relapses at time x. The numbers of relapses at time x in groups 1 and 2 are shown in columns 6 and 7, respectively.

TABLE 12.7 Expected Values and Variances of the (1,1) Cells

Year under care	Expected value	Variance
0–1	110.86	3.03
1–2	127.61	3.27
2–3	134.59	3.28
3–4	190.36	4.44
4–5	172.96	3.88
5–6	217.22	4.58
6–7	200.69	4.13
7–8	206.78	4.04
Total	1361.07	30.65

TABLE 12.8 Comparison of Alcohol-Free Time Distributions for Females and Males

Survival time x_i (1)	Number of subjects: Total $n'_{\cdot i}$ (2)	Female n'_{1i} (3)	Male n'_{2i} (4)	Observed number of relapses: Total $d_{\cdot i}$ (5)	Female d_{1i} (6)	Male d_{2i} (7)	Expected relapses (8)	Variance (9)
4 M	14	7	7	1	0	1	0.50	0.25
6 MF	13	7	6	2	1	1	1.08	0.44
9* M	11	6	5	0	0	0	0	0
10 F	10	6	4	1	1	0	0.60	0.24
14* M	9	5	4	0	0	0	0	0
16 M	8	5	3	1	0	1	0.63	0.24
17* M	7	5	2	0	0	0	0	0
19 F	6	5	1	1	1	0	0.83	0.14
20 F	5	4	1	1	1	0	0.80	0.16
28 M	4	3	1	1	0	1	0.75	0.19
31 F	3	3	0	1	1	0	1.00	0.00
34* F	2	2	0	0	0	0	0	0
47* F	1	1	0	0	0	0	0	0
Total				9	5	4	6.19	1.66

*Censored observation.

The eighth column shows the expected number of relapses at time x_i for females. It is calculated in the same manner as before. For example, at 6 months, two relapses are recorded. The proportion of females under observation at 6 months is 7/13. Therefore, the expected number of relapses for females is 2 * (7/13), or 1.08. The variances of the observed numbers of relapses for females at time x_i are shown in column 9. These calculations are performed only for the uncensored survival times. The values are next summed and the Cochran–Mantel–Haenszel chi-square statistic is calculated as follows:

$$X^2_{\text{CMH}} = \frac{(|O - E| - 0.5)^2}{V} = \frac{(|5 - 6.19| - 0.5)^2}{1.66} = 0.29.$$

As the test statistic is smaller than 3.84 ($= \chi^2_{1,0.95}$), we fail to reject the null hypothesis.

In Chapter 11, we indicated that the CMH test statistic should be used only when the odds ratios are similar across the subtables. The same assumption for its use applies here. This assumption can also be expressed in terms of the plots of the two survival functions instead of in terms of odds ratios. If the plots of the two survival functions cross one another, this means that the odds ratios are not similar across the subtables and the CMH test statistic should not be used. The reason for this is that one group has a better survival experience during part of the study period and the other group has a better experience during another part of the period. Thus it is difficult to say that one group has a better overall experience.

V. CONCLUDING REMARKS

In this chapter, we presented two methods for analyzing survival data: the life-table and product-limit methods. The life-table method is generally used for large data sets and the product-limit method for smaller data sets. In addition, we demonstrated the calculation of the sample median and restricted mean survival times. We also discussed why the median is preferred to the mean as a single summary statistic for use with survival data. To complete the description of survival data, we highly recommended the plotting of the survival distribution. Finally, we showed the use of the Cochran–Mantel–Haenszel test for comparing the equality of two survival distributions.

EXERCISES

12.1. In an effort to understand employment experience of nurses, personnel records of two large hospitals were reviewed (9). A total of 3221 nurses were hired during a 10-year period from 1970 to 1979 and employment records were reviewed 18 months beyond the end of 1979. In this cohort, only 780 nurses worked more than 33 months. The length of employment was presented by 3-month interval as follows:

Month after employment	*Number terminated*	*Number censored*	*Number at beginning of interval*
0–3	582	0	3221
3–6	369	0	
6–9	247	0	
9–12	212	0	
12–15	182	0	
15–18	144	0	
18–21	129	75	
21–24	99	74	
24–27	85	59	
27–30	51	53	
30–33	45	35	
33+			780

a. Prepare a worksheet for a life-table analysis and estimate the cumulative survival probabilities, the restricted mean, and the median length of employment. Also estimate the probability of termination for each of the intervals.

b. Estimate the standard errors of the estimated cumulative survival probabilities and the probability of termination for each interval.

c. Calculate 95 percent confidence intervals for the 24-month cumulative survival probability and the probability of termination during the first 3 months of employment.

d. What additional data, if any, do you need and what further analyses would you perform to assess the nursing employment situation?

12.2. The Hypertension Detection and Follow-up Program collected mortality data for 8 years (8). The following data show the survival experience of two subgroups formed by the level of serum creatinine concentration:

	Serum creatinine concentration (mg/dl)					
	2.00–2.49			≥2.5		
Year care	*Alive*	*Died*	*Censored*	*Alive*	*Died*	*Censored*
0–1	78	3	0	72	8	0
1–2	75	4	0	64	8	0
2–3	71	6	0	56	3	0
3–4	65	3	0	53	3	0
4–5	62	5	0	50	8	0
5–6	57	4	0	42	3	0
6–7	53	2	0	39	5	0
7–8	51	3	3	34	1	1

a. Analyze the survival pattern of each group using the life-table method: estimate the cumulative survival probabilities and their standard errors, and compare the survival curves of these two groups graphically.

b. If it is appropriate, determine whether or not the two survival distributions are equal at the 0.01 significance level.

c. Comment on what factors may have confounded the comparison above and what further analyses you think are necessary before you can draw more defensible conclusions.

12.3. The Systolic Hypertension in the Elderly Program (SHEP) Cooperative Research Group assessed the ability of antihypertensive drug treatment to reduce the risk of stroke (nonfatal and fatal) in a randomized, double-blind, placebo-controlled experiment (10). A total of 4736 persons with systolic hypertension (systolic blood pressure 160 mm Hg and above and diastolic blood pressure less than 90 mm

Hg) were screened from 447,921 elderly persons aged 60 years and above. During the study period, 213 deaths occurred in the treatment group and 242 deaths in the placebo group. The average follow-up period was 4.5 years. Total stroke was the primary endpoint and the following data were reported:

	Treatment group			*Placebo group*		
Year	*Number started*	*Strokes*	*Lost*	*Number started*	*Strokes*	*Lost*
0–1	2365	28	0	2371	34	0
1–2	2316	22	0	2308	42	0
2–3	2264	21	0	2229	22	2
3–4	2153	18	0	2193	34	2
4–5	1438	13	5	1393	24	1
5–6[a]	613	1	0	584	3	0

[a]The last stroke occurred during the 67th month of follow-up.

a. To analyze the above data by the life-table method, how would you set up the worksheet? It is obvious that there were censored observations other than the lost-to-follow-up, such as deaths and withdrawn alive. This can be seen because the difference in the number of persons starting one interval and the number starting the following interval decreased by more than the number of strokes in the interval. Would you include or exclude the data in the last reported interval?

b. If it is appropriate, test the hypothesis of the equality of the two survival distributions at the 0.05 significance level.

12.4. The following data were abstracted from the records of the neonatal intensive care unit (NICU) in a hospital during the month of February 1993 (day and 24-hour clock time are used to describe the timing of events, e.g., 0102 indicates the first day of February, 2 AM):

No.	*Sex*	*Born*	*Last observed*	*Status*
1	Boy	0102	2210	Discharged
2	Girl	0306	1722	Died
3	Boy	0309	1517	Died
4	Boy	0523	2609	Discharged
5	Boy	0918	1001	Died
6	Girl	1004	2411	Died
7	Boy	1107	2512	Discharged
8	Girl	1110	1815	Discharged
9	Boy	1206	1408	Died
10	Girl	1307	2320	Died

No.	Sex	Born	Last observed	Status
11	Girl	1412	2823	Still in NICU
12	Boy	1500	1510	Died
13	Boy	1607	2220	Died
14	Girl	1819	2823	Still in NICU
15	Boy	1903	2009	Died
16	Boy	2009	2711	Discharged
17	Boy	2110	2823	Still in NICU
18	Girl	2208	2329	Died
19	Girl	2321	2823	Still in NICU
20	Girl	2323	2810	Discharged
21	Boy	2402	2823	Still in NICU
22	Girl	2509	2823	Still in NICU
23	Boy	2620	2823	Still in NICU
24	Girl	2701	2822	Died

a. Estimate the neonatal survival function for these NICU infants, estimate the median survival time, and form the 90 percent confidence interval for the 50-hour survival probability.
b. Plot the estimated neonatal survival functions separately for boys and girls and test the equality of the two survival distributions at the 0.10 significance level.

12.5. Quality of care for colorectal cancer was evaluated by comparing the survival experience of patients in two types of health plans (fee-for-service and health maintenance organization) offered by the same health care provider (11). The following data were generated from the reported survival curves:

Practice	*Survival time (months)*											
Fee-for-service	2	5	10	12*	14	14	16	18	23	26*	27	31
	34	37*	39	42*	46	47*	50	53*				
HMO	4	10*	12	15	19	25	30*	35	38	43*	49	54*

Asterisks indicate censored observations.

a. Estimate the survival distributions by the product-limit method and graphically compare the survival curves.
b. Compare the equality of the survival distributions of the two medical services at the 0.01 significance level.

REFERENCES

1. California Tumor Registry (1963). "Cancer Registration and Survival in California." State of California Department of Public Health, Berkeley.

2. National Cancer Institute (1990). "Cancer Statistics Review: 1973–1987," NIH Publ. No. 90-2789. National Institutes of Health, U.S. Public Health Service, Bethesda, MD.
3. Thomas, D. R., and Grunkemeier, G. L. (1975). Confidence interval estimation of survival probabilities for censored data. *J. Am. Stat. Assoc.* **70,** 865–871.
4. Kaplan, E. L., and Meier, P. (1958). Nonparametric estimation from incomplete observations. *J. Am. Stat. Assoc.* **53,** 457–481.
5. Peto, R., Pike, M. C., Armitage, P., Breslow, N. E., Cox, D. R., Howard, S. V., Mantel, N., McPherson, K., Peto, J., and Smith, P. G. (1977). Design and analysis of randomized clinical trials requiring prolonged observation of each patient. II. Analysis and examples. *Br. J. Cancer* **35,** 1–39.
6. Lee, E. T. (1992). "Statistical Methods for Survival Data Analysis," 2nd ed. Wiley, New York.
7. Mantel, N. (1966). Evaluation of survival data and two new rank order statistics arising in its considerations. *Cancer Chemother. Rep.* **50,** 163–170.
8. Shulman, N. B., Ford, C. E., Hall, W. D., Blaufox, M. D., Simon, D., Langford, H. G., and Schneider, K. A. on behalf of the Hypertension Detection and Follow-up Program Cooperative Group. (1989). Prognostic value of serum creatinine and effect of treatment of hypertension on renal function. *Hypertension* **13,** Suppl. I, 80–93.
9. Benedict, M. B., Glasser, J. H., and Lee, E. S. (1989). Assessing hospital nursing staff retention and turnover: A life table approach. *Eval. Health Prof.* **12,** 73–96.
10. Systolic Hypertension in the Elderly Program Cooperative Research Group. (1991). Prevention of stroke by antihypertenstive drug treatment in older persons with isolated systolic hypertension. *JAMA, J. Am. Med. Assoc.* **265,** 3255–3264.
11. Vernon, S. W., Hughes, J. I., Heckel, V. M., and Jackson, G. L. (1992). Quality of care for colorectal cancer in a fee-for-service and health maintenance organization practice. *Cancer (Philadelphia)* **69,** 2418–2425.

CHAPTER 13

Tests of Hypotheses Based on the Normal Distribution

This chapter is similar to and builds on the material in Chapter 7 on confidence intervals that are based on the normal distribution. We first show the equivalence of the confidence interval and the test of hypothesis for the population mean. This equivalence extends to each of the parameters discussed in this chapter. Therefore, we do not repeat the details on the distributions of the test statistics that were already presented in Chapter 7.

Although the confidence interval and the test of hypothesis can be used to reach the same conclusion, their emphases are different. The confidence interval provides limits that are likely to contain the parameter. These limits can also be used to test a hypothesis, but that is not necessarily the reason why they were created. The test of hypothesis aids in reaching a decision about whether we believe that the hypothesized value of the parameter is correct. The use of the test of hypothesis also serves as a reminder to calculate the p value and, occasionally, the power of the test.

We begin with the test about the population mean and demonstrate the use of both one- and two-sided alternatives as well as the calculation of power. In subsequent sections, we usually show only the test for two-sided alternatives.

I. TESTING HYPOTHESES ABOUT THE MEAN

Suppose that we wish to analyze the dietary data shown in Table 4.1; however, before performing the analyses, we wish to determine whether the population represented by the sample of 33 boys differs from the national population of boys as far as caloric intake is concerned. Therefore, we first test the hypothesis that the mean caloric intake for boys in one of the northern suburbs of Houston is the same as the national average. If the caloric intake is different, we may not wish to analyze the data further until we understand why there is a difference.

From the calculations in Chapter 4, we know that the sample mean, based on 33 boys, is 2314 calories. Based on data shown in "Nutrition Monitoring in the United States" (1, Table II-3), we take the national average to be 2400 calories. The test of hypothesis about the population mean, just like the confidence interval, uses the normal distribution if the population variance is known or the t distribution if the variance is unknown. We first assume that the variance is known.

A. Known Variance

In Chapter 7, when we formed the 95 percent confidence interval for the population mean, we assumed that the population standard deviation was 700 calories or that the variance was 490,000 calories^2. We use that value in the test of hypothesis about the population mean. The null and alternative hypotheses are

$$H_0\colon \mu = \mu_0 \qquad \text{and} \qquad H_a\colon \mu \neq \mu_0$$

where μ_0 is 2400 calories in this example. To be able to compare the test results with the confidence interval from Chapter 7, we conduct the test at the 0.05 significance level.

There are two equivalent ways of presenting the test of hypothesis. One method uses z [$= (\bar{x} - \mu_0)/(\sigma/\sqrt{n})$], the standard normal statistic, as the test statistic and the other method uses the sample mean, $\bar{x}$, as the test statistic.

1. Use of the Standard Normal Statistic

The test statistic in this approach is z, the standard normal statistic. If the null hypothesis is true, z will follow the standard normal distribution. The rejection region is thus defined in terms of percentiles of the standard normal distribution. For a two-sided alternative, if z is either less than or equal to $z_{\alpha/2}$ or greater than or equal to $z_{1-\alpha/2}$, we reject the null hypothesis in favor of the alternative hypothesis. In symbols, this is

$$\frac{\bar{x} - \mu_0}{(\sigma/\sqrt{n})} \leq z_{\alpha/2} \quad \text{or} \quad \frac{\bar{x} - \mu_0}{(\sigma/\sqrt{n})} \geq z_{1-\alpha/2}.$$

If the test statistic is not in the rejection region, that is,

$$z_{\alpha/2} < z < z_{1-\alpha/2},$$

we fail to reject the null hypothesis in favor of the alternative hypothesis.

Let us calculate the test statistic for the caloric intakes. The z value is

$$\frac{2314 - 2400}{(700/\sqrt{33})} = -0.706.$$

As -0.706 does not fall in the rejection region, that is, it is not less than -1.96 ($= z_{0.025}$) nor is it greater than 1.96, we fail to reject the null hypothesis. This situation is shown pictorially in Figure 13.1. The p value for this test is the probability of observing a standard normal variable with a value either less than -0.706 or greater than 0.706. This probability is found to be 0.48. The use of MINITAB is illustrated in Box 13.1.

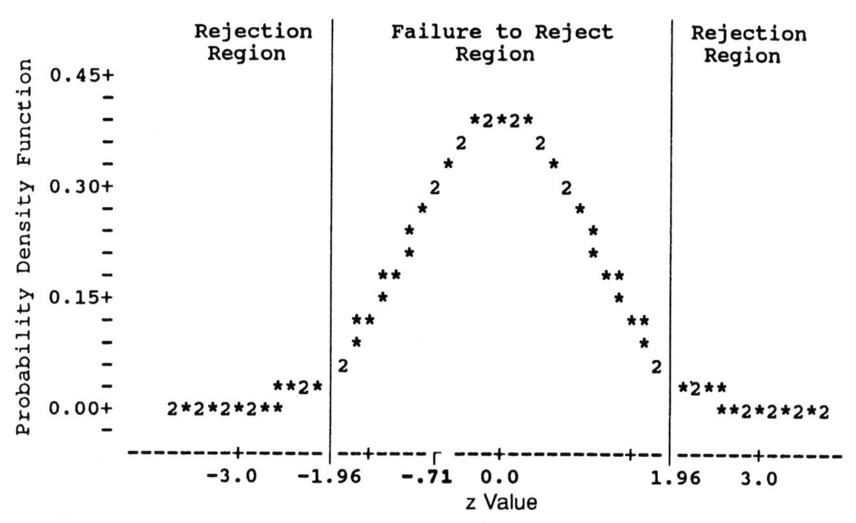

FIGURE 13.1 Representation of the rejection and failure-to-reject regions in terms of the standard normal statistic for $\alpha = 0.05$.

MINITAB BOX 13.1

The command for performing this test with MINITAB is ZTEST and it is shown next. First we read the data from the file in which it is stored and then we specify the values of μ_0 and σ and the column(s) containing the data.

```
MTB > read 'bookch4.dat' c1-c6
MTB > ztest 2400 700 c2

TEST OF MU = 2400.000 VS MU N.E. 2400.000
THE ASSUMED SIGMA = 700

          N      MEAN     STDEV   SE MEAN        Z   P VALUE
C2       33  2313.818   667.895   121.854    -0.71      0.48
```

If no value is specified for μ_0, the value of zero is assumed.

2. Use of the Sample Mean

In this method, the sample mean's value is compared with the values that are based on the hypothesized value of the population mean and the standard error of the sample mean. These values are found by simple manipulation of the above expressions. The values of $\bar{x}$ that are in the rejection region are those that satisfy

$$\bar{x} \leq \mu_0 + z_{\alpha/2}\left(\frac{\sigma}{\sqrt{n}}\right) \quad \text{or} \quad \bar{x} \geq \mu_0 + z_{1-\alpha/2}\left(\frac{\sigma}{\sqrt{n}}\right).$$

For this example, the rejection region in terms of $\bar{x}$ comprises the values of the sample mean such that

$$\bar{x} \leq 2400 - 1.96\left(\frac{700}{\sqrt{33}}\right) \quad \text{or} \quad \bar{x} \geq 2400 + 1.96\left(\frac{700}{\sqrt{33}}\right)$$

which yields the rejection region of

$$\bar{x} \leq 2161.17 \quad \text{calories} \quad \text{or} \quad \bar{x} \geq 2638.83 \quad \text{calories.}$$

Figure 13.2 shows this pictorially. As the observed value of $\bar{x}$ does not fall into the rejection region, we fail to reject the null hypothesis. The choice of which statistic, the z statistic or the sample mean, to use as the test statistic is left to the analyst. As can be seen from the creation of the rejection region, as well as from Figures 13.1 and 13.2, both statistics arrive at the same conclusion.

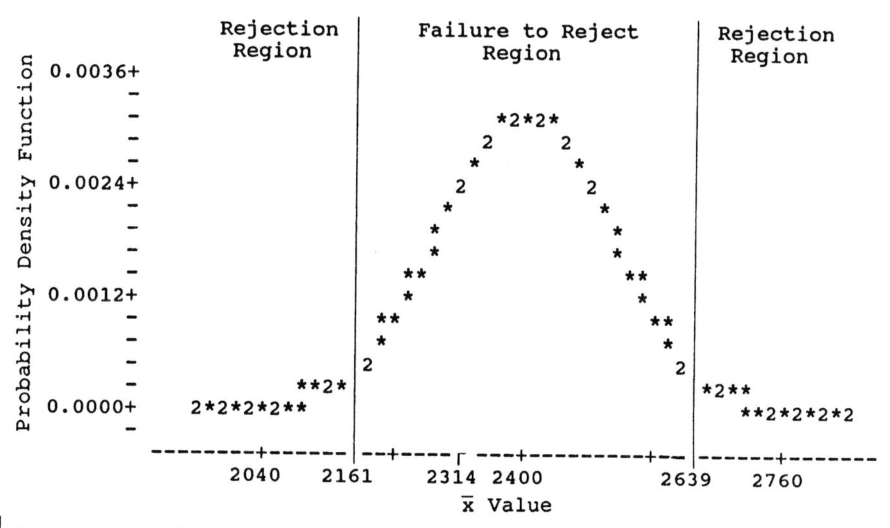

FIGURE 13.2 Representation of the rejection and failure-to-reject regions in terms of the sample mean for $\alpha = 0.05$.

3. Equivalence of Confidence Intervals and Tests of Hypotheses

Recall that in Chapter 7 when we found the $(1 - \alpha) * 100$ percent confidence interval for the population mean, we started with the expression

$$\Pr\left\{-z_{1-\alpha/2} < \frac{\bar{x} - \mu}{(\sigma/\sqrt{n})} < z_{1-\alpha/2}\right\} = 1 - \alpha.$$

We manipulated this expression and obtained the expression

$$\bar{x} - z_{1-\alpha/2}\left(\frac{\sigma}{\sqrt{n}}\right) < \mu < \bar{x} + z_{1-\alpha/2}\left(\frac{\sigma}{\sqrt{n}}\right).$$

If we replace μ in the middle portion of the first expression above by μ_0, the middle portion is the z statistic for testing the hypothesis that μ equals μ_0. The confidence interval was derived from this test statistic; this means that if μ_0 is contained in the confidence interval, then the corresponding z statistic must also be in the failure-to-reject (acceptance) region. If μ_0 is not in the confidence interval, then the z statistic is in the rejection region, that is, it is less than or equal to $-z_{1-\alpha/2}$ or greater than or equal to $z_{1-\alpha/2}$.

In this case, the hypothesized value of 2400 calories is contained in the 95 percent confidence interval for the population mean; we saw in Chapter 7 that the confidence interval ranges from 2075.2 to 2552.8 calories. Therefore we know that the test statistic will be in the failure-to-reject region and we will fail to reject the null hypothesis. In addition, using the same logic, from the confidence interval, we know we would fail to reject the null hypothesis for any μ_0 ranging from 2075.2 to 2552.8 calories.

This same type of argument for the linkage of the test of hypothesis and the corresponding confidence interval can be used with the other tests of hypotheses presented in this chapter. Thus the confidence interval is also very useful from a test of hypothesis perspective; however, the confidence interval does not provide the p value of the test, also a useful statistic.

4. One-Sided Alternative Hypothesis

If we are concerned only when the boys do not receive enough calories, the null and alternative hypotheses are

$$H_0: \mu = \mu_0 \quad \text{and} \quad H_a: \mu < \mu_0.$$

The test statistic does not change, but the rejection region is a one-sided region now. We reject the null hypothesis in favor of the alternative hypothesis if z is less than or equal to z_α or, equivalently, if $\bar{x}$ is less than or equal to $\mu_0 + (z_\alpha * \sigma/\sqrt{n})$. The use of MINITAB is illustrated in Box 13.2. and the corresponding one-sided rejection region is shown in Figure 13.3 in terms of the z test statistic.

MINITAB BOX 13.2

To use MINITAB with a one-sided alternative hypothesis, we must specify the ALTERNATIVE (abbreviated to ALT) subcommand. As mentioned in Chapter 10, a value of 1 means that the alternative hypothesis is greater than, and a value of −1 indicates an alternative hypothesis of less than.

```
MTB > ztest 2400 700 c2;
SUBC> alt -1.

TEST OF MU = 2400.000 VS MU L.T. 2400.000
THE ASSUMED SIGMA = 700

              N      MEAN     STDEV   SE MEAN       Z   P VALUE
C2           33  2313.818   667.895   121.854   -0.71      0.24
```

If we are concerned only when the boys' caloric intake is too high, the null and alternative hypotheses are

$$H_0: \mu = \mu_0 \quad \text{and} \quad H_a: \mu > \mu_0.$$

We now reject if z is greater than or equal to $z_{1-\alpha}$ or, equivalently, if $\bar{x}$ is greater than or equal to $\mu_0 + (z_{1-\alpha} * \sigma/\sqrt{n})$.

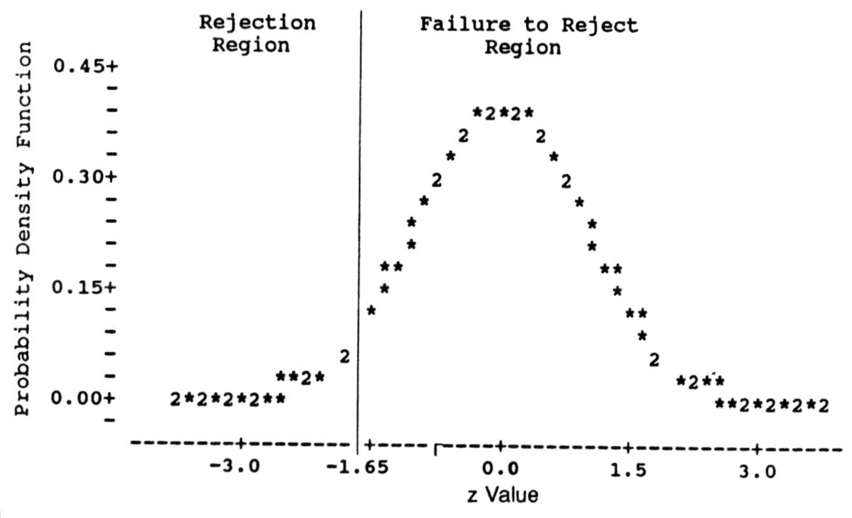

FIGURE 13.3 Representation of the rejection and failure-to-reject regions for a one-sided alternative of less than the mean for $\alpha = 0.05$.

5. Power of the Test

Before collecting the data, suppose that we wanted to be confident that, if the caloric intake in the Houston suburb was substantially less than the national average, we could detect this lower mean intake. By substantially less, we mean 10 percent or more below the national average of 2400 calories. Thus we wish to conclude that there is a difference between the boys in the Houston suburb and the national average if the Houston suburb has a population mean of 2160 calories or less. The use of 10 percent is subjective and other values could be used.

The null and alternative hypotheses for this situation are

$$H_0: \mu = \mu_0 \quad \text{and} \quad H_a: \mu < \mu_0.$$

We again use a significance level of 0.05. Thus the rejection region includes all z less than or equal to $z_{0.05}$, that is, z less than or equal to -1.645. In terms of $\bar{x}$, the rejection region includes all values of $\bar{x}$ less than or equal to

$$\mu_0 + z_{0.05} * \left(\frac{\sigma}{\sqrt{n}}\right) = 2400 + (-1.645) * \left(\frac{700}{\sqrt{33}}\right) = 2199.55.$$

Figure 13.4 shows the rejection and acceptance regions in terms of $\bar{x}$ as well as its distribution under the alternative hypothesis.

The shaded area provides a feel for how large the power—the probability of rejecting the null hypothesis when it should be rejected—of the test is. Power is the proportion of the area under the alternative hypothesis

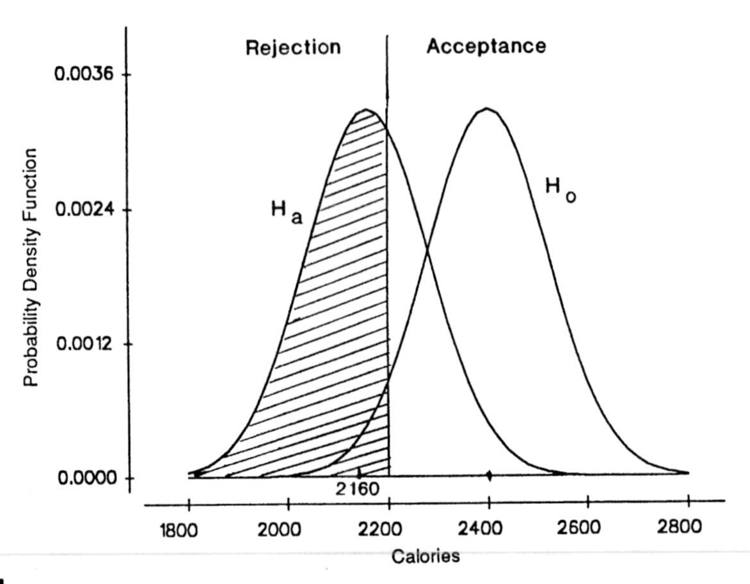

FIGURE 13.4 Rejection and acceptance regions for testing H_0: $\mu = 2400$ versus H_a: $\mu = 2160$ at the 0.05 significance level.

curve that is in the rejection region, that is, less than or equal to 2199.55 calories.

Let us find the power of the test and see if it agrees with our expectations about it based on Figure 13.4. Power is the probability of being in the rejection region, that is, of the sample mean being less than or equal to 2199.55 calories, assuming that the alternative hypothesis ($\mu = 2160$) is true. To find this probability, we convert 2199.55 to a standard normal value by subtracting the mean of 2160 calories and dividing by $\sigma/\sqrt{n}$. Thus the z value is

$$\frac{(2199.55 - 2160)}{(700/\sqrt{33})} = 0.3246.$$

The probability of a standard normal variable being less than or equal to 0.3246 is found from Table B4 to be 0.627.

The power of the test is almost 63 percent. If this value is not large enough, there are several methods of increasing the power. One way is to increase the sample size. For example, let us increase the sample size to 100. Then the z value is

$$\frac{(2199.55 - 2160)}{(700/\sqrt{100})} = 0.565.$$

The probability of a standard normal variable being less than or equal to 0.565 is 0.714, almost 10 percent larger than the power associated with the sample size of 33.

As was discussed in Chapter 9, another way of increasing the power is to increase the significance level, for example, to 0.10. Doing this increases the size of the rejection region. All values of $\bar{x}$ that are less than or equal to

$$2400 + (-1.28) * \left(\frac{700}{\sqrt{33}}\right) = 2244.03$$

now are in the rejection region. Using this significance level and still using a sample size of 33, the z value becomes

$$\frac{(2244.03 - 2160)}{(700/\sqrt{33})} = 0.6896.$$

The probability of a standard normal variable being less than or equal to 0.6896 is 0.755, a value that is even closer to the value of 0.80 which is often used as the desired level for power in the literature.

Another way of increasing the power is to redefine what we consider to be a substantial difference. If our emphasis were on detecting an intake 15 percent less than the national average, instead of an intake of 10 percent less, we would have a higher power. As 15 percent of 2400 calories is 360 calories, the null and alternative hypotheses become

$$H_0\colon \mu = 2400 \quad \text{and} \quad H_a\colon \mu = 2040.$$

The z statistic becomes

$$\frac{(2199.55 - 2040)}{(700/\sqrt{33})} = 1.31$$

and the probability of a standard normal variable being less than or equal to 1.31 is 0.905. The power associated with the alternative that μ equals 2160 has not changed, but our emphasis on what difference is important has changed. We have a much higher chance of detecting this greater difference, from 2040 instead of from 2160, between the null and alternative hypotheses.

Let us consider another example of the calculation of power. Suppose that we have reason to suspect that the systolic blood pressure of 5-year-old boys in Pittsburgh is higher than the national average and we are planning a study to test this. The null and alternative hypotheses for the study are

$$H_0\colon \mu = 94 \text{ mm Hg} \quad \text{and} \quad H_a\colon \mu > 94 \text{ mm Hg}$$

We use a value of 11 mm Hg for the standard deviation of systolic blood pressure for 5-year-old boys.

We must choose a specific value for the mean blood pressure under the alternative hypothesis. We have selected the value of 100 mm Hg, a difference of 6 mm Hg from the national mean, as an important difference that we wish to be able to detect. For this study, our initial plans call for a sample size of 61.

To find the power, we must first find the acceptance and rejection regions. Let us perform the test at the 0.01 significance level. Therefore the rejection region consists of all values of z greater than or equal to $z_{0.99}$ (= 2.326). In terms of the sample mean, the rejection region consists of values of $\bar{x}$ greater than or equal to

$$\mu_0 + z_{0.99} * \left(\frac{\sigma}{\sqrt{n}}\right) = 94 + 2.326 * \left(\frac{11}{\sqrt{61}}\right) = 97.276.$$

Figure 13.5 shows this situation.

Once we know the boundary between the acceptance and rejection regions, we convert the boundary to a z value by subtracting the mean under the alternative hypothesis and dividing by the standard error. For this example, the z value is

$$z = \frac{(97.276 - 100)}{(11/\sqrt{61})} = -1.93.$$

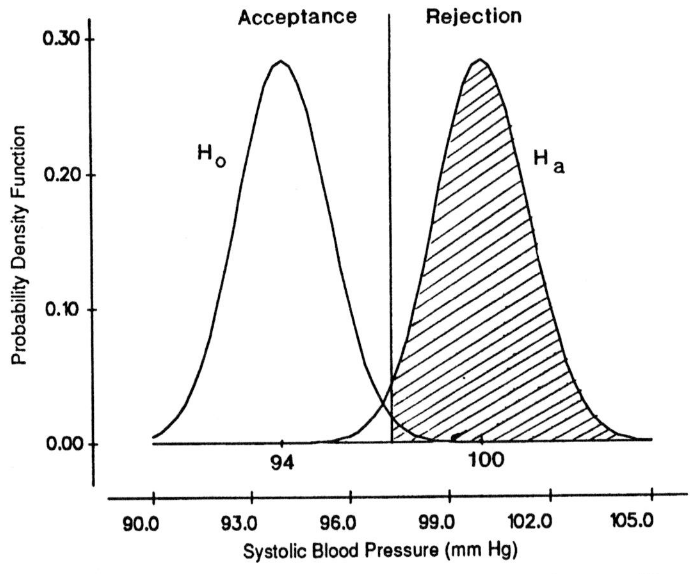

FIGURE 13.5 Rejection and acceptance regions for testing H_0: μ = 94 versus H_a: μ = 100 at the 0.01 significance level.

The power of the test is the probability of observing a z statistic with a value greater than or equal to -1.93. From Table B4, we find the power to be 0.9732. This value is consistent with what we would have expected based on Figure 13.5. A study with 61 boys has an excellent chance of detecting a mean value 6 mm Hg above the national average.

The key point about power is that calculations like these, or like those discussed in the material on confidence intervals, should be performed before any data are collected. These calculations give some indication about whether or not it is worthwhile to conduct an experiment before the resources are actually expended.

B. Unknown Variance

If the variance is unknown, the t statistic is used in place of the z statistic, that is,

$$t = \frac{\bar{x} - \mu_0}{(s/\sqrt{n})}$$

in the test of the null hypothesis that the mean is the particular value, μ_0. The rejection region for a two-sided alternative is $t \leq t_{n-1,\alpha/2}$ or $t \geq t_{n-1,1-\alpha/2}$.

Suppose that we did not know the value of σ in the caloric intake example, or that we were uncomfortable in using the value of 700 for σ. Then we would substitute s for σ and use the t distribution in place of the z distribution. In this case, the value of the t statistic is

$$t = \frac{(2313.8 - 2400)}{(667.9/\sqrt{33})} = -0.74.$$

To be consistent with the test shown above, we also perform this test at the 0.05 significance level. Therefore, t is compared with $t_{32,0.025}$ which is -2.04 and with $t_{32,0.975}$ which is 2.04.[1] As -0.74 is in the failure-to-reject region, we fail to reject the null hypothesis in favor of the alternative. Not surprisingly, this result is very similar to that obtained when the z statistic was used. The results are similar because there was little difference between the values of s and σ, and as the sample size is reasonably large, the critical values of the t and normal distributions are also close in value.

The MINITAB command for performing this test is shown in Box 13.3.

[1]In situations when certain percentiles are not shown in the tables in Appendix B, we have used MINITAB to obtain the exact values of the percentiles shown in the text.

MINITAB BOX 13.3

The MINITAB command for performing this test is TTEST.

```
MTB > ttest 2400 c2

TEST OF MU = 2400.000 VS MU N.E. 2400.000

            N      MEAN     STDEV   SE MEAN       T    P VALUE
C2         33  2313.818   667.895   116.266   -0.74       0.46
```

The TTEST command also has the ALTERNATIVE subcommand which is used with one-sided alternative hypotheses.

II. TESTING HYPOTHESES ABOUT THE PROPORTION

In this section, we focus on situations for which the use of the normal distribution as an approximation for the binomial distribution is appropriate. In general, these are situations in which the sample size is large.

In Chapter 7, one example considered the immunization level of 5-year-olds. The health department took a sample and, based on the sample, wanted to decide whether to provide additional funds for an immunization campaign. For this example, the null and alternative hypotheses are

$$H_0: \pi = \pi_0 = 0.75 \quad \text{and} \quad H_a: \pi \neq \pi_0 = 0.75.$$

The test statistic for this hypothesis is

$$z = \frac{|p - \pi_0| - 1/(2n)}{\sqrt{p * (1 - p)/n}}.$$

If $p - \pi_0$ is positive, a positive sign is assigned to z; if the difference is negative, a minus sign is assigned to z. The rejection region consists of values of z less than or equal to $z_{\alpha/2}$ or z greater than or equal to $z_{1-\alpha/2}$. This framework is very similar to that used with the population mean, the only difference being the use of the continuity correction with the proportion.

The sample proportion, p, had a value of 0.614 based on a sample size of 140. Thus the calculation of z is

$$\frac{|0.614 - 0.75| - 1/(2 * 140)}{\sqrt{0.614 * (1 - 0.614)/140}} = 3.219.$$

As $p - \pi_0$ is negative, the test statistic's value is -3.219. If the test is performed at the 0.01 significance level, values of z less than or equal to -2.576 or greater than or equal to 2.576 form the rejection region. As z is less than -2.576, we reject the null hypothesis in favor of the alternative.

The health department should devote more funds to an immunization effort. This conclusion agrees with that reached based on the confidence interval approach in Chapter 7.

The continuity correction can be eliminated from the calculations for relatively large sample sizes because its effect will be minimal. For example, if we had ignored the continuity correction in this example, the value of the test statistic would be −3.306, not much different from −3.219. MINITAB can be used to analyze these data as shown in Box 13.4.

MINITAB BOX 13.4

For proportions, it is easy to enter the data as is shown below. We enter the data into c1 using the SET command, and then use the ZTEST command. MINITAB does not use the continuity correction, but that poses no problem for relatively large sample sizes. In the ZTEST command, we must provide the value of the standard deviation, that is, the square root of the product of π and $1 - \pi$. In this example, the estimate of the standard deviation, replacing π with p, is 0.4867.

```
MTB > set cl
DATA> 86(1) 54(0)
DATA> end
```

Eighty-six children were immunized and 54 were not immunized; their order of entry is not important.

```
 MTB > ztest 0.75 0.4867 cl

TEST OF MU = 0.7500 VS MU N.E. 0.7500
THE ASSUMED SIGMA = 0.4867

            N      MEAN     STDEV   SE MEAN        Z   P VALUE
C1        140    0.6143    0.4885    0.0411    -3.30    0.0010
```

Except for rounding, the value of the test statistic here would be the same as −3.31, the value that did not involve the continuity correction. Neither value differs much from the continuity corrected value of −3.22.

III. TESTING CRUDE AND ADJUSTED RATES

Just as in Chapter 7, we treat rates as if they were proportions. This provides for a simple approximation to the variance of a rate and also gives a justification for the use of the normal distribution as an approximation to the distribution of the rate. Thus our test statistic has the same form as that used for the proportion.

Suppose we wish to test, at the 0.05 significance level, that the 1986 age-adjusted mortality rate for Harris County, obtained by the direct method of adjustment, is equal to the 1986 mortality rate for the United States. The alternative hypothesis is that the rates differ. In symbols, the null and alternative hypotheses are

$$H_0: \theta = \theta_0 = 0.008732 \qquad \text{and} \qquad H_a: \theta \neq \theta_0$$

where 0.008732 is the rate for the United States in decimal form. The test statistic for this hypothesis is

$$z = \frac{\hat{\theta} - \theta_0}{\text{approximate standard error of } \hat{\theta}}$$

where $\hat{\theta}$, the Harris County rate in decimal form, is 0.008609. In Chapter 7 we found the approximation to the standard error of $\hat{\theta}$ was 0.00007. If this value of z is less than or equal to -1.96 ($= z_{0.025}$) or greater than or equal to 1.96 ($= z_{0.975}$), we reject the null hypothesis in favor of the alternative hypothesis. The value of z is

$$\frac{0.008609 - 0.008732}{0.00007} = -1.757.$$

As -1.757 is not in the rejection region, we fail to reject the null hypothesis in favor of the alternative hypothesis at the 0.05 significance level. There is not sufficient evidence to suggest that the age-adjusted mortality rate for Harris County, obtained by the direct method of adjustment, differs from the national rate. The p value for this test is obtained by taking twice the probability that a z statistic is less than or equal to -1.757; the probability is 0.0394 and thus the p value is 0.0788.

As we have previously discussed, this test makes sense only if we view the Harris County population data as a sample in time or place.

The tests for the crude rate and for the adjusted rate obtained by the indirect method of adjustment have the same form as the above.

IV. TESTING HYPOTHESES ABOUT THE VARIANCE

In Chapter 7 we saw that $(n - 1) * s^2/\sigma^2$ followed the chi-square distribution with $n - 1$ degrees of freedom. Therefore we base the test of hypothesis about σ^2 on this statistic. The null and alternative hypotheses are

$$H_0: \sigma^2 = \sigma_0^2 \qquad \text{and} \qquad H_a: \sigma^2 \neq \sigma_0^2.$$

We define X^2 to be equal to $(n - 1) * s^2/\sigma_0^2$. When X^2 is greater than or equal to $\chi^2_{n-1,1-\alpha/2}$ or when X^2 is less than or equal to $\chi^2_{n-1,\alpha/2}$, we reject H_0 in favor of H_a.

For a one-sided alternative hypothesis, for example, H_a: $\sigma^2 < \sigma_0^2$, the rejection region is $X^2 \leq \chi^2_{n-1,\alpha}$. If the alternative is H_a: $\sigma^2 > \sigma_0^2$, the rejection region is $X^2 \geq \chi^2_{n-1,1-\alpha}$.

Returning to the vitamin D in milk example discussed in Chapter 7, suppose we wish to test the hypothesis that the producer is in compliance with the requirement that the variance be less than 1600. We doubt that the producer is in compliance; therefore, we use the following null and alternative hypotheses:

$$H_0: \sigma^2 = 1600 \qquad \text{and} \qquad H_a: \sigma^2 > 1600.$$

As this test is one-sided, we are implicitly saying that the null hypothesis is that the population variance is less than or equal to 1600 versus the alternative that the variance is greater than 1600. We perform the test at the 0.10 significance level. Thus the test statistic, X^2, which equals

$$\frac{(n-1) * s^2}{\sigma_0^2}$$

is compared with $\chi^2_{29,0.90}$. If X^2 is greater than or equal to 39.09, obtained from Table B7, we reject the null hypothesis in favor of the alternative hypothesis. Using the values of 1700 for s^2 and 30 for n from Chapter 7, the value of X^2 is

$$\frac{29 * 1700}{1600} = 30.81.$$

Because X^2 is not in the rejection region, we fail to reject the null hypothesis. There is not sufficient evidence to suggest that the producer is not in

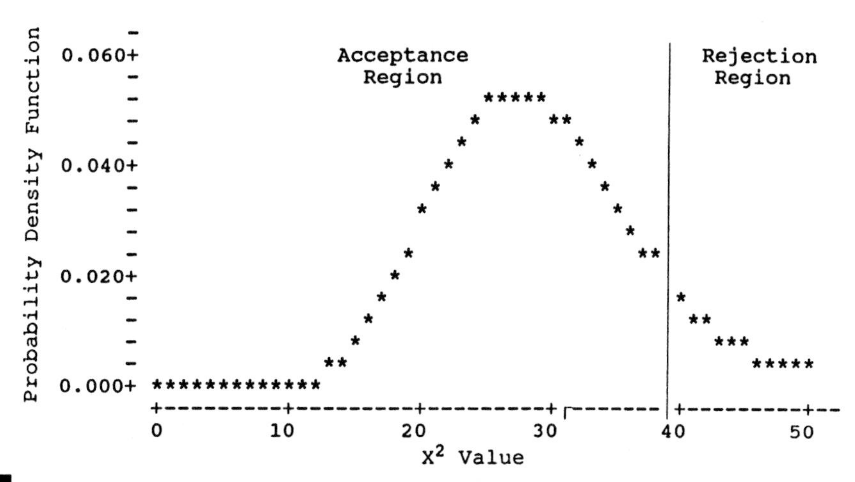

FIGURE 13.6 Rejection and acceptance regions for test of H_0: $\sigma^2 = 1600$ versus H_a: $\sigma^2 > 1600$.

compliance with the variance requirement. This is the same conclusion reached when the confidence interval approach was used in Chapter 7. Figure 13.6 shows the rejection and acceptance regions for this test.

As was mentioned in Chapter 7, the chi-square distribution begins to resemble the normal curve as the degrees of freedom becomes large. This figure is a verification of that fact. From this figure, we also see that the p value for the test statistic is large, approximately 0.40.

V. TESTING HYPOTHESES ABOUT THE PEARSON CORRELATION COEFFICIENT

In Chapter 7, we saw that the z' transformation, $z' = 0.5 * \ln[(1 + r)/(1 - r)]$ approximately followed a normal distribution with a mean of

TABLE 13.1 1988 Infant Mortality Rates and 1987 Health Expenditures as a Percentage of Gross Domestic Product for Selected Countries[a]

Country	1988 Infant mortality rate[b]	1987 Health expenditures as percentage of GDP
Japan	4.8	6.8
Sweden	5.8	9.0
Finland	6.1	7.4
The Netherlands	6.8	8.5
Switzerland	6.8	7.7
Canada	7.2	8.6
West Germany	7.5	8.2
Denmark	7.5	6.0
France	7.8	8.6
Spain	8.1	6.0
Austria	8.1	7.1
Norway	8.3	7.5
Australia	8.7	7.1
Ireland	8.9	7.4
United Kingdom	9.0	6.1
Belgium	9.2	7.2
Italy	9.3	6.9
United States	10.0	11.2
New Zealand	10.8	6.9
Greece	11.0	5.3
Portugal	13.1	6.4

[a]Infant mortality rates are from Table 25 in National Center for Health Statistics (2) and health expenditures are from Table 104 in National Center for Health Statistics (3).

[b]Infant mortality rates are deaths to infants under 1 year of age per 1000 live births.

$0.5 * \ln [(1 + \rho)/(1 - \rho)]$ and a standard error of $1/\sqrt{n-3}$. Therefore to test the null hypothesis of H_0: $\rho = \rho_0$ versus an alternative hypothesis of H_a: $\rho \neq \rho_0$, we use the test statistic λ, defined as $\lambda = (z' - z_0') * \sqrt{n-3}$ where z_0' is $0.5 * \ln[(1 + \rho_0)/(1 - \rho_0)]$. If λ is less than or equal to $z_{\alpha/2}$ or greater than or equal to $z_{1-\alpha/2}$, we reject the null hypothesis in favor of the alternative hypothesis.

There is often interest as to whether or not the Pearson correlation coefficient is zero. If it is zero, then there is no linear association between the two variables. In this case, the test statistic simplifies to

$$\lambda = z' * \sqrt{n-3}.$$

Table 13.1 shows infant mortality rates for 1988 and total health expenditures as a percentage of gross domestic product in 1987 for selected countries. It is thought that there should be some relationship between these two variables. We translate these thoughts into the following null and alternative hypotheses:

$$H_0: \rho = 0.00 \quad \text{and} \quad H_a: \rho \neq 0.00$$

and the null hypothesis will be tested at the 0.10 significance level. The rejection region consists of values of λ that are less than or equal to -1.645 $(= z_{0.05})$ or greater than or equal to 1.645. MINITAB can be used to calculate the value of λ as shown in Box 13.5.

MINITAB BOX 13.5

We first enter the data in columns c1 and c2, next find the correlation between these columns, and then calculate z' and z_0'. After these intermediate statistics are calculated, we are ready to find the value of λ, the test statistic.

```
MTB > set c1
DATA> 4.8 5.8 6.1 6.8 6.8 7.2 7.5 7.5 7.8 8.1 8.1
DATA> 8.3 8.7 8.9 9.0 9.2 9.3 10.0 10.8 11.0 13.1
MTB > set c2
DATA> 6.8 9.0 7.4 8.5 7.7 8.6 8.2 6.0 8.6 6.0 7.1
DATA> 7.5 7.1 7.4 6.1 7.2 6.9 11.2 6.9 5.3 6.4
DATA> end
```

Column c1 contains the infant mortality rates and c2 contains the health expenditures as a percentage of the gross domestic product.

```
MTB > corr c1 c2
Correlation of C1 and C2 = -0.243

MTB > let k1=0.5*loge((1.0-0.243)/(1.0+0.243))
MTB > let k2=k1*sqrt((21-3))
MTB > print k1 k2
K1        -0.247960
K2        -1.05201
```

The constant k1 is z' and k2 contains the value of λ. As -1.05201 is not in the rejection region, we fail to reject the null hypothesis. Based on these data, there is no evidence to claim that there is some nonzero correlation, at the country level, between infant mortality rate and health expenditures as a percentage of gross domestic product.

```
MTB > cdf k2
   -1.0520     0.1464
MTB > let k5=2*0.1461
MTB > print k5
K5          0.292200
```

The constant k5 is the p value of the test.

This procedure can be used with the Spearman correlation coefficient for sample sizes greater than or equal to 10.

The following sections focus on the difference of two population parameters instead of testing a hypothesis about a single parameter.

VI. TESTING HYPOTHESES ABOUT THE DIFFERENCE OF TWO MEANS

A. Independent Means

We begin with the consideration of independent means under various assumptions. The first test assumes that the variances are known, followed by the assumption that the variances are unknown but equal and then unknown and unequal. After these sections, we consider the difference of two dependent means.

1. Known Variances

The null hypothesis of interest for the difference of two independent means is

$$H_0\colon \mu_1 - \mu_2 = \Delta_0$$

where Δ_0 is the hypothesized difference of the two means. Usually Δ_0 is zero; that is, we are testing that the means have the same value. The alternative hypothesis could be either

$$H_a\colon \mu_1 - \mu_2 \neq \Delta_0$$

or that the difference is greater (less) than Δ_0. Regardless of the alternative hypothesis, when the variances are known, the test statistic is

$$z = \frac{(\bar{x}_1 - \bar{x}_2) - \Delta_0}{\sqrt{\sigma_1{}^2/n_1 + \sigma_2{}^2/n_2}}.$$

The rejection region for the two-sided alternative includes values of z less than or equal to $z_{\alpha/2}$ or greater than or equal to $z_{1-\alpha/2}$. The rejection region for the greater than alternative includes values of z greater than or equal to $z_{1-\alpha}$, and the rejection region for the less than alternative includes values of z less than or equal to z_{α}.

We return to the ramipril example from Chapter 7 and test the hypothesis that μ_1, the mean decrease in diastolic blood pressure associated with the 1.25-mg dose, is the same as μ_2, the mean decrease for the 5-mg dose. In practice, we should not initially focus on only two of the three doses; all three doses should be considered together at the start of the analysis. At this stage, however, we do not know how to analyze three means at one time, the topic of the next chapter. Therefore, we are temporarily ignoring the existence of the third dose (2.5 mg) of ramipril that was used in the actual experiment.

As we expect that the higher dose of medication will have the greater effect, the null and alternative hypotheses are

$$H_0\colon \mu_1 - \mu_2 = 0 \quad \text{and} \quad H_a\colon \mu_1 - \mu_2 < 0.$$

We perform the test at the 0.05 significance level; thus if the test statistic is less than -1.645 ($= z_{0.05}$), we reject the null hypothesis in favor of the alternative hypothesis. The sample mean decreases, $\bar{x}_1$ and $\bar{x}_2$, are 10.6 and 14.9 mm Hg, respectively, and both sample means are based on 53 observations. Both σ_1 and σ_2 are assumed to be 9 mm Hg. Therefore the value of z, the test statistic, is

$$z = \frac{(10.6 - 14.9) - 0}{\sqrt{81/53 + 81/53}} = -2.46.$$

As the test statistic is less than -1.645, we reject the null hypothesis in favor of the alternative hypothesis. There appears to be a difference in the effects of the two doses of ramipril, with the higher dose being associated with the greater mean decrease in diastolic blood pressure at the 0.05 significance level.

2. Unknown but Equal Population Variances

The null and alternative hypotheses are the same as in the preceding section; however, the test statistic for the difference of two independent means, when the variances are unknown, but assumed to be equal, changes to

$$t = \frac{(\bar{x}_1 - \bar{x}_2) - \Delta_0}{s_p \sqrt{1/n_1 + 1/n_2}}.$$

For a two-sided alternative hypothesis, the rejection region includes values of t less than or equal to $t_{n-2,\alpha/2}$ or greater than or equal to $t_{n-2,1-\alpha/2}$, where n is the sum of n_1 and n_2.

Let us test, at the 0.05 significance level, the hypothesis that there is no difference in the population mean proportions of total calories coming from fat for fifth and sixth grade boys and seventh and eighth grade boys. The alternative hypothesis is that there is a difference, that is, Δ_0 is not zero. The rejection region includes values of t less than or equal to -2.04 ($= t_{31,0.025}$) or greater than or equal to 2.04.

From Chapter 7, we know that $\bar{x}_1$, the sample mean proportion for the 14 fifth and sixth grade boys, is 0.329 and the corresponding value, $\bar{x}_2$, for the 19 seventh and eighth grade boys is 0.353. The value of s_p is 0.094. Therefore, the test statistic's value is

$$t = \frac{(0.329 - 0.353) - 0}{0.094\sqrt{1/14 + 1/19}} = -0.727.$$

As t is not in the rejection region, we fail to reject the null hypothesis. There does not appear to be a difference in the proportion of calories coming from fat at the 0.01 significance level.

MINITAB can be used to perform this test as shown in Boxes 13.6.1 and 13.6.2 (two parts).

MINITAB BOX 13.6.1

Two MINITAB commands can be used to perform this test. The choice of which command to use depends on how the data have been entered. If the data are stored in separate columns, use the TWOSAMPLE command (shown in this box); if the data are in a single column and another column contains an indicator identifying to which sample the data point belongs, use the TWOT command (shown in the next box).

For the first procedure, the data are entered into two columns. Column c1 contains the data for the fifth and sixth grade boys and c2 has the data for the seventh and eighth grade boys.

```
MTB > set c1
DATA> 0.365 0.437 0.248 0.424 0.403 0.337 0.295 0.319 0.285
DATA> 0.465 0.255 0.125 0.427 0.225
MTB > set c2
DATA> 0.311 0.278 0.282 0.421 0.426 0.345 0.281 0.578 0.383
DATA> 0.299 0.150 0.336 0.425 0.354 0.337 0.289 0.438 0.411
DATA> 0.357
DATA> end
```

We abbreviate the TWOSAMPLE command to TWOS and use the subcommand POOLED, which indicates that we are using s_p, a pooled estimate of variance.

```
MTB > twos 95 c1 c2;
SUBC> pooled.
TWOSAMPLE T FOR C1 VS C2

      N      MEAN      STDEV    SE MEAN
C1   14    0.3293     0.0974      0.026
C2   19    0.3527     0.0894      0.020
```

```
95 PCT CI FOR MU C1 - MU C2: (-0.090, 0.043)
TTEST MU C1 = MU C2 (VS NE): T= -0.72  P=0.48  DF=  31
POOLED STDEV =       0.0928
```

The values shown here differ slightly from those shown in Chapter 7 because we used only three digits after the decimal point in entering the data whereas in Chapter 7, more digits were used. A confidence interval for the difference of the population means is also provided.

MINITAB BOX 13.6.2

We create a single column containing the data by using the STACK command, stacking column c1 on top of c2 and placing all the data in a new column, c3. The subscripts for the data, indicating the sample from which the observation came, are stored in column c4.

```
MTB > stack c1 c2 c3;
SUBC> subscripts c4.
MTB > print c3
C3
0.365   0.437  0.248  0.424  0.403  0.337  0.295  0.319
0.285   0.465  0.255  0.125  0.427  0.225  0.311  0.278
0.282   0.421  0.426  0.345  0.281  0.578  0.383  0.299
0.150   0.336  0.425  0.354  0.337  0.289  0.438  0.411
0.357
MTB > print c4
C4
1  1  1  1  1  1  1  1  1  1  1  1  1  1  2
2  2  2  2  2  2  2  2  2  2  2  2  2  2  2
2  2  2

MTB > twot c3 c4;
SUBC> pooled.
TWOSAMPLE T FOR C3
C4   N      MEAN      STDEV    SE MEAN
1   14    0.3293    0.0974     0.026
2   19    0.3527    0.0894     0.020

95 PCT CI FOR MU 1 - MU 2: (-0.090, 0.043)
TTEST MU 1 = MU 2 (VS NE): T= -0.72  P=0.48  DF=  31
POOLED STDEV =       0.0928
```

If no percentage is indicated for the confidence interval, the 95 percent confidence interval is calculated. Thus it was not necessary to have entered the value 95 in the TWOS command above. These commands both have the ALTERNATIVE subcommand as well for performing one-sided tests.

---

3. Unknown and Unequal Population Variances

The test statistic for testing the null hypothesis of a specified difference in the population means, that is,

$$H_0: \mu_1 - \mu_2 = \Delta_0,$$

assuming that the population variances are unequal, is given by

$$t' = \frac{(\bar{x}_1 - \bar{x}_2) - \Delta_0}{\sqrt{s_1^2/n_1 + s_2^2/n_2}}.$$

The statistic t' approximately follows the t distribution with degrees of freedom, df, given by

$$\text{df} = \frac{(s_1^2/n_1 + s_2^2/n_2)^2}{((s_1^2/n_1)^2/(n_1 - 1)) + ((s_2^2/n_2)^2/(n_2 - 1))}.$$

For a two-sided alternative, if t' is less than or equal to $t_{\text{df},\alpha/2}$ or greater than or equal to $t_{\text{df},1-\alpha/2}$, we reject the null hypothesis in favor of the alternative hypothesis. If the alternative hypothesis is

$$H_a: \mu_1 - \mu_2 < \Delta_0,$$

the rejection region consists of values for t' less than or equal to $t_{\text{df},\alpha}$. If the alternative hypothesis is

$$H_a: \mu_1 - \mu_2 > \Delta_0,$$

the rejection region consists of values for t' greater than or equal to $t_{\text{df},1-\alpha}$.

In Chapter 7, we examined the mean ages of the AML and ALL patients. Suppose that we consider that there is no difference in the population mean ages if the mean age of AML patients minus the mean age of ALL patients is less than or equal to 5 years. Thus the null and alternative hypotheses are

$$H_0: \mu_1 - \mu_2 = 5 \quad \text{and} \quad H_a: \mu_1 - \mu_2 > 5.$$

We perform this test at the 0.01 significance level, which means that we reject the null hypothesis in favor of the alternative hypothesis if t' is greater than or equal to 2.446 ($= t_{33,0.99}$).

Using the values for the sample means, standard deviations, and sample sizes from Chapter 7, we calculate t' to be

$$t' = \frac{(49.86 - 36.65) - 5}{\sqrt{16.51^2/51 + 17.85^2/20}} = 1.781.$$

As t' is less than 2.446, we fail to reject the null hypothesis. There is not sufficient evidence to conclude that the difference in ages is greater than 5 years. Usually one would test the hypothesis of no difference instead of a difference of 5 years; however, by testing the difference of 5 years, we were able to demonstrate the calculations for a nonzero Δ_0. MINITAB can be used to perform this test as is shown in Box 13.7.

MINITAB BOX 13.7

MINITAB can also be used to perform this test by using either the TWOSAMPLE or TWOT command. We demonstrate the TWOSAMPLE command with the following material.

```
MTB > set cl
DATA> 20 25 26 26 27 27 28 28 31 33 33 33 34 36 37 40 40 43
DATA> 45 45 45 45 47 48 50 50 51 52 53 53 56 57 59 59 60 60
DATA> 61 61 61 62 63 65 71 71 73 73 74 74 75 77 80
DATA> end
MTB > set c2
DATA> 18 19 21 22 26 27 28 28 28 28 34 36 37 47 55 56 59 62
DATA> 83 19
DATA> end
```

The sample AML ages are stored in c1 and the sample ALL ages are stored in c2. As TWOSAMPLE is designed to test that Δ_0 is zero, we have to change the data to reflect that we are testing the hypothesis that Δ_0 equals 5. Therefore, we subtract 5 from each of the values in c1.

```
MTB > let c3=cl-5
MTB > twos c3 c2;
SUBC> alternative 1.

TWOSAMPLE T FOR C3 VS C2
     N      MEAN      STDEV    SE MEAN
C3  51      44.9       16.5       2.3
C2  20      36.7       17.8       4.0

95 PCT CI FOR MU C3 - MU C2: (-1.2, 17.6)
TTEST MU C3 = MU C2 (VS GT): T= 1.78  P=0.04  DF=  32
```

The test statistic agrees with that found above, but the degrees of freedom is one less than that used above. This difference could be due to the number of digits used in the calculations. In Chapter 7, we calculated the degrees of freedom to be 32.501 and this was rounded to 33.

As we emphasized in Chapter 7, we seldom know much about the magnitude of the two variances. Therefore, in those situations in which we know little about the variances and have no reason to believe that they are equal, we recommend that the unequal variances assumption should be used.

B. Test about the Difference of Two Dependent Means

The test to be used in this section is the paired t test, one of the more well-known and widely used tests in statistics. The null hypothesis to be tested is that the mean difference of the paired observations has a specified value, that is,

$$H_0\colon \mu_d = \mu_{d0},$$

where μ_{d0} is usually zero. The test statistic is

$$t_d = \frac{\bar{x}_d - \mu_{d0}}{s_d/\sqrt{n}}.$$

The rejection region for a two-sided alternative hypothesis includes values of t_d less than or equal to $t_{\alpha/2}$ or greater than or equal to $t_{1-\alpha/2}$. The rejection region for the alternative of less than includes values of t_d less than or equal to t_α, and the rejection region for the alternative of greater than includes values of t_d greater than or equal to $t_{1-\alpha}$.

We use this method to examine the effect of the 1.25-mg level of ramipril. We analyze the first 6 weeks of observation of the subjects: 4 weeks of run-in followed by 2 weeks of treatment. The null hypothesis is that the mean difference in diastolic blood pressure between the value at the end of the run-in period and the value at the end of the first treatment period is zero. The alternative hypothesis of interest is that there is an effect, that is, the mean difference is greater than zero. In symbols, the hypotheses are

$$H_0\colon \mu_d = 0 \quad \text{and} \quad H_a\colon \mu_d > 0.$$

We perform the test at the 0.10 significance level. Thus the rejection region includes values of t_d greater than or equal to 1.298 ($= t_{52,0.90}$), using 52 degrees of freedom because 53 pairs of observations are being analyzed.

From Chapter 7, we find that the sample mean difference in diastolic blood pressure after the 2 weeks of treatment was 10.6 mm Hg for the 53 subjects. The sample standard deviation of the differences was 8.5 mm Hg. Based on these data, we can calculate the value of t_d:

$$t_d = \frac{(10.6 - 0)}{8.5/\sqrt{53}} = 9.08.$$

As t_d is greater than 1.298, we reject the null hypothesis in favor of the alternative hypothesis. It appears that there is a difference between the values of diastolic blood pressure at the end of the run-in period and the treatment period, with the blood pressure at the end of the treatment period being significantly less than that at the end of the run-in period. Note that we said only that there was a difference; we did not attribute the difference to the medication.

As we have discussed in Chapters 7, 8, and 10, drawing any conclusion from this type of study design is very difficult. Two concerns, the presence of extraneous factors and reversion to the mean, are associated with this design. Without some control group, it is difficult to attribute any effects observed in the study group to the intervention because of the possibility of extraneous factors. In a tightly controlled experiment, the researcher may be able to remove all extraneous factors, but it is difficult. The presence of a control group is also useful in providing an estimate of the reversion-to-the-mean effect if such an effect exists. Thus we are suggesting that the paired *t* test should be used with great caution, that is, only in those situations for which we believe that there are no extraneous factors and no reversion-to-the-mean effect. In other cases, we would randomly assign study subjects either to the control group or to the intervention group and compare the differences of the pre- and postmeasures for both groups.

If we are comfortable with the use of the paired *t* test, it can easily be performed by *Minitab* as is shown in Box 13.8.

MINITAB BOX 13.8

Let the preintervention data be in column c1 and the postintervention data in c2. We take their difference and store it in column c3. Note that the pre- and post-measurements must be arranged in the same order; that is, we are subtracting person A's postvalue from person A's preintervention value, and so forth. We then use the TTEST command on the differences stored in c3.

VII. TESTING HYPOTHESES ABOUT THE DIFFERENCE OF TWO PROPORTIONS

As in Chapter 7, we are considering only the case of two independent proportions. The null hypothesis is

$$H_0\colon \pi_1 - \pi_2 = \Delta_0$$

where Δ_0 usually is taken to be zero. The test statistic for this hypothesis, assuming that the sample sizes are large enough for the use of the normal approximation to the binomial to be appropriate, is

$$z_{\pi_d} = \frac{(p_1 - p_2) - \Delta_0}{\sqrt{p_1(1 - p_1)/n_1 + p_2(1 - p_2)/n_2}}$$

The rejection region for a two-sided alternative includes values of z_{π_d} less than or equal to $z_{\alpha/2}$ or greater than or equal to $z_{1-\alpha/2}$. If the alternative is less than, the rejection region consists of values of z_{π_d} less than or equal to z_{α}; if the alternative is greater than, the rejection region consists of values of z_{π_d} greater than or equal to $z_{1-\alpha}$.

We test the hypothesis, at the 0.01 significance level, that there is no difference in the proportions of milk that contain 80 to 120 percent of the amount of vitamin D stated on the label between the Eastern and Southwestern milk producers. The alternative hypothesis is that there is a difference. From Chapter 7, we find the values of p_1 and p_2 are 0.286 and 0.420, respectively. Thus the test statistic is

$$z_{\pi_d} = \frac{(0.286 - 0.420) - 0}{\sqrt{0.286 * 0.714/42 + 0.420 * 0.580/50}} = -1.358.$$

As z_{π_d} is not in the rejection region, we fail to reject the null hypothesis. MINITAB can be used to perform this test as shown in Box 13.9.

MINITAB BOX 13.9

An approximation to this test can be carried out using the TWOSAMPLE command in MINITAB.

```
MTB > set cl
DATA> 12(1) 30(0)
DATA> set c2
DATA> 21(1) 29(0)
DATA> end
```

The data from the first sample are entered into cl and the data from the second sample are entered into c2.

```
MTB > twosample cl c2

TWOSAMPLE T FOR Cl VS C2
       N        MEAN       STDEV    SE MEAN
Cl    42       0.286       0.457      0.071
C2    50       0.420       0.499      0.071

95 PCT CI FOR MU Cl - MU C2: (-0.333, 0.064)
TTEST MU Cl = MU C2 (VS NE): T= -1.35   P=0.18   DF=  89
```

The estimate of the standard error of the difference differs slightly from that found using the binomial formula because of the division by $n - 1$ used in the TWOS command instead of the division by n used in the binomial calculation. The reported p value is also slightly off as it uses the t distribution instead of the normal distribution in its calculation. For large samples, these differences are very small.

VIII. CONCLUDING REMARKS

In this chapter, we added to the material on confidence intervals presented in Chapter 7, demonstrating the equivalence of the confidence intervals to the test of hypothesis. We showed how to test hypotheses about the more common parameters used with normally distributed data and how to calculate power for a test of hypothesis about the mean when the population variance is known. In addition, we presented statistics to be used in the tests of hypotheses about the difference of two means and two proportions. This latter material prepares us for the next chapter, the analysis of variance, in which we extend the test of hypothesis to comparing two or more means.

EXERCISES

13.1. In a recent study, Hall (4) examined the pulmonary functioning of 135 male Caucasian asbestos product workers. An earlier study had suggested that the development of clinical manifestations of the exposure to asbestos required a minimum of 20 years. Therefore, Hall partitioned his data set into two groups, one with less than 20 years of exposure to asbestos and the other with 20 or more years of exposure. Two of the variables used to examine pulmonary function are the forced vital capacity (FVC) measured in liters and the percentage of the predicted FVC value where the prediction is based on age, height, sex, and race. Age is a particularly important variable to consider as there is a strong positive correlation between FVC and age. The sample means and standard deviations of FVC and the percentage of the predicted FVC for each of the two groups are as follows:

	Length of exposure			
	<20 years (n = 66)		*≥20 years (n = 69)*	
Variable	*Mean*	*SD.*	*Mean*	*SD.*
FVC (liters)	5.19	0.78	4.27	0.63
% Predicted FVC	104	9.7	45	12.8

Choose the more appropriate of these two variables to use in a test of whether there is a difference in the means of the two exposure groups. Perform the test at the 0.05 significance level. Explain your choice for which variable to use and also your choice of a one- or two-sided alternative hypothesis. What assumption did you make about the population variances? Does this study support the idea that there is a difference between those with less than 20 years of exposure and those with 20 or more years of exposure? What is the p value of the test? What, if any, other variable should be taken into account in the analysis?

13.2. Kirklin *et al.* (5) performed a study of infants less than 3 months old who underwent open heart surgery. There were 175 infants in their study based on data from 1967 to 1980. It was suggested that the survival probabilities improved over time. To examine this, the data were broken into two time periods. Test the hypothesis that there is a difference in the survival probabilities over these two periods versus the alternative hypothesis of no difference over time at the 0.01 significance level. Use the following hypothetical data, based on data presented in the study.

Date	*Probability of survival*	*Sample size*
January 1967 to December 1973	0.46	66
January 1974 to July 1980	0.64	109

Provide possible reasons why there might be a difference in the survival probabilities over time.

13.3. Data from the National Institute of Occupational Safety and Health for the period 1980–1988 were used to obtain estimates of the annual workplace fatality rates by state (6). The average annual state rates over the 9-year period are given in Exercise 7.4. There is tremendous variability in the rates, ranging from a low of 1.9 to a high of 33.1 deaths per 100,000 workers. Provide some possible reasons for this variability. For the state of your residence, test the hypothesis of no difference in the crude workplace fatality rate and the national average of 7.2 per 100,000 workers. Exercise 7.7 gives the population total for your state. Perform this test against a two-sided alternative at the 0.05 significance level. What is the p value of the test? Provide possible reasons why there is or is not evidence of a difference between your state and the national average.

13.4. In the study by Reisin *et al.* (7), previously discussed in Chapter 10, one of the goals was to observe the effect of weight loss without salt restriction on blood pressure. We focus on one of the intervention groups, the group that was on a weight reduction program and given no medication. The program consisted of a strict diet with caloric intake reduced to about 50 percent of the usual adult intake for a 2-month period. Before examining the data for an effect on blood pressure, it is necessary to determine whether the diet worked. The summary weight data for the sample of 24 patients was a mean reduction of 8.8 kg and the standard deviation of the weight changes was 4.3 kg. This is a paired t test situation for this single group; however, there was also a control group that was not part of the weight reduction effort. During the period when the study group lost an average of 8.8 kg, the control group showed an average decrease of only 0.7 kg. The results from the control group increase our confidence in the use of the paired t test here. Test the null hypothesis of no weight reduction versus the appropriate one-sided alternative hypothesis at the 0.01 significance level. Did the weight reduction program work?

13.5. There were a number of drug recalls during 1993 because of the failure of the drugs to meet dissolution specifications or content uniformity specifications or because of subpotency (8). Three products from the Parke–Davis Division of the Warner–Lambert Company were recalled. One of the products, Tedral, met neither the dissolution or the content uniformity specifications. Suppose that the content uniformity specifications were expressed in terms of the variance. For example, say that the variance of the amount of phenobarbital in tablets was supposed to be less than or equal to 0.015 g^2. We selected a sample of 30 tablets and found the sample standard deviation of phenobarbital to be 0.14 g. Test the appropriate hypothesis to determine, at the 0.10 level, whether there is compliance with the content uniformity specifications for the amount of phenobarbital in the tablets.

13.6. In Chapter 7, using data from Table 7.6, we saw that there was a statistically significant (at the 0.01 level) difference in the mean ages of the AML and ALL patients. The difference in ages is important, particularly if the length of survival is strongly related to age. Calculate the sample Pearson correlation coefficient between age and length of survival based on all the patients in Table 7.6. Then test the null hypothesis, at the 0.05 level, that the population correlation coefficient is -0.30 versus the alternative hypothesis that the correlation is less (more negative) than -0.30. Here we are using -0.30 or more negative values to indicate a strong inverse correlation. Based on your analysis, is it necessary to control for the effect of age in the comparison of the length of survival of AML and ALL patients?

13.7. In Exercise 7.10, we examined progress toward the Surgeon General's goal of reducing the proportion of 12- to 18-year-old adolescents who smoked to below 6 percent for a hypothetical community. We found that in 1990, of the 12- to 18-year-olds in the sample, 11 of 85 admitted that they smoked. Test the hypothesis that the hypothetical community has already attained the Surgeon General's goal at the 0.05 significance level. Should you use a one- or two-sided alternative hypothesis? Explain your reasoning.

13.8. Opponents of a national health system argue that it will lead to rationing of services, something that is viewed as being unacceptable to people in the United States. To determine how people in the United States really felt about rationing of services, the American Board of Family Practice conducted a survey, some of the results of which are reported by Potter and Porter (9). One question asked whether people would approve of rationing medical attention in the case of a terminal illness. Suppose that we have decided that there is substantial support for rationing if the proportion of the population who would approve of rationing in this case is 40 percent. In the sample of 1007 Americans, 34 percent supported rationing in the case of terminal illness. Test the hypothesis that the population proportion equals 40 percent versus the alternative hypothesis that it is less than 40 percent. Use the 0.01 significance level. It is interesting to note that 43 percent of the physicians surveyed supported rationing in this situation.

13.9. The proportions of caloric intake coming from fat for the 33 boys in the Houston suburb study are shown in MINITAB Box 13.6.1. As was mentioned, the recommended amount of calories coming from fat should be no more than 30 percent. Test the hypothesis that this proportion for boys in the northern suburbs of Houston is equal to 30 percent at the 0.02 level. In performing the test, assume that the population standard deviation is 0.09. Should this test be one- or two-sided? What is the p value of the test? What is the power of the test to detect a value of 35 percent of the calories coming from fat?

13.10. Anderson *et al.* (10) performed a study on the effects of oat bran on serum cholesterol for males with high or borderline high values of serum cholesterol. High values of serum cholesterol are greater than or equal to 240 mg/dl (6.20 mmol/L). We wish to use the data from the study to determine whether or not there is a linear relationship between body mass index and serum cholesterol. The body mass index is defined as weight (in kilograms) divided by square of the height (in meters).

Body mass index	*Serum cholesterol*	*Body mass index*	*Serum cholesterol*
29.0	7.29	26.3	8.04
21.6	8.43	21.8	7.96
27.2	5.43	24.8	5.77
25.2	6.96	24.5	6.23
25.1	6.65	23.5	6.26
27.9	8.20	24.8	6.21
31.9	5.92	24.4	5.92

Test the hypothesis of no correlation between body mass index and serum cholesterol at the 0.10 level. Explain your choice of a one- or two-sided alternative hypothesis. What is the p value of the test?

13.11. For men, overweight is defined as a body mass index greater than 27.8 kg/m^2. Test the hypothesis that the men in Exercise 13.10 come from a population that is overweight. Perform the test at the 0.05 level.

13.12. Exercise 7.6 shows 15 hypothetical serum cholesterol values. For these data, test the hypothesis that the population variance equals 100 (mg/dl)^2 versus the alternative hypothesis that the population variance is greater than 100 (mg/dl)^2. Perform the tesf at the 0.025 level. Discuss the results of this test in relation to the confidence interval obtained in Exercise 7.6. Recall that this test requires that the cholesterol values follow a normal distribution. Examine the assumption of normality of the cholesterol values.

13.13. For the same data from Exercise 7.6, test the hypothesis that the measuring process works; that is, test the hypothesis that the population mean of the values measured by this process equals 190 versus the alternative hypothesis that the population mean is not equal to 190 mg/dl. Perform the test at the 0.02 significance level.

REFERENCES

1. Life Sciences Research Office, Federation of American Societies for Experimental Biology (1989) "Nutrition Monitoring in the United States: An Update Report on Nutrition Monitoring," DHHS Publ. No. (PHS) 89-1255. U.S. Department of Agriculture and the U.S. Department of Health and Human Services. Public Health Service, Washington, U.S. Government Printing Office.
2. National Center for Health Statistics (1992). "Health, United States, 1991 and Prevention Profile," DHHS Publ. No. 92-1232. Public Health Service, Hyattsville, MD.
3. National Center for Health Statistics (1991). "Health, United States, 1990," DHHS Publ. No. 91-1232. Public Health Service, Hyattsville MD.
4. Hall, S. K. (1989). Pulmonary health risk. *J. Environ. Health* **52,** 165–167.
5. Kirklin, J. K., Blackstone, E. H., Kirklin, J. W., McKay, R., Pacifico, A. D., and Bargeron, L. M., Jr. (1981). Intracardiac surgery under 3 months of age: Incremental risk factors for hospitality mortality. *Am. J. Cardiol.* **48,** 500–506.
6. Public Citizen Health Research Group (1992). Work-related injuries reached record level last year. *Public Citizen Health Res. Group Health Lett.* **8**(12), 1–3, 9.
7. Reisin, E., Abel, R., Modan, M., Silverberg, D. S., Eliahou, H. E., and Modan, B. (1978). Effect of weight loss without salt restriction on the reduction of blood pressure in overweight hypertensive patients. *N. Engl. J. Med.* **298,** 1–6.
8. Public Citizen Health Research Group (1993). Drug recalls March 9-June 7, 1993. *Public Citizen Health Res. Group Health Lett.* **9**(7), 9–10.
9. Potter, C., and Porter, J. (1989). American perceptions of the British National Health Service: Five myths. *J. Health Politics, Policy Law* **14,** 341–365.
10. Anderson, J. W., Spencer, D. B., Hamilton, C. C., Smith, S. F., Tietyen, J., Bryant, C. A., and Oeltgen, P. (1990). Oat-bran cereal lowers serum total and LDL cholesterol in hypercholesterolemic men. *Am. J. Clin. Nutr.* **52,** 495–499.

CHAPTER 14

Analysis of Variance

In Chapter 13, we used the t test to test the equality of two population means based on data from two independent samples. In this chapter, we introduce a procedure for testing the equality of two or more means. The two statistical designs discussed in Chapter 8, the completely randomized and the randomized block designs, are considered.

The comparison of two or more means is based on partitioning the variation in the dependent variable into its components, and hence the method is called the analysis of variance (ANOVA). It was introduced by Sir Ronald A. Fisher and has been used in many fields of research. We begin this chapter with a presentation of the assumptions made when the ANOVA is used. This section is followed by an introduction to the one-way ANOVA. In conjunction with this analysis, we present three methods used in multiple comparison analysis. These topics are followed by the analysis of the randomized block design, an example of a two-way ANOVA, and a two-way ANOVA with interaction. Lastly, a linear model representation of the ANOVA is provided.

I. ASSUMPTIONS FOR THE USE OF THE ANOVA

The ANOVA is used to determine whether there is a statistically significant difference among the population means of two or more groups. The theoretical basis of the ANOVA requires that the data being analyzed are independent and normally distributed. We must also assume that the population variances in each of the groups have the same value, σ^2. The ANOVA procedure works reasonably well if there are small departures from the normality assumption; however, if the variances are very different, there is concern about the significance levels reported in the analysis (1, Chapter 10). This concern is consistent with the material presented in Chapter 7. In Chapter 7, we presented different methods for comparing two means, depending on whether we assumed that the population variances were equal. One method for protecting against the effects of different values for the variances is to have approximately equal numbers of observations in each of the groups being analyzed. Another approach involves transformations of the dependent variable (2, Chapter 12; 3). One further assumption is that the groups being compared are the only groups of interest.[1]

There are no firm rules for the number of observations required by the ANOVA. It is possible to perform power calculations or to use the size of confidence intervals to estimate the sample size required. In general, we recommend that there be a minimum of 5 to 10 observations for each of the combinations of levels of the independent variables used in the analysis. For example, with two independent variables, if one variable has three levels and the other variable has four levels, there are 12 combinations of levels.

II. ONE-WAY ANOVA

In a one-way ANOVA, there is only one independent variable. The data to be analyzed are obtained from either (1) a random sample of subjects who belong to different groups, for example, different racial groups, or (2) an experiment in which the subjects are randomly assigned to one of several groups. The latter situation arises when we use the completely randomized design discussed in Chapter 8. In the completely randomized design, subjects are randomly allocated to groups and the groups represent the levels of the independent variable. Observations of the continuous variable of

[1]This assumption means that the factors, the independent variables, are fixed factors. For more information on fixed and random factors and the implications, see Steel and Torrie (4, Chapter 9).

TABLE 14.1 Hypothetical Ages for Control and Surgery Subjects

Group	Ages																	
Surgery	32	28	22	25	20	20	28	28	20	29	22	37	18	29	22	32	21	34
	19	23	23	26	41	20	33											
Control I	32	26	31	39	34	33	29	41	35	33	33	43	25	39	36	37	28	34
	27	45	22	29	51	28	35											
Control II	31	35	26	28	22	29	27	21	22	27	24	44	21	25	27	18	27	36

interest, the dependent variable, are taken on the subjects and the subjects' group membership is also recorded.

Data shown in Table 14.1 are based on an article by Kimball *et al.* (5) and can be analyzed using a one-way ANOVA. In the article, the authors wished to evaluate ventricular performance after surgical correction of congenital coarctation of the aorta. The ventricular performance was compared with that found in two control groups. Because of the possible roles that age and gender play on ventricular performance, the authors wanted the age and sex distributions of the subjects who had undergone the surgery to be similar to those of the members of the two control groups. We wish to examine whether the authors were successful in obtaining groups that were similar on the age variable. The ages shown in Table 14.1 are hypothetical, based on the summary values reported by Kimball *et al.* In this example, the dependent variable is age and the independent variable is the group to which the subjects belong.

The entries in Table 14.1 can be represented symbolically as y_{ij}, where the first subscript indicates the subject's group membership and the second subscript identifies the subject in the ith group. For example, y_{11} is 32 years old, y_{12} is 28 years old, y_{25} is 34 years old, y_{26} is 33 years old, and so on. The first subscript ranges from 1 to 3. When the first subscript has the value of 1, the range of the second subscript is 1 to 25; this is also the case when the first subscript is 2. When the first subscript has the value of 3, the second subscript ranges from 1 to 18. In general, there are r groups and n_i observations in the ith group. We also use the $\cdot$ notation introduced in Chapter 11. For example, $y_{i\cdot}$ is a shorthand notation for $\sum_j y_{ij}$, and $y_{\cdot\cdot}$ is shorthand for $\sum_i \sum_j y_{ij}$. Thus, $y_{1\cdot}$ represents the sum of all the ages for the subjects in the surgery group and $y_{\cdot\cdot}$ is the sum of all the 68 ages in the sample. It follows that $\bar{y}_{i\cdot}$ is the sample mean of the ith group and $\bar{y}_{\cdot\cdot}$ is the overall sample mean.

In the previous chapter, two means were compared using a t test; however, the t-test method used in Chapter 13 does not directly extend to the comparison of more than two group means; hence we must introduce another method of analysis.

A. Sums of Squares and Mean Squares

As was mentioned above, the analysis of variance is based on a partitioning of the variation in the dependent variable. In the one-way ANOVA, there are two possible sources of variation in the dependent variable. One source is variation among (or between) the groups; that is, the groups may have different means that vary about the overall mean. The other possible source is variation within the groups. Not all the subjects in the same group will have exactly the same values and the within-group variation reflects this.

The null hypothesis being tested here is that the population group means are equal to one another. If this hypothesis is true, all the observations come, in effect, from the same population. Thus, any variation that remains among the group means really reflects the random variation among the observations, that is, the within-group variation. Therefore, the adjusted among and within variations should be similar if the null hypothesis is true. If the null hypothesis is false, the adjusted among-group variation should be larger than the adjusted within-group variation because it includes variation between the populations as well as the within-group variation. Thus, we can use the adjusted among- and within-group variations as the basis of a test of the equality of the group means.

We can represent the above idea in symbols as

$$\sum_{i=1}^{r}\sum_{j=1}^{n_i}(y_{ij}-\bar{y}_{..})^2=\sum_{i=1}^{r}\sum_{j=1}^{n_i}(\bar{y}_{i\cdot}-\bar{y}_{..})^2+\sum_{i=1}^{r}\sum_{j=1}^{n_i}(y_{ij}-\bar{y}_{i\cdot})^2.$$

This equation shows the partitioning of the total variation in Y, the dependent variable, about its mean into an among- (or between)-group component and a within-group component. These sums of squares are called the total sum of squares corrected for the mean (SST), the among (or between)-group sum of squares (SSB), and the within-group sum of squares (SSW).

If we adjust these two components for the number of independent observations used in their calculations, that is, divide each component sum of squares by its degrees of freedom, we have the mean square among (or between) and the mean square within. The mean square between is

$$\text{MSB}=\frac{\sum_{i=1}^{r}\sum_{j=1}^{n_i}(\bar{y}_{i\cdot}-\bar{y}_{..})^2}{r-1}=\frac{\sum_{i=1}^{r}n_i(\bar{y}_{i\cdot}-\bar{y}_{..})^2}{r-1}$$

where the second expression reflects the fact that the terms in parentheses do not vary with j. The mean square within is

$$\text{MSW}=\frac{\sum_{i=1}^{r}\sum_{j=1}^{n_i}(y_{ij}-\bar{y}_{i\cdot})^2}{n-r}$$

where n is the total number of observations, that is, the sum of the n_i. The degrees of freedom for the mean square between, $r - 1$, comes from the calculation of the variation in r means. The degrees of freedom for the mean square within, $n - r$, is the result of summing the $n_i - 1$ degrees of freedom associated with the ith group over the r groups.

The mean square within is particularly useful as it also provides an adjusted estimate of the variation within groups, that is, of σ^2, the variance of the dependent variable. It is based on the assumption that the variance of the dependent variable is the same within each group. If there is no difference between the group means, then the mean square between also estimates σ^2.

For the data in Table 14.1, we have the following values of means and sums of squares. First, $\bar{y}_{1\cdot}$, the sample mean of the first group, is 26.08, $\bar{y}_{2\cdot}$ is 33.80, and $\bar{y}_{3\cdot}$ is 27.22. The overall sample mean, $\bar{y}_{\cdot\cdot}$, is 29.22 years. The sum of squares between is

$$\begin{aligned} SSB &= 25 * (26.08 - 29.22)^2 + 25 * (33.80 - 29.22)^2 + 18 * (27.22 - 29.22)^2 \\ &= 842.9 \end{aligned}$$

The sum of squares within involves too many terms to show, but its sum of squares is 2660.8 and the total sum of squares (corrected) is 3503.7.

B. The *F* Statistic

The comparison of these two mean squares provides information about whether the null hypothesis is true. One way of comparing the mean squares is to take their difference. If the null hypothesis were true, then the difference would be zero; however, the probability distribution of the difference is not widely available. Another way of comparing the mean squares is to take the ratio of the mean square between to the mean square within. If the null hypothesis were true, the ratio would equal one. If the null hypothesis were false, the ratio would be larger than one. Fortunately, the probability distribution of the ratio has been worked out, and it is an F distribution with $r - 1$ and $n - r$ degrees of freedom. Tables of the F distribution, named in honor of Sir Ronald Fisher, are shown in Appendix B11 for the 0.01, 0.05, and 0.10 significance levels for values of the numerator (f_1) and denominator (f_2) degrees of freedom parameters.

The F distribution has many different shapes, depending on the values of the degrees of freedom parameters. Figure 14.1 shows the shape of the F distribution for degrees of freedom pairs 1 and 20 and 5 and 20. We can see that the shapes are different, but most of the probability (area) is associated with values of F close to one.

There is also a relationship between the t and F distributions that can be seen from the t and F tables. The relationship is $t^2_{k,1-\alpha/2}$ is equal to $F_{1,k,1-\alpha}$. For example, when k is 10, $t_{10,0.95}$ is 1.8125, and its square is 3.2852.

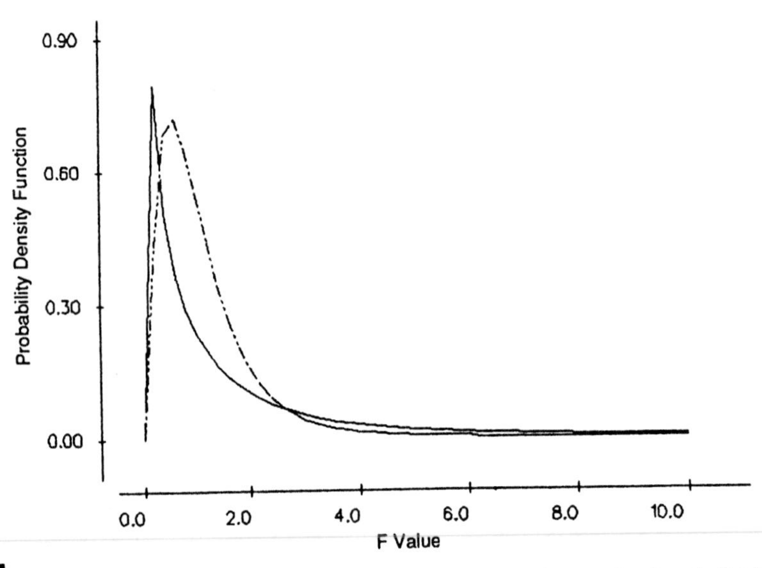

FIGURE 14.1 Plot of the probability density functions of the F distribution for $F_{1,20}$ (solid line) and $F_{5,20}$ (broken line).

Examination of the F tables in Appendix B11 shows that $F_{1,10,0.90}$ is 3.29. This equivalence when there are two groups leads us to think that there may be a relationship between the ANOVA and t-test approaches in the two-group situation.

C. The ANOVA Table

The sums of squares and mean squares just described are usually presented in tabular format as shown in Table 14.2. The degrees of freedom and sums of squares associated with the between and within groups sum to the corresponding total values. If these values do not sum to the total, a mistake has been made in the calculations.

The F statistic is then used to test the null hypothesis that the group means are equal against the alternative hypothesis that the group means

TABLE 14.2 Typical ANOVA Table for a One-Way Analysis

Source of variation	Degrees of freedom	Sum of squares	Mean square	F
Between groups	$r - 1$	SSB	SSB/$(r - 1)$ = MSB	MSB/MSW
Within groups	$n - r$	SSW	SSW/$(n - r)$ = MSW	
Total (corrected)	$n - 1$	SST		

are not all equal. When the null hypothesis is true, the F statistic follows an F distribution with $r-1$ and $n-r$ degrees of freedom. If the calculated F statistic is greater than $F_{r-1,n-r,1-\alpha}$, found in Appendix Table B11, we reject the null hypothesis in favor of the alternative hypothesis at the α significance level. If the calculated F statistic is less than this critical value, we do not have sufficient evidence to reject the null hypothesis.

D. Analysis of the Ages

Let us test the hypothesis of the equality of the mean ages at the 0.01 significance level. Based on the sums of squares presented above, we can complete the ANOVA table for the ages shown in Table 14.1. Table 14.3 is the ANOVA table for the age data.

A computer package can easily perform these calculations. Box 14.1 shows the use of two different MINITAB commands for performing a one-way ANOVA. There are 68 observations in the three groups. Hence there are 2 degrees of freedom for the factor (between groups) variable, 65 degrees of freedom for error (within groups), and 67 degrees of freedom for the total sum of squares. The output shows the sum of squares (SS) and mean squares (MS) as well as the F ratio and its p value. It also shows the mean and standard deviations of the age variable along with the 95 percent confidence intervals for the group means. The pooled standard deviation is the square root of the MS for error and estimates σ.

Based on the ANOVA table, as the p value associated with the F ratio is 0.000, we know that the F ratio is greater than $F_{2,65,0.99}$. Therefore we reject the equality of the mean ages in favor of the alternative hypothesis. It appears that the three groups differ on age. This means that it may be necessary to take age into account in the analysis of ventricular performance.

From the confidence intervals shown in Box 14.1, it appears that the difference is due mainly to the first control group having a mean age much greater than those of the other two groups. When there is a statistically significant difference among the group means, we can perform additional tests to see if we can determine the source of the differences in the means. The next section describes three approaches to this additional testing.

TABLE 14.3 ANOVA Table for the Ages Shown in Table 14.1

Source of variation	Degrees of freedom	Sum of squares	Mean square	F
Between groups	2	842.9	421.4	10.29
Within groups	65	2660.8	40.9	
Total (corrected)	67	3503.7		

MINITAB BOX 14.1

One MINITAB command for performing the analysis is AOVONEWAY (abbreviated AOVO). This command assumes that data from different groups are entered in different columns. The ages shown in Table 14.1 are entered. Ages for the surgery group are in c1, ages for control I are in c2, and ages for control II are in c3. These columns are specified after the command.

```
MTB > aovoneway c1-c3

ANALYSIS OF VARIANCE
SOURCE     DF        SS        MS        F        p
FACTOR      2     842.7     421.4    10.29    0.000
ERROR      65    2661.0      40.9
TOTAL      67    3503.7
                                    INDIVIDUAL 95 PCT CI'S FOR MEAN
                                        BASED ON POOLED STDEV
 LEVEL      N      MEAN     STDEV   ---+---------+---------+---------+---
     1     25    26.080     6.157   (-------*------)
     2     25    33.800     6.708                      (-------*------)
     3     18    27.222     6.283     (--------*-------)
                                    ---+---------+---------+---------+---
POOLED STDEV =    6.398                24.5      28.0      31.5      35.0
```

The other MINITAB command for performing the analysis is ONEWAY. This command assumes that all the data are in one column and that there is a second column that indicates group membership. In the following we stack columns c1, c2, and c3 together in c4. We next create indicators of group membership in c5. The first group is given the code of 1 and there are 25 subjects in that group. The 25 members in the second group are assigned the value of 2, and the 18 members in the third group are assigned the value of 3.

```
MTB > stack c1 c2 c3, c4
MTB > set c5
DATA> 25(1) 25(2) 18(3)
MTB > oneway c4 c5
```

The output is the same as that shown above. The ONEWAY command also has other capabilities that are associated with multiple comparisons and the linear model representation of the ANOVA.

III. MULTIPLE COMPARISONS

If the overall F statistic from the ANOVA is statistically significant, multiple comparison procedures can be used in an attempt to discover the source of the significant differences among the group means. Most of these procedures are designed to examine the pairwise differences in group means, although there are more general procedures. The comparison of the group means is accomplished through the presentation of confidence intervals for pairwise differences of group means. The use of the multiple comparison procedures is generally not recommended when we fail to reject the null hypothesis; however, exceptions may occur when certain comparisons have been planned in the course of the experiment.

There are many different multiple comparison procedures. We present three procedures: the Tukey–Kramer method, Fisher's least significant difference (LSD) method, and Dunnett's method. The Tukey–Kramer method is the procedure recommended when one wishes to estimate simultaneously all pairwise differences among the means in a one-way ANOVA assuming that the variances are equal (6). We present the LSD method because it is frequently used in the literature. Dunnett's procedure is used when we wish to compare several groups with a specific group selected before the data were obtained (or the control group designated in the design). For example, if there were several new treatments and a standard treatment, we would use Dunnett's procedure to compare each of the new treatments with the standard. The multiple comparison procedures presented here use the mean square within as the estimator of σ^2.

Before presenting these methods, we present material on the error rates associated with the methods.

A. Error Rates: Individual and Family

In the pairwise comparison of the group means, many confidence intervals are formed. For example, when there are three groups, we form confidence intervals for the differences between groups 1 and 2, groups 1 and 3, and groups 2 and 3. When there are r groups, there are ${}_rC_2$ confidence intervals for the pairwise comparisons. Thus we see that there are two probabilities of errors in multiple comparison procedures. One probability of error is associated with each individual confidence interval: the individual error rate. The other probability of error is associated with the ${}_rC_2$ intervals, the family of confidence intervals: the family error rate. This is the rate that is usually of primary interest, the rate that we want to be less than or equal to α.

It is clear that if we use the $t_{1-\alpha/2}$ value in the creation of the confidence intervals, the family error rate will be larger than α. If we wish to control the family error rate to be less than or equal to α, then we must use some value other (greater) than $t_{1-\alpha/2}$ in the calculation of the confidence intervals.

B. Tukey–Kramer Method

The *Tukey–Kramer method* focuses on the family error rate. It replaces $t_{n-r,1-\alpha/2}$ in the confidence interval for the difference between two group means by $q_{r,n-r,1-\alpha}/\sqrt{2}$, where q is the upper α value from the studentized range distribution. Table B12 shows the upper 0.05 and 0.01 points of the studentized range distribution. Note that the q value takes the number of possible comparisons into account because its value depends on r, the number of groups.

The confidence interval for the difference between μ_i and μ_j is

$$\bar{y}_{i\cdot} - \bar{y}_{j\cdot} \pm \frac{q_{r,n-r,1-\alpha}}{\sqrt{2}} \sqrt{\text{MSW}} \sqrt{\frac{1}{n_i} + \frac{1}{n_j}}.$$

Let us calculate the confidence intervals for the three pairwise comparisons for the hypothetical age data shown in Table 14.1. We set the family error rate to be 0.05. The value of $q_{3,65,0.95}$ is not found in Table B12. As there is little variation in the value of q as $n - r$ changes from 40 to 60 to 120 in the table, we use 3.40 ($= q_{3,60,0.95}$) as an approximation to the desired value. The confidence interval for the difference between groups 1 and 2 is

$$26.08 - 33.80 \pm \left(\frac{3.40}{\sqrt{2}} * \sqrt{40.9} * \sqrt{\frac{1}{25} + \frac{1}{25}}\right)$$

which yields

$$-7.72 \pm 4.35$$

and the interval ranges from -12.07 to -3.37. The corresponding interval for $\mu_1 - \mu_3$ is -5.89 to 3.61 and the interval for $\mu_2 - \mu_3$ is from 1.83 to 11.33. Both of the intervals involving μ_2 fail to contain zero, suggesting that the first control group differs significantly from both the study group and the second control group.

C. Fisher's Least Significant Difference Method

Fisher's LSD method focuses on the individual error rate. When the n_i are all equal, there is a value, the least significant difference, such that if any of the differences in sample means are greater than that value, the difference is statistically significant. If a difference is greater than that value, the corresponding confidence interval for the difference does not contain zero. If the number of sample observations differ across the groups, there is not a single least significant difference.

The LSD confidence interval looks like the ordinary confidence interval for the difference between two means with one exception. The mean square within is used as the estimator for the population variance instead of an estimator based on only data from the two groups being compared. The LSD confidence interval for $\mu_i - \mu_j$ is

$$\bar{y}_{i\cdot} - \bar{y}_{j\cdot} \pm t_{n-r,1-\alpha/2} \sqrt{\text{MSW}} \sqrt{\frac{1}{n_i} + \frac{1}{n_j}}.$$

Let us calculate the 0.05 individual error rate LSD confidence interval for $\mu_1 - \mu_2$. We have

$$26.08 - 33.80 \pm (2.00 * 6.395 * \sqrt{0.08})$$

which yields

$$-7.72 \pm 3.62$$

and the interval ranges from −11.34 to −4.10. This interval is narrower than the corresponding Tukey–Kramer interval as it must be because it is based on the individual error rate, not the family error rate used by the Tukey–Kramer procedure. The corresponding LSD interval for $\mu_1 - \mu_3$ ranges from −5.10 to 2.82 and the interval for $\mu_2 - \mu_3$ ranges from 2.62 to 10.54.

D. Dunnett's Method

Dunnett's method is used in situations in which we wish to compare the means of several groups with the mean of another group that was selected in advance. For example, we may wish to compare the means of different dosage levels of a new medication with the mean of a placebo group. In our example, there are two control groups and one treatment group. We wish to see if there is a difference between the two control groups and the treatment group (group 1). Thus the comparisons of interest are $\mu_2 - \mu_1$ and $\mu_3 - \mu_1$.

The confidence interval for $\mu_i - \mu_j$ using Dunnett's procedure is given by

$$\bar{y}_{i\cdot} - \bar{y}_{j\cdot} \pm d_{r-1,n-r,1-\alpha/2} \sqrt{\text{MSW}} \sqrt{\frac{1}{n_i} + \frac{1}{n_j}}$$

where the upper 0.005 and 0.025 levels of d are given in Table B13. Let us now calculate the confidence intervals using a family error rate of 0.05 and Dunnett's method.

For the comparison of the first control group with the treatment group, we have

$$33.80 - 26.08 \pm \left(2.27 * 6.395 * \sqrt{\frac{1}{25} + \frac{1}{25}}\right)$$

where 2.27 is the value of $d_{2,60,0.975}$ and this is used as an approximation to $d_{2,65,0.975}$. This calculation yields

$$7.72 \pm 4.11$$

and the interval ranges from 3.61 to 11.83. The corresponding interval for $\mu_3 - \mu_1$ ranges from −3.35 to 5.63. The confidence intervals using Dunnett's procedure are narrower than those provided by the Tukey–Kramer method. This is reasonable because we are doing fewer comparisons with Dunnett's procedure. Based on these intervals, there is a statistically significant difference between the first control group and the treatment group, but no significant difference between the second control group and the treatment group.

E. Analysis by Computer

These calculations can be performed with MINITAB using subcommands to the ONEWAY command. Box 14.2 shows the use of the subcommands to generate these intervals by the three different methods discussed above.

MINITAB BOX 14.2

Multiple comparisons are conducted by using subcommands in ONEWAY. The family error rates are specified for the Tukey–Kramer and Dunnett procedures and the individual error rate is specified for Fisher's LSD procedure. The level against which the other levels are to be compared is also specified for the Dunnett procedure (1 for this example). The ANOVA output is not shown below and we jump directly to the multiple comparison output.

```
MTB > oneway c4 c5;
SUBC> tukey .05;
SUBC> fisher .05;
SUBC> dunnett .05 1.

Tukey's pairwise comparisons
    Family error rate = 0.0500
Individual error rate = 0.0193
Critical value = 3.39
Intervals for (column level mean) - (row level mean)
                 1         2
      2   -12.058
           -3.382

      3    -5.883     1.837
            3.599    11.319

Fisher's pairwise comparisons
    Family error rate = 0.121
Individual error rate = 0.0500
Critical value = 1.997
Intervals for (column level mean) - (row level mean)
                 1         2
      2   -11.334
           -4.106

      3    -5.092     2.628
            2.808    10.528
```

```
Dunnett's intervals for treatment mean minus control mean
    Family error rate = 0.0500
Individual error rate = 0.0268
Critical value = 2.27
Control = level 1 of C5

Level      Lower  Center   Upper   -----+---------+---------+---------+------
    2       3.61    7.72    11.8              (----*----)
    3      -3.35    1.14    5.63    (----*-----)
                                   -----+---------+---------+---------+------
                                       0.0       8.0      16.0      24.0
```

IV. TWO-WAY ANOVA FOR THE RANDOMIZED BLOCK DESIGN WITH *m* REPLICATES

As discussed in Chapter 8, in many situations the same experiment is conducted in several sites or under different conditions. In these situations, the random allocation of subjects takes place separately at each site or for each condition. These experiments are using what is called a *randomized block design*. The random allocation of the subjects to the treatments is performed separately for each block (site or condition) because it is thought that there may be an effect of the blocks on the outcome variable. If the subjects were randomly assigned ignoring the blocks, as in a completely randomized design, there is a chance that the block effects might be confounded with the treatment effects. Hence the random assignment is done separately.

The data in Table 14.4 are from a randomized block design with five replicates per cell. The data are the changes in weight for moderately overweight female employees who participated in weight reduction programs. The women worked at one of two company sites, the headquarters or a manufacturing plant. At each site, after a semiannual health examination, the women were randomly given memberships to a diet clinic, a

TABLE 14.4 Difference between Pre- and Postintervention Weights (Pounds) after 6 Months of Participation by Intervention Program at Two Sites

Program	Office site					Factory site				
Diet clinic	6	2	10	−1	8	3	15	4	8	6
Exercise club	3	4	−2	6	−2	−4	6	8	−2	3
Both programs	8	12	7	10	5	15	8	10	16	3

health club, or both. There was a control for company site because it was thought that there may be a difference in the effects of the weight reduction programs between those who were less physically active—the headquarters group—and the women in the plant. After the next health examination, weight reduction was measured.

In this table, data are classified by program, the row variable, and site, the column variable. The type of intervention program is the treatment variable with three levels, and site is considered to be the blocking variable with two levels. These two independent variables form six cells and the cells all have the same number of observations. When the same number of observations is in each cell, the design is said to be balanced. The analysis of unbalanced data is more complicated and is not discussed here.

The entries in Table 14.4 can be represented symbolically as y_{ijk}, where i is an indicator of the program (the row variable), j represents the site (the column variable), and k indicates the subject number within the ith program and jth site. The first subscript ranges from 1 to 3, the second subscript has the value 1 or 2, and the third subscript ranges from 1 to 5.

We continue to use the $\cdot$ notation. For example, $y_{\cdot 1 \cdot}$ represents $\sum_i \sum_k y_{i1k}$, the sum of weight losses for the female employees at the office site. With this notation, the sample mean of the ith level of the program variable is $\bar{y}_{i \cdot \cdot}$, the sample mean of the jth level of the site variable is $\bar{y}_{\cdot j \cdot}$, and the overall sample mean is $\bar{y}_{\cdot \cdot \cdot}$. The values of these sample means follow:

Program means		*Site means*		*Overall mean*
Diet	6.10	Office	5.07	5.83
Exercise	2.00	Factory	6.60	
Both	9.40			

To analyze this data set, we use a two-way ANOVA. The method of analysis is called two-way because there are now two independent variables: the blocking variable with c levels and the treatment variable with r levels. The total sum of squares of the dependent variable about its mean is now partitioned into a sum of squares between treatment groups, a sum of squares between blocks, and the within-cell (error or residual) sum of squares. This partitioning, based on m observations per cell, is

$$\sum_{i=1}^{r} \sum_{j=1}^{c} \sum_{k=1}^{m} (y_{ijk} - \bar{y}_{\cdot\cdot\cdot})^2 = cm \sum_{i=1}^{r} (\bar{y}_{i\cdot\cdot} - \bar{y}_{\cdot\cdot\cdot})^2 + rm \sum_{j=1}^{c} (\bar{y}_{\cdot j \cdot} - \bar{y}_{\cdot\cdot\cdot})^2 + \text{SSW}.$$

The total variation of Y about its mean (SST) is partitioned into the sum of squares for the row or treatment variable (SSR), the sum of squares for the column or block variable (SSC), and the within or residual sum of squares (SSW). SSW is found by subtracting the sum of SSR and SSC from

SST. The value of SSR is

$$\text{SSR} = 2 * 5\,[(6.10 - 5.83)^2 + (2.00 - 5.83)^2 + (9.40 - 5.83)^2] = 274.9.$$

The value of SSC is similarly found and is

$$\text{SSC} = 3 * 5\,[(5.07 - 5.83)^2 + (6.60 - 5.83)^2] = 17.56.$$

Too many terms are involved to show the calculation of SST, but its value is 768.2 and SSW, found by subtraction, is 475.7.

We use the same approach to the analysis in the two-way ANOVA as was used in the one-way ANOVA. To test the hypothesis of no difference in the treatments, we use the F statistic calculated as the ratio of the mean square for treatment to the residual mean square. If the null hypothesis of no difference in the treatment means, adjusted for the blocking variable, is true, this F statistic follows the F distribution. The mean square for treatments has $r - 1$ degrees of freedom and the residual mean square has $n - r - c + 1\,[= n - (r - 1) - (c - 1) - 1]$ degrees of freedom. Thus the F statistic for the treatment variable will follow an F distribution with $r - 1$ and $n - r - c + 1$ degrees of freedom if there is no difference in the treatment group means. In the same way, we could also test the null hypothesis of no difference in the block means. The F statistic associated with this hypothesis follows the F distribution with $c - 1$ and $n - r - c + 1$ degrees of freedom if this null hypothesis is true. Usually, we are not as interested in the hypothesis about the block means as in the treatment group means.

The ANOVA table for a randomized block design with m replicates per cell is shown in Table 14.5.

Let us perform the test of no treatment effect, that is, of no difference in the population means associated with the three interventions at the 0.05 significance level. The analysis for the change in weight data is shown in Table 14.6. As the calculated F value of 7.51 is greater than the critical value of 3.37 ($= F_{2,26,0.95}$), we reject the null hypothesis and conclude that the intervention programs are significantly different. We are not interested in the site difference.

TABLE 14.5 ANOVA Table for a Randomized Block Design

Source of variation	Degrees of freedom	Sum of squares	Mean square	F
Treatments	$r - 1$	SSR	$\text{SSR}/(r - 1) = \text{MSR}$	MSR/MSW
Blocks	$c - 1$	SSC	$\text{SSC}/(c - 1) = \text{MSC}$	MSC/MSW
Residual	$n - r - c + 1$	SSW	$\text{SSW}/(n - r - c + 1) = \text{MSW}$	
Total (corrected)	$n - 1$	SST		

TABLE 14.6 ANOVA Table for Weight Change Data from Table 14.4: Three Intervention Programs at Two Sites

Source of variation	Degrees of freedom	Sum of squares	Mean square	F
Between programs	2	274.9	137.4	7.51
Between sites	1	17.6	17.6	
Residual	26	475.7	18.3	
Total	29	768.2		

MINITAB BOX 14.3

The weight change data in Table 14.3 are entered in c1 along with the site identification in c2 and the treatment identification in c3.

```
MTB > twoway c1-c3;
SUBC> additive;
SUBC> means c2 c3.
```

The ADDITIVE subcommand means that the only sources of variation being considered are treatment, block, and residual. If we had not used this subcommand, another source of variation, the interaction of the treatment and site variables, would have been included. Interaction is discussed in the next section. The MEANS subcommand causes the dependent variable means and confidence intervals to be created for the levels of the specified independent variables.

```
ANALYSIS OF VARIANCE  C1
SOURCE         DF        SS        MS
C2              1      17.6      17.6
C3              2     274.9     137.4
ERROR          26     475.7      18.3
TOTAL          29     768.2
```

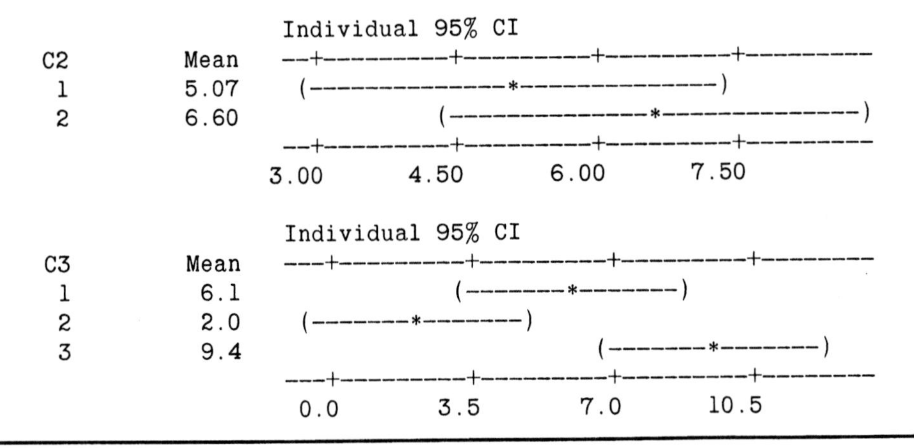

```
                           Individual 95% CI
     C2          Mean     --+---------+---------+---------+--------
      1          5.07      (---------------*---------------)
      2          6.60               (---------------*---------------)
                          --+---------+---------+---------+--------
                         3.00      4.50      6.00      7.50

                           Individual 95% CI
     C3          Mean     ---+---------+---------+---------+-------
      1          6.1                (--------*--------)
      2          2.0       (--------*--------)
      3          9.4                          (--------*--------)
                          ---+---------+---------+---------+-------
                            0.0       3.5       7.0      10.5
```

Box 14.3 shows how to conduct the preceeding analysis using MINITAB. As MINITAB does not know which of the two independent variables is of primary interest, it does not calculate the *F* statistic for us. The confidence intervals for the treatment variable, c3, support the finding of a difference in the treatment group means at the 0.05 significance level. There is no overlap for groups 2 and 3. It appears that using both types of intervention is the most effective intervention.

What would have happened had we ignored the site variable in the above analysis? If we assume that we would have had the same assignment of the subjects to the different treatments, we can examine the effect of the use of the blocking variable. The residual sum of squares in the two-way ANOVA is less than or equal to the residual sum of squares in the corresponding one-way ANOVA, reflecting the removal of the between-blocks sum of squares. If the sum of squares between the blocks is large and its degrees of freedom are small, then the residual mean square is much smaller in the two-way ANOVA. This means that if the blocking variable is important, there is a greater chance of detecting a difference in the treatment group means using the two-way ANOVA than using the corresponding one-way ANOVA.

In the next section, we show a more general two-way analysis of variance that includes the interaction of the two independent variables.

V. TWO-WAY ANOVA WITH INTERACTION

In some instances, a researcher is interested in studying the effects of two factors. In these instances, the experimental subjects are randomly allocated to all combinations of levels of both factors. For example, if both the row and column factors have two levels each, then the subjects are randomly allocated to four groups. This type of experimental design is especially useful when we want to study the effects of each factor as well as the interaction effect of the factors with one another. *Interaction* exists when the differences in responses to the levels of one factor depend on the level of another factor. For example, in a study of byssinosis (brown lung disease) in textile workers in North Carolina (7), two variables of interest were whether the worker smoked and whether the worker was exposed to dust in the workplace. Both of these variables were important; that is, both smoking and exposure to dust were associated with a higher occurrence of byssinosis. In addition, if a worker smoked and also was exposed to the dust, the occurrence of byssinosis was much higher than would have been expected by simply adding the effects of the smoking and exposure variables. In this case, there is a synergistic effect, that is, an interaction of these two independent variables.

TABLE 14.7 Increase in Test Scores after 4 Weeks of Instruction Using Three Textbooks and Two Teaching Methods

Textbook	Method of instruction: Lecture						Discussion					
1	30	43	12	18	22	16	36	34	15	18	40	45
2	21	26	10	14	17	16	33	31	28	15	29	26
3	42	30	18	10	21	18	41	46	19	23	38	48

We have previously been concerned about interaction, although we did not use the term interaction, when we considered the Cochran–Mantel–Haenszel procedure. We said that the procedure should not be used when the odds ratios were not consistent across the subtables. If the odds ratios are not consistent, this means that the relationship between the dependent and independent variables depends on the levels of an extraneous or confounding variable; that is, there is interaction between the independent and extraneous variables. If the interaction exists, it does not make sense to talk about an overall effect of the independent variable because its effect varies with the level of the extraneous or confounding variable.

The data in Table 14.7 are from a two-factor experiment in a health education teacher training program. Three new textbooks (factor A) were tested with two methods of instruction (factor B) and 36 trainees were randomly allocated to the 6 groups with 6 subjects per group. The trainees were tested before and after 4 weeks of instruction and the increases in test scores were recorded as shown in the table. As in the randomized block design, data are classified by textbook, the row factor, and method of instruction, the column factor. In this experiment, the random allocation of subjects was done simultaneously to all combinations of the two sets of levels, whereas the randomization took place separately in each block in the randomized block design.

The entries in this table are also represented symbolically by y_{ijk} as in the randomized block design with replicates. Several means again are used in the analysis. The means here include the cell means ($\bar{y}_{ij\cdot}$); two sets of marginal means, row ($\bar{y}_{i\cdot\cdot}$) and column ($\bar{y}_{\cdot j\cdot}$); and the overall mean ($\bar{y}_{\cdot\cdot\cdot}$). The values of these means follow:

Textbook	*Methods of instruction*: *Lecture*	*Discussion*	*Marginal book means*
1	23.50	31.33	27.42
2	17.33	27.00	22.17
3	23.17	35.83	29.50
Marginal method means	21.33	31.39	26.36 (overall mean)

We analyze this data set using a two-way ANOVA with interaction. For the randomized block design, we used a two-way ANOVA ignoring interaction. The researcher for this experiment could have used two separate completely randomized experiments (one-way ANOVAs): one to compare the three textbooks and the other to compare the two types of instructional methods. Based on these two separate experiments, however, the researcher would not know whether any textbook works better with one instructional method than the other. The effects of the textbooks may differ across the instructional methods. Interaction measures the difference in the textbook effects across the two instructional methods. If the distribution of the mean increase in test scores for the three textbook types for those taught by lecture differs from the corresponding distribution for those taught by discussion, there is interaction. The average effects of textbooks across both types of instruction and the average instructional effects across all textbooks are measures of the main effects of the two independent variables.

If there is an interaction of the two independent variables, then usually the interaction terms are of more interest than the main effects of the two independent variables. This is because, if there is an interaction, the effect of one independent variable depends—it changes—as the level of the other independent variable changes. Hence, in our analysis, we must first examine the test of hypothesis that there is no interaction before considering the test of no main effects of the independent variables.

If there is interaction, we can examine the cell means in an attempt to discover the nature of the interaction. If there is no evidence of an interaction, then we consider the hypotheses about the main effects. In this case, some statisticians would remove the interaction term from the analysis, that is, incorporate its sum of squares and degrees of freedom into the error term before calculating the F statistics for the main effects. The decision to incorporate or not to incorporate the nonsignificant interaction term into the error term usually has little effect on the results.

To include interaction in the analysis, the total sum of squares (SST) of the dependent variable about its mean is now partitioned into a sum of squares for the row factor R (SSR), a sum of squares for the column factor C (SSC), a sum of squares for interaction between factor R and factor C (SSRC), and the error sum of squares (SSE). As before, we use the symbols r and c for the numbers of levels for factors R and C, respectively, and use m to represent the number of replicates in each of the cells formed by the crosstabulation of factors R and C. This partitioning of the total sum of squares is expressed symbolically as

$$\sum_{i=1}^{r}\sum_{j=1}^{c}\sum_{k=1}^{m}(y_{ijk}-\bar{y}_{\cdots})^2 = cm\sum_{i=1}^{r}(\bar{y}_{i\cdot\cdot}-\bar{y}_{\cdots})^2 + rm\sum_{j=1}^{c}(\bar{y}_{\cdot j\cdot}-\bar{y}_{\cdots})^2$$
$$+ m\sum_{i=1}^{r}\sum_{j=1}^{c}(\bar{y}_{ij\cdot}-\bar{y}_{i\cdot\cdot}-\bar{y}_{\cdot j\cdot}+\bar{y}_{\cdots})^2 + \sum_{i=1}^{r}\sum_{j=1}^{c}\sum_{k=1}^{m}(y_{ijk}-\bar{y}_{ij\cdot})^2.$$

TABLE 14.8 ANOVA Table for a Two-Factor Design with Interaction

Source of variation	Degrees of freedom	Sum of squares	Mean square	F
Factor R	$r - 1$	SSR	SSR/$(r - 1)$ = MSR	MSR/MSE
Factor C	$c - 1$	SSC	SSC/$(c - 1)$ = MSC	MSC/MSE
Interaction	$(r - 1)(c - 1)$	SSRC	SSRC/$(r - 1)(c - 1)$ = MSRC	MSRC/MSE
Error	$n - rc$	SSE	SSE/$(n - rc)$ = MSE	
Total (corrected)	$n - 1$	SST		

The rest of the analytic approach is the same as before. The mean squares for the main effects and the interaction are calculated by dividing the sums of squares by appropriate degrees of freedom. The mean squares for factors R and C have $r - 1$ and $c - 1$ degrees of freedom, respectively. The mean square for interaction has $(r - 1)(c - 1)$ degrees of freedom, and the error mean square has $n - rc$ $[= rc(m - 1)]$ degrees of freedom. The error mean square is then used as the denominator in the calculation of the F statistics for the two main effects and interaction. The ANOVA table for a two-factor experimental design with interaction is shown in Table 14.8.

The calculations of the sums of squares, similar to those shown above in the randomized block analysis, are not shown here, but are summarized in Table 14.9.

Let us perform the tests of hypotheses at the 0.05 significance level. The F statistic and its associated p value for interaction indicate that there is no statistically significant interaction of the two independent variables. As this is the case, we can now examine the F statistics associated with the hypotheses of no difference in the test score improvement between the two methods of instruction and among the three textbooks. There is a statistically significant effect for the methods of instruction, a p value less than 0.05, but no significant effect associated with the textbooks.

TABLE 14.9 ANOVA Table for Test Score Increase Data in Table 14.6 by Combinations of Three Textbooks and Two Methods of Instruction

Source	Degrees of freedom	Sum of squares	Mean square	F	p value
Textbooks	2	342.7	171.4	1.66	0.207
Methods of instruction	1	910.0	910.0	8.81	0.006
Interaction	2	35.7	17.9	0.17	0.842
Error	30	3099.8	103.3		
Total	35	4388.3			

If we had removed the interaction term from the analysis after finding that it was not important, the error sum of squares would have been 3135.5 (= 35.7 + 3099.8) and there would have been 32 degrees of freedom associated with this error sum of squares. The error mean square would have been 97.98 instead of 103.3 and the F ratios for textbooks and methods of instruction would have been 1.75 and 9.29, respectively.

Let us explore further the preceding analytical results in relation to the cell means that were calculated above and are repeated here for our convenience.

	Methods of instruction		
Textbook	*Lecture*	*Discussion*	*Marginal book means*
1	23.50	31.33	27.42
2	17.33	27.00	22.17
3	23.17	35.83	29.50
Marginal method means	21.33	31.39	26.36 (overall mean)

The lack of a significant main effect for textbooks is reflected in the marginal means for textbooks. The first and third textbooks appear to be a little more effective than the second book, but the ANOVA results indicated that these differences are not statistically significant. On the other hand, the discussion method was associated with a much greater increase, about 10 points, in test scores than the lecture method and this difference was statistically significant. The lack of an interaction effect is reflected in the cell means that are plotted in Figure 14.2.

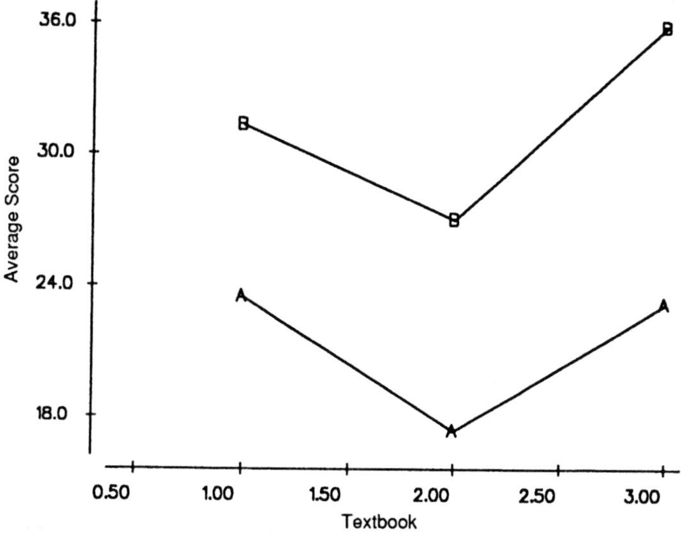

FIGURE 14.2 Plot of mean scores by methods of instruction on three textbooks. A, lecture method; B, Discussion method.

Interaction measures the degree of similarity between the responses to factor A at different levels of factor B. The lines connecting the three cell means for the discussion method are roughly parallel with the lines connecting cell means for the lecture method, reflecting the absence of interaction. If these two lines were not parallel or crossed each other, then the interaction effect would have been statistically significant. If a significant interaction is present, we need to examine the cell means carefully to draw appropriate conclusions.

Box 14.4 shows how to conduct a two-factor ANOVA with interaction by computer software for the above data. As shown in the box, a two-way ANOVA can be used with or without interaction, which suggests that we need to specify the model to be used in the analysis. The choice of a model is dependent on how the data are collected and how we consider each effect to be specified. We consider this modeling aspect of ANOVA in the next section.

MINITAB BOX 14.4

The test score data in Table 14.6 are entered in c1 along with the level designation of factor A (textbooks) in c2 and the level designation of factor B (methods of instruction) in c3. The command TWOWAY is used without the subcommand ADDITIVE in order to include interaction in the analysis.

```
MTB > twoway c1-c3;
SUBC> means c2 c3.

ANALYSIS OF VARIANCE  C1
SOURCE          DF         SS         MS
C2               2        343        171
C3               1        910        910
INTERACTION      2         36         18
ERROR           30       3100        103
TOTAL           35       4388

                          Individual 95% CI
      C2         Mean     --------+---------+---------+---------+---
       1         27.4               (-----------*-----------)
       2         22.2     (-----------*-----------)
       3         29.5                   (-----------*-----------)
                          --------+---------+---------+---------+---
                                20.0      25.0      30.0      35.0

                          Individual 95% CI
      C3         Mean     --------+---------+---------+---------+---
       1         21.3      (---------*---------)
       2         31.4                          (---------*---------)
                          --------+---------+---------+---------+---
                                20.0      25.0      30.0      35.0
```

VI. LINEAR MODEL REPRESENTATION OF THE ANOVA

In the ANOVA, we have partitioned the sum of squares of Y about its mean into within and between components in the completely randomized design or into treatment, blocking, and within components in the randomized block design. Underlying these partitions are linear models that show the relationship between the dependent variable and the independent (treatment and/or blocking) variables. In the following sections, we show these models as well as the model with interaction. From these models, we can also see that it is possible to extend the ANOVA method of analysis to include combinations of the independent variables as well as including more than two independent variables.

A. Linear Model for the Completely Randomized Design

One representation of the linear model underlying the completely randomized design shows the dependent variable being equal to a constant plus a treatment effect plus individual variation, that is,

$$y_{ij} = \mu + \alpha_i + \varepsilon_{ij}$$

for i ranging from 1 to r and j going from 1 to n_i. The value of the jth observation of the dependent variable at the ith treatment level is y_{ij}. There are r levels of the treatment variable and n_i observations of Y at the ith treatment level. The constant is represented by μ and the effect of the ith treatment level is represented by α_i. As not everyone who has received the ith level of treatment will have the same value of the dependent variable, this individual variation, the departure from the sum of μ plus α_i, is represented by ε_{ij}.

Note that this model can be rewritten as

$$y_{ij} = \mu + x_{ij} * \alpha_i + \varepsilon_{ij}$$

where x_{ij} is an indicator variable which has the value of 1 if the ijth subject has received the ith level of the treatment and zero otherwise. The x variable here simply indicates which level of treatment the person has received. We not use this representation of the model here, but refer to it in the next chapter.

In this linear model, there are $r + 1$ population parameters: the constant μ and the r α's; however, there are only r different treatment levels or groups. As we can only estimate the same number of parameters as there are groups, to obtain estimators for r of the parameters, we must make some assumption about them. The appendix on the linear model in Forthofer and Lehnen (8) provides a presentation of a number of assumptions that we could make. In this book, we measure the effect of the treatment levels from the effect of the rth treatment level. This means that α_r is assumed to be zero.

Now let us rewrite the linear model in terms of the population means. The equation for the ith level becomes

$$\mu_i = \mu + \alpha_i$$

and the representation of the model for all r levels is

$$\begin{aligned} \mu_1 &= \mu + \alpha_1 \\ \mu_2 &= \mu + \alpha_2 \\ &\vdots \\ \mu_{r-1} &= \mu + a_{r-1} \\ \mu_r &= \mu. \end{aligned}$$

From these equations, we can see that the constant term is the mean of the rth level, and the effects of the other levels, $\alpha_1, \alpha_2, \ldots, \alpha_{r-1}$, are measured from μ_r (or μ). For example, using the first of these equations to solve for α_1, we have

$$\alpha_1 = \mu_1 - \mu = \mu_1 - \mu_r.$$

This equation makes it clear that we are measuring the effects of the first level relative to the effect of the rth level and the same is true for levels 2 through $r - 1$.

The sample estimator of the ith effect, $\hat{\alpha}_i$, is obtained by substituting the sample means for the population means:

$$\hat{\alpha}_i = \bar{y}_{i\cdot} - \bar{y}_{r\cdot}$$

and the estimator of μ is simply $\bar{y}_{r\cdot}$.

The t test for comparing the means of two populations, assuming equal variances, also fits into the ANOVA framework. In this case, r is 2, and the above model still applies.

B. Linear Model for the Randomized Block Design with *m* Replicates

A linear model underlying the randomized block design has the dependent variable being equal to a constant plus the effect of the ith level of the treatment variable plus the jth block effect plus the individual variation term. In symbols, this is

$$y_{ijk} = \mu + \alpha_i + \beta_j + \varepsilon_{ijk}$$

where i goes from 1 to r, j ranges from 1 to c, and k ranges from 1 to m.

Just as in the completely randomized situation, the effects of the levels of the treatment variable are measured relative to the rth level of the treatment variable. In the same way, the effects of the levels of the blocking variable are measured relative to the cth level of the blocking variable.

The definition of the parameters in terms of the μ_{ij} is complicated and is not shown for this model, but is shown for the model in the next section.

C. Two-Way ANOVA with Interaction

The model for this situation is similar to the above two-way ANOVA model except that it includes the interaction term, denoted by $\alpha\beta_{ij}$, in the model. The model is

$$y_{ijk} = \mu + \alpha_i + \beta_j + \alpha\beta_{ij} + \varepsilon_{ijk}$$

where i goes from 1 to r, j ranges from 1 to c, and k ranges from 1 to m.

The main effect terms in the model, the α_i and the β_j, again are all measured relative to their last level. The representation of this model in terms of the cell means, the μ_{ij}, for the first row is

$$\begin{aligned}
\mu_{11} &= \mu + \alpha_1 + \beta_1 + \alpha\beta_{11} \\
\mu_{12} &= \mu + \alpha_1 + \beta_2 + \alpha\beta_{12} \\
&\cdots \\
&\cdots \\
&\cdots \\
\mu_{1c} &= \mu + \alpha_1.
\end{aligned}$$

Note that there is no β_c term nor any $\alpha\beta_{1c}$ terms in the last equation above. As the cth level is the reference level for the column variable, β_c is taken to be zero. In addition, interaction terms having either an r or a c as a subscript are reference levels and these interaction terms are also assumed to be zero. This pattern is repeated for the other rows except the last one.

$$\begin{aligned}
\mu_{21} &= \mu + \alpha_2 + \beta_1 + \alpha\beta_{21} \\
\mu_{22} &= \mu + \alpha_2 + \beta_2 + \alpha\beta_{22} \\
&\cdots \\
&\cdots \\
&\cdots \\
\mu_{2c} &= \mu + \alpha_2 \\
&\cdots \\
&\cdots \\
&\cdots \\
\mu_{r1} &= \mu \qquad\quad + \beta_1 \\
\mu_{r2} &= \mu \qquad\quad + \beta_2 \\
&\cdots \\
&\cdots \\
&\cdots \\
\mu_{rc} &= \mu
\end{aligned}$$

For the cells in the rth row, no α_r effect is shown as the rth level is the reference level for the row variable and α_r is taken to be zero. There are also no $\alpha\beta_{aj}$ terms in the last row, as the rth level is also a reference level for the interaction terms.

Using these equations, we obtain the following definitions of the parameters (μ, α, β, and $\alpha\beta$) in terms of the cell means. For example, from the last equation above, we see that the constant term in the model is simply the mean of the cell formed by the rth row and cth column, that is, $\mu = \mu_{rc}$.

Once we have expressed μ in terms of the cell means, we can find the estimate of α_i from the equation $\mu_{ic} = \mu + \alpha_i$. This gives the solution that $\alpha_i = \mu_{ic} - \mu_{rc}$ where we have replaced μ_{rc} for μ. This definition for α_i is reasonable as it compares the mean of the cell in the ith row and cth column with the mean of the cell in the rth row and cth column. It is comparing a cell in the ith row with its reference cell in the rth row. The column effect, β_j, is similarly defined as $\beta_j = \mu_{rj} - \mu_{rc}$.

The definition of the interaction term is $\alpha\beta_{ij} = (\mu_{ij} - \mu_{ic}) - (\mu_{rj} - \mu_{rc})$. The rcth cell is the reference cell and the other parameters are defined in terms of it. The ijth interaction parameter focuses on the difference of the jth and cth columns, and compares that difference for the ith and rth rows. If there is no interaction, the difference of the jth and cth columns is the same over all the rows.

VII. CONCLUDING REMARKS

In this chapter we presented several basic models of analysis of variance. One-way ANOVA is used to analyze data from a completely randomized experimental design. Two-way ANOVA can be used for a randomized block design as well as for a two-factor design with interaction. To use these analytical methods properly, we need to be aware of how the data are collected and to make sure that the data collection design warrants the assumptions involved in the ANOVA. In the next chapter, we expand the linear model to regression models.

EXERCISES

14-1. The data shown below, taken from Brogan and Kutner (9), are the change in the maximal rate of urea synthesis (MRUS) level for cirrhotic patients who underwent either a standard operation (a nonselective shunt) or a new procedure (a selective shunt). The purpose of the operations was to improve liver function, measured by MRUS. A low value of MRUS is associated with poor liver function. Patients in the nonselective shunt group are divided into two groups based on their preoperative MRUS values (≤ 40 and > 40).

Perform an analysis of variance of these data at the 0.05 significance level to determine if there is a difference in the three groups. If there is a significant difference, use an appropriate multiple comparison procedure to find the source of the difference.

Group	Change in MRUS (mg urea N/hr/kg body weight) by group							
Selective shunt	−3	20	−6	−5	−3	−3	−6	12
Nonselective shunt I	−18	−4	−18	−18	−6	−18		
Nonselective shunt II	−24	−7	−15	4	−14	−8	−11	

14.2. In Chapter 13, we used the *t* test to compare the proportions of caloric intake from fat for fifth and sixth grade boys with those for seventh and eighth grade boys. The calculated *t*-test statistic was 0.727. Perform a one-way ANOVA on the data in Table 7.5 and compare your results with the *t*-test approach. How does the *t* statistic compare with the *F* statistic?

14.3. For the weight change data shown in Table 14.4, we were concerned about the level of physical activity of the women. Instead of using the site (headquarters or plant) as a way of controlling for physical activity, how else might we have controlled for the physical activity? Do you think that a control group (no intervention) should have been used? Explain your reasoning. Would you do anything to determine whether the women used the memberships? What, if any, other variables should be included in the analysis?

14.4. To investigate publication bias, 75 referees for one journal were randomly assigned to receive one of five versions of a manuscript (10). All versions were identical in the Introduction and Methods sections but varied in either the Results or Discussion sections. The first and second groups received versions with either positive or negative results, respectively. The third and fourth groups received versions with mixed results and either positive or negative discussion. The fifth group was asked to evaluate the manuscript on the basis of the Methods section and no data were provided. The referees used a scale of 0 to 6 (low to high) to rate different aspects of the manuscript. The average scores for three aspects are shown below:

		Mean rating		
Manuscript version	Number of referees	Methods	Scientific contribution	Publication merit
Positive results	12	4.2	4.3	3.2
Negative results	14	2.4	2.4	1.8
Mixed results with positive discussion	13	2.5	1.6	0.5
Mixed results with negative discussion	14	2.7	1.7	1.4
Methods only	14	3.4	4.5	3.4

State an appropriate linear model for this experiment using scientific contribution as the dependent variable. What are the null and alternative hypotheses of interest for this model? Assuming that the standard deviations for the scientific contribution score for the five groups are 1.1, 0.9, 0.7, 0.8, and 1.1, respectively, perform an analysis of variance of these data at the 0.05 significance level to determine if there is a bias in refereeing scientific papers for this journal. If there is a significant difference, use an appropriate multiple comparison procedure to find the source of the bias. State your conclusions clearly.

14.5. In an investigation of the effect of smoking on work performance under different lighting conditions in a large company, a random sample of nine male workers was selected from each of the three smoking status groups: nonsmokers, moderate smokers, and heavy smokers. Each sample was randomly assigned to three working environments with different levels of lighting. The time to complete a standard assembling task was recorded in minutes. The sums of squares were as follows:

Source	*Degrees of freedom*	*Sum of squares*	*Mean square*	*F*
Smoking status		84.90		
Lighting conditions		298.07		
Interaction		2.81		
Error		59.25		
Total		445.03		

Perform an analysis of variance for these data to examine the interaction of the variables at the 0.05 significance level. If there is no significant interaction, test whether the smoking and lighting condition variables have significant effects on the workers' performance and state your conclusions.

REFERENCES

1. Scheffé, H. (1959). "The Analysis of Variance." Wiley, New York.
2. Kleinbaum, D. G., Kupper, L. L., and Muller, K. E. (1988). "Applied Regression Analysis and Other Multivariable Methods," 2nd ed. PWS-Kent, Boston.
3. Lin, L. I., and Vonesh, E. F. (1989). An empirical nonlinear data-fitting approach for transforming data to normality. *Am. Stat.* **43,** 237–243.
4. Steel, R. G. D., and Torrie, J. H. (1980). "Principles and Procedures of Statistics: A Biometrical Approach." McGraw-Hill, New York.

5. Kimball, B. P., Shurvell, B. L., Houle, S., Fulop, J. C., Rakowski, H., and McLaughlin, P. R. (1986). Persistent ventricular adaptations in postoperative coarctation of the aorta. *J. Am. Coll. Cardiol.* **8,** 172–178.
6. Stoline, M. R. (1981). The status of multiple comparisons: Simultaneous estimation of all pairwise comparisons in one-way ANOVA designs. *Am. Stat.* **35,** 134–141.
7. Higgins, J. E., and Koch, G. G. (1977). Variable selection and generalized chi-square analysis of categorical data applied to a large cross-sectional occupational health survey. *Int. Stat. Rev.* **45,** 51–62.
8. Forthofer, R. N., and Lehnen, R. G. (1981). "Public Program Analysis: A New Categorical Data Approach." Lifetime Learning Publications, Belmont, CA.
9. Brogan, D. R., and Kutner, M. H. (1980). Comparative analyses of pretest-posttest research designs. *Am. Stat.* **34,** 229–232.
10. Dickersin, K. (1990). The existence of publication bias and risk factors for its occurrence. *JAMA, J. Am. Med. Assoc.* **263,** 1385–1389.

CHAPTER 15

Linear and Logistic Regression

In this chapter we present methods for examining the relationship between a response or dependent variable and one or more predictor or independent variables. The methods are based on the linear model introduced in Chapter 14. In linear regression, we examine the relationship between a normally distributed response or dependent variable and one or more continuous predictor or independent variables. In a sense, linear regression is an extension of the correlation coefficient. Although linear regression was created for the examination of the relationship between continuous variables, in practice, people often use the term *linear regression* even when continuous and discrete independent variables are used in the analysis. In logistic regression, we examine the relationship between a binary dependent variable and one or more independent variables.

Linear regression and logistic regression are two of the more frequently used techniques used in statistics today. These methods are often used because problems, particularly those concerning humans, usually involve several independent variables. For example, in the creation of norms for lung functioning, age, race, and sex are taken into account. Linear regression is one approach that allows multiple independent variables to be used in the analysis. In the linear regression model, the dependent variable is the observed pulmonary function test value and age, race, and sex are the independent variables. As another example, in an attempt to identify risk factors—variables related to the occurrence of the disease—many variables are considered. Logistic regression is an approach that allows many possible risk factors to be considered simultaneously. In logistic regression, the dependent variable is disease status (presence or absence) and the potential risk factors are included as the independent variables.

I. SIMPLE LINEAR REGRESSION

Simple linear regression is used to examine the relationship between a normally distributed dependent variable and a continuous independent variable. An example of a situation in which simple linear regression is useful is the following.

Some physicians believe that there should be a standard, a value that only a small percentage of the population exceeds, for blood pressure in children (1). When a standard is used, it is desirable that it be easy for the physician to determine quickly and accurately how the patient relates to the standard. Therefore, the standards should be based on a small number of variables that are easy to measure. As it is known that blood pressure is related to maturation, the variables used in the development of the standard should therefore reflect maturation. Two variables that are related to maturation and are easy to measure are age and height. Of these two variables, height appears to be the more appropriate variable for the development of standards (2–4). Because of physiological differences, the standards are developed separately for females and males. In the following, we focus on systolic blood pressure (SBP).

In developing the standards, we are going to assume that the mean SBP for girls increases by a constant amount for each one-unit increase in height. The use of the mean instead of the individual SBP values reflects the fact that there is variation in the SBP of girls of the same height. Not all the girls who are 50 in. tall have the same SBP; their SBPs vary about the mean SBP of girls who are 50 in. tall. The assumption of a constant increase in the mean SBP for each one-unit increase in height is characteristic of a linear relationship. Thus, in symbols, the relationship between Y, the SBP variable, and X, the height variable, can be expressed as

$$\mu_{Y|X} = \beta_0 + X * \beta_1$$

where $\mu_{Y|X}$ is the mean SBP for girls who are X units tall, β_0 is a constant term, and β_1 is the coefficient of the height variable. The β_0 coefficient is the Y intercept and β_1 is the slope of the straight line.

In general, the X variable shown in the above expression may represent the square, the reciprocal, the logarithm, or some other nonlinear transformation of a variable. This is acceptable in linear regression because the expression is really a linear combination of the β_i's, not of the independent variables.

The above equation is similar to the linear model representation of ANOVA. In the ANOVA model, values of the X variables, 1 or 0, indicate which effect should be added in the model. In the regression model, the values of the X variable are the individual observations of the continuous independent variable. The parameters in the ANOVA model are the effects of the different levels of the independent variable. In the regression model, the parameters are the Y intercept and the slope of the line.

Figure 15.1 shows the graph of this simple linear regression equation. The ⊗ symbols show the values of the mean SBP for the different values of height we are considering. As we can see, a straight line does indeed have a rate of increase in the mean SBP that is constant for each one-unit increase in height. The ■ symbols show the projected values of the mean SBP assuming that the relationship holds for very small height values as well. It is usually inappropriate to estimate the values of $\mu_{Y|X}$ for values of X outside the range of observation. The point at which the projected line intersects the $\mu_{Y|X}$ axis is β_0. As β_1 is the amount of increase in $\mu_{Y|X}$ for each one-unit increase in X, the bracketed change in $\mu_{Y|X}$ is 8 β_1 because X has increased 8 units from x_1 to x_2. Note that if the regression line is flat, that is, parallel to the X axis, there is no change in $\mu_{Y|X}$ regardless of how much X changes. Thus, if the regression line is flat, then β_1 is zero and there is no linear relationship between $\mu_{Y|X}$ and X.

If we wish to express this relationship in terms of individual observations, we must take the variation in SBP for each height into account. The model that does this is

$$y_i = \beta_0 + x_i\beta_1 + \varepsilon_i$$

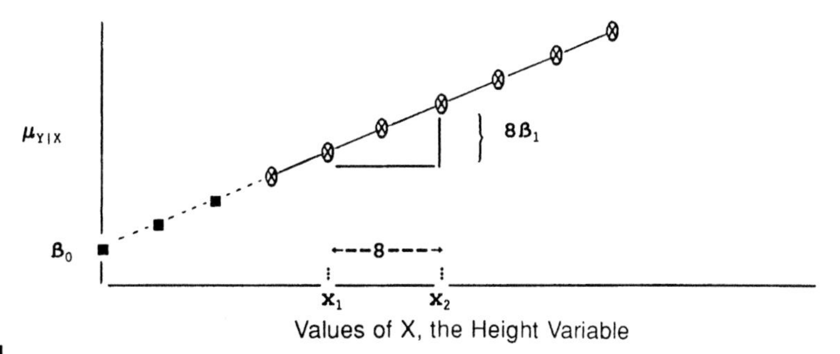

FIGURE 15.1 Line showing the regression of $\mu_{Y|X}$ on X.

where ε_i represents the difference between the mean SBP value at height x_i and the SBP of the ith girl who is also x_i units tall. The ε term is also referred to as the residual or error term.

Knowledge of β_0 and β_1 is necessary in developing the standards for SBP; however, we do not know them and we have to collect data to estimate these values.

II. ESTIMATION OF THE COEFFICIENTS

There are a variety of ways of estimating β_0 and β_1. We must decide on what criterion we will use to find the "best" estimators of these two coefficients. Possible criteria include minimization of the following:

1. The sum of the differences of y_i and $\hat{y}_i$, where y_i is the observed value of the SBP and $\hat{y}_i$ is the estimated value of the SBP for the ith girl. The value of $\hat{y}_i$ is found by substituting the estimates of β_0 and β_1 in the simple linear regression equation, that is, $\hat{y}_i = \hat{\beta}_0 + x_i\hat{\beta}_1$, where x_i is the observed value of height for the ith girl.
2. The sum of the absolute differences of y_i and $\hat{y}_i$.
3. The sum of the squared differences of y_i and $\hat{y}_i$.

Based on considerations similar to those discussed in Chapter 4 in the presentation of the variance, we are going to use the third criterion to determine our "best" estimators.[1]

Thus our estimators of the coefficients will be derived based on the minimization of the sum of squares of the differences of the observed and estimated values of SBP. In symbols, this is the minimization of

$$\sum_i (y_i - \hat{y}_i)^2.$$

The use of this criterion provides estimators that are called *least squares estimators* because they minimize the sum of squares of the differences.

The least squares estimators of the coefficients are given by

$$\hat{\beta}_1 = \frac{\sum_{i=1}^{n} (x_i - \bar{x})(y_i - \bar{y})}{\sum_{i=1}^{n} (x_i - \bar{x})^2} = \frac{\sum_{i=1}^{n} x_i y_i - n\bar{x}\bar{y}}{\sum_{i=1}^{n} x_i^2 - n\bar{x}^2}$$

and

$$\hat{\beta}_0 = \bar{y} - \hat{\beta}_1\bar{x}.$$

[1]The first criterion can be made to equal zero by setting $\hat{\beta}_1$ to zero and letting $\hat{\beta}_0$ equal the sample mean. The use of the absolute value yields interesting estimators, but the testing of hypotheses is more difficult with these estimators.

The second formula for $\hat{\beta}_1$ is provided because it is easier to calculate. Let us use these formulas to calculate the least squares estimates for the data in Table 15.1. The hypothetical values of the SBP and height variables for the 50 girls are based on data from NHANES II (4).

The value of $\hat{\beta}_1$ is found from

$$\hat{\beta}_1 = \frac{\sum_{i=1}^{n} x_i y_i - n\bar{x}\bar{y}}{\sum_{i=1}^{n} x_i^2 - n\bar{x}^2} = \frac{269{,}902 - 50 * 52.5 * 101.5}{142{,}319 - 50 * 52.5^2} = 0.7688.$$

The calculation of $\hat{\beta}_0$ is easier to perform, and its value is found from

$$\hat{\beta}_0 = \bar{y} - \hat{\beta}_1\bar{x} = 101.5 - 0.7688 * 52.5 = 61.138.$$

The estimated coefficient of the height variable is about 0.8. This means that there is an increase of 0.8 mm Hg in SBP for an increase of 1 in. in height for girls between 36 and 69 in. The estimate of the β_0 coefficient is about 60 mm Hg and that is the Y intercept. When the regression line is projected beyond the data values observed, the Y intercept gives the value of SBP for a girl 0 in. tall; however, it does not make sense to talk about the SBP for a girl 0 in. tall, and this shows one of the dangers of extrapolating the regression line beyond the observed data.

TABLE 15.1 Hypothetical Data: SBP and Predicted SBP[a] (mm Hg) and Height (in.) for 50 Girls

SBP	Predicted SBP	Height	SBP	Predicted SBP	Height	SBP	Predicted SBP	Height
105	88.8	36	120	98.0	48	94	106.5	59
90	89.6	37	114	98.8	49	88	107.3	60
82	90.4	38	78	98.8	49	110	107.3	60
96	90.4	38	116	99.6	50	124	107.3	60
82	91.1	39	74	99.6	50	86	108.0	61
74	91.1	39	80	100.3	51	120	108.0	61
104	91.9	40	98	101.1	52	112	108.8	62
100	91.9	40	90	101.9	53	100	109.6	63
80	92.7	41	92	102.7	54	122	110.3	64
98	93.4	42	80	102.7	54	122	110.3	64
96	94.2	43	88	102.7	54	110	111.1	65
86	95.0	44	104	103.4	55	124	111.1	65
88	95.0	44	100	104.2	56	122	111.9	66
128	95.0	44	126	105.0	57	94	112.6	67
118	95.7	45	108	105.7	58	110	112.6	67
90	96.5	46	106	106.5	59	140	114.2	69
108	98.0	48	98	106.5	59			

[a]Predicted using the least squares estimates of the regression coefficients.

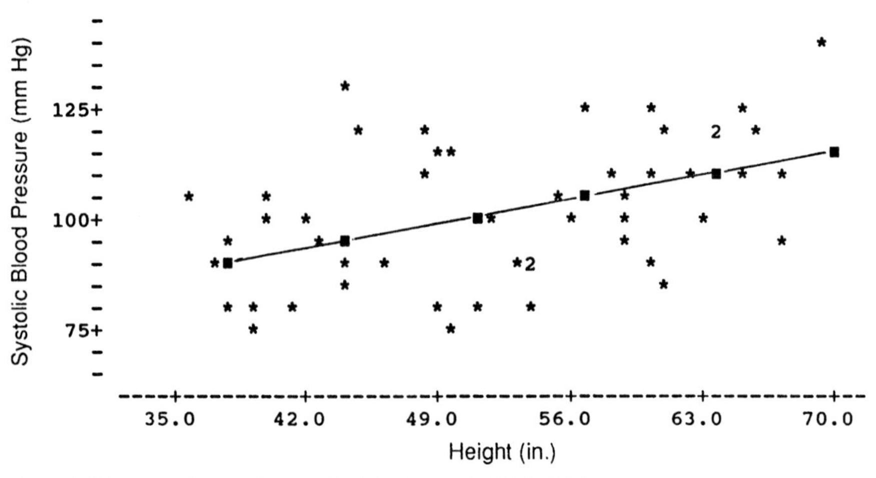

FIGURE 15.2 Plot of SBP versus height for the 50 girls shown in Table 15.1.

Figure 15.2 is a plot of SBP versus height for the data shown in Table 15.1. From this plot, we can see that there is a slight tendency for the larger values of SBP to be associated with the larger values of height, but the relationship is not particularly strong. The squares show the path of the regression line.

We can use the above estimates of the population coefficients in predicting SBP values for the hypothetical data shown in Table 15.1. For example, the predicted value of SBP for the first observation in Table 15.1, a girl 36 in. tall, is

$$61.14 + 36 * 0.7688 = 88.82 \text{ mm Hg}.$$

The other predicted SBP values are found in the same way and they are also shown in Table 15.1.

III. VARIANCE OF $Y|X$

Before going forward with the use of the regression line in the development of the standards, we should examine whether the estimated regression line is an improvement over simply using the sample mean as an estimate of the observed values. One way of obtaining a feel for this is to examine the sum of squares of deviations of Y from $\hat{Y}$, that is,

$$\sum_{i=1}^{n} (y_i - \hat{y}_i)^2.$$

If we subtract and add $\bar{y}$ in this expression, we can rewrite this sum of squares as

$$\sum_{i=1}^{n} [(y_i - \bar{y}) - (\hat{y}_i - \bar{y})]^2$$

and we have not changed the value of the sum of squares; however, this sum of squares can be rewritten as

$$\sum_{i=1}^{n} (y_i - \hat{y}_i)^2 = \sum_{i=1}^{n} (y_i - \bar{y})^2 - \sum_{i=1}^{n} (\hat{y}_i - \bar{y})^2$$

because the crossproduct terms, $(y_i - \bar{y})(\hat{y}_i - \bar{y})$, sum to zero. In regression terminology, the first sum of squares is called the *sum of squares about regression* or the *residual or error sum of squares*. The second sum of squares, about the sample mean, is called the *total sum of squares* (corrected for the mean), and the third sum of squares is called the *sum of squares due to regression*. If we rewrite this equation, putting the total sum of squares (corrected for the mean) on the left side of the equal sign, we have

$$\sum_{i=1}^{n} (y_i - \bar{y})^2 = \sum_{i=1}^{n} (y_i - \hat{y}_i)^2 + \sum_{i=1}^{n} (\hat{y}_i - \bar{y})^2.$$

This equation shows the partition of the total sum of squares into two components, the sum of squares about regression and the sum of squares due to regression.

Figure 15.3 is a graph that shows the differences, $(y_i - \bar{y})$, $(y_i - \hat{y}_i)$, and $(y_i - \bar{y})$, for one y_i. In Figure 15.3, the crosses are points representing the regression line and the horizontal line is the value of the sample mean. We have focused on the last point, the girl who is 69 in. tall and who has a SBP of 140 mm Hg. For this point, the deviation of the observed SBP of 140 from the sample mean of 101.5 can be partitioned into two components. The first component is the difference between the observed value and 114.2, the value predicted from the regression line. The second component is the difference between this predicted value and the sample mean. This partitioning cannot be done for many of the points because, for example, the sample mean may be closer to the observed point than the regression line is.

Ideally, we would like the sum of squares about the regression line to be close to zero. From the last equation above, we see that the sum of the square deviations from the regression line must be less than or equal to the sum of the square deviations from the sample mean; however, the direct comparison of the sum of squares is not fair as they are based on different degrees of freedom. The sum of squares about the sample mean has $n - 1$ degrees of freedom as we discussed in the material about the variance.

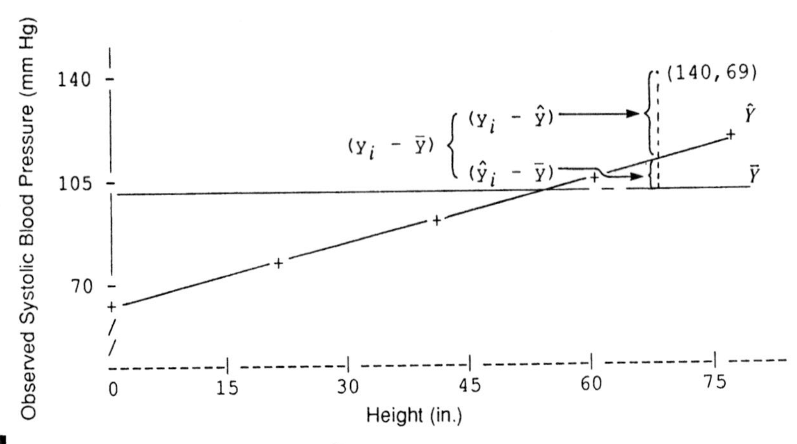

FIGURE 15.3 Blowup of part of the regression line.

Because we estimated two coefficients in obtaining the least squares estimator of Y, there are thus $n - 2$ degrees of freedom associated with sum of squares about $\hat{Y}$. Thus, let us compare s_Y^2 with $s_{Y|X}^2$, that is,

$$\frac{\sum_{i=1}^{n} (y_i - \bar{y})^2}{n - 1} \quad \text{versus} \quad \frac{\sum_{i=1}^{n} (y_i - \hat{y}_i)^2}{n - 2}.$$

If $s_{Y|X}^2$ is much less than s_Y^2, then the regression was worthwhile; if not, then we should use the sample mean as there appears to be little linear relationship between Y and X.

Let us calculate the sample variance of Y taking X into account. The sample variance of Y is

$$\frac{\sum_{i=1}^{50} (y_i - \bar{y})^2}{50 - 1} = \frac{12{,}780}{49} = 260.827$$

and the sample variance of Y given X is

$$\frac{\sum_{i=1}^{50} (y_i - \hat{y}_i)^2}{50 - 2} = \frac{10{,}117}{48} = 210.772.$$

Thus $s_{Y|X}^2$ is less than s_Y^2. The use of the height variable has reduced the sample variance from 260.827 to 210.772, about a 20 percent reduction. It appears that the inclusion of the height variable has allowed for somewhat better estimation of the SBP values.

IV. R^2, THE COEFFICIENT OF DETERMINATION

An additional way of examining whether the regression was helpful is to divide the ratio of the sum of squares due to regression by the sum of squares about the mean, that is,

$$\frac{\sum_{i=1}^{n} (\hat{y}_i - \bar{y})^2}{\sum_{i=1}^{n} (y_i - \bar{y})^2} = \frac{\sum_{i=1}^{n} (y_i - \bar{y})^2 - \sum_{i=1}^{n} (y_i - \hat{y}_i)^2}{\sum_{i=1}^{n} (y_i - \bar{y})^2}.$$

If the regression line provides estimates of the SBP values that closely match the observed SBP values, this ratio will be close to one. If the regression line is close to the mean line, then this ratio will be close to zero. Hence, the ratio provides a measure that varies from 0 to 1, with 0 indicating no linear relationship between Y and X, and 1 indicating a perfect linear relationship between Y and X. This ratio is denoted by R^2 and is called the coefficient of determination. It is a measure of how much of the variation in Y is accounted for by X. R^2 is also the square of the sample Pearson correlation coefficient between Y and X.

For the SBP example, the value of R^2 is

$$\frac{12{,}780 - 10{,}117}{12{,}780} = 0.2084$$

Approximately 21 percent of the variation in SBP is accounted for by height for girls between 36 and 69 in. tall. This is not an impressive amount. Almost 80 percent of the variation in SBP remains to be explained. Even though this measure of the relationship between SBP and height is only 21 percent, it is larger than its corresponding value for the relationship between SBP and age.

The derivation of the R^2 term is based on a linear model which has both a β_0 term and a β_1 term. If the model does not include β_0, then a different expression must be used to calculate R^2.

The sample Pearson correlation coefficient, r, is defined as

$$\frac{\sum_{i=1}^{n} (x_i - \bar{x})(y_i - \bar{y})}{\sqrt{\sum_{i=1}^{n} (x_i - \bar{x})^2 \sum_{i=1}^{n} (y_i - \bar{y})^2}}$$

and its numerical value is

$$r = \frac{3464.5}{\sqrt{4506.6 * 12780}} = 0.4565.$$

If we square r, r^2 is 0.2084 which agrees with R^2 as it must.

Although, symbolically, R^2 is the square of the sample Pearson correlation coefficient, R^2 does not necessarily measure the strength of the linear association between Y and X. In correlation analysis, the observed pairs of values of Y and X are obtained by simple random sampling from a population. Neither variable is thought to be dependent on the other and r measures the strength of the linear association between the two variables. In contrast, linear regression provides a formula that describes the linear relationship between a dependent variable and an independent variable(s). To discover that relationship, we often use stratified random sampling, that is, we select SRSs of Y for specified values of X; however, as Ranney and Thigpen (5) show, the value of R^2 depends on the range of the values of X used in the analysis, the number of repeated observations at given values of X, and the location of the X values. Hence, although symbolically R^2 is the square of the correlation coefficient between two variables, it does not necessarily measure the strength of the linear association between the variables. It does reflect how much of the variation in Y is accounted for by knowledge of X. Korn and Simon provide more on the interpretation of R^2 (6).

There is also a relationship between the sample correlation coefficient and the estimator of β_1. From Chapter 4, we had another form for r than the defining formula given above and it was

$$r = \frac{\Sigma\,(x_i - \bar{x})(y_i - \bar{y})/(n - 1)}{s_x * s_y}.$$

The estimator of β_1 is

$$\hat{\beta}_1 = \frac{\Sigma\,(x_i - \bar{x})(y_i - \bar{y})}{\Sigma\,(x_i - \bar{x})^2} = \frac{\Sigma\,(x_i - \bar{x})(y_i - \bar{y})/(n - 1)}{s_x^{\,2}}.$$

If we multiply r by s_y and divide r by s_x, we have

$$\frac{s_y}{s_x} * r = \frac{s_y}{s_x} * \frac{\Sigma\,(x_i - \bar{x})(y_i - \bar{y})/(n - 1)}{s_x * s_y}$$

or

$$\frac{s_y}{s_x} * r = \frac{\Sigma\,(x_i - \bar{x})(y_i - \bar{y})/(n - 1)}{s_x^{\,2}} = \hat{\beta}_1.$$

As the above relationship shows, if the correlation coefficient is zero, the slope coefficient is also zero and vice versa.

V. INFERENCE ABOUT THE COEFFICIENTS

The parametric approach to testing hypotheses about a parameter requires that we know the probability distribution of the sample estimator of the parameter. The standard approach to finding the probability distributions

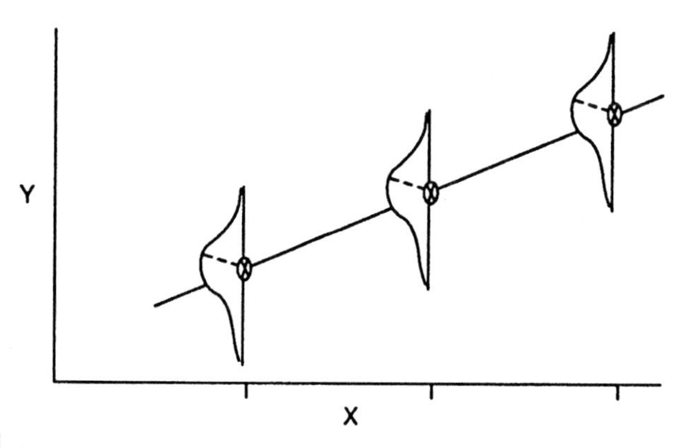

FIGURE 15.4 Distribution of Y at selected values of X.

of the sample estimators of β_0 and β_1 is based on the following assumptions.

A. Assumptions for Inference in Linear Regression

We assume that the y_i's are independent, normally distributed for each value of X, and that the normal distributions at the different values of X all have the same variance, σ^2. Figure 15.4 graphically shows these assumptions. The regression line, showing the relationship between $\mu_{Y|X}$ and X is graphed as are the distributions of Y at the selected values of X. Note that Y is normally distributed at each of the selected X values and that the normal distributions have the same shapes, that is, the same variance, σ^2. The mean of the normal distribution, $\mu_{Y|X}$, is obtained from the regression equation and is $\beta_0 + \beta_1 * X$.

In the following, we consider the values of the X variable to be fixed. This is the conventional approach and it means that the error or residual term, ε, also follows a normal distribution with mean 0 and variance σ^2.[2] Note that the least squares estimation of the regression coefficients did not require this specification of the probability distribution of Y.

Before testing hypotheses about the regression coefficients, we should attempt to determine whether the assumptions stated above are true. We should also examine whether any single data point is exercising a large influence on the estimates of the regression coefficients. These two issues are discussed in the next section.

[2]Two ways X can be viewed as being fixed are the following. First, we may have used a stratified sample, stratified on height, to select girls with the heights shown in Table 15.1. Because we have chosen the values of the height variable, they are viewed as being fixed. In a second way, we consider our results to be conditional on the observed values of X. The conditional approach is usually used with simple random samples in which both Y and X otherwise would be considered to be random variables.

B. Regression Diagnostics

In our brief introduction to regression diagnostics—methods for examining the regression equation—we consider only two of the many methods that exist. More detail on other methods is given in Kleinbaum *et al.* (7). The first method we present involves plotting of the residuals. Plots are used in an attempt to determine whether the residuals or errors are normally distributed or to see if there are any patterns in the residuals. The second method tries to discover the existence of data points that play a major role in the estimation of the regression coefficients.

1. Residuals and Standardized Residuals

The sample estimator of ε_i is the residual e_i, defined as the difference between y_i and $\hat{y}_i$, and the e_i can be used to examine the regression assumptions. As we are used to dealing with standardized variables, people often consider a standardized residual, $e_i/s_{Y|X}$, instead of e_i itself. The standardized residuals should approximately follow a standard normal distribution

TABLE 15.2 Residuals and Leverage for the Data in Table 15.1

Y	Residual	Standardized residual	h_i leverage	Y	Residual	Standardized residual	h_i leverage
105	16.1848	1.16253	0.08041	92	−10.6532	−0.74143	0.02049
90	0.4161	0.02977	0.07331	80	−22.6532	−1.57659	0.02049
82	−8.3527	−0.59552	0.06665	88	−14.6532	−1.01982	0.02049
96	5.6473	0.40264	0.06665	104	0.5781	0.04025	0.02138
82	−9.1215	−0.64818	0.06044	100	−4.1907	−0.29199	0.02271
74	−17.1215	−1.21667	0.06044	126	21.0405	1.46735	0.02449
104	12.1097	0.85790	0.05467	108	2.2717	0.15861	0.02671
100	8.1097	0.57452	0.05467	106	−0.4971	−0.03475	0.02937
80	−12.6590	−0.89430	0.04934	98	−8.4971	−0.59407	0.02937
98	4.5722	0.32218	0.04446	94	−12.4971	−0.87373	0.02937
96	1.8034	0.12678	0.04002	88	−19.2658	−1.34912	0.03248
86	−8.9654	−0.62897	0.03603	110	2.7342	0.19146	0.03248
88	−6.9654	−0.48866	0.03603	124	16.7342	1.17184	0.03248
128	33.0346	2.31756	0.03603	86	−22.0346	−1.54585	0.03603
118	22.2658	1.55920	0.03248	120	11.9654	0.83944	0.03603
90	−6.5029	−0.45465	0.02937	112	3.1966	0.22473	0.04002
108	9.9595	0.69457	0.02449	100	−9.5722	−0.67450	0.04446
120	21.9595	1.53144	0.02449	122	11.6590	0.82366	0.04934
114	15.1907	1.05843	0.02271	122	11.6590	0.82366	0.04934
78	−20.8093	−1.44991	0.02271	110	−1.1097	−0.07862	0.05467
116	16.4219	1.14344	0.02138	124	12.8903	0.91320	0.05467
74	−25.5781	−1.78096	0.02138	122	10.1215	0.71924	0.06044
80	−20.3468	−1.41608	0.02049	94	−18.6473	−1.32950	0.06665
98	−3.1156	−0.21679	0.02005	110	−2.6473	−0.18874	0.06665
90	−11.8844	−0.82693	0.02005	140	25.8152	1.85426	0.08041

if the regression assumptions are met. Thus, values of the standardized residuals larger than 2.5 or less than −2.5 are unusual. Table 15.2 shows these residuals and a quantity called leverage (described in the next section) for the data in Table 15.1.

We use the standardized residuals in our examination of the normality assumption. Other residuals could also be used for this examination (7). The normal scores of the standardized residuals are plotted in Figure 15.5.

The normal scores plot looks reasonably straight; thus the assumption that the error term is normally distributed does not appear to be violated.

If this plot deviates sufficiently from a straight line to cause us to question the assumption of normality, then it may be necessary to consider a transformation of the dependent variable. A number of mathematical functions can be used to transform nonnormally distributed data to normality (7–9).

It is also of interest to plot the standardized residuals against the values of the X variable(s). Observation of any pattern in this plot suggests that another term involving the X variable, for example, X^2, might be needed in the model. Figure 15.6 shows the plot of the standardized residuals versus the height variable. No pattern is immediately obvious from an examination of this plot. Again there is no evidence to cause us to reject this model. If the data have been collected in time sequence, it is also useful to examine a plot of the residuals against time.

2. Leverage

The predicted values of Y are found from

$$\hat{\beta}_0 + \hat{\beta}_1 * X$$

where the estimators of β_0 and β_1 are linear combinations of the observed

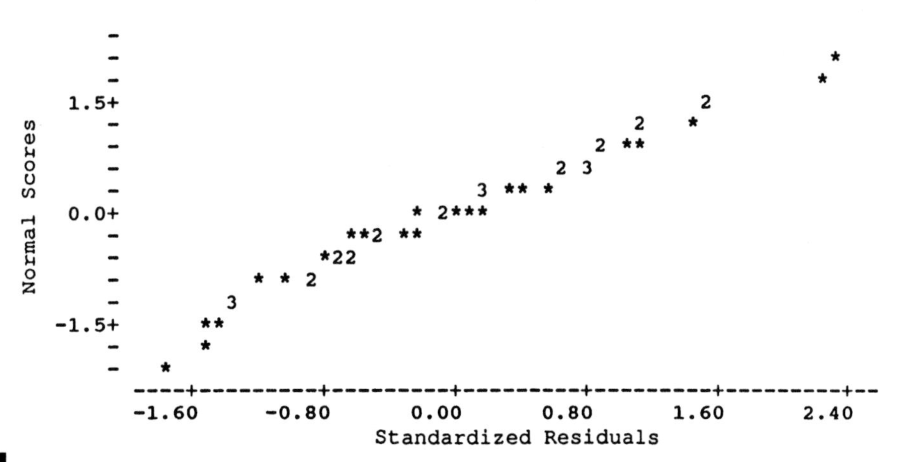

FIGURE 15.5 Normal scores plot of the standardized residuals from the linear regression of SBP on height.

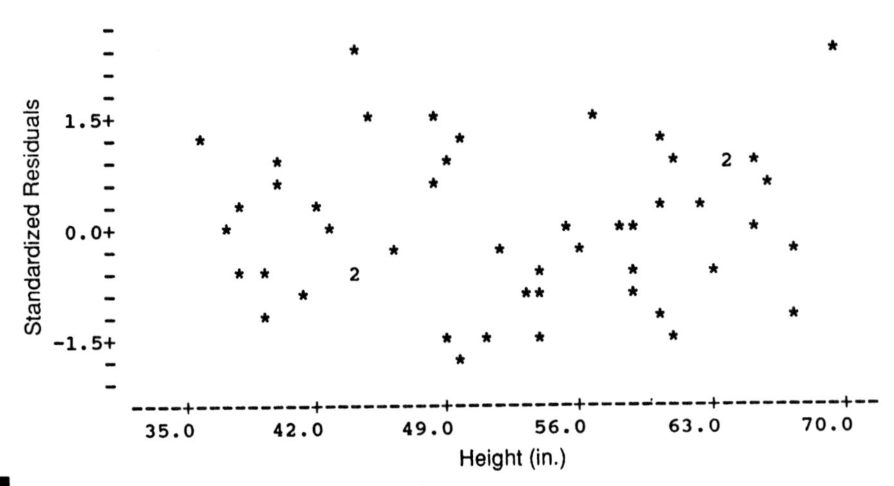

FIGURE 15.6 Plot of standardized residuals versus height.

values of Y. Thus the predicted values of Y are also linear combinations of the observed values of Y. An expression for the predicted value of y_i reflecting this relationship is

$$\hat{y}_i = h_{i1} * y_1 + h_{i2} * y_2 + \ldots + h_{ii} * y_i + \ldots + h_{in} * y_n$$

where h_{ij} is the coefficient of y_j in the expression for $\hat{y}_i$. For simplicity, h_{ii} is denoted by h_i. The effect of y_i on its predicted value is denoted by h_i and this effect is called *leverage*. Leverage shows how much change there is in the predicted value of y_i per unit change in y_i. The possible values of the h_i are greater than or equal to zero and less than or equal to one. The average value of the leverages is the number of estimated coefficients in the regression equation divided by the sample size. In our problem, we estimated two coefficients and there were 50 observations. Thus the average value of the leverages is 0.04 (= 2/50). If any of the leverages are large—some statisticians consider large to be greater than twice the average leverage and others say greater than three times the average—the points with these large leverages should be examined. Perhaps there was a mistake in recording the values or there is something unique about the points that should be examined. If there is nothing wrong or unusual with the points, it is useful to perform the regression again excluding these points. A comparison of the two regression equations can be made, and the effect of the excluded points can be observed.

In our problem, we can see from Table 15.2 that there are two points, the first and the last, with the larger leverages. Both of these points had leverages slightly larger than twice the average leverage value. The first girl had a large SBP value relative to her height and the last girl had the highest SBP value. At this stage, we assume that there was no error in recording or entering the data. We could perform the regression again, and see if there is much difference in the results; however, because the leverages are only slightly larger than twice the average leverage, we do not perform any additional regressions.

Based on these looks at the data, we have no reason to doubt the appropriateness of the regression assumptions and there do not appear to be any really unusual data points that would cause us concern. Therefore, it is appropriate to move into the inferential part of the analysis, that is, to test hypotheses and to form confidence and prediction intervals. We begin the inferential stage with consideration of the slope coefficient.

C. Slope Coefficient

Even though there is an indication of a linear relationship between SBP and height, that is, it appears that β_1 is not zero, we do not know if β_1 is statistically significantly different from zero. To determine this, we must estimate the standard error of $\hat{\beta}_1$ which is used in both confidence intervals and tests of hypotheses about β_1. To form the confidence interval about β_1 or to test a hypothesis about it, we also must know the probability distribution of $\hat{\beta}_1$.

As we are assuming that Y is normally distributed, this means that $\hat{\beta}_1$, a linear combination of the observed Y values, is also normally distributed. Therefore, to form a confidence interval or to test a hypothesis about β_1, we now need to know the standard error of its estimator. The standard error (s.e.) of $\hat{\beta}_1$ is

$$\text{s.e.}(\hat{\beta}_1) = \frac{\sigma}{\sqrt{\sum_{i=1}^{n} (x_i - \bar{x})^2}}$$

and, because σ is usually unknown, the standard error is estimated by substituting $s_{Y|X}$ for σ. From the above equation, we can see that the magnitude of the standard error depends on the variability in the X variable. Larger variability decreases the standard error of β_1. Thus we should be sure to include some values of X at the extremes of X over the range of interest.

To test the hypothesis that β_1 is equal to β_{10}, that is,

$$H_0: \beta_1 = \beta_{10},$$

we use the statistic

$$t = \frac{\hat{\beta}_1 - \beta_{10}}{\text{est. s.e. } (\hat{\beta}_1)} = \frac{(\hat{\beta}_1 - \beta_{10}) \sqrt{\sum (x_i - \bar{x})^2}}{s_{Y|X}}.$$

If σ were known, the test statistic, using σ instead of $s_{Y|X}$, would follow the standard normal distribution; however, σ is usually unknown and the test statistic using $s_{Y|X}$ follows the t distribution with $n - 2$ degrees of freedom. The degrees of freedom parameter has the value of $n - 2$ because we have estimated two coefficients, β_0 and β_1.

If the alternative hypothesis is

$$H_a: \beta_1 \neq \beta_{10},$$

the rejection region consists of values of t less than or equal to $t_{n=2,\alpha/2}$ or greater than or equal to $t_{n-2,1-\alpha/2}$.

The hypothesis usually of interest is that β_{10} is zero; that is, there is no linear relationship between Y and X. If, however, our study is one attempting to replicate previous findings, we may wish to determine if our slope coefficient is the same as that reported in the original work. Then β_{10} will be set equal to the previously reported value. Let us test the hypothesis that there is no linear relationship between SBP and height versus the alternative hypothesis that there is some linear relationship at the 0.05 significance level.

The test statistic, t, is

$$t = \frac{(\hat{\beta}_1 - \beta_{10}) * \sqrt{\sum (x_i - \bar{x})^2}}{s_{Y|X}}$$

which is

$$t = \frac{(0.7688 - 0) * \sqrt{4506.5}}{14.518} = 3.555.$$

This value is compared with -2.01 ($= t_{48,0.025}$) and 2.01 ($= t_{48,0.975}$). As 3.555 is greater than 2.01, we reject the hypothesis of no linear relationship between SBP and height. The p value of this test is approximately 0.001.

The $(1 - \alpha) * 100$ percent confidence interval for β_1 is formed by

$$\hat{\beta}_1 \pm t_{n-2,1-\alpha/2} * \text{est. s.e.}(\hat{\beta}_1)$$

which is

$$\hat{\beta}_1 \pm t_{n-2,1-\alpha/2} * \frac{s_{Y|X}}{\sqrt{\sum (x_i - \bar{x})^2}}.$$

The 95 percent confidence interval for β_1 is found using

$$0.7688 \pm 2.01 * \frac{14.518}{\sqrt{4506.5}} = 0.7688 \pm 0.4347$$

and this gives a confidence interval from 0.3341 to 1.2035. The confidence interval is consistent with the test given above. As zero is not contained in the confidence interval for β_1, there appears to be a linear relationship between SBP and height.

As there is evidence to suggest that β_1 is not zero, this also means that the correlation coefficient between Y and X is not zero.

D. *Y*-Intercept Coefficient

It is also possible to form confidence intervals and to test hypotheses about β_0, although these are usually of less interest than those for β_1. The location of the Y intercept is relatively unimportant compared with determining whether a relationship exists between the dependent and independent variables. Sometimes, however, we wish to compare whether both our coefficients, slope and Y intercept, agree with those presented in the literature. In this case, we are interested in examining β_0 as well as β_1.

As the estimator of β_0 is also a linear combination of the observed values of the normally distributed dependent variable, $\hat{\beta}_0$ also follows a normal distribution. The standard error of $\hat{\beta}_0$ is estimated by

$$\text{est. s.e.}(\hat{\beta}_0) = \sqrt{\frac{\sum x_i^2}{n \sum (x_i - \bar{x})^2}} * s_{Y|X}.$$

The hypothesis of interest is

$$H_0\colon \beta_0 = \beta_{00}$$

versus either a one- or two-sided alternative hypothesis. The test statistic for this hypothesis is

$$t = \frac{\hat{\beta}_0 - \beta_{00}}{\sqrt{\sum x_i^2/[n \sum (x_i - \bar{x})^2]} * s_{Y|X}}$$

and this is compared with $\pm t_{n-2,1-\alpha/2}$ for the two-sided alternative hypothesis. If the alternative hypothesis is that β_0 is greater than β_{00}, we reject the null hypothesis in favor of the alternative when t is greater than $t_{n-2,1-\alpha}$. If the alternative hypothesis is that β_0 is less than β_{00}, we reject the null hypothesis in favor of the alternative when t is less than $-t_{n-2,1-\alpha}$.

The $(1 - \alpha/2) * 100$ percent confidence interval for β_0 is given by

$$\hat{\beta}_0 \pm t_{n-2,1-\alpha/2} \sqrt{\frac{\sum x_i^2}{n \sum (x_i - \bar{x})^2}} * s_{Y|X}.$$

Let us form the 99 percent confidence interval for β_0 for these SBP data. The 0.995 value of the t distribution with 48 degrees of freedom is approximately 2.68. Therefore, the confidence interval is found from the calculations

$$61.14 \pm 2.68 \sqrt{\frac{142{,}319}{50 * 4506.5}} * 14.52 = 61.14 \pm 30.93$$

which gives an interval from 30.21 to 92.07, a wide interval.

E. An ANOVA Table Summary

Table 15.3 shows the information required to test the hypothesis of no relationship between the dependent and independent variables in an ANOVA table similar to that used in Chapter 14. The test statistic for the hypothesis of no linear relationship between the dependent and independent variables is the F ratio, which is distributed as an F variable with 1 and $n - 2$ degrees of freedom. Large values of the F ratio cause us to reject the null hypothesis of no linear relationship in favor of the alternative hypothesis of a linear relationship. The F statistic is the ratio of the mean square due to regression to the mean square about regression (mean square error or residual mean square). The degrees of freedom parameters for the F ratio come from the two mean squares involved in the ratio. The degrees of freedom due to regression is the number of parameters estimated minus one. The degrees of freedom associated with the about regression source of variation is the sample size minus the number of coefficients estimated in the regression model. The ANOVA table for the SBP and height data is shown in Table 15.4. If we perform this test at the 0.05 significance level, we compare the calculated F ratio to $F_{1,48,0.95}$, which is approximately 4.04. As the calculated value, 12.63, is greater than the tabulated value, 4.04, we reject the null hypothesis in favor of the alternative hypothesis. There appears to be a linear relationship between SBP and height at the 0.05 significance level.

Note that if we take the square root of 12.63, we obtain 3.554. With allowance for rounding, we have obtained the value of the t statistic calculated in the section for testing the hypothesis that β_1 is zero. This equality is additional verification of the relationship, pointed out in Chapter 14, between the t and F statistics. An F statistic with 1 and $n - p$ degrees of freedom is the square of the t statistic with $n - p$ degrees of freedom. Examination of the t and F tables shows that $t^2_{n-p,1-\alpha/2}$ equals $F_{1,n-p,1-\alpha}$. Hence we have two equivalent ways of testing whether the dependent and independent variables are linearly related at a given significance level. As we shall see in the multiple regression material, the F statistic directly

TABLE 15.3 An ANOVA Table for the Simple Linear Regression Model

Source of variation	Degrees of freedom	Sum of squares	Mean square	F ratio[a]
Due to regression	1	$\sum (\hat{y}_i - \bar{y})^2$	$\sum (\hat{y}_i - \bar{y})^2/1$	MSR/MSE
About regression or error	$n - 2$	$\sum (y_i - \hat{y}_i)^2$	$\sum (y_i - \hat{y}_i)^2/(n - 2)$	
Corrected total	$n - 1$	$\sum (y_i - \bar{y})^2$		

[a]MSR, mean square due to regression; MSE, mean square error term.

TABLE 15.4 ANOVA Table for the Regression of SBP on Height

Source of variation	Degrees of freedom	Sum of squares	Mean square	*F* ratio
Due to regression	1	2,663	2663	12.63
About regression or error	48	10,117	210.77	
Corrected total	49	12,780		

extends to simultaneously testing several variables, whereas t can be used with only one variable at a time.

These calculations associated with the regression analysis require much time, care, and effort; however, they can be quickly and accurately performed with MINITAB as is shown in Box 15.1.

VI. INTERVAL ESTIMATION FOR $\mu_{Y|X}$ and $Y|X$

Even though the relationship between SBP and height is not impressive, we continue with the idea of developing a height-based standard for SBP for children. We would be much more comfortable doing this if the relationship between height and SBP were stronger. The height-based standards that we shall create are the SBP levels such that 95 percent of the girls of a given height have a lower SBP and 5 percent have a higher SBP. This standard is not based on the occurrence of any disease or other undesirable property. When a standard created in this manner is used, approximately 5 percent of the girls will be said to have undesirably high SBP, regardless of whether that is really a problem.

The standard is based on a one-sided prediction interval for the SBP variable. Also of interest is the confidence interval for the SBP variable and we consider the confidence interval first.

A. Confidence Interval for $\mu_{Y|X}$

The regression line provides estimates of the mean of the dependent variable for different values of the independent variable. How confident are we about these estimates or predicted values? The confidence interval provides one way of answering this question. To create the confidence interval, we require knowledge of the distribution of $\hat{Y}$ and also an estimate of its standard error.

As the predicted value of $\mu_{Y|X}$ at a given value of x, say x_k, is also a linear combination of normal values, it is normally distributed. Its standard

MINITAB BOX 15.1

The REGRESS command (abbreviated REGR) is used to perform linear regression analysis. The command is followed by the specification of a column containing the dependent variable, the number of independent variables to be used, and the specification of the column(s) containing the independent variable(s).

```
MTB > regr c2 1 cl

The regression equation is
C2 = 61.1 + 0.769 C1
Predictor          Coef          Stdev      t-ratio         p
Constant          61.14          11.54         5.30     0.000
C1               0.7688         0.2163         3.55     0.001
s = 14.52          R-sq = 20.8%        R-sq(adj) = 19.2%
Analysis of Variance
SOURCE          DF            SS             MS          F          p
Regression       1        2663.4         2663.4      12.64      0.001
Error           48       10117.1          210.8
Total           49       12780.5

Unusual Observations
Obs.        C1          C2        Fit Stdev.Fit   Residual    St.Resid
 14       44.0      128.00      94.97      2.76      33.03       2.32R

R denotes an obs. with a large st. resid.
```

The output provides a summary of the calculations. The SBP variable is contained in c2 and is regressed on one independent variable, height, stored in c1. The regression equation is shown followed by more details about the estimators of the coefficients. The standard errors are provided, as are the t statistics for testing that the coefficients are equal to zero. The test statistic for the hypothesis that β_0 equals zero is 5.30 and the corresponding t statistic for β_1 is 3.55. The s value shown corresponds to our $s_{Y|X}$ and is 14.52. The R^2 value is shown and agrees with our value of 0.208, or 20.8 percent. The R-sq(adj) value is discussed below. After the ANOVA table, MINITAB shows what it characterizes as unusual observations. An unusual observation either has a large standardized residual (R) or a large leverage value (X). For these unusual observations, the observed height and SBP values are shown, as are the predicted SBP value, its standard error, the residual, and the standardized residual.

To obtain the predicted values, residuals, and leverages shown in Tables 15.1 and 15.2, we used subcommands RESIDUALS and HI. The commands used were the following:

```
MTB > regr c2 1 cl c4 c5;
SUBC> residuals c3;
SUBC> hi c6.
```

The two extra columns, c4 and c5, appended to the REGRESS command after the independent variable column(s), c1, contain the standardized residuals and the predicted SBP values, respectively. The RESIDUALS subcommand puts the residuals in column c3 and the HI subcommand puts the leverages in column c6.

error is estimated by

$$\text{est. s.e.}(\hat{\mu}_{Y|X_k}) = s_{Y|X}\sqrt{\frac{1}{n} + \frac{(x_k - \bar{x})^2}{\sum (x_i - \bar{x})^2}}$$

The estimated standard error increases with increases in the distance between x_k and $\bar{x}$, and there is a unique estimate of the standard error for each x_k.

Because we are using $s_{Y|X}$ to estimate σ, we must use the t distribution in place of the normal in the formation of the confidence interval. The confidence interval for $\mu_{Y|X}$ has the form

$$\hat{\mu}_{Y|X} \pm t_{n-2,1-\alpha/2} * \text{est. s.e.}(\hat{\mu}_{Y|X})$$

Figure 15.7 shows the 95 percent confidence interval for SBP as a function of height. As we can see from the graph, the confidence interval widens as the values of height move away from the mean of the height variable. This is in accord with the expression for the confidence interval which has the term $(x_k - \bar{x})^2$ in the numerator. We are thus less sure of our prediction for the extreme values of the independent variable. The confidence interval is about 17 mm Hg wide for girls 35 or 70 in. tall and narrows to about 8 mm Hg for girls about 50 to 55 in. tall.

B. Prediction Interval for $Y|X$

In the preceding section, we learned how to form the confidence interval for the mean of SBP for a height value. In this section, we form the predic-

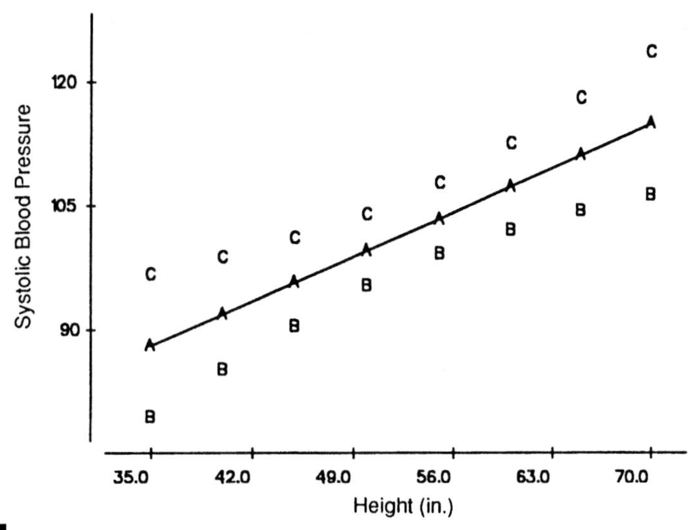

FIGURE 15.7 Ninety-five percent confidence interval for $\mu_{Y|X}$. A represents the regression line. B and C give the 95 percent confidence level.

tion interval, the interval for a single observation. The prediction interval is of interest to a physician because the physician is examining a single person, not an entire community. How does the person's SBP value relate to the standard?

As we saw in Chapter 7 in the material on intervals based on the normal distribution, the prediction interval is wider than the corresponding confidence interval because we must add the individual variation about the mean to the mean's variation. Similarly, the formula for the prediction interval based on the regression equation adds the individual variation to the mean's variation. Thus, the estimated standard error for a single observation is

$$\text{est. s.e.}(\hat{y}_k) = s_{Y|X}\sqrt{1 + \frac{1}{n} + \frac{(x_k - \bar{x})^2}{\Sigma\,(x_i - \bar{x})^2}}.$$

The corresponding two-sided $(1 - \alpha) * 100$ percent prediction interval is

$$\hat{y}_k \pm t_{n-2,1-\alpha/2} * \text{est. s.e.}(\hat{y}_k).$$

Figure 15.8 shows the 95 percent prediction interval for the data in Table 15.1. The prediction interval is much wider than the corresponding confidence interval because of the addition of the individual variation in the standard error term. The prediction interval here is about 60 mm Hg wide. Inclusion of the individual variation term has greatly reduced the effect of the $(x_k - \bar{x})^2$ term in the estimated standard error in this example. The upper and lower limits are essentially straight lines, in contrast to the shape of the upper and lower limits of the confidence interval.

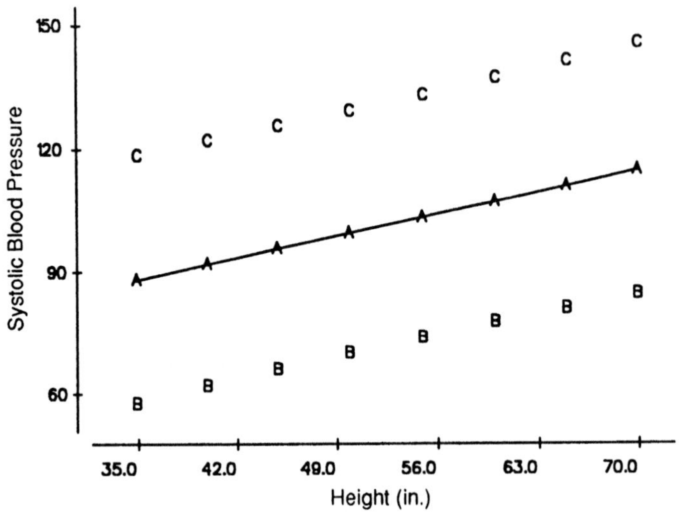

FIGURE 15.8 Ninety-five percent prediction interval for y_k. A represents the regression line. B and C give the 95 percent prediction level.

MINITAB can be used to perform the calculations necessary to create the 95 percent confidence and prediction intervals. The required commands are shown in MINITAB Box 15.2.

MINITAB BOX 15.2

```
MTB > set c3
DATA> 35:70/5
DATA> end
MTB > regr c2 1 c1;
SUBC> pred c3.
```

The values in column c3 contain values of the height variable from 35 to 70 in. in steps of 5 in. These are the values of height that will be used as x_k in the calculation of the intervals. They cover the range of height values of interest and allow us to see the shapes of the intervals when plotted.

The following table is printed after the unusual observations have been pointed out in the regression output.

```
   Fit  Stdev.Fit          95% C.I.              95% P.I.
  88.05       4.31    (  79.39,  96.71)   (  57.59, 118.50)
  91.89       3.39    (  85.06,  98.72)   (  61.91, 121.87)
  95.73       2.62    (  90.47, 101.00)   (  66.07, 125.40)
  99.58       2.12    (  95.31, 103.85)   (  70.07, 129.09)
 103.42       2.12    (  99.15, 107.69)   (  73.91, 132.93)
 107.27       2.62    ( 102.00, 112.53)   (  77.60, 136.93)
 111.11       3.39    ( 104.28, 117.94)   (  81.13, 141.09)
 114.95       4.31    ( 106.29, 123.61)   (  84.50, 145.41)
```

To create the plots of the confidence and prediction intervals, these values have to be entered into columns as is shown next.

```
MTB > set c4
DATA> 88.05 91.89 95.73 99.58 103.42 107.27 111.11 114.95
MTB > set c5
DATA> 79.39 85.06 90.47 95.31 99.15 102.00 104.28 106.29
MTB > set c6
DATA> 96.71 98.72 101.00 103.85 107.69 112.53 117.94 123.61
MTB > set c7
DATA> 57.59 61.91 66.07 70.07 73.91 77.60 81.13 84.50
MTB > set c8
DATA> 118.50 121.87 125.40 129.09 132.93 136.93 141.09 145.41
DATA> end
MTB > gmplot c4 c3, c5 c3, c6 c3, c7 c3, c8 c3;
SUBC> lines c4 c3;
SUBC> xlabel 'Height';
SUBC> ylabel 'Systolic Blood Pressure';
SUBC> footnote 'B & C show the confidence interval';
SUBC> footnote 'D & E give the prediction interval'.
```

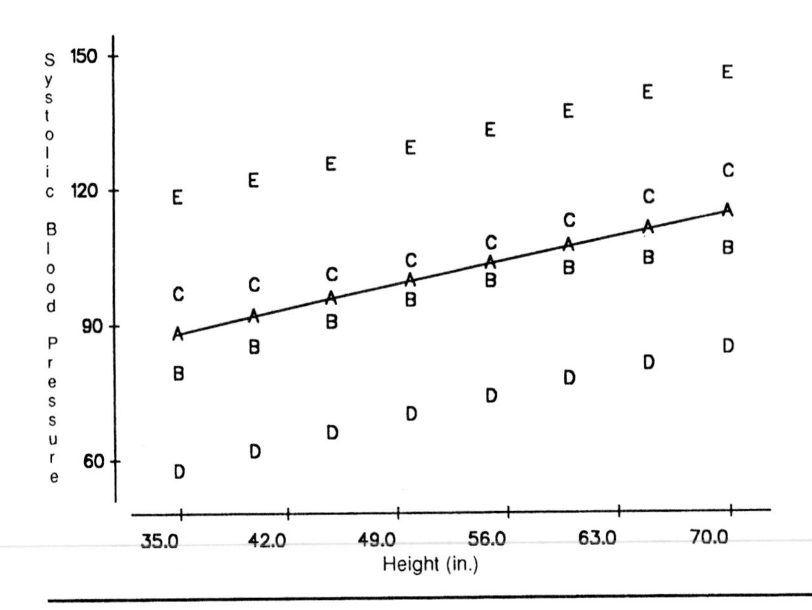

As we are concerned only about systolic blood pressures that may be too high, we use a one-sided prediction interval in the creation of the height-based standard for SBP for girls. The upper $(1 - \alpha) * 100$ percent prediction interval for SBP is found from

$$\hat{y}_k + t_{n-2,1-\alpha} * \text{est. s.e.}(\hat{y}_k).$$

Because the standard is the value such that 95 percent of the SBP values fall below it and 5 percent of the values are greater than it, we use the upper 95 percent prediction interval to obtain the standard.

The data shown in MINITAB Box 15.2 can be used to help create the height-based standards for SBP. The difference between the one- and two-sided intervals is the use of $t_{n-2,1-\alpha}$ in place of $t_{n-2,1-\alpha/2}$. Thus the amount to be added to $\hat{y}_k$ for the upper one-sided interval is simply 0.834 ($= t_{48,0.95}/t_{48,0.975}$) times the amount added for the two-sided interval. To find the amount added for the two-sided interval, we subtract the predicted SBP value shown from the upper limit of the 95 percent prediction interval. For example, for a girl 35 in. tall, the amount added, using the two-sided interval, is found by subtracting 88.05 from 118.50. This yields a difference of 30.45 mm Hg. If we multiply this difference by 0.834, we have the amount to add to the 88.05 value. Thus the standard for a girl 35 in. tall is

$$0.834 * (118.50 - 88.05) + 88.05 = 113.45 \text{ mm Hg}.$$

Table 15.5 shows these calculations and the height-based standards for SBP for girls. As shown above, the calculations in Table 15.5 consist of taking column 2 minus column 3. This is stored in column 4. Column 5 contains 0.834 times column 4. The standard, column 6, is the sum of column 3 and column 5.

The upper one-sided prediction interval is one way of creating height-based standards for SBP. It has the advantage over simply using the observed 95th percentiles of the SBP at the different heights in that it does not require such a large sample size to achieve the same precision. If SBP is really linearly related to height, standards based on the prediction interval also smooth out random fluctuations that may be found in considering each height separately.

The standards developed here are illustrative of the procedure. If one were going to develop standards, a larger sample size would be required. We would also prefer to use additional variables or another variable to increase the amount of variation in the SBP that is accounted for by the independent variable(s). In addition, as we have stated above, the rationale for having standards for blood pressure in children is much weaker than that for having standards in adults. In adults, there is a direct linkage between high blood pressure and disease, whereas in children, no such linkage exists. Additionally, the evidence that relatively high blood pressure in children carries over into adulthood is inconclusive. Use of the 95th percentile or other percentiles as the basis of a standard means that some children will be identified as having a problem when none may exist.

So far we have focused on a single independent variable. In the next section, we consider multiple independent variables.

TABLE 15.5 Creation of Height-Based Standards for SBP (mm Hg) for Girls

x_k inches (1)	Upper limit of prediction interval (2)	y_k (3)	Difference (4)	0.834 * difference (5)	Standard (6)
35	118.50	88.05	30.45	25.40	113.45
40	121.87	91.89	29.98	25.00	116.89
45	125.40	95.93	29.67	24.74	120.67
50	129.09	99.58	29.51	24.61	124.19
55	132.93	103.42	29.51	24.61	128.03
60	136.93	107.27	29.66	24.74	132.01
65	141.09	111.11	29.98	25.00	136.11
70	145.41	114.95	30.46	25.40	140.35

VII. INTRODUCTION TO MULTIPLE LINEAR REGRESSION

For many chronic diseases, no single cause is associated with the occurrence of the disease. Many factors, called risk factors, play a role in the development of the disease. In the study of the occurrence of air pollution, many factors, for example, wind, temperature, and time of day, must be considered. In comparing mortality rates for hospitals, factors such as the mean age of patients, severity of the diseases seen, and percentage of patients admitted from the emergency room must be taken into account in the analysis. As these examples suggest, it is uncommon for an analysis to include only one independent variable. Therefore, in this section, we briefly introduce multiple linear regression, a method for examining the relationship between one normally distributed dependent variable and more than one continuous independent variable.

The equation showing the hypothesized relationship between the dependent and independent variables is

$$y_i = \beta_0 + x_{1i}\,\beta_1 + x_{2i}\beta_2 + \ldots + x_{p-1,i}\,\beta_{p-1} + \varepsilon_i .$$

We are making the same assumptions—independence, normality, and constant variance—about the dependent variable and the error term in this model as we did in the simple linear regression model.

The coefficient β_i describes how much change there is in the dependent variable when the ith independent variable changes by one unit and the other independent variables are held constant. Again, the key hypothesis is whether β_i is equal to zero. If β_i is equal to zero, we probably would drop the corresponding X_i from the equation because there is no linear relationship between X_i and the dependent variable once the other independent variables are taken into account.

A goal of multiple regression is to obtain a small set of independent variables that make sense substantively and that do a reasonable job in accounting for the variation in the dependent variable. Often we have a large number of variables as candidates for the independent variables and our job is to reduce that larger set to a parsimonious set of variables. As was mentioned above, we do not want to retain a variable in the equation if it is not making a contribution. Inclusion of redundant or noncontributing variables increases the standard errors of the other variables and may also make it more difficult to discern the true relationship among the variables. A number of approaches have been developed to aid in the selection of the independent variables, and we show summary output from two of these approaches.

The calculations and the details of multiple linear regression are much more than we can cover in this introductory text. For more information on this topic, see books by Kleinbaum, Kupper, and Muller (7) and Draper and Smith (10), excellent books that focus on linear regression methods. We consider an example using MINITAB showing the use of multiple

TABLE 15.6 Hypothetical Values for Height, SBP, Age, and Weight Based on Data from NHANES II

Height (in.)	SBP (mm Hg)	Age (years)	Weight (pounds)	Height (in.)	SBP (mm Hg)	Age (years)	Weight (pounds)
36	105	7	57	54	92	12	80
37	90	7	46	54	80	12	77
38	82	6	42	54	88	13	86
38	96	7	52	55	104	12	77
39	82	8	56	56	100	11	118
39	74	7	45	57	126	12	97
40	104	8	54	58	108	13	108
40	100	9	73	59	106	14	107
41	80	6	65	59	98	11	97
42	98	7	63	59	94	13	91
43	96	10	74	60	88	14	83
44	86	8	62	60	110	15	95
44	88	8	50	60	124	10	84
44	128	10	68	61	86	13	93
45	118	10	86	61	120	14	75
46	90	10	83	62	112	12	80
48	108	9	86	63	100	13	110
48	120	12	82	64	122	11	102
49	114	9	71	64	122	15	114
49	78	11	65	65	110	15	123
50	116	10	77	65	124	12	108
50	74	8	71	66	122	12	132
51	80	10	65	67	94	16	118
52	98	11	75	67	110	13	142
53	90	12	91	69	140	13	122

linear regression. The hypothetical data used in the example are based on NHANES II and are shown in Table 15.6.

Before starting with the multiple regression analysis, we examine the correlation among these variables. The simple correlation coefficients among these variables can be represented in the format shown in Table 15.7. The correlation between SBP and weight is 0.509, the largest of the correlations between SBP and any of the variables. The correlation between height and weight is 0.867, the largest correlation in this table. These large

TABLE 15.7 Correlations among SBP, Height, Age, and Weight

	SBP	Height	Age
Height	0.457		
Age	0.368	0.864	
Weight	0.509	0.867	0.772

correlations among the height, age, and weight variables suggest that knowledge of one of these variables tells us a lot about the other two variables. Thus, not all three variables are needed as the use of only one of them conveys most of the independent information about SBP. As is clear from these estimates of the correlations among these three independent variables, they are not really independent of one another. We prefer to use the term *predictor variables*, but the term *independent variables* is so widely accepted that it is unlikely to be changed.

There are no firm sample size requirements for performing a multiple regression analysis; however, a reasonable guideline is that the sample size should be at least 10 times as large as the number of independent variables to be used in the final multiple linear regression equation. In our example, there are 50 observations and we will probably use no more than three independent variables in the final regression equation. Hence our sample size meets the guideline, assuming that we do not add interaction terms or higher-order terms of the three independent variables.

In this multiple regression situation, we have three variables that are candidates for inclusion in the multiple linear regression equation to help account for the variation in SBP. As mentioned above, we wish to obtain a parsimonious set of independent variables that account for much of the variation in SBP. We use a stepwise regression procedure and an all-possible-regressions procedure in MINITAB to demonstrate two approaches to selecting the independent variables to be included in the final regression model.

There are many varieties of stepwise regression; we consider forward stepwise regression. In forward stepwise regression, independent variables are added to the equation in steps, one per each step. The first variable to be added to the equation is the independent variable with the highest correlation with the dependent variable, provided that the correlation is high enough. The analyst provides the level that is used to determine whether the correlation is high enough. Instead of actually using the value of the correlation coefficient, the criterion for inclusion into the model is expressed in terms of the F ratio for the test that the regression coefficient is zero.

After the first variable is entered, the next variable to enter the model is the one that has the highest correlation with the residuals from the model that regressed the dependent variable on the first independent variable. This variable must also satisfy the F ratio requirement for inclusion. This process continues in this stepwise fashion, and an independent variable may be added or deleted at each step. An independent variable that had been added previously may be deleted from the model if, after the inclusion of other variables, it no longer meets the required F ratio.

The all-possible-regressions procedure in effect considers all possible regressions with one independent variable, with two independent variables, with three independent variables, and so on, and it provides a summary report of the results for the "best" models. *Best* here is defined in statistical terms but the actual determination of what is best must use substantive knowledge as well as statistical measures. MINITAB Box 15.3 shows the output from both of these procedures.

MINITAB BOX 15.3

Illustration of Stepwise and All Possible Regressions

```
MTB > name c1 'height', c2 'sbp', c3 'age', c4 'weight'
```

In the stepwise (STEP) and all possible regression (BREG) commands, the dependent variable is specified first, followed by the candidates for the independent variables.

```
MTB > step c2 c1 c3 c4

 STEPWISE REGRESSION OF  sbp   ON  3 PREDICTORS, WITH N = 50
     STEP          1
 CONSTANT      72.61
 weight        0.346
 T-RATIO        4.09
 S              14.0
 R-SQ          25.88
  MORE? (YES, NO, SUBCOMMAND, OR HELP)

 SUBC> y
  NO VARIABLES ENTERED OR REMOVED
  MORE? (YES, NO, SUBCOMMAND, OR HELP)
 SUBC> n
```

The subcommand asks whether we wish to consider another step. We responded yes the first time, but after it performed the calculations and said that there were no additions or deletions, there was no need to consider any more steps.

```
MTB > breg c2 c1 c3 c4
Best Subsets Regression of sbp
                                               h   w
                                               e   e
                                               i   i
                                               g a g
                        Adj.                   h g h
Vars     R-sq     R-sq      C-p            s   t e t

   1     25.9     24.3      0.4      14.049        X
   1     20.8     19.2      3.5      14.518    X
   2     26.0     22.9      2.3      14.183      X X
   2     26.0     22.8      2.3      14.188    X   X
   3     26.5     21.7      4.0      14.289    X X X
```

In the stepwise output, we see that the weight variable is the independent variable that entered the model first. It is highly significant with a t value of 4.09, and the R^2 for the model is 0.259. None of the other variables had a large enough F ratio—the default value is 4—to enter the model. Thus, this is the model selected by the forward stepwise process.

In the all-possible-regressions output, four different variables are shown: R^2, adjusted R^2, c_p, and s. Adjusted R^2 is similar to R^2, but it takes the number of variables in the equation into account. If a variable is added to the equation but its associated F ratio is less than one, the adjusted R^2 will decrease. In this sense, the adjusted R^2 is a better measure than R^2. One minor problem with adjusted R^2 is that it can be slightly less than zero. The formula for calculating the adjusted R^2 is

$$\overline{R}_p^2 = 1 - (1 - R_p^2)\left(\frac{n}{n - p}\right)$$

where R_p^2 is the coefficient of determination for a model with p coefficients. We do not use c_p. The quantity s is our $s_{Y|X}$.

From the all-possible-regressions output, we also see that the model including weight was the best model with one independent variable. The second best one-independent-variable model used the height variable. There is little to choose from between the two best two-independent-variable models. Note that the adjusted R^2 for both of these models is less than the adjusted R^2 for the best one-variable model, indicating that we would probably prefer the one-variable model. In the only three-independent-variable model, the adjusted R^2 has again decreased compared with the best one- and two-variable models. Thus, on statistical grounds, we should select the model with weight as the only independent variable. It has the highest adjusted R^2 and the lowest value of $s_{Y|X}$.

Let us now examine the use of the REGRESS command with the full three-independent-variable model and compare it with the model with weight as the only independent variable. MINITAB Box 15.4 shows the regression with the three independent variables.

In the model with the three independent variables, none of the independent variables are statistically significant as is shown by the three t values. Weight is closest to being significant. The F ratio of 5.53 is the value of the test statistic for the hypothesis that all the coefficients are simultaneously zero. As its associated p value is 0.003, we reject the hypothesis in favor of the alternative hypothesis that at least one of the coefficients is not zero. This situation—overall significance as indicated by the F statistic, but none of the variables significant when considered by themselves—results from the inclusion of unnecessary or nearly redundant variables. As we will see below, the weight variable is highly significant when considered in a model with no redundant terms. The height and age variables would also be significant if they were the only variables in the model.

MINITAB BOX 15.4

Multiple Regression with Three Independent Variables

```
MTB > regr c2 3 cl c3 c4

The regression equation is
sbp = 69.1 + 0.298 height - 0.91 age + 0.318 weight

Predictor        Coef       Stdev     t-ratio        p
Constant        69.11       12.55        5.51    0.000
height         0.2978      0.5410        0.55    0.585
age            -0.908       1.570       -0.58    0.566
weight         0.3181      0.1730        1.84    0.072
s = 14.29        R-sq = 26.5%     R-sq(adj) = 21.7%

Analysis of Variance
SOURCE        DF          SS          MS         F        p
Regression     3      3388.0      1129.3      5.53    0.003
Error         46      9392.5       204.2
Total         49     12780.5

SOURCE        DF      SEQ SS
height         1      2663.4
age            1        34.4
weight         1       690.1

Unusual Observations
Obs.  height       sbp       Fit  Stdev.Fit  Residual    St.Resid
 14     44.0    128.00     94.76       3.28     33.24       2.39R
R denotes an obs. with a large st. resid.
```

The sequential sum of squares can be generally ignored. These sums of squares show the added contribution of the variables when they are entered in the order specified in the REGRESS statement. MINITAB Box 15.5 shows the model with weight alone.

In this model, the coefficient for the weight variable is highly significant with a t value of 4.09 and an F ratio of 16.76. The estimated coefficient for the weight variable (0.346) is not that different from its value in the three-independent-variable model (0.318), but its standard error has decreased to 0.0845 from 0.1730 in the previous model. Inclusion of the unnecessary terms in the three-independent-variable model has caused the increase in the estimated standard errors and thus makes it hard to discern the significance of any of the independent variables.

As is clear from the significance of the weight variable in the model in which it is the only variable, compared with its nonsignificance in the

MINITAB BOX 15.5

Model with Weight as the Only Independent Variable

```
MTB > regr c2 1 c4

The regression equation is
sbp = 72.6 + 0.346 weight

Predictor        Coef        Stdev     t-ratio        p
Constant       72.606        7.333        9.90    0.000
weight        0.34579      0.08447        4.09    0.000
s = 14.05        R-sq = 25.9%      R-sq(adj) = 24.3%

Analysis of Variance
SOURCE        DF          SS          MS          F         p
Regression     1      3307.1      3307.1      16.76     0.000
Error         48      9473.4       197.4
Total         49     12780.5

Unusual Observations
Obs.      C4        C2        Fit   Stdev.Fit   Residual    St.Resid
 14       68    128.00      96.12        2.38      31.88       2.30R
 49      142    110.00     121.71        5.32     -11.71      -0.90X
R denotes an obs. with a large st. resid.
X denotes an obs. whose X value gives it large influence.
```

three-independent-variable model, the conclusions that can be drawn about the importance of an independent variable depend on the model that is being considered. Because these predictor variables are not independent of one another, all conclusions are model dependent.

The consideration of the extra variables did not greatly increase the R^2 or adjusted R^2 statistics. We have not been able to account for the great majority of the variation in SBP even by considering other variables. Unfortunately, R^2 of this magnitude, 0.2 to 0.3, are not uncommon when analyzing human data. There are still many sources of variation that remain a mystery to us. Much work needs to be done to discover these additional sources of variation before standards are created.

VIII. INTRODUCTION TO LOGISTIC REGRESSION

As was mentioned in the introduction to this chapter, logistic regression is a method that allows one to examine the relationship between a dependent variable with two levels and one or more independent variables. We are forced to consider this new method because the linear regression approach was based on the assumption that the dependent variable was a continuous and normally distributed variable. A variable with only two levels does not meet this assumption. In logistic regression, the independent variables may be continuous or discrete. The following example shows a situation in which both independent variables are discrete.

Suppose that we wish to determine whether a relationship exists between a male's pulmonary function test (PFT) results and air pollution at his residence, lead in the air serving as a proxy for overall air pollution. The data for this situation are shown in Table 15.8 [taken from Table 7.1 in Forthofer and Lehnen (11)]. We have categorized the PFT results as normal or not normal. This is a 2 by 2 table and we already know several ways of analyzing it. We are considering this simple table initially because we can show the logistic regression results in terms of statistics with which we are already familiar.

We could analyze the proportion of normal PFT results as a function of the lead level. Methods for doing this are shown in Forthofer and Lehnen (11); however, if we do that, sometimes the estimated proportions are less than 0 or greater than 1, impossible values for proportions. In addition, the methods in Forthofer and Lehnen (11) do not apply when the independent variables are continuous. Logistic regression provides an alternative method of analysis that avoids these problems.

In logistic regression, the underlying model is that the natural logarithm, ln, of the odds of a normal (or nonnormal) PFT is a linear function of the constant and the effect of lead pollution. The logarithm of the odds is also referred to as the *log odds* or *logit*. In this example, a larger logit value indicates a more favorable outcome because it indicates a greater proportion of males having a normal PFT. Hence those with low exposures to lead have a more favorable outcome than those with higher exposure to lead for this sample.

Using symbols for the logit, this model is

$$\ln\left(\frac{\pi_{i1}}{\pi_{i2}}\right) = \text{constant} + \text{lead pollution}_i$$

TABLE 15.8 PFT Results by Ambient Air Pollution

Lead level	PFT results		Logit (normal)
	Normal	Not normal	
Low	368	19	2.964
High	82	10	2.104

Source: Forthofer and Lehnen (11, Table 7.1).

where π_{i1} is the probability of a normal PFT, π_{i2} is the probability of a nonnormal PFT for the ith lead level, and lead pollution$_i$ is the effect of the ith lead level. The ratio of π_{i1} to π_{i2} is the odds of a normal PFT for the ith lead level.

Substituting symbols for all the terms in the above equation, we have

$$\ln\left(\frac{\pi_{i1}}{\pi_{i2}}\right) = \mu + \alpha_i$$

where μ represents the constant and α_i is the effect of the ith level of lead. This model has the same structure as the linear model representation of the ANOVA shown in Chapter 14. Just as in Chapter 14, we measure the effects of the levels of a variable from a reference level. For the lead variable, we consider the high level of pollution to be the reference level. This means that α_2 is taken to be 0 and that μ is the logit for the high lead level as can be seen from the following.

As there are only two lead levels, we have the following two equations:

$$\ln\left(\frac{\pi_{11}}{\pi_{12}}\right) = \mu + \alpha_1$$

$$\ln\left(\frac{\pi_{21}}{\pi_{22}}\right) = \mu$$

It is clear from the second of these two equations that μ is the logit of a normal PFT for those exposed to the high lead pollution level. If we subtract the second of these two equations from the first, we see that α_1 is simply the difference of the two logits, that is,

$$\ln\left(\frac{\pi_{11}}{\pi_{12}}\right) - \ln\left(\frac{\pi_{21}}{\pi_{22}}\right) = \alpha_1.$$

Because, as we saw in Chapter 7, the difference of two logarithms is the logarithm of the ratio, we have

$$\alpha_1 = \ln\left(\frac{\pi_{11}\pi_{22}}{\pi_{12}\pi_{21}}\right).$$

That is, α_1 is also the natural logarithm of the odds ratio.

It is beyond the scope of this book to provide the details of the estimation and the testing methods used in logistic regression. For more information on logistic regression, see the excellent book by Hosmer and Lemeshow (12). In the following we show some of the statistics usually provided by different logistic regression programs.

The estimates of μ and α_1 are 2.104 and 0.860, respectively. The effect of the low lead level is to increase the estimate of the logit by 0.860 over the corresponding logit for those who were exposed to the high level of lead. As was shown above, because of the model we are using, the estimate of α_1

is simply the natural logarithm of the odds ratio for Table 15.8. If we take the exponential of the estimate of α_1, we obtain 2.362, the odds ratio. This odds ratio is much larger than one, and it strongly supports the idea that those with the lower lead exposure have the greater proportion of a normal PFT. The estimate of the constant term is the logit for the high level of lead, and the exponential of $\hat{\mu}$ is 8.2, the odds of a normal result for those with high lead exposures. Thus the logistic regression model leads to parameters that are readily interpretable.

It is surprising that there appears to be a lead effect as lead has not been shown to have a negative impact on the respiratory system in other studies; however, during the period 1974–1975 when this study was performed, automobile emissions were a major source of lead pollution. Thus, a possible explanation for this finding is that lead pollution is serving as a proxy for nitrogen dioxide or other pollutants that have adverse respiratory effects. Another possible explanation is that we have not controlled for possible confounding variables. Smoking status is a key variable that has been ignored in the analysis so far. Table 15.9 shows the inclusion of the smoking status variable.

We begin by considering a model containing the main effects of lead and smoking. The symbolic representation of this model is

$$\ln\left(\frac{\pi_{ij1}}{\pi_{ij2}}\right) = \mu + \alpha_i + \beta_j$$

where π_{ij1} is the probability of a normal PFT and π_{ij2} is the probability of a nonnormal PFT for the subgroup formed by the ith level of the lead variable and the jth level of the smoking variable. The constant term is represented by μ, the effect of the ith lead level is α_i, and the effect of the jth smoking level is β_j. The reference level for the smoking variable is the heavy smoking level which means that β_4 is taken to be zero.

TABLE 15.9 PFT Results by Smoking Status and Ambient Air Pollution

		PFT results		
Lead level	**Smoking status**	**Normal**	**Not normal**	**Logit (normal)**
Low	Never	160	4	3.69
	Former	49	6	2.10
	Light	75	6	2.53
	Heavy	84	3	3.33
High	Never	33	3	2.40
	Former	12	2	1.79
	Light	21	2	2.35
	Heavy	16	3	1.67

The estimated values of the logistic regression parameters are the following:

$$\hat{\mu} = 2.18, \quad \hat{\alpha}_1 \text{ (low lead level)} = 0.84$$
$$\hat{\beta}_1 \text{ (never smoked)} = 0.50$$
$$\hat{\beta}_2 \text{ (former smoker)} = -0.77$$
$$\hat{\beta}_3 \text{ (light smoker)} = -0.29.$$

The addition of the smoking variable has not changed the parameter estimates much as the constant previously was estimated to be 2.104 and the low lead effect was previously estimated to be 0.860. In this multiple logistic regression situation, α_1 is the natural logarithm of the odds ratio that would have been obtained if the high and low lead levels had had the same distributions for the smoking status variable. Examination of Table 15.9 shows that the distributions of the smoking status variable are similar for the high and low lead levels. Hence it is not surprising that the estimates of α_1 based on Tables 15.8 and 15.9 are similar.

There are usually one or more chi-square statistics for assessing the goodness of fit of the model as well as test statistics associated with the variables included in the model. In our case, the model provides a reasonable fit to the data, and we can proceed to the test of primary interest, the test for no lead effect. The test statistic for the null hypothesis of no lead effect is 4.29 and this is asymptotically distributed as a chi-square statistic with one degree of freedom if the null hypothesis is true. The p value for this test statistic is 0.038. If we use a 0.05 significance level, there is still a significant lead effect.

Logistic regression is an important method, particularly in epidemiology, as it allows the investigator to examine the relationship between a binary dependent variable and a set of continuous and discrete independent variables. The binary variable may indicate the presence or absence of a disease or a variable showing the survival status of a person. Interpretation of the parameters in terms of the odds and odds ratios is a key attraction of the logistic regression procedure. Many of the procedures for multiple linear regression have also been adapted to logistic regression, making it an even more attractive method.

IX. CONCLUDING REMARKS

In this chapter, we showed how to examine the relationship between a normally distributed dependent variable and a continuous independent variable. This method can be extended to include many independent variables and a brief introduction to this topic was provided. In linear models analysis, the independent variables may be continuous or discrete. Often we wish to use linear regression or ANOVA, but the dependent variable is a binary variable, for example, the occurrence of a disease. In this case, the logistic regression method can be used, and a brief introduction to this topic was also provided in this chapter.

EXERCISES

15.1. Restenosis, narrowing of the blood vessels, frequently occurs after coronary angioplasty, but accurate prediction of which individuals will have this problem is problematic. In their study, Simons *et al.* (13) hypothesized that restenosis is more likely to occur if activated smooth muscle cells are present in coronary lesions at the time of surgery. They used the number of reactive nuclei in the coronary lesions as an indicator of the presence of activated smooth muscle cells. The number of reactive nuclei in the lesions and the degree of stenosis at follow-up for 16 patients who underwent a second angiography are:

Patient	*Degree of stenosis (%) at follow-up*	*Number of reactive nuclei at initial surgery*
1	28	5
2	15	3
3	22	2
4	93	10
5	60	12
6	90	25
7	42	8
8	53	3
9	72	15
10	0	13
11	79	17
12	28	0
13	82	13
14	28	14
15	100	17
16	21	1

Are you suspicious of any of these data points? If so, why? Does there appear to be a linear relationship between the degree of stenosis and the number of reactive nuclei? If there is, describe the relationship. Are there any points that have a large influence on the estimated regression line? If there are, eliminate the point with the greatest leverage and refit the equation. Is there much difference between the two regression equations? Are there any points that have a large standardized residual? Explain why the residuals are large for these points. Do you think that Simons *et al.* have a promising lead for predicting which patients will undergo restenosis?

15.2. The estimated age-adjusted percentages of persons 18 years of age and older who smoke cigarettes are shown for females and males for selected years [taken from Table 62 in the National Center for Health Statistics (14)].

	Estimated age-adjusted percentage smoking cigarettes	
Year	*Female*	*Male*
1965	34.0	51.6
1974	32.5	42.9
1979	30.3	37.2
1983	29.9	34.7
1985	28.2	32.1
1987	26.7	31.0
1988	26.0	30.1
1990	23.1	28.0

Describe the linear relationship between the estimated age-adjusted percentage smoking and time for females and males separately. How much of the variation in the percentages is accounted for by time for females and for males? Do females and males appear to have the same rate of decrease in the estimated age-adjusted percentage smoking? Provide an estimate of when the age-adjusted percentage of males who smoke will equal the corresponding percentage for females. What assumption(s) have you made in coming up with the estimate of this time point? Do you think this assumption is reasonable? Explain your answer.

15.3. Use the following data taken from Table 112 in the National Center for Health Statistics (14) to determine whether there is a linear relationship between the U.S. national health expenditures as a percentage of gross national product (GNP) and time.

Year	*National health expenditures as percentage of GNP*	*Year*	*National health expenditures as percentage of GNP*
1929	3.5	1981	9.5
1935	4.0	1982	10.3
1940	4.0	1983	10.5
1950	4.5	1984	10.3
1955	4.4	1985	10.5
1960	5.3	1986	10.7
1965	5.9	1987	10.9
1970	7.3	1988	11.2
1975	8.3	1989	11.6
1980	9.2	1990	12.2

What is your predicted value for national health expenditures as a percentage of GNP for 1995? For 2000? What are the 95 percent confidence intervals for your estimates? What data have you used as the basis of your predictions? What assumptions have you made?

15.4. Use the data in Table 15.1 to construct height-based standards for systolic blood pressure for girls. In constructing these standards, you should be concerned about values that may be too low as well as too high.

15.5. Data from an article by Madsen (15) are used here to examine the relationship between survival status—less than 10 years or 10 years or longer—and type of operation—extensive (total removement of the ovaries and the uterus) and not extensive—for 299 patients with cancer of the ovary. Other factors could be included, for example, stage of the tumor, whether radiation was used, and the tumor had spread, in a logistic regression analysis; however, we begin our consideration with only the one independent variable. The data are

	Survival status	
Type of operation	*<10 years*	*≥10 years*
Extensive	129	122
Not extensive	20	28

In a logistic regression analysis, using the logit for ≥ 10 years of survival and the not extensive type of operation as the base level, the estimates of the constant term and the regression coefficient for the type of operation (extensive) are 0.3365 and 0.3920, respectively. Provide an interpretation for these estimates. Demonstrate that your interpretations are correct by relating these estimates to the above table.

15.6. The following data are a sample of observations from NHANES II. We wish to determine whether diastolic blood pressure (DBP) of adults can be predicted based on knowledge of the person's body mass index (BMI: weight in kilograms divided by square of height in meters); age; sex (females are coded as 0 and males as 1); smoking status (not currently a smoker is coded as 0 and currently a smoker as 1); race (0 represents non-African-American and 1 represents African-American); years of education; poverty status (household income expressed as a multiple of the poverty level for households of the same size); and vitamin status (0 indicates not taking supplements and 1 taking supplements).

Vitamin status	*BMI*	*Sex*	*Race*	*Education*	*Age*	*Poverty index*	*DBP*	*Smoke*
1	18.46	0	0	13	24	1.93	50	0
0	32.98	1	0	14	24	3.97	98	0
1	29.48	1	0	12	39	1.71	80	1
1	19.20	0	0	12	29	1.62	62	1
0	24.76	0	0	12	45	5.49	90	0
1	20.60	0	0	14	24	4.78	70	0
0	24.80	1	0	8	65	3.63	80	0
1	24.24	0	0	12	25	4.55	56	1
0	29.95	1	0	16	24	2.77	90	0
0	21.80	1	0	17	29	2.15	78	0
0	23.19	1	0	13	29	1.09	56	0
0	28.34	0	0	12	18	1.71	78	0
0	22.00	1	0	12	28	5.49	70	1
0	24.60	1	0	8	65	3.35	70	1
1	21.83	0	0	16	26	0.77	74	0
0	30.50	0	0	3	73	1.10	70	0
1	19.63	0	0	13	33	5.48	62	1
0	27.92	0	0	12	65	3.83	78	0
1	26.77	1	0	12	59	3.57	90	0
1	21.02	1	0	15	21	1.25	64	0
1	19.40	0	0	16	26	3.25	70	0
0	31.12	0	0	12	58	1.91	100	0
0	20.68	0	0	7	57	4.63	74	0
0	22.48	0	0	12	28	1.75	75	0
0	24.89	0	0	14	23	3.25	74	0
1	21.08	0	0	12	56	5.04	68	0
1	23.67	1	0	14	23	4.47	86	1
1	28.19	1	0	12	24	3.38	82	1
0	22.09	0	1	7	58	1.73	80	0
0	23.46	1	0	14	66	5.12	70	0
1	21.11	1	0	13	18	0.64	70	1
1	21.35	0	1	12	20	0.26	60	1
0	20.36	0	1	14	23	2.85	78	0
1	25.00	0	1	4	36	0.72	80	0
1	20.47	0	0	17	37	3.97	88	1
0	24.73	0	1	8	44	1.36	82	0
0	27.87	0	0	12	50	3.31	70	1
0	28.22	1	0	15	50	3.41	112	0
0	26.05	1	0	13	33	5.85	80	0
0	24.51	0	0	12	42	3.17	92	0
1	28.09	0	1	16	46	2.39	92	0
1	18.85	0	1	11	36	1.62	56	1
0	25.99	0	1	12	74	1.40	80	0
1	23.47	1	0	16	35	1.97	96	1
0	26.57	0	0	12	55	6.11	86	0
0	25.09	1	0	12	33	2.15	104	1
0	30.78	0	0	12	38	1.37	74	0
0	28.89	1	0	14	49	1.82	90	1
1	23.82	1	0	17	35	2.85	70	0
0	28.29	1	0	12	62	6.89	60	0

Select an appropriate multiple regression model that shows the relationship between DBP and the set or a subset of the independent variables shown here. Note that the independent variables include both continuous and discrete variables. Provide an interpretation of the estimated regression coefficients for each discrete independent variable used in the model. From these independent variables, are we able to do a good job of predicting DBP? What other independent variables, if any, should be included to improve the prediction of DBP?

15.7. Anderson *et al.* (16) provide serum cholesterol and body mass index (BMI) values for subjects who participated in a study to examine the effects of oat-bran cereal on serum cholesterol. The values of serum cholesterol and BMI for the 12 subjects included in the analysis are:

Subject	*Serum cholesterol (mmol/liter)*	*Body mass index*
1	7.29	29.0
2	8.04	26.3
3	8.43	21.6
4	7.96	21.8
5	5.43	27.2
6	5.77	24.8
7	6.96	25.2
8	6.23	24.5
9	6.65	25.1
10	6.26	23.5
11	8.20	27.9
12	6.21	24.8

Plot serum cholesterol versus BMI. Calculate the correlation coefficient between serum cholesterol and BMI. Regress serum cholesterol on BMI. Does there appear to be any linear relationship between these two variables? Form a new variable that is BMI minus its mean. Square this new variable. Include this new independent variable in the regression equation along with the BMI variable. Does there appear to be any linear relationship between these two independent variables and serum cholesterol? Why do you think we suggested that this new variable be added to the regression equation?

REFERENCES

1. NHLBI Task Force on Blood Pressure Control in Children (1987). The report of the second task force on blood pressure control in children, 1987. *Pediatrics* **79**, 1–25.

2. Voors, A., Webber, L., Frerichs, R., Berenson, G. S. (1977). Body height and body mass as determinants of basal blood pressure in children: The Bogalusa heart study. *Am. J. Epidemiol.* **106,** 101–108.
3. Gillum, R., Prineas, R., and Horibe, H. (1982). Maturation vs age: Assessing blood pressure by height. *J. Natl. Med. Assoc.* **74,** 43–46.
4. Forthofer, R. N. (1991). Blood pressure standards in children. Paper presented at the American Statistical Association meeting, August, 1991.
5. Ranney, G. B., and Thigpen, C. C. (1981). The sample coefficient of determination in simple linear regression. *Am. Stat.* **35,** 152–153.
6. Korn, E. L., and Simon, R. (1991). Explained residual variation, explained risk, and goodness of fit. *Am. Stat.* **45,** 201–206.
7. Kleinbaum, D. G., Kupper, L. L., and Muller, K. E. (1988). "Applied Regression Analysis and Other Multivariable Methods," 2nd ed. PWS-Kent, Boston.
8. Lin, L. I., and Vonesh, E. F. (1989). An empirical nonlinear data-fitting approach for transforming data to normality. *Am. Stat.* **43,** 237–243.
9. Miller, D. M. (1984). Reducing transformation bias in curve fitting. *Am. Stat.* **38,** 124–126.
10. Draper, N. R., and Smith, H. (1981). "Applied Regression Analysis." Wiley, New York.
11. Forthofer, R. N., and Lehnen, R. G. (1981). "Public Program Analysis: A New Categorical Data Approach." Lifetime Learning Publications, Belmont, CA.
12. Hosmer, D. W., and Lemeshow, S. (1989). "Applied Logistic Regression." Wiley, New York.
13. Simons, M., Leclerc, G., Safian, R. D., Isner, J. M., Weir, L., and Baim, D. S. (1993). Relation between activated smooth-muscle cells in coronary-artery lesions and restenosis after atherectomy. *N. Engl. J. Med.* **328,** 608–613.
14. National Center for Health Statistics (1992). "Health, United States, 1991 and Prevention Profile," DHHS Publ. No. 92-1232. Public Health Service, Hyattsville, MD.
15. Madsen, M. (1976). Statistical analysis of multiple contingency tables. Two examples. *Scand. J. Stat.* **3,** 97–106.
16. Anderson, J. W., Spencer, D. B., Hamilton, C. C., Smith, S. F., Tietyen, J., Bryant, C. A., and Oeltgen, P. (1990). Oat-bran cereal lowers serum total and LDL cholesterol in hypercholesterolemic men. *Am. J. Clin. Nutr.* **52,** 495–499.

APPENDIX A

SAS and Stata Commands

SAS and Stata are two other widely used statistical software packages. SAS was developed on the mainframe computer and, as a result, can be used with small to very large data sets. Stata is a much newer package and is usually used with small to moderate-size data sets. This appendix shows commands for both of these packages that can be used to perform analyses comparable to those in many of the MINITAB boxes shown in the text. As many of the boxes are repetitive in that they show several plots or calculations of the pdf for a probability distribution, we show only one plot or pdf from the probability distribution, not all the ones shown in the text. As with the MINITAB statements in the text, the statements shown here are illustrative. For a complete description and details on the various commands, one should consult the SAS (1–4) and Stata (5, 6) documentation. In addition, all three of the computer packages used in the text and in this appendix show many of the features of the packages that were available when the book was being written. New versions of each of these packages are continually being developed, and the new releases will have greater capabilities than those illustrated here.

I. SAS COMMANDS

The material here shows the SAS commands for carrying out the analyses in the text; however, the material does not show all the capabilities of the various commands. We are deliberately attempting to keep the SAS material relatively simple, and the price we pay for the simplicity is some inefficiency in our commands. Even though we are attempting to keep things relatively simple, SAS requires that some programming be done in the creation of some of the pdf and cdf plots. This is reflected primarily in the boxes in Chapter 6. We are also using SAS graphical procedures, for example, GPLOT and GCHART. If you do not have SAS Graphics, then PLOT and CHART can be used in their place, but the options are slightly different. In the following, we are considering that each box represents a separate SAS session.

SAS uses PROCEDURES (PROCs) for carrying out the desired analyses and the name of a PROC provides a clue to its purpose. SAS also performs some of the calculations in DATA steps as is shown below. SAS differs from MINITAB and Stata in that there are not individual commands for many of the descriptive tools, for example, box plots. Instead, these tools are contained in PROCs and are invoked as options within the PROC.

Box 4.1: Data Entry from the Keyboard

We are first creating a library that will store our permanent SAS data files. This is done by using the LIBNAME statement. The name of the library follows the LIBNAME statement, followed by its location in quotes. If you wish to use data files in the library or to enter data files into the library, you must use the LIBNAME statement when you begin the SAS session. The name of the SAS data file that is to contain the data entered from the keyboard is specified in a DATA statement. The name is limited to eight characters, excluding the name of the library that is used. The names of the variables to be entered are listed in their order of input in an INPUT statement. Following the INPUT statement is a CARDS statement (the name is a relic of an earlier era in computers) that indicates the data are being entered from the keyboard. The data are entered after the CARDS statement. The following shows the entry of the data in Table 4.1, which is stored in a data set named NUTRIENT. The characters BOOK. before NUTRIENT indicate that this data set is to be a permanent SAS file and is stored in a library named BOOK. BOOK is an example of a library name. Once we have created a SAS file, we refer to the variables by their names instead of referring to columns. There is a carriage return (Enter) at the end of each line.

```
LIBNAME BOOK 'C:\';
DATA BOOK.NUTRIENT;
  INPUT DAY GRADE CALORIES PROTEIN FAT VITA;
CARDS;
3 8 1823  83  63  4876
4 8 2007  64  62  6202
4 8 1053  23  33   964
5 8 4322 128 202  6761
    .     .    .
    .     .    .
    .     .    .
3 5 1723  45  43  5703
;
RUN;
```

Note that there is a semicolon at the end of each SAS statement; these semicolons must be there. The last line above before the RUN statement is also a semicolon and it indicates that all the data have been entered. The RUN statement tells SAS to execute the above statements. If we wish to view the data, we use PROC PRINT followed by the name of the SAS data set to be printed.

```
PROC PRINT DATA=BOOK.NUTRIENT;
RUN;
```

Box 4.2: Data Entry from a Non-SAS File

We assume that the data are stored in a file named CALORIES that looks exactly like the listing of the data shown above. Again the DATA statement is required and the name after DATA is the name of the file that will contain the SAS data set we are creating. The name of the non-SAS file, CALORIES, that contains the data is specified using an INFILE statement. The INPUT statement is again used to specify the location of the variables. The numbers after the variable names give the column locations containing the variables. We again create a permanent SAS file with the same name as used above.

```
LIBNAME BOOK 'C:\';
DATA BOOK.NUTRIENT;
  INFILE CALORIES;
    INPUT  DAY  1  GRADE  3  CALORIES 5-8  PROTEIN 10-12  FAT 14-16
           VITA 18-22;
RUN;
```

Box 4.3: Creation of One- and Two-Way Tables

We wish to create a one-way table showing the distribution of days and a two-way table showing the distribution of days with calories (grouped into less than 2500 calories and greater than or equal to 2500 calories). We first need to create a variable showing the grouping of calories. We create a temporary SAS data set, NUTRTEMP, containing this information and use that temporary data set in the analysis. In the DATA step, the SET statement identifies the SAS data file that is used for input. The variable GROUP will be 0 if the caloric intake is less than 2500 calories and will be 1 if the intake is greater than or equal to 2500 calories. PROC FREQ creates contingency tables and is also used in the analysis of the tables. The LIST option tells SAS not to put the tables in tabular format, but simply to list the frequency of the different cells. This saves space in the output. TABLES is the statement that tells SAS what tables are to be created. The * between two variable names means that these variables are to be crosstabulated with one another.

```
LIBNAME BOOK 'C:\';
DATA NUTRTEMP;
SET BOOK.NUTRIENT;
  GROUP = 0;
  IF CALORIES >= 2500 THEN GROUP=1;
PROC FREQ  DATA=NUTRTEMP LIST;
  TABLES DAY  DAY*GROUP;
RUN;
```

Boxes 4.4 and 4.5: Creation of a Plot Showing Three Line Graphs

To create a plot with the three line graphs, we first must enter the data into a SAS data set. The $ after the word COUNTRY in the INPUT statement tells SAS that the variable COUNTRY will contain some alphabetic characters. The @@ symbols at the end of the INPUT statement tell SAS that more than one set of variables will be entered per line.

```
DATA EXPEND;
  INPUT COUNTRY $ PERCENT YEAR @@;
CARDS;
GB  3.9  60 GB  4.1  65  GB  4.5  70
GB  5.5  75 GB  5.8  80  GB  6.0  85
GB  6.1  87
US  5.2  60 US  6.0  65  US  7.4  70
US  8.4  75 US  9.2  80  US 10.6  85
US 11.2  87
GER 4.7  60 GER 5.1  65  GER 5.5  70
GER 7.8  75 GER 7.9  80  GER 8.2  85
GER 8.2  87
;
```

We use the TITLE statement to identify what is being graphed. SAS prints what is in the single quotes at the top of the page. In the following, we are using three different symbols to represent the values for the three countries and these symbols are specified by the VALUE option. As we wish to have the points for each country connected, we use the JOIN value for the INTERPOL option. We want the vertical axis, AXIS2, to have the label of % of GDP instead of simply the variable name. Therefore, we use the AXIS2 statement and its LABEL option. Now we are ready to use the PROC GPLOT to display the line graphs. In the PLOT statement, the first variable specified is the vertical axis variable and the second variable is the horizontal axis variable. If there are to be multiple graphs (a graph for each level of a third variable) in the same figure, the third variable is specified after the = sign.

```
TITLE 'Health Expenditures as % of GDP over Time';
SYMBOL1 INTERPOL=JOIN VALUE=CIRCLE;
SYMBOL2 INTERPOL=JOIN VALUE=SQUARE;
SYMBOL3 INTERPOL=JOIN VALUE=DIAMOND;
AXIS2 LABEL=('% of GDP');
GOPTIONS DEVICE = VGA;
PROC GPLOT  DATA=EXPEND;
  PLOT PERCENT*YEAR=COUNTRY;
RUN;
```

Box 4.6: Creation of a Histogram

SAS uses PROC GCHART to produce bar charts and histograms. We are going to create a histogram for the systolic blood pressure data shown in Table 4.5. First the data are stored in a permanent SAS data set called BOOK.SBP.

```
LIBNAME BOOK 'C:\';
DATA BOOK.SBP;
 INPUT SBP @@;
CARDS;
130  100  125   92   98  108  104  100  100  102  120  110  100
112  110  110  100  128  122  110  120  108   94  130  110  104
120  118   84  115  102  100  112  104  100  120  110  110  106
130  120  108  104  106  114   96  112  114  100  112   80  100
110  126   95  100  100   94  102   95  140  124   98  110   90
 80  102  116  102   90  116  110  128  140   90  104  130  104
105   80  116  106  100   95  105   90  108   88  105  112  134
116  108  108  100  105  110   90   95  125
;
```

Next PROC GCHART is used to create the histogram shown in Figure 4.8. We have used the TITLE command to provide a title for the histogram. Note that if we do not change the TITLE statement, all subsequent analyses performed in the same SAS session will have this title. The VBAR subcommand causes a vertical histogram to be created. If a horizontal histogram is desired, the HBAR subcommand is used. As this is a histogram, there should be no space between the bars. Thus we set the SPACE variable equal to zero. In addition, because we wanted a histogram with the same midpoints as those in Figure 4.8, we had to specify the values of the interval midpoints. If we do not specify the midpoints, SAS will create what it thinks is the most appropriate histogram.

```
TITLE  'Histogram of Systolic Blood Pressure Values';
GOPTIONS DEVICE=VGA;
PROC GCHART DATA=BOOK.SBP;
  VBAR SBP / SPACE=0 MIDPOINTS=77 84 91 98 105 112 119 126 133 140;
RUN;
```

Boxes 4.7 to 4.9: Creation of Stem-and-Leaf Plots

A stem-and-leaf plot of the systolic blood pressure data in the SAS data set BOOK.SBP is created using PROC UNIVARIATE. The PLOT option listed causes the stem-and-leaf plot to be created for the blood pressure variable. The SAS user can not specify the size of the increment for the stem and leaf plot.

```
TITLE 'Description of the SBP Variable';
LIBNAME BOOK 'C:\';
PROC UNIVARIATE DATA=BOOK.SBP PLOT;
  VAR SBP;
```

A stem-and-leaf plot of vitamin A is also created using the values in the SAS data set BOOK.NUTRIENT.

```
TITLE 'Stem and Leaf Plot of Vitamin A';
PROC UNIVARIATE DATA=BOOK.NUTRIENT PLOT;
 VAR VITA;
RUN;
```

Box 4.10: Creation of a Scatter Plot

The data used to create the scatter plot are the fat and protein values from BOOK.NUTRIENT. Hence we do not have to enter any additional data and can immediately use PROC GPLOT to create the scatter plot.

```
LIBNAME BOOK 'C:\';
TITLE 'Scatter Plot of Total Fat with Vitamin A';
GOPTIONS DEVICE = VGA;
```

```
PROC GPLOT DATA=BOOK.NUTRIENT;
 PLOT FAT*VITA;
RUN;
```

Boxes 4.11 to 4.13: Creation of Box Plots and Univariate Statistics

The first box plot uses the vitamin A data in BOOK.NUTRIENT and PROC UNIVARIATE is used to create it. When the PLOT option is included, besides obtaining the descriptive statistics and a stem-and-leaf plot from PROC UNIVARIATE, we also obtain a box plot for the variables indicated in the VAR statement. Because we want the descriptive statistics for the four dietary variables in BOOK.NUTRIENT, we make only one call to PROC UNIVARIATE. Thus we also create stem-and-leaf plots and box plots for CALORIES, FAT, and PROTEIN as well as those for VITA.

Included in the descriptive statistics are the sample size, mean, standard deviation, variance, standard error of the mean, coefficient of variation, quartiles including the median, minimum and maximum, range, interquartile range, mode, and 1st, 5th, 10th, 90th, 95th, and 99th percentiles. The five smallest and largest values are also printed.

```
LIBNAME BOOK 'C:\';
TITLE 'Creation of Box Plots';
PROC UNIVARIATE DATA=BOOK.NUTRIENT PLOT;
 VAR CALORIES FAT PROTEIN VITA;
RUN;
```

Next the box plots are created for the SBP variable after it has been split into two groups. We use the variable GROUP to indicate group membership. The SAS variable _N_ is the observation number. In this case, _N_ ranges from 1 to 100 as there are 100 observations of SBP. The OUTPUT command tells SAS to write each observation to the data file TEMPSBP.

```
DATA TEMPSBP;
SET BOOK.SBP;
 GROUP=0;
 IF _N_ > 50 THEN GROUP=1;
 OUTPUT;
```

A separate box plot will be created for each level of the GROUP variable because we are using the BY command in PROC UNIVARIATE. If the data were not ordered according to the variable GROUP (50 0's followed by 50 1's), then it would have been necessary to use PROC SORT, shown below, to sort the data set based on GROUP.

```
PROC UNIVARIATE DATA=TEMPSBP PLOT;
 BY GROUP;
 VAR SBP;
RUN;
```

Box 4.14: Sorting the Data

We wish to sort the CALORIES variable in ascending order. This is easily accomplished in SAS with the use of PROC SORT.

```
LIBNAME BOOK 'C:\';
PROC SORT DATA=BOOK.NUTRIENT; BY CALORIES;
PROC PRINT;
RUN;
```

The data have all been ordered on the basis of the CALORIES variable and have been printed to show the sorting. If no SAS data set is listed with a PROC statement, the last SAS data set created during this SAS computer session is used.

Box 4.15: Calculation of the Coefficient of Variation

There is no need to perform any additional calculations here as the coefficient of variation is part of the output from PROC UNIVARIATE.

Box 4.16: Calculation of the Geometric Mean

In Box 4.16 in the text, MINITAB was, in effect, used as a desk calculator to calculate the geometric mean. SAS can be used in that fashion as well, but it is inefficient to do so. We show the use of SAS to perform these calculations in this case, but we do not use SAS in this fashion for the other desk calculator applications shown in the text. The calculations are performed in the DATA step. The values used are the number of microbes in six different areas. In the following, we have not taken advantage of SAS ARRAYS and DO statements in an effort not to introduce too much of the SAS language too soon. The KEEP statement tells SAS that it can delete the *A* and *B* variables; only LOGGM and GM will be printed by PROC PRINT. As no data set is specified below, PROC PRINT uses the most recently created SAS data set. Here we have used the natural logarithm to the base *e*, not the base 10. As the answer shows, the choice of which base to use does not make any difference in the value of the geometric mean (GM).

```
DATA GEOMMEAN;
 KEEP LOGGM GM;
 A1=100; A2=100; A3=1000; A4=1000; A5=10000; A6=1000000;
 B1=LOG(A1); B2=LOG(A2); B3=LOG(A3);
 B4=LOG(A4); B5=LOG(A5); B6=LOG(A6);
 LOGGM = (B1 + B2 + B3 + B4 + B5 + B6) / 6;
 GM = EXP(LOGGM);
PROC PRINT;
 VAR GM;
RUN;
```

Box 4.18: Calculation of the Correlation Coefficient

We use PROC CORR to obtain the correlation coefficient between variables. In this case, we want the correlation between FAT and PROTEIN, variables in the BOOK.NUTRIENT data set.

```
LIBNAME BOOK 'C:\';
PROC CORR DATA=BOOK.NUTRIENT;
 VAR FAT PROTEIN;
RUN;
```

Box 5.1: Creation of 10 Samples of 30 Birthdates

SAS will create these samples in a DATA step. This section is more complicated than those shown above and may be skipped if desired. We use the ARRAY statement to indicate to SAS that the variable *B* has locations set up to hold the 30 birthdates. The RETAIN command tells SAS not to lose the values stored in *B* or the SEED value as we process one observation to the next. The DO statement indicates how many times the statements included between the DO statement and its closing END statement are to be performed. We are using the function RANUNI to obtain a random number between 0 and 1. The value of the SEED variable must be initialized before RANUNI is called. By multiplying the number created by RANUNI by 1000, we obtain a number between 0 and 1000. The INT function truncates the number to an integer. If the integer is 0 or greater than 365, we delete it and draw another random number. If the integer is between 1 and 365, we store it in the *j*th location of *B*. After obtaining 30 numbers, we use the OUTPUT statement to write them to the file named BIRTH. We repeat the process nine more times and then print the resultant file.

```
DATA BIRTH;
  RETAIN SEED;
  ARRAY B(30) B1-B30;
  SEED=0;
  DO I=1 TO 10;
     DO J=1 TO 30;
       REPL:CALL RANUNI(SEED,X);
       X=1000*X;
       X=INT(X);
       IF X = 0 OR X > 365 THEN GOTO REPL;
       B(J)=X;
     END;
     OUTPUT;
  END;
RUN;
PROC PRINT;
  ID I;
  VAR I B1-B30;
RUN;
```

Box 6.1: Drawing 10 Binomial Variates

In this section, we wish to obtain the values of binomial variables in 10 samples from a binomial distribution with a sample size of 4 and a population proportion of 0.25. RANBIN generates random numbers from a binomial distribution and stores the value in X. Instead of using PROC PRINT to print the results, we can also use a PUT statement to print the results. The PUT statement prints the results in the LOG window, not in the OUTPUT window.

```
DATA BINOM;
  RETAIN SEED;
  SEED=5;
  DO I = 1 TO 10;
  CALL RANBIN(SEED,4,0.25,X);
  PUT ' The value of a binomial variable from B(4,.25) is ' X;
  END;
RUN;
```

Box 6.2: Generation of the Binomial pdf

SAS uses a command, PROBBNML, to obtain the values of the binomial (BNML) cumulative distribution function. By subtracting the cdf evaluated at x from the cdf evaluated at $x + 1$, we can obtain the pdf of the binomial distribution. These calculations are performed in the DATA step. We show the calculations for n equal to 4 and π equal to 0.25.

```
DATA BNMLPDF;
  ARRAY P(5) P1-P5;
  P(1) = PROBBNML(.25,4,0);
  DO X = 0 TO 3;
    P(X+2) = PROBBNML(.25,4,X+1) - PROBBNML(.25,4,X);
  END;
PROC PRINT;
  VAR P1-P5;
RUN;
```

Box 6.4: Finding the cdf for a *B*(62,0.235) Variable

We wish to find the probability that a *B*(62,0.235) variable is less than 22. The following DATA step shows this calculation.

```
DATA BNMLCDF;
  P=PROBBNML(.235,62,21);
  PUT ' The probability that a B(62,0.235) variate is < 22 is ' P;
RUN;
```

Boxes 6.5 and 6.6: Plot of a Binomial pdf

We plot the binomial pdf for n equal to 10 and π equal to 0.1. The OUTPUT statement tells SAS to write a record.

```
DATA BNMLPLOT;
  RETAIN X J;
  J=0;
  P = PROBBNML(.10,10,0);
  OUTPUT;
  DO X = 0 TO 9;
    P = PROBBNML(.10,10,X+1) - PROBBNML(.10,10,X);
    J=J+1;
    OUTPUT;
  END;
GOPTIONS DEVICE = VGA;
PROC GPLOT;
  PLOT P*J;
RUN;
```

Boxes 6.7 and 6.8: Calculation of the pdf and cdf and Their Plots for the Poisson Distribution with a Mean Equal to 2.0

The pdf and cdf for the Poisson distribution and their plots with a mean of 2 are obtained below using the POISSON statement.

```
DATA POISSON;
  RETAIN X J PDF CDF;
  J=0;
  CDF = POISSON(2.0,0);
  PDF = CDF;
  PUT J ' PDF = ' PDF ' CDF = ' CDF;
  OUTPUT;
  DO X=0 TO 9;
  PDF = POISSON(2.0,X+1) - CDF;
  CDF = POISSON(2.0,X+1);
  J = J + 1;
  PUT J ' PDF = ' PDF ' CDF = ' CDF;
  OUTPUT;
  END;
GOPTIONS DEVICE = VGA;
PROC GPLOT;
  PLOT (PDF CDF)*J;
RUN;
```

Box 6.10: Creation of a Poissonness Plot

We use the DATA statement for performing the necessary transformations for the Poissonness plot.

```
DATA PPLOT;
  INPUT FREQS I IFACT;
  Y = LOG(FREQS); Z = LOG(IFACT);
  YAXIS = Y + Z;
CARDS;
103 0   1
143 1   1
 98 2   2
 42 3   6
  8 4  24
  4 5 120
  2 6 720
;
GOPTIONS DEVICE = VGA;
PROC GPLOT;
  PLOT YAXIS*I;
RUN;
```

Box 6.15: Plot of the Standard Normal cdf

The PROBNORM statement is used in the DATA step to obtain the values of the standard normal cdf for a number of values. PROC GPLOT is then used to plot these values.

```
DATA NORMCDF;
  DO X = -3.8 TO 3.8 BY .1;
    Y = PROBNORM(X);
    OUTPUT;
  END;
GOPTIONS DEVICE = VGA;
PROC GPLOT;
  PLOT Y*X;
RUN;
```

Box 6.16: Finding the cdf for a *N*(80, 10) Variable

PROBNORM is used to find the cdf for a $N(0,1)$ variable. In particular, we wish to find the probability that a $N(80,10)$ variable is greater than 95. We find the probability that the variable is less than or equal to 95 and then subtract that probability from 1.

```
DATA NORMCDF;
  Z=(95-80)/10;
  PLTE=PROBNORM(Z);
  PGT=1-PLTE;
  PUT ' PROBABILITY OF BEING LESS THAN 95 IS ' PGT;
RUN;
```

Box 6.17: Finding the Inverse cdf for a *N*(80,10) Variable

The PROBIT statement provides the inverse cdf for a standard normal variable.

```
DATA INVCDF;
  Z=PROBIT(0.95);
  X=Z*10+80;
  PUT ' 95-TH PERCENTILE IS ' X;
RUN;
```

Box 6.19: A Normal Probability Plot

We wish to create a normal probability plot, not the normal scores plot shown in the text, for the vitamin A data. This plot is part of the PROC UNIVARIATE output when the option PLOT is specified.

```
LIBNAME BOOK 'C:\';
PROC UNIVARIATE DATA=BOOK.NUTRIENT PLOT;
  VAR VITA;
RUN;
```

Box 6.20: Normal Probability Plot for 100 Observations from a *N*(80,10) Distribution

We first must generate the 100 observations in the DATA step using RANNOR and then use PROC UNIVARIATE to create the normal probability plot.

```
DATA NORPLOT;
RETAIN SEED;
SEED=3;
  DO I=1 TO 100;
    Z=RANNOR(SEED);
    X=Z*10+80;
    OUTPUT;
  END;
PROC UNIVARIATE PLOT;
  VAR X;
RUN;
```

Box 6.23: Generation of 100 Samples of Size 5 from a Poisson Distribution; Description and Plot of the Means from the 100 Samples

The function RANPOI is used in a DATA step to create the 100 samples of size 5 from a Poisson distribution with a mean of 1. The MEAN function is used to calculate the means for each of the 100 samples.

```
DATA POISSAMP;
  RETAIN SEED; ARRAY X(5) X1-X5;
  SEED=-7;
  DO I = 1 TO 100;
    DO J=1 TO 5;
      X(J)=RANPOI(SEED,1);
    END;
    XMEAN=MEAN(OF X1-X5);
    OUTPUT;
  END;
PROC UNIVARIATE;
  VAR XMEAN;
GOPTIONS DEVICE = VGA;
PROC GCHART;
  VAR XMEAN;
RUN;
```

Boxes 7.5 and 7.6: Creation of Histogram of Sample Variances Based on 200 Samples of Size 3 from a *N*(0,5) Distribution

RANNOR is used in a DATA step to generate samples from a standard normal distribution. The function STD is used to calculate the sample standard deviation of each sample. PROC MEANS can also be used to describe the sample variances.

```
DATA STDHIST;
  RETAIN SEED;
  ARRAY X(3) X1-X3;
  SEED=-9;
  DO I = 1 TO 200;
    DO J = 1 TO 3;
      Z=RANNOR(SEED);
      X(J)=5*Z;
    END;
    STANDDEV=STD(OF X1-X3);
    SAMPVAR=STANDDEV*STANDDEV;
    OUTPUT;
  END;
PROC MEANS;
  VAR SAMPVAR;
GOPTIONS DEVICE = VGA;
PROC GCHART;
  VAR SAMPVAR;
RUN;
```

Boxes 7.9 and 7.10: Calculation of Confidence Interval for the Population Correlation Coefficient

These calculations are performed in the DATA step using the LOG, SQRT, and EXP functions.

```
DATA CORRCI;
  X=0.648; Y=(1+X)/(1-X);
  X=0.5*LOG(Y);
  Y=1/SQRT(30);
  A=1.96*Y;
  B=X-A; C=X+A;
  D=2*B; E=2*C;
  F=EXP(D);
  G=EXP(E);
  H=(F-1)/(F+1);
  I=(G-1)/(G+1);
PUT ' THE LOWER LIMIT IS ' H ' AND THE UPPER LIMIT IS ' I;
RUN;
```

Boxes 10.1 and 10.4: Data Entry of Seven Smallest and Seven Largest Day 1 Values, Descriptive Statistics, and Sign Test

The following shows the entry of the 14 extreme values (7 smallest and 7 largest) for calories for the first day of recording as well as the corresponding second day's values. The variable CODE indicates whether or not the observation is one of the smallest values (CODE = 1) or one of the largest values (CODE = 0). The variable DIFF is the day 1 value minus the day 2 value for the 7 largest day 1 observations. It is the day 2 value minus the day 1 value for the 7 smallest day 1 observations.

```
LIBNAME BOOK 'C:\';
DATA BOOK.EXTREME;
  INPUT DAY1 DAY2 CODE;
  DIFF=DAY1-DAY2;
  IF CODE=1 THEN DIFF=DAY2-DAY1;
CARDS;
1053 2484 1
4322 2926 0
1753 1054 1
3532 3289 0
2842 2849 0
1505 1925 1
3076 2431 0
1292  810 1
3049 2573 0
3277 2185 0
```

```
1781 1844 1
2773 3236 0
1645 2269 1
1723 3163 1
;
```

The statistic labeled M in the PROC UNIVARIATE output is the number of positive sample values minus the expected number of values greater than zero if the null hypothesis is true. (Note that some releases of SAS do not have M as part of the PROC UNIVARIATE output. PROBBNML can be used if M is not part of the PROC UNIVARIATE output.) In our example, there are 10 difference values greater than zero and we would expect 7 (= 14/2) positive values under the null hypothesis. Therefore $M = 3$ in this case. The p value associated with this test, "Prob > $|M|$ 0.xxxx," is the p value for a two-sided alternative hypothesis. Therefore, for our one-sided alternative, we must divide the indicated probability by 2 to obtain the desired p value. Usually M is of little interest, but in this case, it provides what we want.

```
PROC UNIVARIATE;
  VAR DIFF;
RUN;
```

Box 10.5: Wilcoxon Signed Rank Test

PROC UNIVARIATE also performs the Wilcoxon signed rank test. The test statistic is labeled Sgn Rank and the corresponding p value for a two-sided alternative hypothesis is given by "Prob > $|S|$." The signed rank statistic, S, that SAS uses is the sum of the ranks of the positive values minus the sum expected under the null hypothesis, $n * (n + 1)/4$. In this example, S is 29.5 (= 82 − 52.5). As we are performing a one-sided test, the p value is 1/2 the value that SAS has reported; that is, it is 0.0338. SAS uses a t-distribution approximation to calculate the p value for n, the sample size, greater than 20. SAS adjusts its calculation of the t statistic for ties in the data.

```
PROC UNIVARIATE  DATA=BOOK.EXTREME;
  VAR DIFF;
RUN;
```

Box 10.6: Wilcoxon Rank Sum Test

PROC NPAR1WAY is used to perform the Wilcoxon rank sum test. The data to be used in this section are the percentages of calories coming from fat for the 33 boys whose data are shown in Table 4.1. We use the BOOK.NUTRIENT data file to create these percentages for the boys. After the new data file is created, we use PROC NPAR1WAY to carry out the analysis. The test statistic is the smallest sum of ranks and the p value is calculated using a normal and t-distribution approximation as well as a chi-square approximation. As SAS carried out the calculations to four decimal places, there are no ties in the data. Hence its test statistic, S, is 225, not the value 224.5 that is shown in the text. The p value from the normal distribution approximation is 0.6489; that from the t-distribution approximation is 0.6520; and that from the chi-square approximation is 0.6358. All of these are very close to the value calculated by MINITAB.

```
LIBNAME BOOK 'C:\';
DATA BOOK.FAT;
SET BOOK.NUTRIENT;
  PERCTFAT=9*FAT/CALORIES;
  GRADEVAR=1;
  IF GRADE > 6 THEN GRADEVAR=2;
PROC NPAR1WAY DATA=BOOK.FAT WILCOXON;
  CLASS GRADEVAR;
  VAR PERCTFAT;
RUN;
```

Box 10.7: Kruskal–Wallis Test

PROC NPAR1WAY is also used to perform the Kruskal–Wallis test. The data to be used here are the weight reductions shown in Table 10.8. The entry of the data is shown next.

```
DATA WEIGHT;
  INPUT  REDUCT  GROUP @@;
CARDS;
  38  1  10  1  10  1   28  1   6  1   8  1   33  1   8  1
  19  2  36  2  16  2   36  2  38  2  28  2   36  2  22  2
  42  2  24  2  40  2   34  2   6  2  16  2   30  2
  12  3  16  3   0  3  -12  3  14  3  16  3  -10  3   4  3
 -20  3  -6  3  18  3   16  3 -14  3   6  3  -16  3   6  3
;
```

The same commands used in Box 10.6 are used here because the Kruskal–Wallis test is an extension of the Wilcoxon rank sum test. SAS shows the sum of the ranks, the expected rank sums, and the mean rank for each of the groups. The test statistic is labeled CHISQ and its p value is given by 'Prob > CHISQ ='.

```
PROC NPAR1WAY WILCOXON;
  CLASS GROUP;
  VAR REDUCT;
RUN;
```

Box 11.1: Chi-Square Test for 2 by 2 Contingency Tables

PROC FREQ, used earlier to form contingency tables, can also be used to test the hypotheses of no association between the row and column variables. The data to be analyzed are the frequency counts for the crosstabulation of education and iron status for 100 women shown in Table 11.6. We first use a DATA step to create the sample data. The variable COUNT shows the number of women in each of the four cells created by the crosstabulation of education and iron status. We are using alpha variables as one way of making the output from PROC FREQ more understandable. The alpha values shown will be printed with the table. This is more useful than having levels indicated by a 1 and 2 printed.

```
DATA IRON;
  INPUT EDUC $  IRONST $  COUNT;
CARDS;
LT12  DEF   4
LT12  ACC  26
GE12  DEF   4
GE12  ACC  66
;
```

In PROC FREQ, we specified ORDER=DATA to tell SAS that we wanted the table shown in the same order as we entered it in the DATA step. If we had not specified ORDER=DATA, SAS would have used alphabetical order, placing the GE12 response before the LT12 response because G precedes L in the alphabet. It would also have reversed the order of DEF and ACC because A precedes D. The variable listed in the WEIGHT statement gives the frequency associated with each of the cells. The use of the CHISQ option requests that several different test statistics for the hypothesis of no association of the row and column variables be calculated. The test statistics we use are 'Chi-Square' (the Pearson chi-square) and 'Continuity Adj. Chi-Square' (the Yates chi-square) and its value is 0.783. PROC FREQ also calculates the value of Fisher's exact test which we mentioned. The use of MEASURES provides the value of the sample odds ratio as well as a number of other statistics that we have not discussed, but that are discussed in the documentation for PROC FREQ. The sample odds ratio is the 'Case–Control' estimate of the 'Relative Risk' and its value is 2.538. The corresponding confidence interval for the population odds ratio is also provided. The use of EXPECTED requests that the cell frequencies expected under the hypothesis of no association be printed.

```
PROC FREQ ORDER=DATA;
 WEIGHT COUNT;
 TABLES EDUC*IRONST / CHISQ MEASURES EXPECTED;
RUN;
```

Box 11.2: Chi-Square Test for *r* by *c* Contingency Tables

The data entry and the use of PROC FREQ are similar to that shown for the 2 by 2 table. The contingency table results from the crosstabulation of whether a woman knew someone with breast cancer with her opinion about mammography. The value of the chi-square test statistic is 6.648 and its p value is 0.036.

One additional feature is presented here that was not presented above. Because we also perform a test for no linear trend for these data, we have used the CMH option to generate an approximate test statistic for this hypothesis. The output associated with this hypothesis is found in the section on the 'Cochran–Mantel–Haenszel Statistics' and the test statistic similar to the one shown in the text for no trend is given by 'Nonzero Correlation.' Its value is 6.056, close to the value of 6.100 that we had calculated. The scores that PROC FREQ uses for the columns are 1, 2, and 3.

```
DATA CANCER;
  INPUT  KNOWL $  OPINION $  CT;
CARDS;
YES   POSITIVE  120
YES   NEUTRAL    45
YES   NEGATIVE   28
NO    POSITIVE   77
NO    NEUTRAL    15
NO    NEGATIVE    8
;
PROC FREQ ORDER=DATA;
 WEIGHT CT;
 TABLES KNOWL*OPINION / CHISQ EXPECTED CMH;
RUN;
```

Box 11.3: CMH Test Statistic and Estimate of the Odds Ratio

There is no Box 11.3 in the text but, because SAS calculates these statistics, we show this analysis. The data are on the relationship between outdoor pollution status and occurrence of upper respiratory infection while controlling for passive smoke in the home.

```
DATA URI;
  INPUT  SMOKE $  POLLUT $  URI $  WT;
CARDS;
YES HIGH SOME 100
YES HIGH NONE  20
YES LOW  SOME 124
YES LOW  NONE  40
NO  HIGH SOME 128
NO  HIGH NONE  62
```

```
NO  LOW  SOME 166
NO  LOW  NONE 119
;
```

The Cochran–Mantel–Haenszel test statistic used by SAS does not include the continuity correction. Its value is given by the 'General Association' test statistic. The Mantel–Haenszel estimate of the common odds ratio is given by the 'Case–Control' estimate of the common relative risk.

```
PROC FREQ ORDER=DATA;
 WEIGHT WT;
 TABLES SMOKE*POLLUT*URI / CMH;
RUN;
```

Box 12.1: Life-Table Method of Calculating Survival Probabilities

PROC LIFETEST is the SAS procedure used in the nonparametric analysis of life tables. In this section, we show its use in performing the life-table method of calculating survival probabilities for the data in Table 12.1. When the data are already in summary form, PROC LIFETEST requires a variable to indicate the frequency of those who have died or withdrawn. In the following, this is the variable named COUNT. It also requires a variable that indicates the censor status: died or withdrawn. This is the variable labeled C below. When C is 0, it is a death and when C is 1, it indicates a withdrawal. The option FREQ uses COUNT to indicate the frequencies of death and withdrawals.

```
DATA LTMETHOD;
  INPUT DIED WITHDREW YEAR @@;
  COUNT=DIED; C=0;
  OUTPUT;
  COUNT=WITHDREW; C=1;
  OUTPUT;
CARDS;
1421  68  0     335  56  1     132  101  2     64  57  3
  44  60  4      20  51  5      19   45  6     14  33  7
   7  35  8       7  28  9       5   18 10      5  21 11
   1  15 12       3  16 13       1   13 14      0   7 15
   0   9 16
;
```

In the TIME statement, the first variable is the variable representing time and the variable following the * is the variable representing the censor status. The value in parentheses after C is the code used to indicate withdrawal. It must be a numeric value. The PLOTS option asks SAS to plot the survival and hazard curves. The INTERVALS option specifies the time intervals to be used in the analysis. The METHOD option tells SAS to use the actuarial (life-table) method of analysis. If METHOD is not specified, then the product-limit method is used. In the output, SAS creates Tables 12.2 and 12.3 as well as providing an estimate of the pdf for the time variable. A crude estimated survival curve corresponding to Figure 12.3 is provided and an estimated hazard function is also plotted.

```
PROC LIFETEST PLOTS=(S,H) INTERVALS=(0 TO 16 BY 1) METHOD=ACT;
  TIME YEAR*C(1);
  FREQ COUNT;
RUN;
```

Box 12.2: Product-Limit Method of Calculating Survival Probabilities

Box 12.2 was not shown in the text, although it would have been similar to Box 12.1. PROC LIFETEST also can perform the product-limit method of analysis of survival tables. We use PROC LIFETEST in the analysis of the 14 alcohol-dependent subjects shown in Table 12.4.

```
DATA PLMETHOD;
  INPUT MONTHS CENSOR GENDER $;
CARDS;
 4 1 M
 6 1 M
 6 1 F
 9 2 M
10 1 F
14 2 M
16 1 M
17 2 M
19 1 F
20 1 F
28 1 M
31 1 F
34 2 F
47 2 F
;
```

As above, the TIME statement identifies the time variable and the variable following the * is the variable indicating whether the time is right-censored. The value in parentheses is the value used to indicate that the time is censored, and it must be numeric. We have run the analysis twice, once without and once with the STRATA option. STRATA is used to indicate group membership. The first analysis ignores the gender variable, whereas the second analysis takes gender into account. SAS creates most of the information in Table 12.5 as part of its output as well as providing a crude plot corresponding to Figure 12.4. In calculating its estimate of the standard error of the survival probabilities, SAS did not use the simplified approximation that we used in the text. In the output above the estimated survival curve, SAS also provides estimates of the median and the restricted mean survival times.

```
PROC LIFETEST PLOTS=(S);
  TIME MONTHS*CENSOR(2);
```

When STRATA is used, the survival probabilities and medians and restricted means are estimated separately for females and males. There is a plot of both of their estimated survival curves. There are three different test statistics for the null hypothesis of no difference in survival times for females and males. These tests are different from that discussed in the text, but we can recognize their form. If the null hypothesis is true, each of these three tests asymptotically follows a chi-square distribution with the indicated degrees of freedom. Their p values are also shown. All three tests agree with the test that we used in the text. There does not appear to be a statistically significant difference in the survival distributions for females and males.

```
PROC LIFETEST PLOTS=(S);
  TIME MONTHS*CENSOR(2);
  STRATA GENDER;
RUN;
```

Box 13.3: Testing a Hypothesis about the Mean Assuming σ is Unknown

PROC UNIVARIATE can be used to test a hypothesis about the mean assuming that σ is unknown. Part of the output from PROC UNIVARIATE is a test that the population mean is zero. We can convert that to a test that the population mean is μ_0, different from zero, by subtracting the hypothesized value of μ_0 from all the sample observations for that variable. To test the null hypothesis that μ is equal to 2400 calories, we subtract 2400 from all the observations of CALORIES in BOOK.NUTRIENT. The test statistic is labeled '*T*: Mean = 0' and the test statistic's value is -0.74125. The corresponding two-sided p value is labeled 'Prob > $|T|$ 0.xxxx' and its value is 0.4639.

```
LIBNAME BOOK 'C:\';
DATA TTEST;
SET BOOK.NUTRIENT;
  NEWCAL = CALORIES - 2400;
PROC UNIVARIATE;
  VAR NEWCAL;
RUN;
```

Box 13.4: Testing a Hypothesis about a Population Proportion

PROC UNIVARIATE can be used to test a hypothesis about the population proportion. The test uses a t statistic to test the null hypothesis that π equals π_0 instead of the z statistic shown in the text. In addition, the estimated standard error uses $n - 1$ in its denominator instead of n and the continuity correction term is not used. Hence there will be slight differences between the test statistic provided by PROC UNIVARIATE and that shown in the text. Additionally, the p value is calculated from the t distribution, not the normal. For large sample sizes, there will be little difference between the t and z test statistics and their corresponding p values. In the DATA step, we also must subtract the value of π_0 (= 0.75) from each of the observations. There were 86 children said to be immunized and 54 who were reported not to have been fully immunized.

```
DATA PROP;
  INPUT IMMUN @@;
  IMMUN = IMMUN - 0.75;
CARDS;
1 1 1 1 1 1 1 1 1 1 1 1 1 1 1 1 1 1 1 1 1 1
1 1 1 1 1 1 1 1 1 1 1 1 1 1 1 1 1 1 1 1 1 1
1 1 1 1 1 1 1 1 1 1 1 1 1 1 1 1 1 1 1 1 1 1
1 1 1 1 1 1 1 1 1 1 1 1 1 1 1 1 1 1 1 1
0 0 0 0 0 0 0 0 0 0 0 0 0 0 0 0 0 0 0 0 0 0
0 0 0 0 0 0 0 0 0 0 0 0 0 0 0 0 0 0 0 0 0 0
0 0 0 0 0 0 0 0 0 0
;
PROC UNIVARIATE;
 VAR IMMUN;
RUN;
```

Box 13.5: Testing a Hypothesis about the Correlation Coefficient

PROC CORR is used to test a hypothesis about the correlation coefficient. The data are the infant mortality rates and health expenditures shown in Table 13.1. The p value shown is for the test that the population correlation coefficient is zero. If the null hypothesis involves a nonzero value, then PROC CORR does not provide the appropriate test statistic. The p value differs slightly from that in the text because we rounded the correlation coefficient to three decimal places in the text, whereas PROC CORR used at least five decimal places in its calculations.

```
DATA IMF;
  INPUT  RATE  EXPEND @@;
CARDS;
4.8 6.8  5.8 9.0  6.1 7.4  6.8 8.5  6.8 7.7  7.2 8.6
7.5 8.2  7.5 6.0  7.8 8.6  8.1 6.0  8.1 7.1  8.3 7.5
```

```
8.7 7.1  8.9 7.4  9.0 6.1  9.2 7.2  9.3 6.9  10.0 11.2
10.8 6.9  11.0 5.3  13.1 6.4
;
PROC CORR;
  VAR RATE EXPEND;
RUN;
```

Boxes 13.6.1 and 13.6.2: Testing the Hypothesis of No Difference in Two Population Means Assuming Equal Variances

PROC TTEST is used to test the hypothesis of no difference in two population means. PROC TTEST provides a test statistic and p value assuming that the population variances are equal as well as assuming that they are unequal. In this section, we are assuming that the variances are equal. The data are the proportions of calories coming from fat for the boys whose data were originally displayed in Table 4.1. The variable specified in the CLASS statement identifies the two groups used in the analysis. We are not using the information in the test of the equality of the population variances.

```
LIBNAME BOOK 'C:\';
PROC TTEST DATA=BOOK.FAT;
  CLASS GRADEVAR;
  VAR PERCTFAT;
RUN;
```

Box 13.7: Testing the Hypothesis of a Difference of 5 Years in Two Population Means Assuming Unequal Variances

As was stated above, PROC TTEST is used to test the hypothesis of no difference in two population means. We now use the test statistic associated with the assumption that the variances are unequal. The data are the ages of the AML and ALL patients from Chapter 7. We must subtract 5 years from all the observations in the AML group.

```
DATA AGES;
  INPUT GROUP $  AGE  @@;
  IF GROUP = 'AML' THEN AGE=AGE-5;
  OUTPUT;
CARDS;
AML 20  AML 25  AML 26  AML 26  AML 27  AML 27  AML 28
AML 28  AML 31  AML 33  AML 33  AML 33  AML 34  AML 36
AML 37  AML 40  AML 40  AML 43  AML 45  AML 45  AML 45
AML 45  AML 47  AML 48  AML 50  AML 50  AML 51  AML 52
AML 53  AML 53  AML 56  AML 57  AML 59  AML 59  AML 60
AML 60  AML 61  AML 61  AML 61  AML 62  AML 63  AML 65
AML 71  AML 71  AML 73  AML 73  AML 74  AML 74  AML 75
```

```
AML 77   AML 80
ALL 18   ALL 19   ALL 21   ALL 22   ALL 26   ALL 27   ALL 28
ALL 28   ALL 28   ALL 28   ALL 34   ALL 36   ALL 37   ALL 47
ALL 55   ALL 56   ALL 59   ALL 62   ALL 83   ALL 19
;
PROC TTEST;
  CLASS GROUP;
  VAR AGE;
RUN;
```

Box 13.8: Paired *t* Test

PROC UNIVARIATE can be used to test for no difference in two dependent population means. The variable to be used in the analysis is the difference of the sample observations and this difference is created in a DATA step.

Box 13.9: Testing a Hypothesis about the Difference of Two Proportions

PROC TTEST can be used to provide an approximate test statistic for the test of the equality of two population proportions. The estimate of the standard error of the difference differs slightly from that found using the binomial formula because of the division by $n - 1$ in PROC TTEST instead of n used in the binomial calculation. The reported p value is also slightly off as the t distribution instead of the normal distribution is used in its calculation. For large samples, these differences are small.

```
DATA MILK;
  INPUT REGION $  COMPLY  @@;
CARDS;
NE 1  NE 1  NE 1  NE 1  NE 1  NE 1  NE 1  NE 1  NE 1  NE 1
NE 1  NE 1  NE 0  NE 0  NE 0  NE 0  NE 0  NE 0  NE 0  NE 0
NE 0  NE 0  NE 0  NE 0  NE 0  NE 0  NE 0  NE 0  NE 0  NE 0
NE 0  NE 0  NE 0  NE 0  NE 0  NE 0  NE 0  NE 0  NE 0  NE 0
NE 0  NE 0
SW 1  SW 1  SW 1  SW 1  SW 1  SW 1  SW 1  SW 1  SW 1  SW 1
SW 1  SW 1  SW 1  SW 1  SW 1  SW 1  SW 1  SW 1  SW 1  SW 1
SW 1  SW 0  SW 0  SW 0  SW 0  SW 0  SW 0  SW 0  SW 0  SW 0
SW 0  SW 0  SW 0  SW 0  SW 0  SW 0  SW 0  SW 0  SW 0  SW 0
SW 0  SW 0  SW 0  SW 0  SW 0  SW 0  SW 0  SW 0  SW 0  SW 0
;
PROC TTEST;
  CLASS REGION;
  VAR COMPLY;
RUN;
```

Boxes 14.1 and 14.2: One-Way ANOVA and Multiple Comparisons

PROC ANOVA can be used to analyze the age data shown in Table 14.1. PROC ANOVA uses a MODEL statement to identify the dependent and independent variables. The variable to the left of the equal sign is the dependent variable and the variable to the right of the equal sign is the independent variable. The independent variable is also identified in the CLASS statement. The MEANS statement tells SAS that we wish to see the mean of the dependent variable for each level of the variable shown in the MEANS statement. The words after the / symbol indicate which types of multiple comparisons we wish to use in the analysis. For the Dunnett procedure, we must specify which level of the independent variable is to be used in the comparisons with the other levels.

```
DATA AGES;
  INPUT GROUP $  AGE @@;
CARDS;
SURG 32  SURG 28  SURG 22  SURG 25  SURG 20  SURG 20  SURG 28
SURG 28  SURG 20  SURG 29  SURG 22  SURG 37  SURG 18  SURG 29
SURG 22  SURG 32  SURG 21  SURG 34  SURG 19  SURG 23  SURG 23
SURG 26  SURG 41  SURG 20  SURG 33
CON1 32  CON1 26  CON1 31  CON1 39  CON1 34  CON1 33  CON1 29
CON1 41  CON1 35  CON1 33  CON1 33  CON1 43  CON1 25  CON1 39
CON1 36  CON1 37  CON1 28  CON1 34  CON1 27  CON1 45  CON1 22
CON1 29  CON1 51  CON1 28  CON1 35
CON2 31  CON2 35  CON2 26  CON2 28  CON2 22  CON2 29  CON2 27
CON2 21  CON2 22  CON2 27  CON2 24  CON2 44  CON2 21  CON2 25
CON2 27  CON2 18  CON2 27  CON2 36
;
PROC ANOVA DATA=AGES;
  CLASS GROUP;
  MODEL AGE=GROUP;
  MEANS GROUP / TUKEY LSD DUNNETT ('SURG');
RUN;
```

Box 14.3: ANOVA for Randomized Block with *k* Replicates per Cell

PROC ANOVA can also be used with the two-way ANOVA (or other more general ANOVAs as well). The model shown here does not include any interaction terms. The MODEL statement again has the dependent variable to the left of the equal sign and the independent variables to the right of the equal sign. The MEANS statement indicates that we wish to see the mean of the dependent variable shown for the levels of both of the independent variables.

```
DATA WEIGHT;
  INPUT PROG $  SITE $  WTCHANGE  @@;
CARDS;
DIET OFFICE 6  DIET OFFICE 2  DIET OFFICE 10  DIET OFFICE -1
DIET OFFICE 8
DIET FACTORY 3  DIET FACTORY 15  DIET FACTORY 4  DIET FACTORY 8
DIET FACTORY 6
EXERCISE OFFICE 3  EXERCISE OFFICE 4  EXERCISE OFFICE -2
EXERCISE OFFICE 6  EXERCISE OFFICE -2
EXERCISE FACTORY -4  EXERCISE FACTORY 6  EXERCISE FACTORY 8
EXERCISE FACTORY -2  EXERCISE FACTORY 3
BOTH OFFICE 8  BOTH OFFICE 12  BOTH OFFICE 7  BOTH OFFICE 10
BOTH OFFICE 5
BOTH FACTORY 15  BOTH FACTORY 8  BOTH FACTORY 10
BOTH FACTORY 16  BOTH FACTORY 3
;
PROC ANOVA;
  CLASS PROG SITE;
  MODEL WTCHANGE = PROG SITE;
  MEANS PROG SITE;
RUN;
```

Box 14.4: Balanced Two-Way ANOVA with Interaction

As you might have guessed, PROC ANOVA can also be used here to analyze the data shown in Table 14.6. Now that we are familiar with the simple, but very tedious, way of entering the data that we have used throughout, we complicate the input section a little by using DO and ARRAY statements. The DO statement indicates how many times the statements included between the DO statement and its closing END statement are to be performed. One form of the ARRAY statement provides the name of the variable that will store multiple values, and this name is followed by the names of each of the variables storing one of the values. In the following, the variable REP is the name of the array, and there are six values to be stored in REP1 through REP6. The first value of the variable METHOD is lecture and the value of BOOK is 1. We next read the six values, REP1 through REP6. The DO statement is used to create six records containing the values for METHOD, BOOK, and INCREASE. The first value of INCREASE is REP1, the second value is REP2, and so on through REP6. Thus for each of the six lines of INPUT, we have created six data points that are used in PROC ANOVA.

```
DATA SCORES;
 ARRAY REP REP1 - REP6;
  INPUT  METHOD $ BOOK REP1 - REP6;
    DO OVER REP;
      INCREASE = REP;
      OUTPUT;
    END;
CARDS;
LECTURE 1 30 43 12 18 22 16
LECTURE 2 21 26 10 14 17 16
LECTURE 3 42 30 18 10 21 18
DISCUSS 1 36 34 15 18 40 45
DISCUSS 2 33 31 28 15 29 26
DISCUSS 3 41 46 19 23 38 48
;
PROC ANOVA;
  CLASS METHOD BOOK;
  MODEL INCREASE = METHOD BOOK METHOD*BOOK;
  MEANS METHOD BOOK;
RUN;
```

In the MODEL statement, we have the term METHOD*BOOK which means to include the interaction of the two independent variables in the analysis as well as the two main effects. It was not necessary to write the terms METHOD and BOOK in the MODEL statement, as the use of the METHOD*BOOK term also tells SAS to include the main effect terms.

```
PROC ANOVA;
  CLASS METHOD BOOK;
  MODEL INCREASE = METHOD*BOOK;
  MEANS METHOD BOOK;
RUN;
```

Box 15.1: Simple Linear Regression

PROC REG is one of the procedures that can be used to perform linear regression analyses in SAS. The following shows the SAS statements to perform the simple linear regression of SBP on height. The data are shown in Table 15.1.

```
DATA BP;
  INPUT SBP HEIGHT @@;
CARDS;
105 36   90 37   82 38   96 38   82 39   74 39  104 40 100 40
 80 41   98 42   96 43   86 44   88 44  128 44  118 45  90 46
108 48  120 48  114 49   78 49  116 50   74 50   80 51  98 52
 90 53   92 54   80 54   88 54  104 55  100 56  126 57 108 58
```

```
106 59   98 59   94 59   88 60  110 60  124 60   86 61 120 61
112 62  100 63  122 64  122 64  110 65  124 65  122 66  94 67
110 67  140 69
;
```

In the regression model, the dependent variable is the variable to the left of the equal sign in the MODEL statement and the independent variable is to the right of the equal sign. The options specified in the MODEL statement, *R* and INFLUENCE, cause the predicted, observed, and residual values to be printed as well as the leverage values, labeled as Hat Diag, *H*, for each observation. Several other statistics are also printed, but we do not discuss them. The calculations performed in Box 15.2 are part of this SAS session.

```
PROC REG DATA=BP;
  MODEL SBP = HEIGHT / R INFLUENCE;
RUN;
```

Box 15.2: 95 Percent Confidence and Prediction Intervals

PROC REG also produces the 95 percent confidence and prediction intervals as is shown below. The option CLI prints the 95 percent prediction intervals and CLM prints the 95 percent confidence interval for the mean of *Y* for each value of *X*. If we choose not to print the statistics from PROC REG, and focus only on the plot of the intervals, we can also use the NOPRINT option of the MODEL statement. In the following, U95. and L95. are keywords in SAS that refer to the upper and lower 95 percent prediction interval values. P. is another SAS keyword that refers to the predicted value of the dependent variable and U95M. and L95M. are keywords that refer to the upper and lower 95 percent confidence interval values for the mean of *Y* at each value of the independent variable. All five of these variables are plotted versus HEIGHT, the independent variable. The OVERLAY option means that all three of these plots are shown in a single figure. The symbol *P* will represent the upper and lower 95 percent prediction intervals, *R* shows the predicted regression line, and *C* shows the upper and lower 95 percent confidence intervals. The plot that is produced is very crude and cluttered. It is probably better not to include the prediction interval and the intervals for the mean in the same graph. To improve the quality of the plot, it is also possible to create an output file containing the points to be plotted and then to use PROC GPLOT to create the plots. This could be done with some of the other plots that are built into some of the other procedures as well.

```
PROC REG DATA=BP;
  MODEL SBP = HEIGHT / P CLI CLM;
  PLOT (U95. L95.)*HEIGHT='P' P.*HEIGHT='R'
       (U95M. L95M.)*HEIGHT='C' / OVERLAY;
RUN;
```

Box 15.3: Stepwise and All Possible Regressions

PROC REG can also be used to perform stepwise and all possible regressions in the multiple regression setting. The data to be analyzed are those shown in Table 15.6. The CORR option of PROC REG requests that the correlation matrix of all the variables listed in the MODEL statement be printed. It is printed as a square matrix with 1's—the correlation of a variable with itself—printed down the diagonal. The SELECTION option tells SAS what method of analysis should be used. The selection of STEPWISE tells SAS to use a forward stepwise regression method, which allows for a variable to be entered or deleted at each step. SAS prints the *F* test statistic, the square of the *t* value, instead of the *t* statistic printed by MINITAB as part of the STEPWISE output. When SELECTION is ADJRSQ, SAS performs all possible regressions using the adjusted *r*-square measure as the criterion for determining the best model. By including CP, we have requested that its values also be printed as part of the output. Calculations in Boxes 15.4 and 15.5 are part of this SAS session.

```
DATA BOOK.MULTIPLE;
  INPUT  HEIGHT  SBP  AGE  WEIGHT  @@;
CARDS;
36  105   7  57        37   90   7   46        38  82  6  42
38   96   7  52        39   82   8   56        39  74  7  45

       .                    .                        .
       .                    .                        .
       .                    .                        .

67  110  13  142       69  140  13  122
;
PROC REG  CORR;
  MODEL SBP = HEIGHT  AGE  WEIGHT / SELECTION=STEPWISE;
  MODEL SBP = HEIGHT  AGE  WEIGHT / SELECTION=ADJRSQ CP;
RUN;
```

Boxes 15.4 and 15.5: Multiple Linear Regression

PROC REG can also be used to perform the multiple regression analysis. The SAS statements for analyzing the data in Table 15.6 are the following.

```
PROC REG DATA=BOOK.MULTIPLE;
  MODEL SBP = HEIGHT AGE WEIGHT;
  MODEL SPB = WEIGHT;
RUN;
```

Box 15.6: Logistic Regression Analysis

There is no Box 15.6 in the text as the current version of MINITAB does not have a logistic regression command; however, SAS has PROC LOGISTIC for performing logistic regression and its use is shown here for the lead, smoking, and pulmonary function test results in Table 15.8. We are using 0 to represent a normal PFT result and 1 to indicate a result that was not normal. We must create our own coding of the smoking variable. The codes 1, 2, 3, and 4 represent the smoking levels of never, former, light, and heavy, respectively. In the DATA step, we create three smoking status variables. The variable SMOK1 has the value of 1 if SMOKING is 1 and 0 otherwise; the variable SMOK2 will has the value of 1 if SMOKING is 2 and 0 otherwise; and the variable SMOK3 has the value of 1 if SMOKING is 3 and 0 otherwise. This means that the heavy level of smoking is the reference level.

```
DATA PFT;
  INPUT LEAD  SMOKING  PFT  COUNT;
  SMOK1=0;  SMOK2=0;  SMOK3=0;
  IF SMOKING = 1 THEN SMOK1=1;
  IF SMOKING = 2 THEN SMOK2=1;
  IF SMOKING = 3 THEN SMOK3=1;
  OUTPUT;
CARDS;
1 1  0  160
1 1  1    4
1 2  0   49
1 2  1    6
1 3  0   75
1 3  1    6
1 4  0   84
1 4  1    3
0 1  0   33
0 1  1    3
0 2  0   12
0 2  1    2
0 3  0   21
0 3  1    2
0 4  0   16
0 4  1    3
;
PROC LOGISTIC;
  WEIGHT COUNT;
  MODEL PFT = LEAD SMOK1 SMOK2 SMOK3;
RUN;
```

II. STATA COMMANDS

Stata is similar to MINITAB in that it is command oriented instead of having many functions grouped into a procedure as is done in SAS. As stated above, we are demonstrating the use of Stata; we are not showing all the features available. In addition, just as with SAS, we do not show the boxes in which MINITAB was used simply as a desk calculator or those boxes in which MINITAB simply provided values of pdf's or cdf's. We are assuming that each box below was created during a separate Stata session unless we indicate otherwise. This means that the **use** command (discussed below) is specified in each box that uses a Stata data set that we previously created. Stata uses a . to indicate when it is ready for an additional command. We enter a carriage return (Enter) at the end of each line or, if there is more information than fits on a line, we continue typing until we have entered all the required information and then we enter the carriage return. The **label** definition statement in Box 4.1 provides an example of how Stata handles information that takes more than one line. Information shown in parentheses after a Stata command tells what action Stata took in response to the command.

Box 4.1: Data Entry from the Keyboard

To enter data from the keyboard, we use the **input** command. The names of the variables being entered are listed after the word **input.** For the data in Table 4.1, we use the following format. We enter the values of the variables after Stata provides the line numbers. There is a carriage return (Enter) at the end of each line. The **end** statement tells Stata that all the data have been entered.

```
input day grade calories protein fat vita
        day      grade      calories      protein      fat      vita
  1. 3 8 1823  83  63  4876
  2. 4 8 2007  64  62  6202
  3. 4 8 1053  23  33   964
  4. 5 8 4322 128 202  6761
       .     .     .
       .     .     .
       .     .     .
 33. 3 5 1723  45  43  5703
 34. end
```

We use the **label define** statement to name the levels of the variables. For example, for the day of the week variable, we can assign the labels shown in Table 4.2 as follows. The dayfmt variable now contains the labels, and the use of the **label values** statement connects the day variable with the labels in daylab.

```
. label define daylab 1 "Sun" 2 "Mon" 3 "Tue" 4 "Wed" 5 "Thu" 6
> "Fri" 7  "Sat"
. label values day daylab
```

When the day variable is used, the labels assigned to it will appear instead of the numeric codes that were entered originally.

To save this file, we use the **save** command. Stata appends **.dta** to the file name, indicating that this is a Stata data file. We are saving the file on drive c in subdirectory book, hence the **c:\book** before the file name.

```
. save c:\book\nutrient
(file c:\book\nutrient.dta saved)
```

To print some of the values to see if they have been entered correctly, we use the **list** command with the **in** range which specifies which records are to be printed. Suppose that we wished to print only the first 10 records. The command for doing this is the following.

```
. list in 1/10
```

If we wish to see only the values for calories in records 5 to 20, the list command would be the following.

```
. list calories in 5/20
```

If we have made an error in data entry, we can use the **replace** command to correct the error. For example, suppose that we entered the wrong value for vita for the second boy in the data set. To **replace** the incorrect value, we do the following.

```
. replace vita=6202 in 2
(1 change made)
```

Box 4.2: Data Entry from a Non-Stata File

Again we assume that the non-Stata file looks exactly like the file shown in Box 4.1, except that the observation numbers are not included. Suppose that the file is named nutrient.dat. The file is entered into Stata using the **infile** and **using** commands. The variable names follow the **infile** statement and the file location and name follow the **using** statement.

```
. infile day grade calories protein fat vita using c:\book\nutrient.dat
(33 observations read)
```

We can use the **desc** command to describe the file.

```
. desc
```

We wish to save this file as a Stata file, so we use the **save** command. Note that the input file was not a Stata file. Its name was nutrient.dat, which is different from nutrient.dta, a Stata file.

```
. save c:\book\nutrient
(file c:\book\nutrient.dta saved)
```

Once a file is saved as a Stata file, we access it by the **use** command as is shown in Box 4.3.

Box 4.3: Creation of One- and Two-Way Tables

The **tabulate (tab)** command is used to create one- and two-way tables in Stata. The one-way table for day of the week is created by the following commands. Note that in the **use** command, we do not have to include the **.dta** modifier for Stata files.

```
. use c:\book\nutrient
. tabulate day
```

In the two-way table, we wish to crosstab day of the week by a variable that indicates whether the calories are less than 2500. Therefore, we first must create this new indicator variable. We **generate (gen)** a new variable whose values are all 0 and then we use the **replace** command and the **if** statement to change the value of the new variable to 1 if calories are greater than or equal to 2500.

```
. gen codecal=0
. replace codecal=1 if calories > 2499
(11 changes made)

. tab day codecal
```

Boxes 4.4 and 4.5: Creation of a Plot Showing Three Line Graphs

The data to be graphed are the health expenditures as a percentage of GDP by year for Great Britain, West Germany, and the United States. The data are entered as shown below.

```
. input GB US WGERM year
                GB          US        WGERM         year

 1.  3.9  5.2  4.7  1960
 2.  4.1  6.0  5.1  1965
 3.  4.5  7.4  5.5  1970
 4.  5.5  8.4  7.8  1975
 5.  5.8  9.2  7.9  1980
 6.  6.0 10.6  8.2  1985
 7.  6.1 11.2  8.2  1987
 8. end
```

We use the **graph (gr)** command to create the plots. Each of the countries' health expenditures as a percentage of GDP are to be plotted on the same graph versus the year variable. The last variable listed is the variable for the horizontal axis. The **t1title (t1)** is the title at the top of the plot. The **l1title (l1)** is the title on the left axis. If there are no special symbols in the title, the " symbols are not necessary. The **connect()** option tells Stata what method to use to connect the points for GB, US, and WGERM. Below, we have entered **connect(lll)**. The code of **l** tells Stata to use straight lines to connect the points. As there are three 1's, we wish to connect the points by straight lines for all three countries. The **symbol (s)** option tells Stata which symbols to use for the points to be plotted. The **O** indicates that a large circle is to be used for GB values, **S** means that a large square is to be used for US values, and **T** indicates that a large triangle is to be used for WGERM values. Stata does not plot its graphs within Stata, but requires that the file be saved. The **saving** command saves a file named expend.gph. Once we exit from Stata using the **exit, clear** statement, we can then plot the graph on the printer. If we do not wish to have a printout of the plot, then we do not need to use the **saving** option.

```
. gr GB US WGERM year, tl("Health Exp. as a % of GDP by Year") l1("% of
> GDP") connect(lll) s(OST) saving(expend)
. exit, clear
```

In response to the DOS prompt, we use the **gphdot** command and provide the file name.

```
c:\STATA> gphdot c:expend.gph
```

Box 4.6: Creation of a Histogram

The **graph** command is also used to create histograms in Stata. The data are entered from the keyboard and saved in a file bp.dta. **Histogram** is the option that causes the histogram to be created. **Bin(x)** specifies that x intervals are to be used. If **bin(x)** is not specified, Stata uses 5 intervals. We have specified that 11 intervals be used for this histogram. **Freq** tells Stata that the vertical axis is to be labeled in frequency units rather than fractional units. Because we are unable to specify the midpoints of the intervals or the starting point in Stata, this histogram may differ from that shown in Figure 4.8. We can also use the **xlabel()** and **ylabel()** statements to provide more information on the x and y axes. The labels that we want shown are included in the parentheses. In this and all future demonstrations of the graph command, we are not showing the exiting from Stata to draw the graphs as was shown in the box immediately above.

```
. input sbp
                   sbp

  1. 130
  2. 100
  3. 125
     .
     .
     .
100. 125
101. end
```

```
. save c:\book\bp
(file c:\book\bp.dta saved)
. graph sbp, histogram bin(11) freq xlabel(80,92,104,116,128,140)
> ylabel(0,4,8,12,16,20,24) saving(histsbp)
```

Box 4.10: Creation of a Scatter Plot

The **graph** command is also used to create scatter plots. The data to be used are the total fat and protein values in the file nutrient.dta. Therefore, we access that file by the **use** command. **Twoway** is the option that causes the creation of the scatter plot. When there are two variables and no option is specified, Stata assumes that twoway is the intended option.

```
. use c:\book\nutrient
. gr fat protein, twoway xlabel(0,30,60,90,120,150,180)
> ylabel (0,70,140,210) saving(scat)
```

Boxes 4.11 to 4.13: Creation of Box Plots and Univariate Statistics

The box plot for the vitamin A data in the nutrient.dta file is created using the **graph** command and the **box** option as follows.

```
. use c:\book\nutrient
. gr vita, box ylabel(0,2500,5000,7500,10000,12500)
saving(boxvita)
. exit, clear
```

We also wish to create two box plots for the systolic blood pressure values in Table 4.5. This is easily accomplished in Stata by creating an indicator variable that has the value of 0 for the first 50 observations and the value of 1 for the next 50 observations. The symbol **_n** is a Stata variable that is the observation number. The **by** option tells Stata to draw box plots of sbp for each level of the variable in parentheses following the **by** statement.

```
. use c:\book\bp
. gen dummy=1
. replace dummy=0 if _n < 51
(50 changes made)

. graph sbp, box by(dummy) ylabel(80,100,120,140)
saving(boxsbp)
. exit, clear
```

We also want to have the descriptive statistics for calories, protein, total fat, and vitamin A. Stata uses the **summarize (summ)** command with the **detail** option to have the descriptive statistics, including the key percentiles calculated.

```
. use c:\book\nutrient
. summ calories protein fat vita, detail
```

Box 4.14: Sorting the Data

Sorting is easy with Stata. For example, to sort the nutrient.dta file based on calories, we use the **sort** statement. We use the **list** command to show that the calories have been sorted.

```
. use c:\book\nutrient
. sort calories
. list calories
```

If we do not save this file, the next time we access it with the **use** statement, it will be in its original unsorted order.

Box 4.18: Calculation of the Correlation Coefficient

Stata uses the **correlate (corr)** statement to calculate the Pearson correlation coefficient.

```
. use c:\book\nutrient
. corr protein fat
```

Box 4.19: Scatter Plot of a Quadratic Relationship

We can easily show the plot of a quadratic relationship with Stata using the **graph** command and the **twoway** option. We first enter the data to be used and then plot them. The **connect(s)** statement tells Stata to connect the points, and **s** means to smoothe the line.

```
. input y x
                    y                    x

  1.  4  -2
  2.  1  -1
  3.  0   0
  4.  1   1
  5.  4   2
  6. end

. graph y x, twoway c(s) saving(quad)
```

Box 4.20: Spearman Rank Correlation Coefficient

We can use the **genrank** statement along with the **correlation** command to calculate the Spearman correlation coefficient; however, it is easier to use the **spearman** command which calculates the correlation in one step as is also shown below.

```
. use c:\book\nutrient
. genrank rankprot = protein
. genrank rankfat = fat
. corr rankprot rankfat
. spearman protein fat
```

Box 5.1: Creation of 10 Samples of 30 Birthdates

Stata uses the **uniform()** statement in the creation of 10 samples of 30 birthdates. The **uniform()** statement produces random numbers in the range 0 to 1. We convert them to integers in the following **generate** statement. We must also **set** the **seed** to obtain different sets of uniform random numbers and **set** the number of **observations** to 1500. We only need 300 birthdates, but because we will obtain many numbers outside the range of 1 to 365, we create extra numbers to give a high probability of obtaining 300 numbers in the range 1 to 365. The **int()** function returns the integer portion of the number contained in parentheses. We use the **drop** command to delete observations not in the desired range. The == sign tells Stata to drop values of y if they equal to zero. The double equal sign is also used in the **generate** command. Stata requires the == sign when it wishes to test for equality in an **if** statement. If only one equal sign is used, Stata is confused and gives an error message. We use the **sort** command to sort the observations in sets of 30 and then we use the **list** statement to list the set. We repeat the **sort** and **list** statements 10 times to list the first 300 birthdates in sets of 30.

```
. set seed -3
. set obs 1500
(obs was 0, now 1500)

. gen x=1000*uniform()
. gen y=int(x)
. drop if y > 365
(931 observations deleted)

. drop if y==0
(3 observations deleted)

. sort y in 1/30
. list y in 1/30
. sort y in 31/60
. list y in 31/60
.     .
.     .
.     .
. sort y in 271/300
. list y in 271/300
```

Box 6.16: Finding the cdf for a *N*(80,10) Variable

Normprob() is used to find the cumulative probability for a standard normal variable. We **set** the number of **observations** equal to 1, transform the observation to a standard normal, and then use the **normprob()** statement.

```
. set obs 1
(obs was 0, now 1)

. gen z=(95-80)/10
. gen prob=normprob(z)
. list prob in 1
```

Box 6.17: Finding the Inverse cdf for a *N*(80,10) Variable

The **invnorm()** command is used to find the inverse cdf value for a standard normal variable.

```
. set obs 1
(obs was 0, now 1)

. gen z=invnorm(0.95)
. gen x=z*10+80
. list x in 1
```

Box 6.19: Normal Probability Plot

We can create a normal probability plot by using the **qnorm** command. The horizontal and vertical axes have the values of the random variable as the labels instead of the cdf values. We are creating the plot for the vitamin A variable in the nutrient data set. The plot provided by Stata has the observed variable, vita, on the vertical axis, instead of the horizontal axis as shown in the text.

```
. use c:\book\nutrient
. qnorm vita ylabel(0,2500,5000,7500,10000,12500)
> xlabel(0,2500,5000,7500,10000,12500)
```

Boxes 10.1 and 10.4: Data Entry of Seven Smallest and Seven Largest Day 1 Values, Descriptive Statistics, and Sign Test

The data are entered as shown below and the **summarize** command provides the summary descriptive statistics. The **signtest** is used to test the hypothesis that π equals 0.50. The rejection region is specified in the output and the two-sided p value based on the normal approximation is also provided.

```
. input day1 day2
                    day1                    day2

  1. 1053 2484
  2. 4322 2926
```

```
 3. 1753 1054
 4. 3532 3289
 5. 2842 2849
 6. 1505 1925
 7. 3076 2431
 8. 1292  810
 9. 3049 2573
10. 3277 2185
11. 1781 1844
12. 2773 3236
13. 1645 2269
14. 1723 3163
15. end

. sort day1
. gen d1=day1 if _n < 8
(7 missing values generated)

. gen d2=day2 if _n < 8
(7 missing values generated)

. gen d3=day1 if _n > 7
(7 missing values generated)

. gen d4=day2 if _n > 7
(7 missing values generated)

. summ d1-d4, detail
```

The **replace** command used with the **if** statement is used to change some of the values for a variable. In the following, we wish to change the order of subtraction for the seven smallest day 1 values.

```
. gen diff=day1-day2
. replace diff=day2-day1 if _n < 8
(7 changes made)

. gen zero=0
. signtest diff=zero
```

Box 10.5: The Wilcoxon Signed Rank Test

The command **signrank** is used to perform this test. A two-sided p value based on the normal approximation is provided. We are assuming here that we have not exited Stata, but are continuing with the Stata session from Box 10.4.

```
. signrank diff=zero
```

Box 10.6: The Wilcoxon Rank Sum Test

The data to be analyzed are the percentages of calories coming from fat for the fifth and sixth grade boys compared with seventh and eighth grade boys. The **wilcoxon** command is used to perform the test. The data in c:\book\nutrient are ordered with the 19 seventh and eighth grade boys appearing first followed by the 14 fifth and sixth grade boys. The two-sided p value based on the normal approximation is provided.

```
. use c:\book\nutrient
. gen fatcal=9*fat
. gen perfat=fatcal/calories
. gen gradecod=0
. replace gradecod=1 if _n > 19
(14 changes made)

. save c:\book\fatdata
(file c:\book\fatdata.dta saved)
. wilcoxon perfat, by(gradecod)
```

Box 10.7: Kruskal–Wallis Test

The data to be analyzed are the weight reductions from Table 10.8. The command **kwallis** is used to perform the test of no difference in location across the three groups. The sum of the group ranks and the chi-square test statistic and its p value are provided.

```
. input reduct group
           reduct              group

  1. 38 1
  2. 10 1
  3. 10 1
  4. 28 1
  .  .  .
  .  .  .
  .  .  .
 39.  6 3
 40. end

. kwallis reduct, by(group)
```

Box 11.1: Chi-Square Test for 2 by 2 Contingency Tables

We wish to analyze the education and iron status data to determine if there is a statistically significant relationship between these variables. The **tabulate** command with the **chi2** option is used to test the hypothesis of no relationship. The test statistic is the uncorrected Pearson chi-square statistic. If we also specify **exact,** we obtain the p value from Fisher's exact test. The variables to be crosstabulated are specified, and the variable containing the cell frequencies follows the equal sign.

```
. input educ ironst count
           educ           ironst   count

  1. 1   1  4
  2. 1   2 26
  3. 2   1  4
  4. 2   2 66
  5. end

. tab educ ironst = count, chi2 exact
```

Box 11.2: Chi-Square Test for *r* by *c* Contingency Tables

The data entry and the use of **tabulate** are similar to that shown for the 2 by 2 table. The contingency table results from the crosstabulation of whether a woman knew someone with breast cancer with her opinion about mammography. There is no directly calculated test for a linear trend.

```
. input knowl  opinion  ct
           knowl       opinion              ct

  1. 1 1 120
  2. 1 2  45
  3. 1 3  28
  4. 2 1  77
  5. 2 2  15
  6. 2 3   8
  7. end

. tab knowl opinion = ct, chi2
```

Box 12.2: Product-Limit Method of Calculating Survival Probabilities

The data are the times to relapse of the 14 alcohol-dependent patients shown in Table 12.4. We use several commands in Stata to summarize the survival (time to relapse) experience of the 14 patients. The command **kapmeier** is used to graph the survival curve. The command **gwood** adds confidence bands to the survival curve.

The command **survsum** displays summary statistics for the survival data. The command **mantel** is used when we wish to test for a difference in two survival curves. The data include the time to relapse or to a censoring variable, a dead variable that is 0 if the subject's time is censored and 1 if the subject died (relapsed). If there is more than one group, there is also a group variable. We use a 1 to indicate that the subject is female and a 2 to indicate a male.

```
. input time relapse sex
                      time            relapse            sex

  1.  4 1 2
  2.  6 1 1
  3.  6 1 2
  4.  9 0 2
  5. 10 1 1
  6. 14 0 2
  7. 16 1 2
  8. 17 0 2
  9. 19 1 1
 10. 20 1 1
 11. 28 1 2
 12. 31 1 1
 13. 34 0 1
 14. 47 0 1
 15. end
```

We first provide summary statistics: the estimated mean and median survival times for each group.

```
. survsum time relapse, by(sex)
```

Next we create a graph of the survival curves for females and males. As we have told Stata not to use any special symbols, it uses the numeric codes of 1 and 2 for females and males to indicate the group.

```
. kapmeier time relapse, by(sex)
```

We add a confidence band to the overall survival curve.

```
. gwood time relapse
```

We now test the hypothesis of no difference in the survival distributions for females and males. Stata uses a normal statistic in the test of hypothesis. If we square the value Stata calculates, we have the chi-square statistic shown in the text. Stata also shows the expected and observed values.

```
. mantel time relapse, by(sex)
```

Box 13.3: Testing a Hypothesis about the Mean Assuming σ Is Unknown

Stata uses the **ttest** command to test a hypothesis about the mean when σ is unknown. To test the hypothesis that the mean of calories is 2400, we use the following format. Stata provides the mean and standard deviation, the value of the *t* statistic, its degrees of freedom, and its two-sided *p* value.

```
. use c:\book\nutrient
. ttest calories=2400
```

Box 13.4: Testing a Hypothesis about a Population Proportion

The **ttest** command can be used to test a hypothesis about the population proportion. The test uses a t statistic to test the null hypothesis that π equals π_0 instead of the z statistic shown in the text. In addition, the estimated standard error uses $n - 1$ in its denominator instead of n and the continuity correction term is not used. Hence there will be slight differences between the test statistic provided by **ttest** and that shown in the text. Additionally, the p value is calculated from the t distribution, not the normal. For large sample sizes, there will be little difference between the t and z test statistics and their corresponding p values. We use the **generate** command to create the data. We create 140 observations with the value of 1 and then replace 54 of them with the value of 0. It is easier to do this than to enter 140 observations.

```
. set obs 140
(obs was 0, now 140)

. gen immun=1
. replace immun=0 in 87/140
(54 changes made)

. ttest immun=0.75
```

Boxes 13.6.1 and 13.6.2: Testing the Hypothesis of No Difference in Two Population Means Assuming Equal Variances

The **ttest** command is again used to test the hypothesis of no difference in two population means. The **ttest** command provides a test statistic and p value assuming that the population variances are equal. If we do not wish to assume equal variances, we use the **unequal** option. In this section, we are assuming that the variances are equal. The data are the proportions of calories coming from fat for the boys whose data were originally displayed in Table 4.1 and saved in the file c:\book\fatdata.dta. We must convert the perfat variable into two variables, fat1 and fat2, corresponding to the gradecod variable. The == here is used to test if gradecod is equal to 1 and if gradecod is equal to 0.

```
. use c:\book\fatdata
. gen fat1=perfat if gradecod==1
(19 missing values generated)

. gen fat2=perfat if gradecod==0
(14 missing values generated)

. ttest fat1=fat2
```

Box 13.7: Testing the Hypothesis of a Difference of 5 Years in Two Population Means Assuming Unequal Variances

As was stated above, **ttest** is used to test the hypothesis of no difference in two population means. We now use the test statistic associated with the assumption that the variances are unequal. The data are the ages of the AML and ALL patients from Chapter 7. We wish to test that the mean AML age is less than or equal to 5 years greater than the mean ALL age versus the alternative that the difference is greater than 5 years. The first 51 observations are from the AML patients and the last 20 are from the ALL patients.

```
. input age
                  age

  1. 20
  2. 25
  3. 26
      .
      .
      .
 71. 19
 72. end

. gen age1=age-5 in 1/51
(20 missing values generated)

. gen age2=age in 52/71
(51 missing values generated)

. ttest age1=age2, unequal
```

Box 13.8: Paired *t* Test

The **ttest** command can be used to test for no difference in two dependent population means. The variable to be used in the analysis can be the difference of the sample observations and this difference is created with the **generate** command. As an alternative, we could use the two dependent means in **ttest** with the **paired** option.

Box 13.9: Testing a Hypothesis about the Difference of Two Proportions

The **ttest** command can be used to provide an approximate test statistic for the test of the equality of two population proportions. The estimate of the standard error of the difference differs slightly from that found using the binomial formula because of the division by $n - 1$ in **ttest** instead of n used in the binomial calculation. The reported p value is also slightly off because the t distribution instead of the normal distribution is used in its calculation. For large samples, these differences are small. The data are the compliance status of the 42 milk producers in the East and the 50 milk producers in the Southwest. We use the **generate** and **replace** statements to create the data instead of entering 140 lines with the compliance status and region.

```
. set obs 92
(obs was 0, now 92)

. gen comply = 0
. replace comply=1 in 1/12
(12 changes made)

. replace comply=1 in 43/63
(21 changes made)

. gen region=1
. replace region=2 in 43/92
(50 changes made)

. label define reglab 1 "East" 2 "SW"
. label values region reglab
. gen compl = comply if region==1
(50 missing values generated)

. gen comp2 = comply if region==2
(42 missing values generated)

. ttest compl=comp2
```

Boxes 14.1 and 14.2: One-Way ANOVA and Multiple Comparisons

The **oneway** command can be used to analyze the age data shown in Table 14.1. In the **oneway** command, the dependent variable is the first variable listed after the command and the independent variable is listed next. The **tabulate** option produces a table of means, standard deviations, and frequencies for each level of the independent variable. Stata reports the results of three different multiple comparison procedures. These three procedures are different from the methods shown in the text. The procedures are **Bonferroni, Scheffe,** and **Sidak.** All three of these procedures focus on the family error rate. The **Scheffe** procedure is the most general of the three procedures used in Stata. The method of presentation of the multiple comparison results differs from that shown in the text. The difference in the sample means is shown and the corresponding *p* value associated with the difference and method is shown beneath the difference in the sample means. Note that in the data entry, we do not show the group variable as it is easier to have Stata create it for us by using **generate** and **replace** statements.

```
. input age
              age

  1. 32
  2. 28
  3. 22
     .
     .
     .
 68. 36
 69. end

. gen group=1
. replace group=2 in 26/50
(25 changes made)

. replace group=3 in 51/68
(18 changes made)

. label define grouplab 1 "Surg" 2 "Contl" 3 "Cont2"
. label values group grouplab

. oneway age group, tabulate bonferroni scheffe sidak
```

Box 14.3: ANOVA for Randomized Block with *k* Replicates per Cell

The **anova** command is used with the two-way ANOVA (can be used for the one-way ANOVA as well). The model shown here does not include any interaction terms. The dependent variable immediately follows the **anova** command and the independent variables are given after the dependent variable. If we wish to see the sample means of the dependent variable by the levels of the independent variables, we must use the **tabulate** command and the **summarize** command with the **mean** option separate from the **anova** command. By specifying the **mean** option, we are telling Stata that we wish to see only the means, not the standard deviations.

```
. input reduct program site
             reduct            program            site

  1.   6 1 1
  2.   2 1 1
  3.  10 1 1
  4.  -1 1 1
  5.   8 1 1
  6.   3 1 2
  7.  15 1 2
  8.   4 1 2
  9.   8 1 2
 10.   6 1 2
 11.   3 2 1
 12.   4 2 1
 13.  -2 2 1
 14.   6 2 1
 15.  -2 2 1
 16.  -4 2 2
 17.   6 2 2
 18.   8 2 2
 19.  -2 2 2
 20.   3 2 2
 21.   8 3 1
 22.  12 3 1
 23.   7 3 1
 24.  10 3 1
 25.   5 3 1
 26.  15 3 2
 27.   8 3 2
 28.  10 3 2
 29.  16 3 2
 30.   3 3 2
 31. end

. label define proglab 1 "diet" 2 "exer" 3 "both"
. label values program proglab
. label define sitelab 1 "office" 2 "factory"
. label values site sitelab
. tabulate program site, summarize(reduct) means
. anova reduct program site
```

Box 14.4: Balanced Two-Way ANOVA with Interaction

The **anova** command can also be used here to analyze the data shown in Table 14.6. The only difference from the previous analysis is that we include the interaction term. The interaction of the two independent variables is specified by the * between the two variables.

```
. input increase text method
          increase        text          method

  1. 30 1 1
  2. 43 1 1
  3. 12 1 1
       .
       .
       .
 36. 48 3 2
 37. end

. label define methlab 1 "lecture" 2 "discuss"
. label values method methlab
. tabulate text method, summarize(increase) means
. anova increase text method text*method
```

Box 15.1: Simple Linear Regression

The **regress** command is one of the commands that can be used to perform linear regression analyses in Stata. The following shows the Stata statements to perform the simple linear regression of SBP on height. The data are shown in Table 15.1. The name of the dependent variable immediately follows the **regress** command and it is in turn followed by the name of the independent variable. The **means** option causes Stata to summarize the variables before displaying the regression results. The *F* table is provided as are the regression coefficients, their standard errors, the *t* statistic for testing that the regression coefficient is zero, the associated *p* value for the test, and the sample mean of the independent variable. The *R*-square and adjusted *R*-square values are also printed as are the root MSE which is the value of $s_{Y|X}$.

```
. input sbp height
          sbp             height

  1. 105 36
  2.  90 37
  3.  82 38
       .
       .
       .
 50. 140 69
 51. end

save c:\book\girlssbp
(file c:\book\girlssbp.dta saved)

. regress sbp height, means
(obs=50)
```

To obtain the predicted values shown in Table 15.1, we use the **predict** command followed by the name of a variable that will store the predicted values. The variable named fitted will contain the predicted values.

```
. predict fitted
```

To obtain the residuals shown in Table 15.2, we again use the **predict** command followed by the name of the variable that will store the residuals and the **residuals** option. The variable that will contain the residuals is resid.

```
. predict resid, residuals
```

To obtain the standardized residuals shown in Table 15.2, we use the **predict** command with the **rstandard** option. We use stresid for the name of the variable that will store the standardized residuals.

```
. predict stresid, rstandard
```

To obtain the leverages shown in Table 15.2, we use the **predict** command with the **hat** option. We use leverage for the name of the variable that will store the leverages.

```
. predict leverage, hat
```

To print the values of the dependent variable, its predicted value, the residuals, standardized residuals, and leverages, we can use the **list** command.

```
. list sbp fitted resid stresid leverage
```

We can also use the **qnorm** command to obtain a normal probability plot of the standardized residuals as is shown next.

```
. qnorm stresid
```

To plot the standardized residuals against height, we use the **graph** command.

```
. graph stresid height
```

Box 15.2: 95 Percent Confidence and Prediction Intervals

To obtain the confidence interval for $\mu_{Y|X}$, we use the **predict** command with the option **stdp** to obtain the estimate of the standard error of the estimate of $\mu_{Y|X}$. The name of the variable that will contain these estimated standard errors is semeanyx. We then use the **generate** command to create the lower and upper bounds for the interval. In the creation of the lower and upper bounds, we are using the multiplier of 2 as an approximation to $t_{n-2,0.975}$. We are also assuming that we are still in the same Stata session in which the regression analysis was performed.

```
. predict semeanyx, stdp
. gen lowmean=fitted - 2*semeanyx
. gen uppmean=fitted + 2*semeanyx
```

To obtain the prediction interval for a single observation, we use the **predict** command with the option **stdf** to obtain the estimated standard error of Y_X. The name of the variable to contain these estimated standard errors is seyx.

```
. predict seyx, stdf
. gen lowy=fitted - 2*seyx
. gen uppy=fitted + 2*seyx
```

To obtain the plot of the confidence and prediction intervals, we use the **graph** command with the **connect()** option to draw lines connecting the points. In the **c(.llll)** option used below, the **.** tells Stata not to connect the values of the first variable, sbp. The **llll** tells Stata to use straight lines to connect the values of the second through fifth variables.

```
. graph sbp fitted lowmean uppmean lowy uppy height,
c(.llll) xlabel(3,5,44,53,62,71)
> ylabel(70,85,100,115,130,145) saving(interv)
```

Box 15.3: Stepwise Regression

We use the **stepwise** command to perform stepwise regression. Stata performs either forward, backward, or stepwise forward or backward regressions. We discussed the forward stepwise regression in the text and that is the method we demonstrate here. Therefore, we use the **forward** and **stepwise** options. Stata uses default values of 0.5 and 0.1 for the *F* values for a variable to enter and to stay in the equation. This contrasts with default values of 4 or more used in some other packages. We can change those values by using the **fenter()** and **fstay()** options if we desire. The *F* values that we wish to use are entered in parentheses. We use the default values. The data are those shown in Table 15.6 which are stored in a non-Stata file. Therefore, we use the **infile** command with the **using** statement to enter the data.

```
. infile height sbp age weight using c:\book\girlssbp.dat
(50 observations read)

. stepwise sbp height age weight, forward stepwise
```

Boxes 15.4 and 15.5: Multiple Linear Regression

The **regress** command can also be used to perform the multiple regression analysis. The Stata statements for analyzing the data in Table 15.6 are the following. We are also assuming that we are still in the same Stata session in which the stepwise regression analysis was performed.

```
. regress sbp height age weight, means
. regress sbp weight, means
(obs=50)
```

Box 15.6: Logistic Regression Analysis

There is no Box 15.6 in the text as the current version of MINITAB does not have a logistic regression command; however, Stata has the **logit** command for performing logistic regression and its use is shown here for the lead, smoking, and pulmonary function test results in Table 15.8. We use the **generate** and **replace** commands to create Table 15.8 for Stata. We are using 1 to represent a normal PFT result and 0 to indicate a result that was not normal. Note that this is the reverse of some programs. The low level of lead is coded as 1 and the high level is coded as 0. The high lead level is thus the reference level. We create three smoking status variables. The variable smok1 will have the value of 1 if the smoking status is never and 0 otherwise, the variable smok2 will have the value of 1 if the smoking status is former and 0 otherwise, and the variable smok3 will have the value of 1 if the smoking status is light and 0 otherwise. This means that the heavy level of smoking is the reference level. In the following, we assume that the 479 observations are in the same order as shown in Table 15.8. Thus the first 160 observations refer to people with a normal PFT exposed to low levels of lead who have never smoked. The next 4 people have PFT results that are not normal. Hence we can take advantage of the order in creating the observations below.

```
. set obs 479
(obs was 0, now 479)

. gen pft=1
. replace pft=0 in 161/164
(4 changes made)

. replace pft=0 in 214/219
(6 changes made)

. replace pft=0 in 295/300
(6 changes made)

. replace pft=0 in 385/387
(3 changes made)

. replace pft=0 in 421/423
(3 changes made)

. replace pft=0 in 436/437
(2 changes made)

. replace pft=0 in 459/460
(2 changes made)

. replace pft=0 in 477/479
(3 changes made)
```

The dependent variable has been created. Now we can create the lead variable.

```
. gen lead=1
. replace lead=0 in 388/479
(92 changes made)
```

The lead level variable has been created. Next we create the three smoking status variables.

```
. gen smok1=1
. replace smok1=0 in 165/387
(223 changes made)

. replace smok1=0 in 424/479
(56 changes made)

. gen smok2=0
. replace smok2=1 in 165/219
(55 changes made)

. replace smok2=1 in 424/437
(14 changes made)

. gen smok3=0
. replace smok3=1 in 220/300
(81 changes made)

. replace smok3=1 in 438/460
(23 changes made)
```

The variables are all created now. We are ready to use the **logit** command. The dependent variable is specified immediately after **logit** and it is followed by the independent variables.

```
. logit pft lead smok1 smok2 smok3
```

References

1. SAS Institute Inc. (1985). "SAS® Introductory Guide for Personal Computers," Version 6 ed. SAS Institute Inc., Cary, NC.
2. SAS Institute Inc. (1985). "SAS® Language Guide for Personal Computers," Version 6 ed. SAS Institute Inc., Cary, NC.
3. SAS Institute Inc. (1987). "SAS/STAT® Guide for Personal Computers," Version 6 ed. SAS Institute Inc., Cary, NC.
4. SAS Institute Inc. (1987). "SAS/GRAPH® Guide for Personal Computers," Version 6 ed. SAS Institute Inc., Cary, NC.
5. Computing Resource Center (1988). "STATA® Release 2 Reference Manual." Computing Resource Center, Los Angeles, CA.
6. Computing Resource Center (1990). "STATA® Release 2.1 Update Manual." Computing Resource Center, Los Angeles, CA.

Statistical Tables

List of Tables

TABLE B1 Random Digits[a]

Line										
1	17174	75908	43306	77061	97755	26780	07446	34836	47656	22475
2	26580	68460	18051	95528	78196	91824	10696	09283	06525	13586
3	24041	33800	09976	36785	11529	19948	21497	94665	54600	51793
4	74838	79323	43962	50531	30826	76623	04007	72395	03544	37575
5	72862	50965	29962	37114	73007	36615	83463	01021	56940	56615
6	82274	94537	52039	68725	06163	47388	62564	46097	71644	00108
7	77586	89168	04043	31926	83333	99957	22204	96361	79770	42561
8	17802	16697	96288	24603	36345	17063	05251	68206	71113	19390
9	10271	06180	39740	01903	01539	59476	83991	07954	83098	01486
10	07780	55451	05276	87719	42723	33685	66024	14236	96801	45797
11	05751	92219	44689	92084	10025	73998	12863	55026	09230	05881
12	14324	44563	13269	88172	47751	64408	86355	16960	72794	30842
13	12869	51161	96952	01895	35785	40807	88980	56656	88839	94521
14	36891	94679	18832	02471	98216	51769	57593	52247	65271	73641
15	22899	37988	68991	28990	87701	99578	06381	33877	45714	45227
16	58556	91925	66542	12852	57203	25725	19844	92696	56861	51882
17	08520	26078	78485	74072	60421	89379	55514	92898	17894	67682
18	31466	97330	39266	06800	32679	37443	53245	81738	73843	64176
19	43780	49375	20055	79095	79987	96005	44296	29004	25059	95752
20	15875	68956	37126	69074	68076	85098	23707	03965	52477	52517
21	22002	20395	72174	70897	00337	70238	19154	77878	33456	89624
22	28968	92168	79825	50945	99479	03121	43217	97297	47547	12201
23	19446	40211	48163	91237	78166	00421	09652	37508	75560	48279
24	98339	39146	76425	55658	60259	59368	49751	44492	99846	07142
25	42746	66199	44160	87627	31369	59756	91765	64760	46878	57467
26	25544	61063	35953	30319	61982	24629	78600	70075	64922	65913
27	22776	62299	05281	92046	98422	95316	20720	90877	01922	32294
28	22578	20732	18421	77419	75391	20665	60627	29382	37782	13163
29	51580	99897	58983	01745	37488	56543	99580	74823	80339	31931
30	63403	74610	23839	69171	52030	91661	18486	83805	62578	67212
31	77353	80198	26674	72839	09944	51278	99333	97341	87588	01655
32	68849	86194	61771	39583	40760	54492	14279	85621	67459	82681
33	50190	86021	96163	18245	58245	41974	05243	66966	07246	09569
34	91239	72671	10759	17927	38958	40672	06409	21979	87813	11939
35	23457	17487	93379	41738	87628	28721	07582	36969	09161	66801
36	60016	28539	40587	27737	50626	22101	74564	65628	11076	75953
37	37076	96887	07002	14535	70186	84065	57590	94324	14132	25879
38	66454	08589	05977	82951	77907	88931	44828	24952	68021	48766
39	14921	18264	69297	84783	83152	82360	46620	53243	56694	17183
40	79201	63127	02632	42083	23715	95916	66794	52598	84195	45420
41	73735	41872	55392	78688	46013	78470	12915	41744	27769	83002
42	67931	75825	80931	07475	06189	88500	36417	35724	65641	35527
43	40580	67626	06630	79770	08154	12159	11322	84871	53591	77690
44	44858	33801	13691	54744	55641	36758	96949	26400	00505	59016
45	84835	40044	86334	34812	35222	20327	71467	37874	51288	95802
46	88089	35765	87473	22457	56445	18890	60892	53132	87424	71714
47	64102	14894	13441	06584	23270	04518	94560	81582	69858	42800
48	62020	92065	06863	58852	84988	81613	53313	58765	27750	71533
49	36121	29901	65962	49271	09970	00719	72935	35598	53014	50036
50	73007	65445	42898	86105	55352	37128	56141	11222	16718	25885

[a]Generated from MINITAB.

Line										
51	32220	61646	87732	07598	05465	68584	64790	56416	21824	61643
52	12782	34043	30801	64642	62329	85019	22481	70105	38254	57186
53	66400	03051	40583	75130	88348	50303	03657	47252	18090	35891
54	76763	78376	40249	52103	36769	53552	55846	61963	86763	67257
55	11767	46380	25290	59073	91662	89160	94869	71368	90732	33583
56	61292	87282	79921	20936	56304	81358	94966	54748	25865	48333
57	64169	56790	91323	29070	49567	86422	13878	42058	53470	22312
58	86741	20680	18422	64127	88381	27590	99659	47854	12163	41801
59	23215	07774	49216	77376	83893	37631	44332	54941	11038	09157
60	72324	05050	52212	82330	10707	92439	33220	11634	35942	09534
61	18209	60272	95944	64495	09247	61000	52564	99690	52055	70716
62	26568	12545	07291	30737	11449	36252	70323	80141	17833	48502
63	66895	34490	95682	44956	39491	54269	07867	84505	05578	91088
64	28908	21020	84646	17475	40539	62981	93042	38181	35279	21843
65	03091	10135	85594	86222	36342	07903	97933	53548	56768	77881
66	69948	54947	28724	33966	90529	16339	40152	06517	18221	53248
67	80774	71613	41590	18430	99863	70872	41549	89671	63628	82167
68	84702	95823	83712	55061	89773	63242	97952	24027	95176	95129
69	18067	54980	38542	86549	43966	92989	87768	16267	47616	63546
70	76825	11257	34842	26130	91870	37116	90770	42369	09614	16645
71	59759	28041	48498	94968	02759	29884	87231	17899	21157	91094
72	67377	59310	86243	30374	18340	58630	21092	62426	37022	40022
73	86655	18980	13739	12234	50705	68189	02212	64653	39716	29953
74	84073	53993	78016	77751	31457	18155	97944	27295	90526	57958
75	58999	77251	84274	15777	66045	84364	62165	24700	00055	06668
76	11308	03979	68271	51776	55915	67970	52691	19073	82178	66031
77	24585	78224	96506	77936	97772	65814	46162	58603	24666	49133
78	22369	34622	75780	67276	06726	07734	48849	60918	83256	17099
79	24914	45155	66234	00460	86700	72578	57617	82212	50104	34094
80	88320	48338	70689	05856	91247	29214	21807	77100	74896	24592
81	69848	33544	50065	69910	15783	76852	25025	37762	49049	21666
82	77987	45152	89425	81350	10697	90522	10496	86753	75366	83410
83	97709	78833	69516	05969	98796	60938	90201	99875	37430	87145
84	05209	88924	10458	20004	65788	91299	41139	76993	47040	15777
85	68616	23573	66693	83674	34890	57000	07586	39661	23774	50682
86	18260	40283	35008	94377	47286	93322	68092	92858	99829	59997
87	29121	89864	44444	03931	34222	49057	49713	50972	23191	29933
88	36834	59756	46105	01156	40367	50950	43614	70178	93359	77431
89	10757	21796	12219	39415	32020	04178	69733	83093	58039	74845
90	99465	88838	45530	96133	66529	57600	52060	98052	72613	32354
91	59157	66024	86610	70068	29879	30664	87190	98772	76243	62043
92	63489	17951	66279	69460	03659	53135	79535	05034	26052	75480
93	08723	61325	57652	18876	08976	51276	12793	60467	11655	04069
94	75883	23261	03050	36180	38486	47570	72493	92403	06412	10039
95	95560	45085	03464	79493	25121	04125	86957	16042	63551	40774
96	81329	74272	70097	05615	91212	73956	43022	64078	77377	14160
97	13536	31170	91648	67487	95149	17890	50223	82906	59466	01721
98	28778	55892	59449	53815	84565	62568	79771	00793	19324	10150
99	39757	44482	21115	01607	93177	26324	66403	91660	62073	34237
00	54595	87336	08030	30633	83752	04706	96494	71064	19061	84919

TABLE B2 Binomial Probabilities[a]

n	x	.01	.05	.10	.15	.20	.25	.30	.35	.40	.45	.50
2	0	.9801	.9025	.8100	.7225	.6400	.5625	.4900	.4225	.3600	.3025	.2500
	1	.0198	.0950	.1800	.2550	.3200	.3750	.4200	.4550	.4800	.4950	.5000
	2	.0001	.0025	.0100	.0225	.0400	.0625	.0900	.1225	.1600	.2025	.2500
3	0	.9703	.8574	.7290	.6141	.5120	.4219	.3430	.2746	.2160	.1664	.1250
	1	.0294	.1354	.2430	.3251	.3840	.4219	.4410	.4436	.4320	.4084	.3750
	2	.0003	.0071	.0270	.0574	.0960	.1406	.1890	.2389	.2880	.3341	.3750
	3		.0001	.0010	.0034	.0080	.0156	.0270	.0429	.0640	.0911	.1250
4	0	.9606	.8145	.6561	.5220	.4096	.3164	.2401	.1785	.1296	.0915	.0625
	1	.0388	.1715	.2916	.3685	.4096	.4219	.4116	.3845	.3456	.2995	.2500
	2	.0006	.0135	.0486	.0975	.1536	.2109	.2646	.3105	.3456	.3675	.3750
	3		.0005	.0036	.0115	.0256	.0469	.0756	.1115	.1536	.2005	.2500
	4			.0001	.0005	.0016	.0039	.0081	.0150	.0256	.0410	.0625
5	0	.9510	.7738	.5905	.4437	.3277	.2373	.1681	.1160	.0778	.0503	.0313
	1	.0480	.2036	.3281	.3915	.4096	.3955	.3602	.3124	.2592	.2059	.1563
	2	.0010	.0214	.0729	.1382	.2048	.2637	.3087	.3364	.3456	.3369	.3125
	3		.0011	.0081	.0244	.0512	.0879	.1323	.1811	.2304	.2757	.3125
	4			.0005	.0022	.0064	.0146	.0284	.0488	.0768	.1128	.1563
	5				.0001	.0003	.0010	.0024	.0053	.0102	.0185	.0313
6	0	.9415	.7351	.5314	.3771	.2621	.1780	.1176	.0754	.0467	.0277	.0156
	1	.0571	.2321	.3543	.3993	.3932	.3560	.3025	.2437	.1866	.1359	.0938
	2	.0014	.0305	.0984	.1762	.2458	.2966	.3241	.3280	.3110	.2780	.2344
	3		.0021	.0146	.0415	.0819	.1318	.1852	.2355	.2765	.3032	.3125
	4		.0001	.0012	.0055	.0154	.0330	.0595	.0951	.1382	.1861	.2344
	5			.0001	.0004	.0015	.0044	.0102	.0205	.0369	.0609	.0938
	6					.0001	.0002	.0007	.0018	.0041	.0083	.0156
7	0	.9321	.6983	.4783	.3206	.2097	.1335	.0824	.0490	.0280	.0152	.0078
	1	.0659	.2573	.3720	.3960	.3670	.3115	.2471	.1848	.1306	.0872	.0547
	2	.0020	.0406	.1240	.2097	.2753	.3115	.3177	.2985	.2613	.2140	.1641
	3		.0036	.0230	.0617	.1147	.1730	.2269	.2679	.2903	.2918	.2734
	4		.0002	.0026	.0109	.0287	.0577	.0972	.1442	.1935	.2388	.2734
	5			.0002	.0012	.0043	.0115	.0250	.0466	.0774	.1172	.1641
	6				.0001	.0004	.0013	.0036	.0084	.0172	.0320	.0547
	7							.0002	.0006	.0016	.0037	.0078
8	0	.9227	.6634	.4305	.2725	.1678	.1001	.0576	.0319	.0168	.0084	.0039
	1	.0746	.2793	.3826	.3847	.3355	.2670	.1977	.1373	.0896	.0548	.0313
	2	.0026	.0515	.1488	.2376	.2936	.3115	.2965	.2587	.2090	.1569	.1094
	3	.0001	.0054	.0331	.0839	.1468	.2076	.2541	.2786	.2787	.2568	.2188
	4		.0004	.0046	.0185	.0459	.0865	.1361	.1875	.2322	.2627	.2734
	5			.0004	.0026	.0092	.0231	.0467	.0808	.1239	.1719	.2188
	6				.0002	.0011	.0038	.0100	.0217	.0413	.0703	.1094
	7					.0001	.0004	.0012	.0033	.0079	.0164	.0313
	8							.0001	.0002	.0007	.0017	.0039

n	x	.01	.05	.10	.15	.20	.25	.30	.35	.40	.45	.50
9	0	.9135	.6302	.3874	.2316	.1342	.0751	.0404	.0207	.0101	.0046	.0020
	1	.0830	.2985	.3874	.3679	.3020	.2253	.1557	.1004	.0605	.0339	.0176
	2	.0034	.0629	.1722	.2597	.3020	.3003	.2668	.2162	.1612	.1110	.0703
	3	.0001	.0077	.0446	.1069	.1762	.2336	.2668	.2716	.2508	.2119	.1641
	4		.0006	.0074	.0283	.0661	.1168	.1715	.2194	.2508	.2600	.2461
	5			.0008	.0050	.0165	.0389	.0735	.1181	.1672	.2128	.2461
	6			.0001	.0006	.0028	.0087	.0210	.0424	.0743	.1160	.1641
	7					.0003	.0012	.0039	.0098	.0212	.0407	.0703
	8						.0001	.0004	.0013	.0035	.0083	.0176
	9								.0001	.0003	.0008	.0020
10	0	.9044	.5987	.3487	.1969	.1074	.0563	.0282	.0135	.0060	.0025	.0010
	1	.0914	.3151	.3874	.3474	.2684	.1877	.1211	.0725	.0403	.0207	.0098
	2	.0042	.0746	.1937	.2759	.3020	.2816	.2335	.1757	.1209	.0763	.0439
	3	.0001	.0105	.0574	.1298	.2013	.2503	.2668	.2522	.2150	.1665	.1172
	4		.0010	.0112	.0401	.0881	.1460	.2001	.2377	.2508	.2384	.2051
	5		.0001	.0015	.0085	.0264	.0584	.1029	.1536	.2007	.2340	.2461
	6			.0001	.0012	.0055	.0162	.0368	.0689	.1115	.1596	.2051
	7				.0001	.0008	.0031	.0090	.0212	.0425	.0746	.1172
	8					.0001	.0004	.0014	.0043	.0106	.0229	.0439
	9							.0001	.0005	.0016	.0042	.0098
	10									.0001	.0003	.0010
11	0	.8953	.5688	.3138	.1673	.0859	.0422	.0198	.0088	.0036	.0014	.0005
	1	.0995	.3293	.3835	.3248	.2362	.1549	.0932	.0518	.0266	.0125	.0054
	2	.0050	.0867	.2131	.2866	.2953	.2581	.1998	.1395	.0887	.0513	.0269
	3	.0002	.0137	.0710	.1517	.2215	.2581	.2568	.2254	.1774	.1259	.0806
	4		.0014	.0158	.0536	.1107	.1721	.2201	.2428	.2365	.2060	.1611
	5		.0001	.0025	.0132	.0388	.0803	.1321	.1830	.2207	.2360	.2256
	6			.0003	.0023	.0097	.0268	.0566	.0985	.1471	.1931	.2256
	7				.0003	.0017	.0064	.0173	.0379	.0701	.1128	.1611
	8					.0002	.0011	.0037	.0102	.0234	.0462	.0806
	9						.0001	.0005	.0018	.0052	.0126	.0269
	10								.0002	.0007	.0021	.0054
	11										.0002	.0005
12	0	.8864	.5404	.2824	.1422	.0687	.0317	.0138	.0057	.0022	.0008	.0002
	1	.1074	.3413	.3766	.3012	.2062	.1267	.0712	.0368	.0174	.0075	.0029
	2	.0060	.0988	.2301	.2924	.2835	.2323	.1678	.1088	.0639	.0339	.0161
	3	.0002	.0173	.0852	.1720	.2362	.2581	.2397	.1954	.1419	.0923	.0537
	4		.0021	.0213	.0683	.1329	.1936	.2311	.2367	.2128	.1700	.1209
	5		.0002	.0038	.0193	.0532	.1032	.1585	.2039	.2270	.2225	.1934
	6			.0005	.0040	.0155	.0401	.0792	.1281	.1766	.2124	.2256
	7				.0006	.0033	.0115	.0291	.0591	.1009	.1489	.1934
	8				.0001	.0005	.0024	.0078	.0199	.0420	.0762	.1209
	9					.0001	.0004	.0015	.0048	.0125	.0277	.0537
	10							.0002	.0008	.0025	.0068	.0161
	11								.0001	.0003	.0010	.0029
	12										.0001	.0002

(*continued*)

n	x	.01	.05	.10	.15	.20	.25	.30	.35	.40	.45	.50
						p						
13	0	.8775	.5133	.2542	.1209	.0550	.0238	.0097	.0037	.0013	.0004	.0001
	1	.1152	.3512	.3672	.2774	.1787	.1029	.0540	.0259	.0113	.0045	.0016
	2	.0070	.1109	.2448	.2937	.2680	.2059	.1388	.0836	.0453	.0220	.0095
	3	.0003	.0214	.0997	.1900	.2457	.2517	.2181	.1651	.1107	.0660	.0349
	4		.0028	.0277	.0838	.1535	.2097	.2337	.2222	.1845	.1350	.0873
	5		.0003	.0055	.0266	.0691	.1258	.1803	.2154	.2214	.1989	.1571
	6			.0008	.0063	.0230	.0559	.1030	.1546	.1968	.2169	.2095
	7			.0001	.0011	.0058	.0186	.0442	.0833	.1312	.1775	.2095
	8				.0001	.0011	.0047	.0142	.0336	.0656	.1089	.1571
	9					.0002	.0009	.0034	.0101	.0243	.0495	.0873
	10						.0001	.0006	.0022	.0065	.0162	.0349
	11							.0001	.0003	.0012	.0036	.0095
	12									.0001	.0005	.0016
	13											.0001
14	0	.8687	.4877	.2288	.1028	.0440	.0178	.0068	.0024	.0008	.0002	.0001
	1	.1229	.3593	.3559	.2539	.1539	.0832	.0407	.0181	.0073	.0027	.0009
	2	.0081	.1229	.2570	.2912	.2501	.1802	.1134	.0634	.0317	.0141	.0056
	3	.0003	.0259	.1142	.2056	.2501	.2402	.1943	.1366	.0845	.0462	.0222
	4		.0037	.0349	.0998	.1720	.2202	.2290	.2022	.1549	.1040	.0611
	5		.0004	.0078	.0352	.0860	.1468	.1963	.2178	.2066	.1701	.1222
	6			.0013	.0093	.0322	.0734	.1262	.1759	.2066	.2088	.1833
	7			.0002	.0019	.0092	.0280	.0618	.1082	.1574	.1952	.2095
	8				.0003	.0020	.0082	.0232	.0510	.0918	.1398	.1833
	9					.0003	.0018	.0066	.0183	.0408	.0762	.1222
	10						.0003	.0014	.0049	.0136	.0312	.0611
	11							.0002	.0010	.0033	.0093	.0222
	12								.0001	.0006	.0019	.0056
	13									.0001	.0002	.0009
	14											.0001
15	0	.8601	.4633	.2059	.0874	.0352	.0134	.0047	.0016	.0005	.0001	
	1	.1303	.3658	.3432	.2312	.1319	.0668	.0305	.0126	.0047	.0016	.0005
	2	.0092	.1348	.2669	.2856	.2309	.1559	.0916	.0476	.0219	.0090	.0032
	3	.0004	.0308	.1285	.2184	.2501	.2252	.1700	.1110	.0634	.0318	.0139
	4		.0049	.0428	.1156	.1876	.2252	.2186	.1792	.1268	.0780	.0417
	5		.0006	.0105	.0449	.1032	.1651	.2061	.2123	.1859	.1404	.0916
	6			.0019	.0132	.0430	.0917	.1472	.1906	.2066	.1914	.1527
	7			.0003	.0030	.0138	.0393	.0811	.1319	.1771	.2013	.1964
	8				.0005	.0035	.0131	.0348	.0710	.1181	.1647	.1964
	9				.0001	.0007	.0034	.0116	.0298	.0612	.1048	.1527
	10					.0001	.0007	.0030	.0096	.0245	.0515	.0916
	11						.0001	.0006	.0024	.0074	.0191	.0417
	12							.0001	.0004	.0016	.0052	.0139
	13								.0001	.0003	.0010	.0032
	14										.0001	.0005
	15											

TABLE B2 —*Continued*

n	*x*	.01	.05	.10	.15	.20	.25	.30	.35	.40	.45	.50
16	0	.8515	.4401	.1853	.0743	.0281	.0100	.0033	.0010	.0003	.0001	
	1	.1376	.3706	.3294	.2097	.1126	.0535	.0228	.0087	.0030	.0009	.0002
	2	.0104	.1463	.2745	.2775	.2111	.1336	.0732	.0353	.0150	.0056	.0018
	3	.0005	.0359	.1423	.2285	.2463	.2079	.1465	.0888	.0468	.0215	.0085
	4		.0061	.0514	.1311	.2001	.2252	.2040	.1553	.1014	.0572	.0278
	5		.0008	.0137	.0555	.1201	.1802	.2099	.2008	.1623	.1123	.0667
	6		.0001	.0028	.0180	.0550	.1101	.1649	.1982	.1983	.1684	.1222
	7			.0004	.0045	.0197	.0524	.1010	.1524	.1889	.1969	.1746
	8			.0001	.0009	.0055	.0197	.0487	.0923	.1417	.1812	.1964
	9				.0001	.0012	.0058	.0185	.0442	.0840	.1318	.1746
	10					.0002	.0014	.0056	.0167	.0392	.0755	.1222
	11						.0002	.0013	.0049	.0142	.0337	.0667
	12							.0002	.0011	.0040	.0115	.0278
	13								.0002	.0008	.0029	.0085
	14									.0001	.0005	.0018
	15											.0002
	16											
17	0	.8429	.4181	.1668	.0631	.0225	.0075	.0023	.0007	.0002		
	1	.1447	.3741	.3150	.1893	.0957	.0426	.0169	.0060	.0019	.0005	.0001
	2	.0117	.1575	.2800	.2673	.1914	.1136	.0581	.0260	.0102	.0035	.0010
	3	.0006	.0415	.1556	.2359	.2393	.1893	.1245	.0701	.0341	.0144	.0052
	4		.0076	.0605	.1457	.2093	.2209	.1868	.1320	.0796	.0411	.0182
	5		.0010	.0175	.0668	.1361	.1914	.2081	.1849	.1379	.0875	.0472
	6		.0001	.0039	.0236	.0680	.1276	.1784	.1991	.1839	.1432	.0944
	7			.0007	.0065	.0267	.0668	.1201	.1685	.1927	.1841	.1484
	8			.0001	.0014	.0084	.0279	.0644	.1134	.1606	.1883	.1855
	9				.0003	.0021	.0093	.0276	.0611	.1070	.1540	.1855
	10					.0004	.0025	.0095	.0263	.0571	.1008	.1484
	11					.0001	.0005	.0026	.0090	.0242	.0525	.0944
	12						.0001	.0006	.0024	.0081	.0215	.0472
	13							.0001	.0005	.0021	.0068	.0182
	14								.0001	.0004	.0016	.0052
	15									.0001	.0003	.0010
	16											.0001
	17											

(*continued*)

TABLE B2 —*Continued*

		p										
n	*x*	.01	.05	.10	.15	.20	.25	.30	.35	.40	.45	.50
18	0	.8345	.3972	.1501	.0536	.0180	.0056	.0016	.0004	.0001		
	1	.1517	.3763	.3002	.1704	.0811	.0338	.0126	.0042	.0012	.0003	.0001
	2	.0130	.1683	.2835	.2556	.1723	.0958	.0458	.0190	.0069	.0022	.0006
	3	.0007	.0473	.1680	.2406	.2297	.1704	.1046	.0547	.0246	.0095	.0031
	4		.0093	.0700	.1592	.2153	.2130	.1681	.1104	.0614	.0291	.0117
	5		.0014	.0218	.0787	.1507	.1988	.2017	.1664	.1146	.0666	.0327
	6		.0002	.0052	.0301	.0816	.1436	.1873	.1941	.1655	.1181	.0708
	7			.0010	.0091	.0350	.0820	.1376	.1792	.1892	.1657	.1214
	8			.0002	.0022	.0120	.0376	.0811	.1327	.1734	.1864	.1669
	9				.0004	.0033	.0139	.0386	.0794	.2844	.1694	.1855
	10				.0001	.0008	.0042	.0149	.0385	.0771	.1248	.1669
	11					.0001	.0010	.0046	.0151	.0374	.0742	.1214
	12						.0002	.0012	.0047	.0145	.0354	.0708
	13							.0002	.0012	.0045	.0134	.0327
	14								.0002	.0011	.0039	.0117
	15									.0002	.0009	.0031
	16										.0001	.0006
	17											.0001
	18											
19	0	.8262	.3774	.1351	.0456	.0144	.0042	.0011	.0003	.0001		
	1	.1586	.3774	.2852	.1529	.0685	.0268	.0093	.0029	.0008	.0002	
	2	.0144	.1787	.2852	.2428	.1540	.0803	.0358	.0138	.0046	.0013	.0003
	3	.0008	.0533	.1796	.2428	.2182	.1517	.0869	.0422	.0175	.0062	.0018
	4		.0112	.0798	.1714	.2182	.2023	.1491	.0909	.0467	.0203	.0074
	5		.0018	.0266	.0907	.1637	.2023	.1916	.1468	.0933	.0497	.0222
	6		.0002	.0069	.0374	.0955	.1574	.1916	.1844	.1451	.0949	.0518
	7			.0014	.0122	.0443	.0974	.1525	.1844	.1797	.1443	.0961
	8			.0002	.0032	.0166	.0487	.0981	.1489	.1797	.1771	.1442
	9				.0007	.0051	.0198	.0514	.0980	.1464	.1771	.1762
	10				.0001	.0013	.0066	.0220	.0528	.0976	.1449	.1762
	11					.0003	.0018	.0077	.0233	.0532	.0970	.1442
	12						.0004	.0022	.0083	.0237	.0529	.0961
	13						.0001	.0005	.0024	.0085	.0233	.0518
	14							.0001	.0006	.0024	.0082	.0222
	15								.0001	.0005	.0022	.0074
	16									.0001	.0005	.0018
	17										.0001	.0003
	18											
	19											

TABLE B2 —*Continued*

		p										
n	*x*	.01	.05	.10	.15	.20	.25	.30	.35	.40	.45	.50
20	0	.8179	.3585	.1216	.0388	.0115	.0032	.0008	.0002			
	1	.1652	.3774	.2702	.1368	.0576	.0211	.0068	.0020	.0005	.0001	
	2	.0159	.1887	.2852	.2293	.1369	.0669	.0278	.0100	.0031	.0008	.0002
	3	.0010	.0596	.1901	.2428	.2054	.1339	.0716	.0323	.0124	.0040	.0011
	4		.0133	.0898	.1821	.2182	.1897	.1304	.0738	.0350	.0139	.0046
	5		.0022	.0319	.1028	.1746	.2023	.1789	.1272	.0746	.0365	.0148
	6		.0003	.0089	.0454	.1091	.1686	.1916	.1712	.1244	.0746	.0370
	7			.0020	.0160	.0546	.1124	.1643	.1844	.1659	.1221	.0739
	8			.0004	.0046	.0222	.0609	.1144	.1614	.1797	.1623	.1201
	9			.0001	.0011	.0074	.0271	.0654	.1158	.1597	.1771	.1602
	10				.0002	.0020	.0099	.0308	.0686	.1171	.1593	.1762
	11					.0005	.0030	.0120	.0336	.0710	.1185	.1602
	12					.0001	.0008	.0039	.0136	.0355	.0727	.1201
	13						.0002	.0010	.0045	.0146	.0366	.0739
	14							.0002	.0012	.0049	.0150	.0370
	15								.0003	.0013	.0049	.0148
	16									.0003	.0013	.0046
	17										.0002	.0011
	18											.0002
	19											

[a]Calculated by MINITAB.

TABLE B3 Poisson Probabilities[a]

x	μ = .2	.4	.6	.8	1.0	1.2	1.4	1.6	x
0	.818731	.670320	.548812	.449329	.367879	.301194	.246597	.201896	0
1	.163746	.268128	.329287	.359463	.367879	.361433	.345236	.323034	1
2	.016375	.053626	.098786	.143785	.183940	.216860	.241665	.258428	2
3	.001092	.007150	.019757	.038343	.061313	.086744	.112777	.137828	3
4	.000055	.000715	.002964	.007669	.015328	.026023	.039472	.055131	4
5	.000002	.000057	.000356	.001227	.003066	.006246	.011052	.017642	5
6		.000004	.000036	.000164	.000511	.001249	.002579	.004705	6
7			.000003	.000019	.000073	.000214	.000516	.001075	7
8				.000002	.000009	.000032	.000090	.000215	8
9					.000001	.000004	.000014	.000038	9
10						.000001	.000002	.000006	10
11								.000001	11

x	μ = 1.8	2.0	2.5	3.0	3.5	4.0	4.5	5.0	x
0	.165299	.135335	.082085	.049787	.030197	.018316	.011109	.006738	0
1	.297538	.270671	.205213	.149361	.105691	.073263	.049990	.033690	1
2	.267784	.270671	.256516	.224042	.184959	.146525	.112479	.084224	2
3	.160671	.180447	.213763	.224042	.215785	.195367	.168718	.140374	3
4	.072302	.090224	.133602	.168031	.188812	.195367	.189808	.175467	4
5	.026029	.036089	.066801	.100819	.132169	.156293	.170827	.175467	5
6	.007809	.012030	.027834	.050409	.077098	.104196	.128120	.146223	6
7	.002008	.003437	.009941	.021604	.038549	.059540	.082363	.104445	7
8	.000452	.000859	.003106	.008102	.016865	.029770	.046329	.065278	8
9	.000090	.000191	.000863	.002701	.006559	.013231	.023165	.036266	9
10	.000016	.000038	.000216	.000810	.002296	.005292	.010424	.018133	10
11	.000003	.000007	.000049	.000221	.000730	.001925	.004264	.008242	11
12		.000001	.000010	.000055	.000213	.000642	.001599	.003434	12
13			.000002	.000013	.000057	.000197	.000554	.001321	13
14				.000003	.000014	.000056	.000178	.000472	14
15				.000001	.000003	.000015	.000053	.000157	15
16					.000001	.000004	.000015	.000049	16
17						.000001	.000004	.000014	17
18							.000001	.000004	18
19								.000001	19

[a]Calculated by MINITAB.

TABLE B3 —*Continued*

	μ								
x	5.5	6.0	6.5	7.0	8.0	9.0	10.0	11.0	x
0	.004087	.002479	.001503	.000912	.000335	.000123	.000045	.000017	0
1	.022477	.014873	.009772	.006383	.002684	.001111	.000454	.000184	1
2	.061812	.044618	.031760	.022341	.010735	.004998	.002270	.001010	2
3	.113323	.089235	.068814	.052129	.028626	.014994	.007567	.003705	3
4	.155819	.133853	.111822	.091226	.057252	.033737	.018917	.010189	4
5	.171401	.160623	.145369	.127717	.091604	.060727	.037833	.022415	5
6	.157117	.160623	.157483	.149003	.122138	.091090	.063055	.041095	6
7	.123449	.137677	.146234	.149003	.139587	.117116	.090079	.064577	7
8	.084871	.103258	.118815	.130377	.139587	.131756	.112599	.088794	8
9	.051866	.068838	.085811	.101405	.124077	.131756	.125110	.108526	9
10	.028526	.041303	.055777	.070983	.099262	.118580	.125110	.119378	10
11	.014263	.022529	.032959	.045171	.072190	.097020	.113736	.119378	11
12	.006537	.011264	.017853	.026350	.048127	.072765	.094780	.109430	12
13	.002766	.005199	.008926	.014188	.029616	.050376	.072908	.092595	13
14	.001087	.002228	.004144	.007094	.016924	.032384	.052077	.072753	14
15	.000398	.000891	.001796	.003311	.009026	.019431	.034718	.053352	15
16	.000137	.000334	.000730	.001448	.004513	.010930	.021699	.036680	16
17	.000044	.000118	.000279	.000596	.002124	.005786	.012764	.023734	17
18	.000014	.000039	.000101	.000232	.000944	.002893	.007091	.014504	18
19	.000004	.000012	.000034	.000085	.000397	.001370	.003732	.008397	19
20	.000001	.000004	.000011	.000030	.000159	.000617	.001866	.004618	20
21		.000001	.000003	.000010	.000061	.000264	.000889	.002419	21
22			.000001	.000003	.000022	.000108	.000404	.001210	22
23				.000001	.000008	.000042	.000176	.000578	23
24					.000003	.000016	.000073	.000265	24
25					.000001	.000006	.000029	.000117	25
26						.000002	.000011	.000049	26
27						.000001	.000004	.000020	27
28							.000001	.000008	28
29							.000001	.000003	29
30								.000001	30

(*continued*)

TABLE B3 —*Continued*

x	μ 12.0	13.0	14.0	15.0	16.0	17.0	x
0	.000006	.000002	.000001				0
1	.000074	.000029	.000012	.000005	.000002	.000001	1
2	.000442	.000191	.000081	.000034	.000014	.000006	2
3	.001770	.000828	.000380	.000172	.000077	.000034	3
4	.005309	.002690	.001331	.000645	.000307	.000144	4
5	.012741	.006994	.003727	.001936	.000983	.000490	5
6	.025481	.015153	.008696	.004839	.002622	.001388	6
7	.043682	.028141	.017392	.010370	.005994	.003371	7
8	.065523	.045730	.030436	.019444	.011988	.007163	8
9	.087364	.066054	.047344	.032407	.021311	.013529	9
10	.104837	.085870	.066282	.048611	.034098	.023000	10
11	.114368	.101483	.084359	.066287	.049597	.035545	11
12	.114368	.109940	.098418	.082859	.066129	.050355	12
13	.105570	.109940	.105989	.095607	.081389	.065849	13
14	.090489	.102087	.105989	.102436	.093016	.079960	14
15	.072391	.088475	.098923	.102436	.099218	.090621	15
16	.054293	.071886	.086558	.096034	.099218	.096285	16
17	.038325	.054972	.071283	.084736	.093381	.096285	17
18	.025550	.039702	.055442	.070613	.083006	.090936	18
19	.016137	.027164	.040852	.055747	.069899	.081363	19
20	.009682	.017657	.028597	.041810	.055920	.069159	20
21	.005533	.010930	.019064	.029865	.042605	.055986	21
22	.003018	.006459	.012132	.020362	.030986	.043262	22
23	.001574	.003651	.007385	.013280	.021555	.031976	23
24	.000787	.001977	.004308	.008300	.014370	.022650	24
25	.000378	.001028	.002412	.004980	.009197	.015402	25
26	.000174	.000514	.001299	.002873	.005660	.010070	26
27	.000078	.000248	.000674	.001596	.003354	.006341	27
28	.000033	.000115	.000337	.000855	.001917	.003850	28
29	.000014	.000052	.000163	.000442	.001057	.002257	29
30	.000005	.000022	.000076	.000221	.000564	.001279	30
31	.000002	.000009	.000034	.000107	.000291	.000701	31
32	.000001	.000004	.000015	.000050	.000146	.000373	32
33		.000001	.000006	.000023	.000071	.000192	33
34		.000001	.000003	.000010	.000033	.000096	34
35			.000001	.000004	.000015	.000047	35
36				.000002	.000007	.000022	36
37				.000001	.000003	.000010	37
38					.000001	.000005	38
39					.000001	.000002	39
40						.000001	40

TABLE B4 Cumulative Distribution Function for Standard Normal Distribution[a]

z	.09	.08	.07	.06	.05	.04	.03	.02	.01	.00
−3.7	.0001	.0001	.0001	.0001	.0001	.0001	.0001	.0001	.0001	.0001
−3.6	.0001	.0001	.0001	.0001	.0001	.0001	.0001	.0001	.0002	.0002
−3.5	.0002	.0002	.0002	.0002	.0002	.0002	.0002	.0002	.0002	.0002
−3.4	.0002	.0003	.0003	.0003	.0003	.0003	.0003	.0003	.0003	.0003
−3.3	.0003	.0004	.0004	.0004	.0004	.0004	.0004	.0005	.0005	.0005
−3.2	.0005	.0005	.0005	.0006	.0006	.0006	.0006	.0006	.0007	.0007
−3.1	.0007	.0007	.0008	.0008	.0008	.0008	.0009	.0009	.0009	.0010
−3.0	.0010	.0010	.0011	.0011	.0011	.0012	.0012	.0013	.0013	.0013
−2.9	.0014	.0014	.0015	.0015	.0016	.0016	.0017	.0018	.0018	.0019
−2.8	.0019	.0020	.0021	.0021	.0022	.0023	.0023	.0024	.0025	.0026
−2.7	.0026	.0027	.0028	.0029	.0030	.0031	.0032	.0033	.0034	.0035
−2.6	.0036	.0037	.0038	.0039	.0040	.0041	.0043	.0044	.0045	.0047
−2.5	.0048	.0049	.0051	.0052	.0054	.0055	.0057	.0059	.0060	.0062
−2.4	.0064	.0066	.0068	.0069	.0071	.0073	.0075	.0078	.0080	.0082
−2.3	.0084	.0087	.0089	.0091	.0094	.0096	.0099	.0102	.0104	.0107
−2.2	.0110	.0113	.0116	.0119	.0122	.0125	.0129	.0132	.0136	.0139
−2.1	.0143	.0146	.0150	.0154	.0158	.0162	.0166	.0170	.0174	.0179
−2.0	.0183	.0188	.0192	.0197	.0202	.0207	.0212	.0217	.0222	.0228
−1.9	.0233	.0239	.0244	.0250	.0256	.0262	.0268	.0274	.0281	.0287
−1.8	.0294	.0301	.0307	.0314	.0322	.0329	.0336	.0344	.0351	.0359
−1.7	.0367	.0375	.0384	.0392	.0401	.0409	.0418	.0427	.0436	.0446
−1.6	.0455	.0465	.0475	.0485	.0495	.0505	.0516	.0526	.0537	.0548
−1.5	.0559	.0571	.0582	.0594	.0606	.0618	.0630	.0643	.0655	.0668
−1.4	.0681	.0694	.0708	.0721	.0735	.0749	.0764	.0778	.0793	.0808
−1.3	.0823	.0838	.0853	.0869	.0885	.0901	.0918	.0934	.0951	.0968
−1.2	.0985	.1003	.1020	.1038	.1056	.1075	.1093	.1112	.1131	.1151
−1.1	.1170	.1190	.1210	.1230	.1251	.1271	.1292	.1314	.1335	.1357
−1.0	.1379	.1401	.1423	.1446	.1469	.1492	.1515	.1539	.1562	.1587
−0.9	.1611	.1635	.1660	.1685	.1711	.1736	.1762	.1788	.1814	.1841
−0.8	.1867	.1894	.1922	.1949	.1977	.2005	.2033	.2061	.2090	.2119
−0.7	.2148	.2177	.2206	.2236	.2266	.2296	.2327	.2358	.2389	.2420
−0.6	.2451	.2483	.2514	.2546	.2578	.2611	.2643	.2676	.2709	.2743
−0.5	.2776	.2810	.2843	.2877	.2912	.2946	.2981	.3015	.3050	.3085
−0.4	.3121	.3156	.3192	.3228	.3264	.3300	.3336	.3372	.3409	.3446
−0.3	.3483	.3520	.3557	.3594	.3632	.3669	.3707	.3745	.3783	.3821
−0.2	.3859	.3897	.3936	.3974	.4013	.4052	.4090	.4129	.4168	.4207
−0.1	.4247	.4286	.4325	.4364	.4404	.4443	.4483	.4522	.4562	.4602
−0.0	.4641	.4681	.4721	.4761	.4801	.4840	.4880	.4920	.4960	.5000

(*continued*)

TABLE B4 —*Continued*

z	.00	.01	.02	.03	.04	.05	.06	.07	.08	.09
0.0	.5000	.5040	.5080	.5120	.5160	.5199	.5239	.5279	.5319	.5359
0.1	.5398	.5438	.5478	.5517	.5557	.5596	.5636	.5675	.5714	.5753
0.2	.5793	.5832	.5871	.5910	.5948	.5987	.6026	.6064	.6103	.6141
0.3	.6179	.6217	.6255	.6293	.6331	.6368	.6406	.6443	.6480	.6517
0.4	.6554	.6591	.6628	.6664	.6700	.6736	.6772	.6808	.6844	.6879
0.5	.6915	.6950	.6985	.7019	.7054	.7088	.7123	.7157	.7190	.7224
0.6	.7257	.7291	.7324	.7357	.7389	.7422	.7454	.7486	.7517	.7549
0.7	.7580	.7611	.7642	.7673	.7704	.7734	.7764	.7794	.7823	.7852
0.8	.7881	.7910	.7939	.7967	.7995	.8023	.8051	.8078	.8106	.8133
0.9	.8159	.8186	.8212	.8238	.8264	.8289	.8315	.8340	.8365	.8389
1.0	.8413	.8438	.8461	.8485	.8508	.8531	.8554	.8577	.8599	.8621
1.1	.8643	.8665	.8686	.8708	.8729	.8749	.8770	.8790	.8810	.8830
1.2	.8849	.8869	.8888	.8907	.8925	.8944	.8962	.8980	.8997	.9015
1.3	.9032	.9049	.9066	.9082	.9099	.9115	.9131	.9147	.9162	.9177
1.4	.9192	.9207	.9222	.9236	.9251	.9265	.9279	.9292	.9306	.9319
1.5	.9332	.9345	.9357	.9370	.9382	.9394	.9406	.9418	.9429	.9441
1.6	.9452	.9463	.9474	.9484	.9495	.9505	.9515	.9525	.9535	.9545
1.7	.9554	.9564	.9573	.9582	.9591	.9599	.9608	.9616	.9625	.9633
1.8	.9641	.9649	.9656	.9664	.9671	.9678	.9686	.9693	.9699	.9706
1.9	.9713	.9719	.9726	.9732	.9738	.9744	.9750	.9756	.9761	.9767
2.0	.9772	.9778	.9783	.9788	.9793	.9798	.9803	.9808	.9812	.9817
2.1	.9821	.9826	.9830	.9834	.9838	.9842	.9846	.9850	.9854	.9857
2.2	.9861	.9864	.9868	.9871	.9875	.9878	.9881	.9884	.9887	.9890
2.3	.9893	.9896	.9898	.9901	.9904	.9906	.9909	.9911	.9913	.9916
2.4	.9918	.9920	.9922	.9925	.9927	.9929	.9931	.9932	.9934	.9936
2.5	.9938	.9940	.9941	.9943	.9945	.9946	.9948	.9949	.9951	.9952
2.6	.9953	.9955	.9956	.9957	.9959	.9960	.9961	.9962	.9963	.9964
2.7	.9965	.9966	.9967	.9968	.9969	.9970	.9971	.9972	.9973	.9974
2.8	.9974	.9975	.9976	.9977	.9977	.9978	.9979	.9979	.9980	.9981
2.9	.9981	.9982	.9982	.9983	.9984	.9984	.9985	.9985	.9986	.9986
3.0	.9987	.9987	.9987	.9988	.9988	.9989	.9989	.9989	.9990	.9990
3.1	.9990	.9991	.9991	.9991	.9992	.9992	.9992	.9992	.9993	.9993
3.2	.9993	.9993	.9994	.9994	.9994	.9994	.9994	.9995	.9995	.9995
3.3	.9995	.9995	.9995	.9996	.9996	.9996	.9996	.9996	.9996	.9997
3.4	.9997	.9997	.9997	.9997	.9997	.9997	.9997	.9997	.9997	.9998
3.5	.9998	.9998	.9998	.9998	.9998	.9998	.9998	.9998	.9998	.9998
3.6	.9998	.9998	.9999	.9999	.9999	.9999	.9999	.9999	.9999	.9999
3.7	.9999	.9999	.9999	.9999	.9999	.9999	.9999	.9999	.9999	.9999

[a]Calculated by MINITAB.

TABLE B5 Critical Values for the *t* Distribution[a]

df	.50	.60	.70	.80	.90	.95	.98	.99
	Probabilites between ± *t* values (two-sided)							
1	1.000	1.376	1.963	3.078	6.314	12.706	31.821	63.657
2	0.816	1.061	1.386	1.886	2.920	4.303	6.965	9.925
3	0.765	0.978	1.250	1.638	2.353	3.182	4.541	5.841
4	0.741	0.941	1.190	1.533	2.132	2.776	3.747	4.604
5	0.727	0.920	1.156	1.476	2.015	2.571	3.365	4.032
6	0.718	0.906	1.134	1.440	1.943	2.447	3.143	3.707
7	0.711	0.896	1.119	1.415	1.895	2.365	2.998	3.499
8	0.706	0.889	1.108	1.397	1.860	2.306	2.896	3.355
9	0.703	0.883	1.100	1.383	1.833	2.262	2.821	3.250
10	0.700	0.879	1.093	1.372	1.812	2.228	2.764	3.169
11	0.697	0.876	1.088	1.363	1.796	2.201	2.718	3.106
12	0.695	0.873	1.083	1.356	1.782	2.179	2.681	3.055
13	0.694	0.870	1.079	1.350	1.771	2.160	2.650	3.012
14	0.692	0.868	1.076	1.345	1.761	2.145	2.624	2.977
15	0.691	0.866	1.074	1.341	1.753	2.131	2.602	2.947
16	0.690	0.865	1.071	1.337	1.746	2.120	2.583	2.921
17	0.689	0.863	1.069	1.333	1.740	2.110	2.567	2.898
18	0.688	0.862	1.067	1.330	1.734	2.101	2.552	2.878
19	0.688	0.861	1.066	1.328	1.729	2.093	2.539	2.861
20	0.687	0.860	1.064	1.325	1.725	2.086	2.528	2.845
21	0.686	0.859	1.063	1.323	1.721	2.080	2.518	2.831
22	0.686	0.858	1.061	1.321	1.717	2.074	2.508	2.819
23	0.685	0.858	1.060	1.319	1.714	2.069	2.500	2.807
24	0.685	0.857	1.059	1.318	1.711	2.064	2.492	2.797
25	0.684	0.856	1.058	1.316	1.708	2.060	2.485	2.787
26	0.684	0.856	1.058	1.315	1.706	2.056	2.479	2.779
27	0.684	0.855	1.057	1.314	1.703	2.052	2.473	2.771
28	0.683	0.855	1.056	1.313	1.701	2.048	2.467	2.763
29	0.683	0.854	1.055	1.311	1.699	2.045	2.462	2.756
30	0.683	0.854	1.055	1.310	1.697	2.042	2.457	2.750
35	0.682	0.852	1.052	1.306	1.690	2.030	2.438	2.724
40	0.681	0.851	1.050	1.303	1.684	2.021	2.423	2.704
45	0.680	0.850	1.049	1.301	1.679	2.014	2.412	2.690
50	0.679	0.849	1.047	1.299	1.676	2.009	2.403	2.678
55	0.679	0.848	1.046	1.297	1.673	2.004	2.396	2.668
60	0.679	0.848	1.045	1.296	1.671	2.000	2.390	2.660
65	0.678	0.847	1.045	1.295	1.669	1.997	2.385	2.654
70	0.678	0.847	1.044	1.294	1.667	1.994	2.381	2.648
75	0.678	0.846	1.044	1.293	1.665	1.992	2.377	2.643
80	0.678	0.846	1.043	1.292	1.664	1.990	2.374	2.639
90	0.677	0.846	1.042	1.291	1.662	1.987	2.369	2.632
100	0.677	0.845	1.042	1.290	1.660	1.984	2.364	2.626
150	0.676	0.844	1.040	1.287	1.655	1.976	2.351	2.609
200	0.676	0.843	1.039	1.286	1.653	1.972	2.345	2.601
500	0.675	0.842	1.038	1.283	1.648	1.965	2.334	2.586
1000	0.675	0.842	1.037	1.282	1.646	1.962	2.330	2.581
∞	0.674	0.842	1.036	1.282	1.645	1.960	2.326	2.576
	.75	.80	.85	.90	.95	.975	.99	.995
	Probability below *t* value (one-sided)							

[a]Calculated by MINITAB.

TABLE B6 Graphs for Binomial Confidence Interval

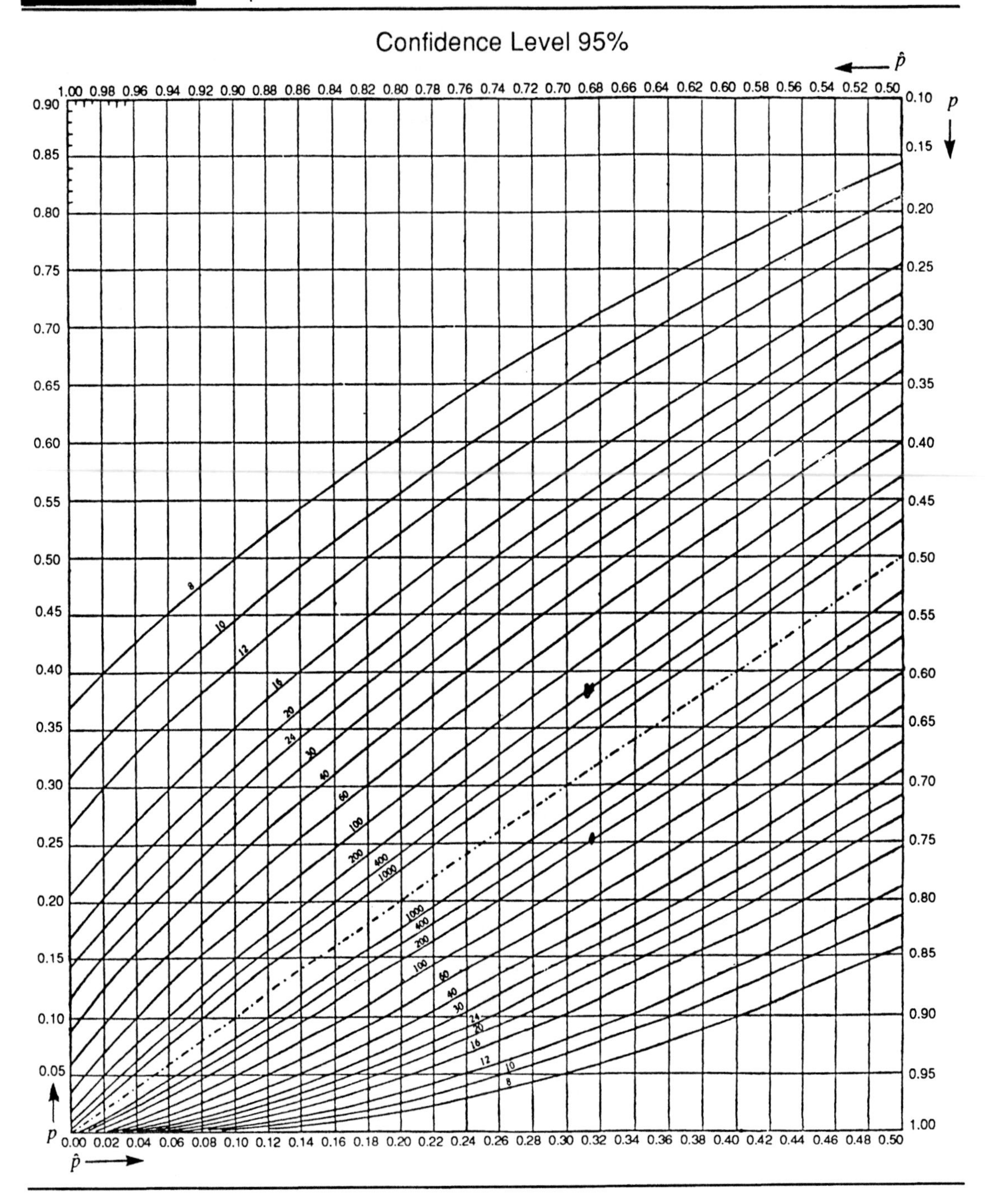

Source: Reprinted from "Biometrika Tables for Statisticians," 3rd ed., Vol. 1, Table 41, Bentley House, London; 1966, with the permission of the Biometrika Trustees. Use the bottom axis with the left axis and the top axis with the right axis.

TABLE B6 —*Continued*

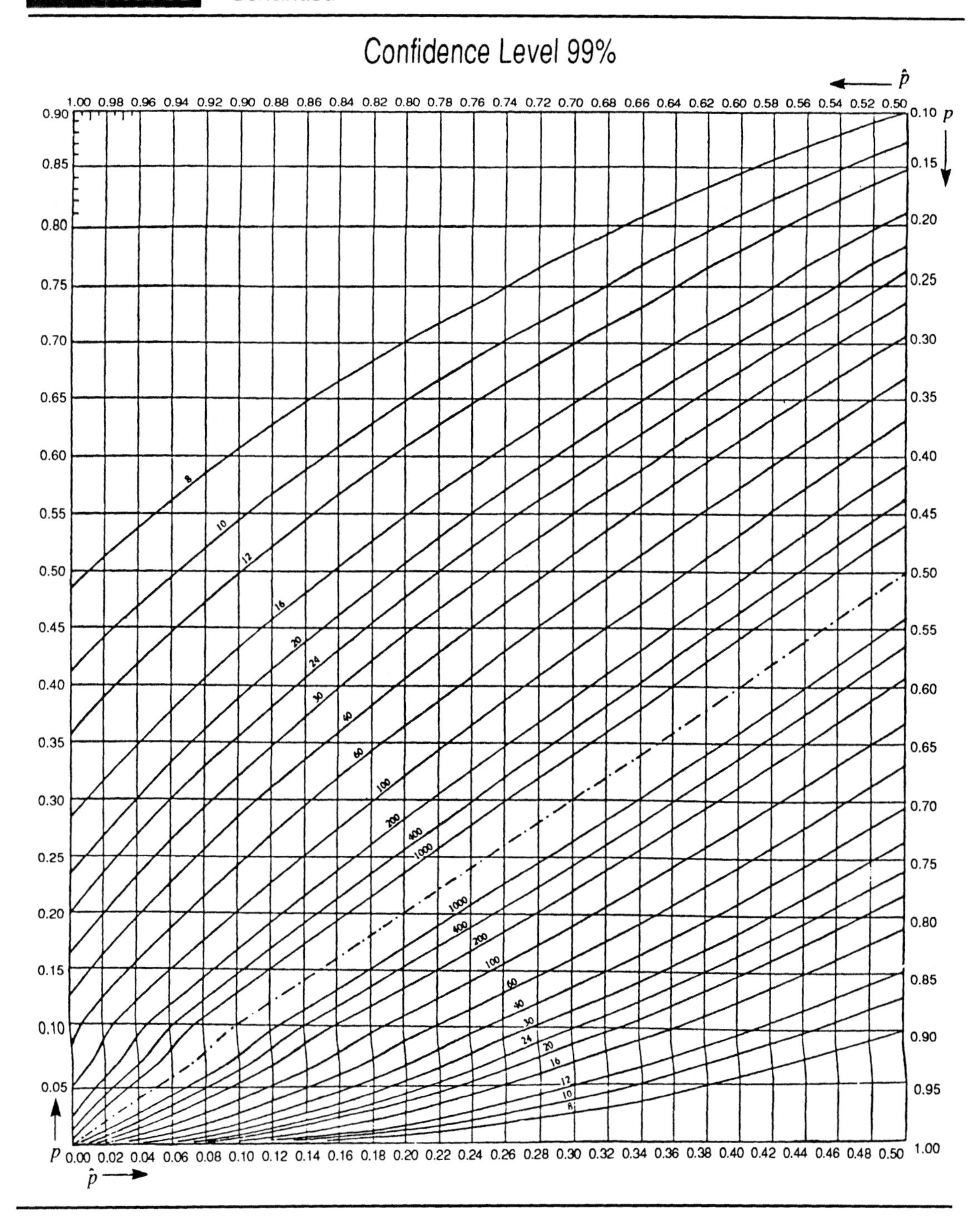

TABLE B7 Critical Values for the Chi-Square (χ^2) Distribution[a]

df	Probability below table value									
	.005	.01	.025	.05	.10	.90	.95	.975	.99	.995
1	0.00	0.00	0.00	0.00	0.02	2.71	3.84	5.02	6.64	7.88
2	0.01	0.02	0.05	0.10	0.21	4.61	5.99	7.38	9.21	10.60
3	0.07	0.12	0.22	0.35	0.58	6.25	7.82	9.35	11.35	12.84
4	0.21	0.30	0.48	0.71	1.06	7.78	9.49	11.14	13.28	14.86
5	0.41	0.55	0.83	1.15	1.61	9.24	11.07	12.83	15.09	16.75
6	0.68	0.87	1.24	1.64	2.20	10.65	12.59	14.45	16.81	18.55
7	0.99	1.24	1.69	2.17	2.83	12.02	14.07	16.01	18.48	20.28
8	1.34	1.65	2.18	2.73	3.49	13.36	15.51	17.54	20.09	21.96
9	1.74	2.09	2.70	3.33	4.17	14.68	16.92	19.02	21.67	23.59
10	2.16	2.56	3.25	3.94	4.86	15.99	18.31	20.48	23.21	2[illegible].19
11	2.60	3.05	3.82	4.57	5.58	17.28	19.67	21.92	24.73	2[illegible].76
12	3.07	3.57	4.40	5.23	6.30	18.55	21.03	23.34	26.22	28.30
13	3.57	4.11	5.01	5.89	7.04	19.81	22.36	24.74	27.69	29.82
14	4.07	4.66	5.63	6.57	7.79	21.06	23.69	26.12	29.14	31.32
15	4.60	5.23	6.26	7.26	8.55	22.31	25.00	27.49	30.58	32.80
16	5.14	5.81	6.91	7.96	9.31	23.54	26.30	28.85	32.00	34.27
17	5.70	6.41	7.56	8.67	10.09	24.77	27.59	30.19	33.41	35.72
18	6.27	7.02	8.23	9.39	10.87	25.99	28.87	31.53	34.81	37.16
19	6.84	7.63	8.91	10.12	11.65	27.20	30.14	32.85	36.19	38.58
20	7.43	8.26	9.59	10.85	12.44	28.41	31.41	34.17	37.57	40.00
21	8.03	8.90	10.28	11.59	13.24	29.62	32.67	35.48	38.93	41.40
22	8.64	9.54	10.98	12.34	14.04	30.81	33.92	36.78	40.29	42.80
23	9.26	10.20	11.69	13.09	14.85	32.01	35.17	38.08	41.64	44.18
24	9.89	10.86	12.40	13.85	15.66	33.20	36.42	39.36	42.98	45.56
25	10.52	11.52	13.12	14.61	16.47	34.38	37.65	40.65	44.31	46.93
26	11.16	12.20	13.84	15.38	17.29	35.56	38.88	41.92	45.64	48.29
27	11.81	12.88	14.57	16.15	18.11	36.74	40.11	43.20	46.96	49.65
28	12.46	13.57	15.31	16.93	18.94	37.92	41.34	44.46	48.28	50.99
29	13.12	14.26	16.05	17.71	19.77	39.09	42.56	45.72	49.59	52.34
30	13.79	14.95	16.79	18.49	20.60	40.26	43.77	46.98	50.89	53.67
35	17.19	18.51	20.57	22.47	24.80	46.06	49.80	53.20	57.34	60.28
40	20.71	22.16	24.43	26.51	29.05	51.81	55.76	59.34	63.69	66.77
45	24.31	25.90	28.37	30.61	33.35	57.51	61.66	65.41	69.96	73.17
50	27.99	29.71	32.36	34.76	37.69	63.17	67.50	71.42	76.15	79.49
55	31.74	33.57	36.40	38.96	42.06	68.80	73.31	77.38	82.29	85.75
60	35.54	37.49	40.48	43.19	46.46	74.40	79.08	83.30	88.38	91.96
65	39.38	41.44	44.60	47.45	50.88	79.97	84.82	89.18	94.42	98.10
70	43.28	45.44	48.76	51.74	55.33	85.53	90.53	95.02	100.42	104.21
75	47.21	49.48	52.94	56.05	59.80	91.06	96.22	100.84	106.39	110.29
80	51.17	53.54	57.15	60.39	64.28	96.58	101.88	106.63	112.33	116.32
90	59.20	61.75	65.65	69.13	73.29	107.57	113.15	118.14	124.12	128.30
100	67.33	70.07	74.22	77.93	82.36	118.50	124.34	129.56	135.81	140.18
120	83.85	86.92	91.57	95.71	100.62	140.23	146.57	152.21	158.95	163.65
140	100.66	104.03	109.14	113.66	119.03	161.83	168.61	174.65	181.84	186.85
160	117.68	121.35	126.87	131.76	137.55	183.31	190.52	196.92	204.54	209.84
180	134.88	138.82	144.74	149.97	156.15	204.70	212.30	219.05	227.06	232.62
200	152.24	156.43	162.73	168.28	174.84	226.02	234.00	241.06	249.46	255.28

[a]Calculated by MINITAB.

TABLE B8 Factors, *k*, for Two-Sided Tolerance Limits for Normal Distributions

	$1-\alpha = 0.75$					$1-\alpha = 0.90$				
n Π:	0.75	0.90	0.95	0.99	0.999	0.75	0.90	0.95	0.99	0.999
2	4.498	6.301	7.414	9.531	11.920	11.407	15.978	18.800	24.167	30.227
3	2.501	3.538	4.187	5.431	6.844	4.132	5.847	6.919	8.974	11.309
4	2.035	2.892	3.431	4.471	5.657	2.932	4.166	4.943	6.440	8.149
5	1.825	2.599	3.088	4.033	5.117	2.454	3.494	4.152	5.423	6.879
6	1.704	2.429	2.889	3.779	4.802	2.196	3.131	3.723	4.870	6.188
7	1.624	2.318	2.757	3.611	4.593	2.034	2.902	3.452	4.521	5.750
8	1.568	2.238	2.663	3.491	4.444	1.921	2.743	3.264	4.278	5.446
9	1.525	2.178	2.593	3.400	4.330	1.839	2.626	3.125	4.098	5.220
10	1.492	2.131	2.537	3.328	4.241	1.775	2.535	3.018	3.959	5.046
11	1.465	2.093	2.493	3.271	4.169	1.724	2.463	2.933	3.849	4.906
12	1.443	2.062	2.456	3.223	4.110	1.683	2.404	2.863	3.758	4.792
13	1.425	2.036	2.424	3.183	4.059	1.648	2.355	2.805	3.682	4.697
14	1.409	2.013	2.398	3.148	4.016	1.619	2.314	2.756	3.618	4.615
15	1.395	1.994	2.375	3.118	3.979	1.594	2.278	2.713	3.562	4.545
16	1.383	1.977	2.355	3.092	3.946	1.572	2.246	2.676	3.514	4.484
17	1.372	1.962	2.337	3.069	3.917	1.552	2.219	2.643	3.471	4.430
18	1.363	1.948	2.321	3.048	3.891	1.535	2.194	2.614	3.433	4.382
19	1.355	1.936	2.307	3.030	3.867	1.520	2.172	2.588	3.399	4.339
20	1.347	1.925	2.294	3.013	3.846	1.506	2.152	2.564	3.368	4.300
21	1.340	1.915	2.282	2.998	3.827	1.493	2.135	2.543	3.340	4.264
22	1.334	1.906	2.271	2.984	3.809	1.482	2.118	2.524	3.315	4.232
23	1.328	1.898	2.261	2.971	3.793	1.471	2.103	2.506	3.292	4.203
24	1.322	1.891	2.252	2.959	3.778	1.462	2.089	2.489	3.270	4.176
25	1.317	1.883	2.244	2.948	3.764	1.453	2.077	2.474	3.251	4.151
26	1.313	1.877	2.236	2.938	3.751	1.444	2.065	2.460	3.232	4.127
27	1.309	1.871	2.229	2.929	3.740	1.437	2.054	2.447	3.215	4.106
30	1.297	1.855	2.210	2.904	3.708	1.417	2.025	2.413	3.170	4.049
35	1.283	1.834	2.185	2.871	3.667	1.390	1.988	2.368	3.112	3.974
40	1.271	1.818	2.166	2.846	3.635	1.370	1.959	2.334	3.066	3.917
45	1.262	1.805	2.150	2.826	3.609	1.354	1.935	2.306	3.030	3.871
50	1.255	1.794	2.138	2.809	3.588	1.340	1.916	2.284	3.001	3.833
55	1.249	1.785	2.127	2.795	3.571	1.329	1.901	2.265	2.976	3.801
60	1.243	1.778	2.118	2.784	3.556	1.320	1.887	2.248	2.955	3.774
65	1.239	1.771	2.110	2.773	3.543	1.312	1.875	2.235	2.937	3.751
70	1.235	1.765	2.104	2.764	3.531	1.304	1.865	2.222	2.920	3.730
75	1.231	1.760	2.098	2.757	3.521	1.298	1.856	2.211	2.906	3.712
80	1.228	1.756	2.092	2.749	3.512	1.292	1.848	2.202	2.894	3.696
85	1.225	1.752	2.087	2.743	3.504	1.287	1.841	2.193	2.882	3.682
90	1.223	1.748	2.083	2.737	3.497	1.283	1.834	2.185	2.872	3.669
95	1.220	1.745	2.079	2.732	3.490	1.278	1.828	2.178	2.863	3.657
100	1.218	1.742	2.075	2.727	3.484	1.275	1.822	2.172	2.854	3.646
110	1.214	1.736	2.069	2.719	3.473	1.268	1.813	2.160	2.839	3.626
120	1.211	1.732	2.063	2.712	3.464	1.262	1.804	2.150	2.826	3.610
130	1.208	1.728	2.059	2.705	3.456	1.257	1.797	2.141	2.814	3.595
140	1.206	1.724	2.054	2.700	3.449	1.252	1.791	2.134	2.804	3.582
150	1.204	1.721	2.051	2.695	3.443	1.248	1.785	2.127	2.795	3.571
160	1.202	1.718	2.047	2.691	3.437	1.245	1.780	2.121	2.787	3.561
170	1.200	1.716	2.044	2.687	3.432	1.242	1.775	2.116	2.780	3.552

(continued)

TABLE B8 —*Continued*

n Π:	1 − α = 0.75 0.75	0.90	0.95	0.99	0.999	1 − α = 0.90 0.75	0.90	0.95	0.99	0.999
180	1.198	1.713	2.042	2.683	3.427	1.239	1.771	2.111	2.774	3.543
190	1.197	1.711	2.039	2.680	3.423	1.236	1.767	2.106	2.768	3.536
200	1.195	1.709	2.037	2.677	3.419	1.234	1.764	2.102	2.762	3.529
250	1.190	1.702	2.028	2.665	3.404	1.224	1.750	2.085	2.740	3.501
300	1.186	1.696	2.021	2.656	3.393	1.217	1.740	2.073	2.725	3.481
400	1.181	1.688	2.012	2.644	3.378	1.207	1.726	2.057	2.703	3.453
500	1.177	1.683	2.006	2.636	3.368	1.201	1.717	2.046	2.689	3.434
600	1.175	1.680	2.002	2.631	3.360	1.196	1.710	2.038	2.678	3.421
700	1.173	1.677	1.998	2.626	3.355	1.192	1.705	2.032	2.670	3.411
800	1.171	1.675	1.996	2.623	3.350	1.189	1.701	2.027	2.663	3.402
900	1.170	1.673	1.993	2.620	3.347	1.187	1.697	2.023	2.658	3.396
1000	1.169	1.671	1.992	2.617	3.344	1.185	1.695	2.019	2.654	3.390
∞	1.150	1.645	1.960	2.576	3.291	1.150	1.645	1.960	2.576	3.291

n Π:	1 − α = 0.95 0.75	0.90	0.95	0.99	0.999	1 − α = 0.99 0.75	0.90	0.95	0.99	0.999
2	22.858	32.019	37.674	48.430	60.573	114.363	160.193	188.491	242.300	303.054
3	5.922	8.380	9.916	12.861	16.208	13.378	18.930	22.401	29.055	36.616
4	3.779	5.369	6.370	8.299	10.502	6.614	9.398	11.150	14.527	18.383
5	3.002	4.275	5.079	6.634	8.415	4.643	6.612	7.855	10.260	13.015
6	2.604	3.712	4.414	5.775	7.337	3.743	5.337	6.345	8.301	10.548
7	2.361	3.369	4.007	5.248	6.676	3.233	4.613	5.488	7.187	9.142
8	2.197	3.136	3.732	4.891	6.226	2.905	4.147	4.936	6.468	8.234
9	2.078	2.967	3.532	4.631	5.899	2.677	3.822	4.550	5.966	7.600
10	1.987	2.839	3.379	4.433	5.649	2.508	3.582	4.265	5.594	7.129
11	1.916	2.737	3.259	4.277	5.452	2.378	3.397	4.045	5.308	6.766
12	1.858	2.655	3.162	4.150	5.291	2.274	3.250	3.870	5.079	6.477
13	1.810	2.587	3.081	4.044	5.158	2.190	3.130	3.727	4.893	6.240
14	1.770	2.529	3.012	3.955	5.045	2.120	3.029	3.608	4.737	6.043
15	1.735	2.480	2.954	3.878	4.949	2.060	2.954	3.507	4.605	5.876
16	1.705	2.437	2.903	3.812	4.865	2.009	2.872	3.421	4.492	5.732
17	1.679	2.400	2.858	3.754	4.791	1.965	2.808	3.345	4.393	5.607
18	1.655	2.366	2.819	3.702	4.725	1.926	2.753	3.279	4.307	5.497
19	1.635	2.337	2.784	3.656	4.667	1.891	2.703	3.221	4.230	5.399
20	1.616	2.310	2.752	3.615	4.614	1.860	2.659	3.168	4.161	5.312
21	1.599	2.286	2.723	3.577	4.567	1.833	2.620	3.121	4.100	5.234
22	1.584	2.264	2.697	3.543	4.523	1.808	2.584	3.078	4.044	5.163
23	1.570	2.244	2.673	3.512	4.484	1.785	2.551	3.040	3.993	5.098

TABLE B8 —*Continued*

	$1-\alpha=0.95$					$1-\alpha=0.99$				
n Π:	0.75	0.90	0.95	0.99	0.999	0.75	0.90	0.95	0.99	0.999
24	1.557	2.225	2.651	3.483	4.447	1.764	2.522	3.004	3.947	5.039
25	1.545	2.208	2.631	3.457	4.413	1.745	2.494	2.972	3.904	4.985
26	1.534	2.193	2.612	3.432	4.382	1.727	2.469	2.941	3.865	4.935
27	1.523	2.178	2.595	3.409	4.353	1.711	2.446	2.914	3.828	4.888
30	1.497	2.140	2.549	3.350	4.278	1.668	2.385	2.841	3.733	4.768
35	1.462	2.090	2.490	3.272	4.179	1.613	2.306	2.748	3.611	4.611
40	1.435	2.052	2.445	3.213	4.104	1.571	2.247	2.677	3.518	4.493
45	1.414	2.021	2.408	3.165	4.042	1.539	2.200	2.621	3.444	4.399
50	1.396	1.996	2.379	3.126	3.993	1.512	2.162	2.576	3.385	4.323
55	1.382	1.976	2.354	3.094	3.951	1.490	2.130	2.538	3.335	4.260
60	1.369	1.958	2.333	3.066	3.916	1.471	2.103	2.506	3.293	4.206
65	1.359	1.943	2.315	3.042	3.886	1.455	2.080	2.478	3.257	4.160
70	1.349	1.929	2.299	3.021	3.859	1.440	2.060	2.454	3.225	4.120
75	1.341	1.917	2.285	3.002	3.835	1.428	2.042	2.433	3.197	4.084
80	1.334	1.907	2.272	2.986	3.814	1.417	2.026	2.414	3.173	4.053
85	1.327	1.897	2.261	2.971	3.795	1.407	2.012	2.397	3.150	4.024
90	1.321	1.889	2.251	2.958	3.778	1.398	1.999	2.382	3.130	3.999
95	1.315	1.881	2.241	2.945	3.763	1.390	1.987	2.368	3.112	3.976
100	1.311	1.874	2.233	2.934	3.748	1.383	1.977	2.355	3.096	3.954
110	1.302	1.861	2.218	2.915	3.723	1.369	1.958	2.333	3.066	3.917
120	1.294	1.850	2.205	2.898	3.702	1.358	1.942	2.314	3.041	3.885
130	1.288	1.841	2.194	2.883	3.683	1.349	1.928	2.298	3.019	3.857
140	1.282	1.833	2.184	2.870	3.666	1.340	1.916	2.283	3.000	3.833
150	1.277	1.825	2.175	2.859	3.652	1.332	1.905	2.270	2.983	3.811
160	1.272	1.819	2.167	2.848	3.638	1.326	1.896	2.259	2.968	3.792
170	1.268	1.813	2.160	2.839	3.527	1.320	1.887	2.248	2.955	3.774
180	1.264	1.808	2.154	2.831	3.616	1.314	1.879	2.239	2.942	3.759
190	1.261	1.803	2.148	2.823	3.606	1.309	1.872	2.230	2.931	3.744
200	1.258	1.798	2.143	2.816	3.597	1.304	1.865	2.222	2.921	3.731
250	1.245	1.780	2.121	2.788	3.561	1.286	1.389	2.191	2.880	3.678
300	1.236	1.767	2.106	2.767	3.535	1.273	1.820	2.169	2.850	3.641
400	1.223	1.749	2.084	2.739	3.499	1.255	1.794	2.138	2.809	3.589
500	1.215	1.737	2.070	2.721	3.475	1.243	1.777	2.117	2.783	3.555
600	1.209	1.729	2.060	2.707	3.458	1.234	1.764	2.102	2.763	3.530
700	1.204	1.722	2.052	2.697	3.445	1.227	1.755	2.091	2.748	3.511
800	1.201	1.717	2.046	2.688	3.434	1.222	1.747	2.082	2.736	3.495
900	1.198	1.712	2.040	2.682	3.426	1.218	1.741	2.075	2.726	3.483
1000	1.195	1.709	2.036	2.676	3.418	1.214	1.736	2.068	2.718	3.472
∞	1.150	1.645	1.960	2.576	3.291	1.150	1.645	1.960	2.576	3.291

Source: Abstracted from C. Eisenhart, M. W. Hastay, and W. A. Wallis, "Techniques of Statistical Analysis," Table 2.1, pp. 102–107. McGraw-Hill, New York, 1947.

TABLE B9 Critical Values for Wilcoxon Signed Rank Test[a]

	Two-sided Comparisons			
n	$\alpha \leq 0.10$	$\alpha \leq 0.05$	$\alpha \leq 0.02$	$\alpha \leq 0.01$
5	0, 15			
6	2, 19	0, 21		
7	3, 25	2, 26	0, 28	
8	5, 31	3, 33	1, 35	0, 36
9	8, 37	5, 40	3, 42	1, 44
10	10, 45	8, 47	5, 50	3, 52
11	13, 53	10, 56	7, 59	5, 61
12	17, 61	13, 65	9, 69	7, 71
13	21, 70	17, 74	12, 79	9, 82
14	25, 80	21, 84	15, 90	12, 93
15	30, 90	25, 95	19,101	15,105
16	35,101	29,107	23,113	19,117
17	41,112	34,119	28,125	23,130
18	47,124	40,131	32,139	27,144
19	53,137	46,144	37,153	33,158
20	60,150	52,158	43,167	37,173
21	67,164	58,173	49,182	42,189
22	75,178	66,187	55,198	48,205
23	83,193	73,203	62,214	54,222
24	91,209	81,210	69,231	61,239
25	100,225	89,236	76,249	68,257
26	110,241	98,253	84,267	75,276
27	119,259	107,271	93,285	83,295
28	130,276	114,278	101,305	91,315
29	140,295	126,309	122,313	100,335
30	151,314	137,328	132,333	109,356
	One-sided Comparisons			
n	$\alpha \leq 0.05$	$\alpha \leq 0.025$	$\alpha \leq 0.01$	$\alpha \leq 0.005$

[a]Extracted from "Critical Values and Probability Levels for the Wilcoxon Rank Sum Test and the Wilcoxon Signed Rank Test," by Frank Wilcoxon, S. K. Katti, and Roberta A. Wilcox, *Selected Tables in Mathematical Statistics*, Vol. 1, 1973, Table II, pp. 237-259, by permission of the American Mathematics Society.

TABLE B10 Critical Values for Wilcoxon Rank Sum Test[a]

$N_2 \backslash N_1$	3	4	5	6	7	8	9	10	11	12	13	14	15
					Two-sided: $\alpha \leq 0.10$		One-sided: $\alpha \leq 0.05$						
3	—	10, 22	16, 29	23, 37	30, 47	39, 57	48, 69	59, 81	71, 94	83,109	97,124	112,140	127,158
4	6, 18	**11, 25**	17, 33	24, 42	32, 52	41, 63	51, 75	62, 88	74,102	87,117	101,133	116,150	132,168
5	7, 20	12, 28	**19, 36**	26, 46	34, 57	44, 68	54, 81	66, 94	78,109	91,125	106,141	121,159	138,177
6	8, 22	13, 31	20, 40	**28, 50**	36, 62	46, 74	57, 87	69,101	82,116	95,133	110,150	126,168	143,187
7	8, 25	14, 34	21, 44	29, 55	**39, 66**	49, 79	60, 93	72,108	85,124	99,141	115,158	131,177	148,197
8	9, 27	15, 37	23, 47	31, 59	41, 71	**51, 85**	63, 99	75,115	89,131	104,148	119,167	136,186	153,207
9	9, 30	16, 40	24, 51	33, 63	43, 76	54, 90	**66,105**	79,121	93,138	108,156	124,178	141,195	159,216
10	10, 32	17, 43	26, 54	35, 67	45, 81	56, 96	69,111	**82,128**	97,145	112,164	128,184	146,204	164,226
11	11, 34	18, 46	27, 58	37, 71	47, 86	59,101	72,117	86,134	**100,153**	116,172	133,192	151,231	170,235
12	11, 37	19, 49	28, 62	38, 76	49, 91	62,106	75,123	89,141	104,160	**120,180**	138,200	156,222	175,245
13	12, 39	20, 52	30, 65	40, 80	52, 95	64,112	78,129	92,148	108,167	125,187	**142,209**	161,231	181,254
14	13, 41	21, 55	31, 69	42, 84	54,100	67,117	81,135	96,154	112,174	129,195	147,217	**166,240**	186,264
15	13, 44	22, 58	33, 72	44, 88	56,105	69,123	84,141	99,161	116,181	133,203	152,225	171,249	**192,273**
16	14, 46	24, 60	34, 76	46, 92	58,110	72,128	87,147	103,167	120,188	138,210	156,234	176,258	197,283
17	15, 48	25, 63	35, 80	47, 97	61,114	75,133	90,153	106,174	123,196	142,218	161,242	182,266	203,292
18	15, 51	26, 66	37, 83	49,101	63,119	77,139	93,159	110,180	127,203	146,226	166,250	187,275	208,302
19	16, 53	27, 69	38, 87	51,105	65,124	80,144	96,165	113,187	131,210	150,234	171,258	192,284	214,311
20	17, 55	28, 72	40, 90	53,109	67,129	83,149	99,171	117,193	135,217	155,241	175,267	197,293	220,320
21	17, 58	29, 75	41, 94	55,113	69,134	85,155	102,177	120,200	139,224	159,249	180,275	202,302	225,330
22	18, 60	30, 78	43, 97	57,117	72,138	88,160	105,183	123,207	143,231	163,257	185,283	207,311	231,339
23	19, 62	31, 81	44,101	58,122	74,143	90,166	108,189	127,213	147,238	168,264	189,292	212,320	236,349
24	19, 65	32, 84	45,105	60,126	76,148	93,171	111,195	130,220	151,245	172,272	194,300	218,328	242,358
25	20, 67	33, 87	47,108	62,130	78,153	96,176	114,201	134,226	155,252	176,280	199,308	223,337	248,367
26	21, 69	34, 90	48,112	64,134	81,157	98,182	117,207	137,233	158,260	181,287	204,316	228,346	253,377
27	21, 72	35, 93	50,115	66,138	83,162	101,187	120,213	141,239	162,267	185,295	208,325	233,355	259,386
28	22, 74	36, 96	51,119	67,143	85,167	104,192	123,219	144,246	166,274	189,303	213,333	238,364	264,396
29	23, 76	37, 99	53,122	69,147	87,172	106,198	127,224	148,252	170,281	194,310	218,341	243,373	270,405
30	23, 79	38,102	54,126	71,151	89,177	109,203	130,230	151,259	174,288	198,318	223,349	249,381	276,414

(continued)

TABLE B10 —*Continued*

$N_2 \backslash N_1$	3	4	5	6	7	8	9	10	11	12	13	14	15
					Two-sided: $\alpha \le 0.05$		One-sided: $\alpha \le 0.025$						
3	—	—	15, 30	22, 38	29, 48	38, 58	47, 70	58, 82	70, 96	82,110	95,126	110,142	125,160
4	—	**10, 26**	16, 34	23, 43	31, 53	40, 64	49, 77	60, 90	72,104	85,119	99,135	114,152	130,170
5	6, 21	11, 29	**17, 38**	24, 48	33, 58	42, 70	52, 83	63, 97	75,112	89,127	103,144	118,162	134,181
6	7, 23	12, 32	18, 42	**26, 52**	34, 64	44, 76	55, 89	66,104	79,119	92,136	107,153	122,172	139,191
7	7, 26	13, 35	20, 45	27, 57	**36, 69**	46, 82	57, 96	69,111	82,127	96,144	111,162	127,181	144,201
8	8, 28	14, 38	21, 49	29, 61	38, 74	**49, 87**	60,102	72,118	85,135	100,152	115,171	131,191	149,211
9	8, 31	14, 42	22, 53	31, 65	40, 79	51, 93	**62,109**	75,125	89,142	104,160	119,180	136,200	154,221
10	9, 33	15, 45	23, 57	32, 70	42, 84	53, 99	65,115	**78,132**	92,150	107,169	124,188	141,209	159,231
11	9, 36	16, 48	24, 61	34, 74	44, 89	55,105	68,121	81,139	**96,157**	111,177	128,197	145,219	164,241
12	10, 38	17, 51	26, 64	35, 79	46, 94	58,110	71,127	84,146	99,165	**115,185**	132,206	150,228	169,251
13	10, 41	18, 54	27, 68	37, 83	48, 99	60,116	73,143	88,152	103,172	119,193	**136,215**	155,237	174,261
14	11, 43	19, 57	28, 72	38, 88	50,104	62,122	76,140	91,159	106,180	123,201	141,223	**160,246**	179,271
15	11, 46	20, 60	29, 76	40, 92	52,109	65,127	79,146	94,166	110,187	127,209	145,232	164,256	**184,281**
16	12, 48	21, 63	30, 80	42, 96	54,114	67,133	82,152	97,173	113,195	131,217	150,240	169,265	190,290
17	12, 51	21, 67	32, 83	43,101	56,119	70,138	84,159	100,180	117,202	135,225	154,249	174,274	195,300
18	13, 53	22, 70	33, 87	45,105	58,124	72,144	87,165	103,187	121,209	139,233	158,258	179,283	200,310
19	13, 56	23, 73	34, 91	46,110	60,129	74,150	90,171	107,193	124,217	143,241	163,266	183,293	205,320
20	14, 58	24, 76	35, 95	48,114	62,134	77,155	93,177	110,200	128,224	147,249	167,275	188,302	210,330
21	14, 61	25, 79	37, 98	50,118	64,139	79,161	95,184	113,207	131,232	151,257	171,284	193,311	216,339
22	15, 63	26, 82	38,102	51,123	66,144	81,167	98,190	116,214	135,239	155,265	176,292	198,320	221,349
23	15, 66	27, 85	39,106	53,127	68,149	84,172	101,196	119,221	139,246	159,273	180,301	203,329	226,359
24	16, 68	27, 89	40,110	54,132	70,154	86,178	104,202	122,228	142,254	163,281	185,309	207,339	231,369
25	16, 71	28, 92	42,113	56,136	72,159	89,183	107,208	126,234	146,261	167,289	189,318	212,348	237,378
26	17, 73	29, 95	43,117	58,140	74,164	91,189	109,215	129,241	149,269	171,297	193,327	217,357	242,388
27	17, 76	30, 98	44,121	59,145	76,169	93,195	112,221	132,248	153,276	175,305	198,335	222,366	247,398
28	18, 78	31,101	45,125	61,149	78,174	96,200	115,227	135,255	156,284	179,313	202,344	227,375	252,408
29	19, 80	32,104	47,128	63,153	80,179	98,206	118,233	138,262	160,291	183,321	207,352	232,384	258,417
30	19, 83	33,107	48,132	64,158	82,184	101,211	121,239	142,268	164,298	187,329	211,361	236,394	263,427

Two-sided: $\alpha \leq 0.02$ One-sided: $\alpha \leq 0.01$

3	—	—	—	—	28, 49	36, 60	46, 71	56, 84	67, 98	80,112	93,128	107,145	123,162
4	—	—	15, 35	22, 44	29, 55	38, 66	48, 78	58, 92	70,106	83,121	96,138	111,155	127,173
5	—	10, 30	**16, 39**	23, 49	31, 60	40, 72	50, 85	61, 99	73,114	86,130	100,147	115,165	131,184
6	—	11, 33	17, 43	**24, 54**	32, 66	42, 78	52, 92	63,107	75,123	89,139	103,157	118,176	135,195
7	6, 27	11, 37	18, 47	25, 59	**34, 71**	43, 85	54, 99	66,114	78,131	92,148	107,166	122,186	139,206
8	6, 30	12, 40	19, 51	27, 63	35, 77	**45, 91**	56,106	68,122	81,139	95,157	111,175	127,195	144,216
9	7, 32	13, 43	20, 55	28, 68	37, 82	47, 97	**59,112**	71,129	84,147	99,165	114,185	131,205	148,227
10	7, 35	13, 47	21, 59	29, 73	39, 87	49,103	61,119	**74,136**	88,154	102,174	118,194	135,215	153,237
11	7, 38	14, 50	22, 63	30, 78	40, 93	51,109	63,126	77,143	**91,162**	106,182	122,203	139,225	157,248
12	8, 40	15, 53	23, 67	32, 82	42, 98	53,115	66,132	79,151	94,170	**109,191**	126,212	143,235	162,258
13	8, 43	15, 57	24, 71	33, 87	44,103	56,120	68,139	82,158	97,178	113,199	**130,221**	148,244	167,268
14	8, 46	16, 60	25, 75	34, 92	45,109	58,126	71,145	85,165	100,186	116,208	134,230	**152,254**	171,279
15	9, 48	17, 63	26, 79	36, 96	47,114	60,132	73,152	88,172	103,194	120,216	138,239	156,264	**176,289**
16	9, 51	17, 67	27, 83	37,101	49,119	62,138	76,158	91,179	107,201	124,224	142,248	161,273	181,299
17	10, 53	18, 70	28, 87	39,105	51,124	64,144	78,165	93,187	110,209	127,233	146,257	165,283	186,309
18	10, 56	19, 73	29, 91	40,110	52,130	66,150	81,171	96,194	113,217	131,241	150,266	170,292	190,320
19	10, 59	19, 77	30, 95	41,115	54,135	68,156	83,178	99,201	116,225	134,250	154,275	174,302	195,330
20	11, 61	20, 80	31, 99	43,119	56,140	70,162	85,185	102,208	119,233	138,258	158,284	178,312	200,340
21	11, 64	21, 83	32,103	44,124	58,145	72,168	88,191	105,215	123,240	142,266	162,293	183,321	205,350
22	11, 67	21, 87	33,107	45,129	59,151	74,174	90,198	108,222	126,248	145,275	166,302	187,331	210,360
23	12, 69	22, 90	34,111	47,133	61,156	76,180	93,204	110,230	129,256	149,283	170,311	192,340	214,371
24	12, 72	23, 93	35,115	48,138	63,161	78,186	95,211	113,237	132,264	153,291	174,320	196,350	219,381
25	13, 74	23, 97	36,119	50,142	64,167	81,191	98,217	116,244	136,271	156,300	178,329	200,360	224,391
26	13, 77	24,100	37,123	51,147	66,172	83,197	100,224	119,251	139,279	160,308	182,338	205,369	229,401
27	13, 80	25,103	38,127	52,152	68,177	85,203	103,230	122,258	142,287	163,317	186,347	209,379	234,411
28	14, 82	26,106	39,131	54,156	70,182	87,209	105,237	125,265	145,295	167,325	190,356	214,388	239,421
29	14, 85	26,110	40,135	55,161	71,188	89,215	108,243	128,272	149,302	171,333	194,365	218,398	243,432
30	15, 87	27,113	41,139	56,166	73,193	91,221	110,250	131,279	152,310	174,342	198,374	223,407	248,442

(continued)

TABLE B10 —*Continued*

$N_2 \backslash N_1$	3	4	5	6	7	8	9	10	11	12	13	14	15
					Two-sided: $\alpha \leq 0.01$		One-sided: $\alpha \leq 0.005$						
3	—	—	—	—	—	—	45, 72	55, 85	66, 99	79,113	92,129	106,146	122,163
4	—	—	—	21, 45	28, 56	37, 67	46, 80	57, 93	68,108	81,123	94,140	109,157	125,175
5	—	—	**15, 40**	22, 50	29, 62	38, 74	48, 87	59,101	71,116	84,132	98,149	112,168	128,187
6	—	10, 34	16, 44	**23, 55**	31, 67	40, 80	50, 94	61,109	73,125	87,141	101,159	116,178	132,198
7	—	10, 38	16, 49	24, 60	**32, 73**	42, 86	52,101	64,116	76,133	90,150	104,169	120,188	136,209
8	—	11, 41	17, 53	25, 65	34, 78	**43, 93**	54,108	66,124	79,141	93,159	108,178	123,199	140,220
9	6, 33	11, 45	18, 57	26, 70	35, 84	45, 99	**56,115**	68,132	82,149	96,168	111,188	127,209	144,231
10	6, 36	12, 48	19, 61	27, 75	37, 89	47,105	58,122	**71,139**	84,158	99,177	115,197	131,219	149,241
11	6, 39	12, 52	20, 65	28, 80	38, 95	49,111	61,128	73,147	**87,166**	102,186	118,207	135,229	153,252
12	7, 41	13, 55	21, 69	30, 84	40,100	51,117	63,135	76,154	90,174	**105,195**	122,216	139,239	157,263
13	7, 44	13, 59	22, 73	31, 89	41,106	53,123	65,142	79,161	93,182	109,203	**125,226**	143,249	162,273
14	7, 47	14, 62	22, 78	32, 94	43,111	54,130	67,149	81,169	96,190	112,212	129,235	**147,259**	166,284
15	8, 49	15, 65	23, 82	33, 99	44,117	56,136	69,156	84,176	99,198	115,221	133,244	151,269	**171,294**
16	8, 52	15, 69	24, 86	34,104	46,122	58,142	72,162	86,184	102,206	119,229	136,254	155,279	175,305
17	8, 55	16, 72	25, 90	36,108	47,128	60,148	74,169	89,191	105,214	122,238	140,263	159,289	180,315
18	8, 58	16, 76	26, 94	37,113	49,133	62,154	76,176	92,198	108,222	125,247	144,272	163,299	184,326
19	9, 60	17, 79	27, 98	38,118	50,139	64,160	78,183	94,206	111,230	129,255	148,281	168,308	189,336
20	9, 63	18, 82	28,102	39,123	52,144	66,166	81,189	97,213	114,238	132,264	151,291	172,318	193,347
21	9, 66	18, 86	29,106	40,128	53,150	68,172	83,196	99,221	117,246	136,272	155,300	176,328	198,357
22	10, 68	19, 89	29,111	42,132	55,155	70,178	85,203	102,228	120,254	139,281	159,309	180,338	202,368
23	10, 71	19, 93	30,115	43,137	57,160	71,185	88,209	105,235	123,262	142,290	163,318	184,348	207,378
24	10, 74	20, 96	30,119	44,142	58,166	78,191	90,216	107,243	126,270	146,298	166,328	188,358	211,389
25	11, 76	20,100	32,123	45,147	60,171	75,197	92,223	110,250	129,278	149,307	170,337	192,368	216,399
26	11, 79	21,103	33,127	46,152	61,177	77,203	94,230	113,257	132,286	152,316	174,346	197,377	220,410
27	11, 82	22,106	34,131	48,156	63,182	79,209	97,236	115,265	135,294	156,324	178,355	201,387	225,420
28	11, 85	22,110	35,135	49,161	64,188	81,215	99,243	118,272	138,302	159,333	182,364	205,397	229,431
29	12, 87	23,113	36,139	50,166	66,193	83,221	101,250	121,279	141,310	163,341	185,374	209,407	234,441
30	12, 90	23,117	37,143	51,171	68,198	85,227	103,257	123,287	144,318	166,350	189,383	213,417	239,451

*Extracted from "Critical Values and Probability Levels for the Wilcoxon Rank Sum Test and Wilcoxon Signed Rank Test," by Frank Wilcoxon, S. K. Katti, and Roberta A. Wilcox, *Selected Tables in Mathematical Statistics*, Vol. 1, 1973, Table I, pp. 177-235, by permission of the American Mathematics Society.

TABLE B11 Critical Values for the *F* Distribution[a]

$F_{.90}$

df in the denominator	df in the numerator 1	2	3	4	5	6	8	10	20	50	100
1	39.86	49.50	53.59	55.83	57.24	58.20	59.44	60.19	61.74	62.69	63.01
2	8.53	9.00	9.16	9.24	9.29	9.33	9.37	9.39	9.44	9.47	9.48
3	5.54	5.46	5.39	5.34	5.31	5.28	5.25	5.23	5.18	5.15	5.14
4	4.54	4.32	4.19	4.11	4.05	4.01	3.95	3.92	3.84	3.80	3.78
5	4.06	3.78	3.62	3.52	3.45	3.40	3.34	3.30	3.21	3.15	3.13
6	3.78	3.46	3.29	3.18	3.11	3.05	2.98	2.94	2.84	2.77	2.75
7	3.59	3.26	3.07	2.96	2.88	2.83	2.75	2.70	2.59	2.52	2.50
8	3.46	3.11	2.92	2.81	2.73	2.67	2.59	2.54	2.42	2.35	2.32
9	3.36	3.01	2.81	2.69	2.61	2.55	2.47	2.42	2.30	2.22	2.19
10	3.29	2.92	2.73	2.61	2.52	2.46	2.38	2.32	2.20	2.12	2.09
11	3.23	2.86	2.66	2.54	2.45	2.39	2.30	2.25	2.12	2.04	2.01
12	3.18	2.81	2.61	2.48	2.39	2.33	2.24	2.19	2.06	1.97	1.94
13	3.14	2.76	2.56	2.43	2.35	2.28	2.20	2.14	2.01	1.92	1.88
14	3.10	2.73	2.52	2.39	2.31	2.24	2.15	2.10	1.96	1.87	1.83
15	3.07	2.70	2.49	2.36	2.27	2.21	2.12	2.06	1.92	1.83	1.79
16	3.05	2.67	2.46	2.33	2.24	2.18	2.09	2.03	1.89	1.79	1.76
17	3.03	2.64	2.44	2.31	2.22	2.15	2.06	2.00	1.86	1.76	1.73
18	3.01	2.62	2.42	2.29	2.20	2.13	2.04	1.98	1.84	1.74	1.70
19	2.99	2.61	2.40	2.27	2.18	2.11	2.02	1.96	1.81	1.71	1.67
20	2.97	2.59	2.38	2.25	2.16	2.09	2.00	1.94	1.79	1.69	1.65
21	2.96	2.57	2.36	2.23	2.14	2.08	1.98	1.92	1.78	1.67	1.63
22	2.95	2.56	2.35	2.22	2.13	2.06	1.97	1.90	1.76	1.65	1.61
23	2.94	2.55	2.34	2.21	2.11	2.05	1.95	1.89	1.74	1.64	1.59
24	2.93	2.54	2.33	2.19	2.10	2.04	1.94	1.88	1.73	1.62	1.58
25	2.92	2.53	2.32	2.18	2.09	2.02	1.93	1.87	1.72	1.61	1.56
26	2.91	2.52	2.31	2.17	2.08	2.01	1.92	1.86	1.71	1.59	1.55
27	2.90	2.51	2.30	2.17	2.07	2.00	1.91	1.85	1.70	1.58	1.54
28	2.89	2.50	2.29	2.16	2.06	2.00	1.90	1.84	1.69	1.57	1.53
29	2.89	2.50	2.28	2.15	2.06	1.99	1.89	1.83	1.68	1.56	1.52
30	2.88	2.49	2.28	2.14	2.05	1.98	1.88	1.82	1.67	1.55	1.51
40	2.84	2.44	2.23	2.09	2.00	1.93	1.83	1.76	1.61	1.48	1.43
50	2.81	2.41	2.20	2.06	1.97	1.90	1.80	1.73	1.57	1.44	1.39
60	2.79	2.39	2.18	2.04	1.95	1.87	1.77	1.71	1.54	1.41	1.36
100	2.76	2.36	2.14	2.00	1.91	1.83	1.73	1.66	1.49	1.35	1.29
200	2.73	2.33	2.11	1.97	1.88	1.80	1.70	1.63	1.46	1.31	1.24
1000	2.71	2.31	2.09	1.95	1.85	1.78	1.68	1.61	1.43	1.27	1.20

(continued)

TABLE B11 —*Continued*

$F_{.95}$

df in the denominator	df in the numerator: 1	2	3	4	5	6	8	10	20	50	100
1	161.5	199.5	215.7	224.6	230.2	234.0	238.9	241.9	248.0	251.8	253.0
2	18.51	19.00	19.16	19.25	19.30	19.33	19.37	19.40	19.45	19.48	19.49
3	10.13	9.55	9.28	9.12	9.01	8.94	8.85	8.79	8.66	8.58	8.55
4	7.71	6.94	6.59	6.39	6.26	6.16	6.04	5.96	5.80	5.70	5.66
5	6.61	5.79	5.41	5.19	5.05	4.95	4.82	4.74	4.56	4.44	4.41
6	5.99	5.14	4.76	4.53	4.39	4.28	4.15	4.06	3.87	3.75	3.71
7	5.59	4.74	4.35	4.12	3.97	3.87	3.73	3.64	3.44	3.32	3.27
8	5.32	4.46	4.07	3.84	3.69	3.58	3.44	3.35	3.15	3.02	2.97
9	5.12	4.26	3.86	3.63	3.48	3.37	3.23	3.14	2.94	2.80	2.76
10	4.96	4.10	3.71	3.48	3.33	3.22	3.07	2.98	2.77	2.64	2.59
11	4.84	3.98	3.59	3.36	3.20	3.09	2.95	2.85	2.65	2.51	2.46
12	4.75	3.89	3.49	3.26	3.11	3.00	2.85	2.75	2.54	2.40	2.35
13	4.67	3.81	3.41	3.18	3.03	2.92	2.77	2.67	2.46	2.31	2.26
14	4.60	3.74	3.34	3.11	2.96	2.85	2.70	2.60	2.39	2.24	2.19
15	4.54	3.68	3.29	3.06	2.90	2.79	2.64	2.54	2.33	2.18	2.12
16	4.49	3.63	3.24	3.01	2.85	2.74	2.59	2.49	2.28	2.12	2.07
17	4.45	3.59	3.20	2.96	2.81	2.70	2.55	2.45	2.23	2.08	2.02
18	4.41	3.55	3.16	2.93	2.77	2.66	2.51	2.41	2.19	2.04	1.98
19	4.38	3.52	3.13	2.90	2.74	2.63	2.48	2.38	2.16	2.00	1.94
20	4.35	3.49	3.10	2.87	2.71	2.60	2.45	2.35	2.12	1.97	1.91
21	4.32	3.47	3.07	2.84	2.68	2.57	2.42	2.32	2.10	1.94	1.88
22	4.30	3.44	3.05	2.82	2.66	2.55	2.40	2.30	2.07	1.91	1.85
23	4.28	3.42	3.03	2.80	2.64	2.53	2.37	2.27	2.05	1.88	1.82
24	4.26	3.40	3.01	2.78	2.62	2.51	2.36	2.25	2.03	1.86	1.80
25	4.24	3.39	2.99	2.76	2.60	2.49	2.34	2.24	2.01	1.84	1.78
26	4.23	3.37	2.98	2.74	2.59	2.47	2.32	2.22	1.99	1.82	1.76
27	4.21	3.35	2.96	2.73	2.57	2.46	2.31	2.20	1.97	1.81	1.74
28	4.20	3.34	2.95	2.71	2.56	2.45	2.29	2.19	1.96	1.79	1.73
29	4.18	3.33	2.93	2.70	2.55	2.43	2.28	2.18	1.94	1.77	1.71
30	4.17	3.32	2.92	2.69	2.53	2.42	2.27	2.16	1.93	1.76	1.70
40	4.08	3.23	2.84	2.61	2.45	2.34	2.18	2.08	1.84	1.66	1.59
50	4.03	3.18	2.79	2.56	2.40	2.29	2.13	2.03	1.78	1.60	1.52
60	4.00	3.15	2.76	2.53	2.37	2.25	2.10	1.99	1.75	1.56	1.48
100	3.94	3.09	2.70	2.46	2.31	2.19	2.03	1.93	1.68	1.48	1.39
200	3.89	3.04	2.65	2.42	2.26	2.14	1.98	1.88	1.62	1.41	1.32
1000	3.85	3.00	2.61	2.38	2.22	2.11	1.95	1.84	1.58	1.36	1.26

TABLE B11—*Continued*

	$F_{.99}$										
df in the denominator	df in the numerator										
	1	2	3	4	5	6	8	10	20	50	100
1	4052	4500	5403	5625	5764	5859	5981	6056	6209	6302	6334
2	98.50	99.00	99.17	99.25	99.30	99.33	99.37	99.40	99.45	99.48	99.49
3	34.12	30.82	29.46	28.71	28.24	27.91	27.49	27.23	26.69	26.35	26.24
4	21.20	18.00	16.69	15.98	15.52	15.21	14.80	14.55	14.02	13.69	13.58
5	16.26	13.27	12.06	11.39	10.97	10.67	10.29	10.05	9.55	9.24	9.13
6	13.75	10.92	9.78	9.15	8.75	8.47	8.10	7.87	7.40	7.09	6.99
7	12.25	9.55	8.45	7.85	7.46	7.19	6.84	6.62	6.16	5.86	5.75
8	11.26	8.65	7.59	7.01	6.63	6.37	6.03	5.81	5.36	5.07	4.96
9	10.56	8.02	6.99	6.42	6.06	5.80	5.47	5.26	4.81	4.52	4.41
10	10.04	7.56	6.55	5.99	5.64	5.39	5.06	4.85	4.41	4.12	4.01
11	9.65	7.21	6.22	5.67	5.32	5.07	4.74	4.54	4.10	3.81	3.71
12	9.33	6.93	5.95	5.41	5.06	4.82	4.50	4.30	3.86	3.57	3.47
13	9.07	6.70	5.74	5.21	4.86	4.62	4.30	4.10	3.66	3.38	3.27
14	8.86	6.51	5.56	5.04	4.69	4.46	4.14	3.94	3.51	3.22	3.11
15	8.68	6.36	5.42	4.89	4.56	4.32	4.00	3.80	3.37	3.08	2.98
16	8.53	6.23	5.29	4.77	4.44	4.20	3.89	3.69	3.26	2.97	2.86
17	8.40	6.11	5.19	4.67	4.34	4.10	3.79	3.59	3.16	2.87	2.76
18	8.29	6.01	5.09	4.58	4.25	4.01	3.71	3.51	3.08	2.78	2.68
19	8.19	5.93	5.01	4.50	4.17	3.94	3.63	3.43	3.00	2.71	2.60
20	8.10	5.85	4.94	4.43	4.10	3.87	3.56	3.37	2.94	2.64	2.54
21	8.02	5.78	4.87	4.37	4.04	3.81	3.51	3.31	2.88	2.58	2.48
22	7.95	5.72	4.82	4.31	3.99	3.76	3.45	3.26	2.83	2.53	2.42
23	7.88	5.66	4.76	4.26	3.94	3.71	3.41	3.21	2.78	2.48	2.37
24	7.82	5.61	4.72	4.22	3.90	3.67	3.36	3.17	2.74	2.44	2.33
25	7.77	5.57	4.68	4.18	3.85	3.63	3.32	3.13	2.70	2.40	2.29
26	7.72	5.53	4.64	4.14	3.82	3.59	3.29	3.09	2.66	2.36	2.25
27	7.68	5.49	4.60	4.11	3.78	3.56	3.26	3.06	2.63	2.33	2.22
28	7.64	5.45	4.57	4.07	3.75	3.53	3.23	3.03	2.60	2.30	2.19
29	7.60	5.42	4.54	4.04	3.73	3.50	3.20	3.00	2.57	2.27	2.16
30	7.56	5.39	4.51	4.02	3.70	3.47	3.17	2.98	2.55	2.25	2.13
40	7.31	5.18	4.31	3.83	3.51	3.29	2.99	2.80	2.37	2.06	1.94
50	7.17	5.06	4.20	3.72	3.41	3.19	2.89	2.70	2.27	1.95	1.82
60	7.08	4.98	4.13	3.65	3.34	3.12	2.82	2.63	2.20	1.88	1.75
100	6.90	4.82	3.98	3.51	3.21	2.99	2.69	2.50	2.07	1.74	1.60
200	6.76	4.71	3.88	3.41	3.11	2.89	2.60	2.41	1.97	1.63	1.48
1000	6.66	4.63	3.80	3.34	3.04	2.82	2.53	2.34	1.90	1.54	1.38

[a]Calculated by MINITAB.

TABLE B12 Upper Percentage Points of the Studentized Range, $q_\alpha = \frac{\bar{X}_{max} - \bar{X}_{min}}{S_{\bar{x}}}$

Error df	α	p = number of treatment means																			α	Error df
		2	3	4	5	6	7	8	9	10	11	12	13	14	15	16	17	18	19	20		
5	.05	3.64	4.60	5.22	5.67	6.03	6.33	6.58	6.80	6.99	7.17	7.32	7.47	7.60	7.72	7.83	7.93	8.03	8.12	8.21	.05	5
	.01	5.70	6.97	7.80	8.42	8.91	9.32	9.67	9.97	10.24	10.48	10.70	10.89	11.08	11.24	11.40	11.55	11.68	11.81	11.93	.01	
6	.05	3.46	4.34	4.90	5.31	5.63	5.89	6.12	6.32	6.49	6.65	6.79	6.92	7.03	7.14	7.24	7.34	7.43	7.51	7.59	.05	6
	.01	5.24	6.33	7.03	7.56	7.97	8.32	8.61	8.87	9.10	9.30	9.49	9.65	9.81	9.95	10.08	10.21	10.32	10.43	10.54	.01	
7	.05	3.34	4.16	4.68	5.06	5.36	5.61	5.82	6.00	6.16	6.30	6.43	6.55	6.66	6.76	6.85	6.94	7.02	7.09	7.17	.05	7
	.01	4.95	5.92	6.54	7.01	7.37	7.68	7.94	8.17	8.37	8.55	8.71	8.86	9.00	9.12	9.24	9.35	9.46	9.55	9.65	.01	
8	.05	3.26	4.04	4.53	4.89	5.17	5.40	5.60	5.77	5.92	6.05	6.18	6.29	6.39	6.48	6.57	6.65	6.73	6.80	6.87	.05	8
	.01	4.74	5.63	6.20	6.63	6.96	7.24	7.47	7.68	7.87	8.03	8.18	8.31	8.44	8.55	8.66	8.76	8.85	8.94	9.03	.01	
9	.05	3.20	3.95	4.42	4.76	5.02	5.24	5.43	5.60	5.74	5.87	5.98	6.09	6.19	6.28	6.36	6.44	6.51	6.58	6.64	.05	9
	.01	4.60	5.43	5.96	6.35	6.66	6.91	7.13	7.32	7.49	7.65	7.78	7.91	8.03	8.13	8.23	8.32	8.41	8.49	8.57	.01	
10	.05	3.15	3.88	4.33	4.65	4.91	5.12	5.30	5.46	5.60	5.72	5.83	5.93	6.03	6.11	6.20	6.27	6.34	6.40	6.47	.05	10
	.01	4.48	5.27	5.77	6.14	6.43	6.67	6.87	7.05	7.21	7.36	7.48	7.60	7.71	7.81	7.91	7.99	8.07	8.15	8.22	.01	
11	.05	3.11	3.82	4.26	4.57	4.82	5.03	5.20	5.35	5.49	5.61	5.71	5.81	5.90	5.99	6.06	6.14	6.20	6.26	6.33	.05	11
	.01	4.39	5.14	5.62	5.97	6.25	6.48	6.67	6.84	6.99	7.13	7.25	7.36	7.46	7.56	7.65	7.73	7.81	7.88	7.95	.01	
12	.05	3.08	3.77	4.20	4.51	4.75	4.95	5.12	5.27	5.40	5.51	5.62	5.71	5.80	5.88	5.95	6.03	6.09	6.15	6.21	.05	12
	.01	4.32	5.04	5.50	5.84	6.10	6.32	6.51	6.67	6.81	6.94	7.06	7.17	7.26	7.36	7.44	7.52	7.59	7.66	7.73	.01	
13	.05	3.06	3.73	4.15	4.45	4.69	4.88	5.05	5.19	5.32	5.43	5.53	5.63	5.71	5.79	5.86	5.93	6.00	6.05	6.11	.05	13
	.01	4.26	4.96	5.40	5.73	5.98	6.19	6.37	6.53	6.67	6.79	6.90	7.01	7.10	7.19	7.27	7.34	7.42	7.48	7.55	.01	
14	.05	3.03	3.70	4.11	4.41	4.64	4.83	4.99	5.13	5.25	5.36	5.46	5.55	5.64	5.72	5.79	5.83	5.92	5.97	6.03	.05	14
	.01	4.21	4.89	5.32	5.63	5.88	6.08	6.26	6.41	6.54	6.66	6.77	6.87	6.96	7.05	7.12	7.20	7.27	7.33	7.39	.01	
15	.05	3.01	3.67	4.08	4.37	4.60	4.78	4.94	5.08	5.20	5.31	5.40	5.49	5.58	5.65	5.72	5.79	5.85	5.90	5.96	.05	15
	.01	4.17	4.83	5.25	5.56	5.80	5.99	6.16	6.31	6.44	6.55	6.66	6.76	6.84	6.93	7.00	7.07	7.14	7.20	7.26	.01	

16	.05	3.00	3.65	4.05	4.33	4.56	4.74	4.90	5.03	5.15	5.26	5.35	5.44	5.52	5.59	5.66	5.72	5.79	5.84	5.90	.05	16
	.01	4.13	4.78	5.19	5.49	5.72	5.92	6.08	6.22	6.35	6.46	6.56	6.66	6.74	6.82	6.90	6.97	7.03	7.09	7.15	.01	
17	.05	2.98	3.63	4.02	4.30	4.52	4.71	4.86	4.99	5.11	5.21	5.31	5.39	5.47	5.55	5.61	5.68	5.74	5.79	5.84	.05	17
	.01	4.10	4.74	5.14	5.43	5.66	5.85	6.01	6.15	6.27	6.38	6.48	6.57	6.66	6.73	6.80	6.87	6.94	7.00	7.05	.01	
18	.05	2.97	3.61	4.00	4.28	4.49	4.67	4.82	4.96	5.07	5.17	5.27	5.35	5.43	5.50	5.57	5.63	5.69	5.74	5.79	.05	18
	.01	4.07	4.70	5.09	5.38	5.60	5.79	5.94	6.08	6.20	6.31	6.41	6.50	6.58	6.65	6.72	6.79	6.85	6.91	6.96	.01	
19	.05	2.96	3.59	3.98	4.25	4.47	4.65	4.79	4.92	5.04	5.14	5.23	5.32	5.39	5.46	5.53	5.59	5.65	5.70	5.75	.05	19
	.01	4.05	4.67	5.05	5.33	5.55	5.73	5.89	6.02	6.14	6.25	6.34	6.43	6.51	6.58	6.65	6.72	6.78	6.84	6.89	.01	
20	.05	2.95	3.58	3.96	4.23	4.45	4.62	4.77	4.90	5.01	5.11	5.20	5.28	5.36	5.43	5.49	5.55	5.61	5.66	5.71	.05	20
	.01	4.02	4.64	5.02	5.29	5.51	5.69	5.84	5.97	6.09	6.19	6.29	6.37	6.45	6.52	6.59	6.65	6.71	6.76	6.82	.01	
24	.05	2.92	3.53	3.90	4.17	4.37	4.54	4.68	4.81	4.92	5.01	5.10	5.18	5.25	5.32	5.38	5.44	5.50	5.54	5.59	.05	24
	.01	3.96	4.54	4.91	5.17	5.37	5.54	5.69	5.81	5.92	6.02	6.11	6.19	6.26	6.33	6.39	6.45	6.51	6.56	6.61	.01	
30	.05	2.89	3.49	3.84	4.10	4.30	4.46	4.60	4.72	4.83	4.92	5.00	5.08	5.15	5.21	5.27	5.33	5.38	5.43	5.48	.05	30
	.01	3.89	4.45	4.80	5.05	5.24	5.40	5.54	5.65	5.76	5.85	5.93	6.01	6.08	6.14	6.20	6.26	6.31	6.36	6.41	.01	
40	.05	2.86	3.44	3.79	4.04	4.23	4.39	4.52	4.63	4.74	4.82	4.91	4.98	5.05	5.11	5.16	5.22	5.27	5.31	5.36	.05	40
	.01	3.82	4.37	4.70	4.93	5.11	5.27	5.39	5.50	5.60	5.69	5.77	5.84	5.90	5.96	6.02	6.07	6.12	6.17	6.21	.01	
60	.05	2.83	3.40	3.74	3.98	4.16	4.31	4.44	4.55	4.65	4.73	4.81	4.88	4.94	5.00	5.06	5.11	5.16	5.20	5.24	.05	60
	.01	3.76	4.28	4.60	4.82	4.99	5.13	5.25	5.36	5.45	5.53	5.60	5.67	5.73	5.79	5.84	5.89	5.93	5.98	6.02	.01	
120	.05	2.80	3.36	3.69	3.92	4.10	4.24	4.36	4.48	4.56	4.64	4.72	4.78	4.84	4.90	4.95	5.00	5.05	5.09	5.13	.05	120
	.01	3.70	4.20	4.50	4.71	4.87	5.01	5.12	5.21	5.30	5.38	5.44	5.51	5.56	5.61	5.66	5.71	5.75	5.79	5.83	.01	
∞	.05	2.77	3.31	3.63	3.86	4.03	4.17	4.29	4.39	4.47	4.55	4.62	4.68	4.74	4.80	4.85	4.89	4.93	4.97	5.01	.05	∞
	.01	3.64	4.12	4.40	4.60	4.76	4.88	4.99	5.08	5.16	5.23	5.29	5.35	5.40	5.45	5.49	5.54	5.57	5.61	5.65	.01	

Source: This table is extracted from Table 29, "Biometrika Tables for Statisticians," 3rd Ed. Vol. I, London, Bentley House, 1966, with the permission of the Biometrika Trustees. The original work appeared in a paper by J. M. May, Extended and corrected tables of the upper percentage points of the 'Studentized' range. *Biometrika* **39**, 192–193 (1952).

TABLE B13 t For Comparisons between p Treatment Means and a Control for a Joint Confidence Coefficient of $P = .95$ and $P = .99$

		One-sided comparisons											Two-sided comparisons								
		p = number of treatment means, excluding control											p = number of treatment means, excluding control								
Error df	P	1	2	3	4	5	6	7	8	9	Error df	P	1	2	3	4	5	6	7	8	9
5	.95	2.02	2.44	2.68	2.85	2.98	3.08	3.16	3.24	3.30	5	.95	2.57	3.03	3.39	3.66	3.88	4.06	4.22	4.36	4.49
	.99	3.37	3.90	4.21	4.43	4.60	4.73	4.85	4.94	5.03		.99	4.03	4.63	5.09	5.44	5.73	5.97	6.18	6.36	6.53
6	.95	1.94	2.34	2.56	2.71	2.83	2.92	3.00	3.07	3.12	6	.95	2.45	2.86	3.18	3.41	3.60	3.75	3.88	4.00	4.11
	.99	3.14	3.61	3.88	4.07	4.21	4.33	4.43	4.51	4.59		.99	3.71	4.22	4.60	4.88	5.11	5.30	5.47	5.61	5.74
7	.95	1.89	2.27	2.48	2.62	2.73	2.82	2.89	2.95	3.01	7	.95	2.36	2.75	3.04	3.24	3.41	3.54	3.66	3.76	3.86
	.99	3.00	3.42	3.66	3.83	3.96	4.07	4.15	4.23	4.30		.99	3.50	3.95	4.28	4.52	4.71	4.87	5.01	5.13	5.24
8	.95	1.86	2.22	2.42	2.55	2.66	2.74	2.81	2.87	2.92	8	.95	2.31	2.67	2.94	3.13	3.28	3.40	3.51	3.60	3.68
	.99	2.90	3.29	3.51	3.67	3.79	3.88	3.96	4.03	4.09		.99	3.36	3.77	4.06	4.27	4.44	4.58	4.70	4.81	4.90
9	.95	1.83	2.18	2.37	2.50	2.60	2.68	2.75	2.81	2.86	9	.95	2.26	2.61	2.86	3.04	3.18	3.29	3.39	3.48	3.55
	.99	2.82	3.19	3.40	3.55	3.66	3.75	3.82	3.89	3.94		.99	3.25	3.63	3.90	4.09	4.24	4.37	4.48	4.57	4.65
10	.95	1.81	2.15	2.34	2.47	2.56	2.64	2.70	2.76	2.81	10	.95	2.23	2.57	2.81	2.97	3.11	3.21	3.31	3.39	3.46
	.99	2.76	3.11	3.31	3.45	3.56	3.64	3.71	3.78	3.83		.99	3.17	3.53	3.78	3.95	4.10	4.21	4.31	4.40	4.47
11	.95	1.80	2.13	2.31	2.44	2.53	2.60	2.67	2.72	2.77	11	.95	2.20	2.53	2.76	2.92	3.05	3.15	3.24	3.31	3.38
	.99	2.72	3.06	3.25	3.38	3.48	3.56	3.63	3.69	3.74		.99	3.11	3.45	3.68	3.85	3.98	4.09	4.18	4.26	4.33
12	.95	1.78	2.11	2.29	2.41	2.50	2.58	2.64	2.69	2.74	12	.95	2.18	2.50	2.72	2.88	3.00	3.10	3.18	3.25	3.32
	.99	2.68	3.01	3.19	3.32	3.42	3.50	3.56	3.62	3.67		.99	3.05	3.39	3.61	3.76	3.89	3.99	4.08	4.15	4.22
13	.95	1.77	2.09	2.27	2.39	2.48	2.55	2.61	2.66	2.71	13	.95	2.16	2.48	2.69	2.84	2.96	3.06	3.14	3.21	3.27
	.99	2.65	2.97	3.15	3.27	3.37	3.44	3.51	3.56	3.61		.99	3.01	3.33	3.54	3.69	3.81	3.91	3.99	4.06	4.13
14	.95	1.76	2.08	2.25	2.37	2.46	2.53	2.59	2.64	2.69	14	.95	2.14	2.46	2.67	2.81	2.93	3.02	3.10	3.17	3.23
	.99	2.62	2.94	3.11	3.23	3.32	3.40	3.46	3.51	3.56		.99	2.98	3.29	3.49	3.64	3.75	3.84	3.92	3.99	4.05

15	.95	1.75	2.07	2.24	2.36	2.44	2.51	2.57	2.62	2.67
	.99	2.60	2.91	3.08	3.20	3.29	3.36	3.42	3.47	3.52
16	.95	1.75	2.06	2.23	2.34	2.43	2.50	2.56	2.61	2.65
	.99	2.58	2.88	3.05	3.17	3.26	3.33	3.39	3.44	3.48
17	.95	1.74	2.05	2.22	2.33	2.42	2.49	2.54	2.59	2.64
	.99	2.57	2.86	3.03	3.14	3.23	3.30	3.36	3.41	3.45
18	.95	1.73	2.04	2.21	2.32	2.41	2.48	2.53	2.58	2.62
	.99	2.55	2.84	3.01	3.12	3.21	3.27	3.33	3.38	3.42
19	.95	1.73	2.03	2.20	2.31	2.40	2.47	2.52	2.57	2.61
	.99	2.54	2.83	2.99	3.10	3.18	3.25	3.31	3.36	3.40
20	.95	1.72	2.03	2.19	2.30	2.39	2.46	2.51	2.56	2.60
	.99	2.53	2.81	2.97	3.08	3.17	3.23	3.29	3.34	3.38
24	.95	1.71	2.01	2.17	2.28	2.36	2.43	2.48	2.53	2.57
	.99	2.49	2.77	2.92	3.03	3.11	3.17	3.22	3.27	3.31
30	.95	1.70	1.99	2.15	2.25	2.33	2.40	2.45	2.50	2.54
	.99	2.46	2.72	2.87	2.97	3.05	3.11	3.16	3.21	3.24
40	.95	1.68	1.97	2.13	2.23	2.31	2.37	2.42	2.47	2.51
	.99	2.42	2.68	2.82	2.92	2.99	3.05	3.10	3.14	3.18
60	.95	1.67	1.95	2.10	2.21	2.28	2.35	2.39	2.44	2.48
	.99	2.39	2.64	2.78	2.87	2.94	3.00	3.04	3.08	3.12
120	.95	1.66	1.93	2.08	2.18	2.26	2.32	2.37	2.41	2.45
	.99	2.36	2.60	2.73	2.82	2.89	2.94	2.99	3.03	3.06
∞	.95	1.64	1.92	2.06	2.16	2.23	2.29	2.34	2.38	2.42
	.99	2.33	2.56	2.68	2.77	2.84	2.89	2.93	2.97	3.00

15	.95	2.13	2.44	2.64	2.79	2.90	2.99	3.07	3.13	3.19
	.99	2.95	3.25	3.45	3.59	3.70	3.79	3.86	3.93	3.99
16	.95	2.12	2.42	2.63	2.77	2.88	2.96	3.04	3.10	3.16
	.99	2.92	3.22	3.41	3.55	3.65	3.74	3.82	3.88	3.93
17	.95	2.11	2.41	2.61	2.75	2.85	2.94	3.01	3.08	3.13
	.99	2.90	3.19	3.38	3.51	3.62	3.70	3.77	3.83	3.89
18	.95	2.10	2.40	2.59	2.73	2.84	2.92	2.99	3.05	3.11
	.99	2.88	3.17	3.35	3.48	3.58	3.67	3.74	3.80	3.85
19	.95	2.09	2.39	2.58	2.72	2.82	2.90	2.97	3.04	3.09
	.99	2.86	3.15	3.33	3.46	3.55	3.64	3.70	3.76	3.81
20	.95	2.09	2.38	2.57	2.70	2.81	2.89	2.96	3.02	3.07
	.99	2.85	3.13	3.31	3.43	3.53	3.61	3.67	3.73	3.78
24	.95	2.06	2.35	2.53	2.66	2.76	2.84	2.91	2.96	3.01
	.99	2.80	3.07	3.24	3.36	3.45	3.52	3.58	3.64	3.69
30	.95	2.04	2.32	2.50	2.62	2.72	2.79	2.86	2.91	2.96
	.99	2.75	3.01	3.17	3.28	3.37	3.44	3.50	3.55	3.59
40	.95	2.02	2.29	2.47	2.58	2.67	2.75	2.81	2.86	2.90
	.99	2.70	2.95	3.10	3.21	3.29	3.36	3.41	3.46	3.50
60	.95	2.00	2.27	2.43	2.55	2.63	2.70	2.76	2.81	2.85
	.99	2.66	2.90	3.04	3.14	3.22	3.28	3.33	3.38	3.42
120	.95	1.98	2.24	2.40	2.51	2.59	2.66	2.71	2.76	2.80
	.99	2.62	2.84	2.98	3.08	3.15	3.21	3.25	3.30	3.33
∞	.95	1.96	2.21	2.37	2.47	2.55	2.62	2.67	2.71	2.75
	.99	2.58	2.79	2.92	3.01	3.08	3.14	3.18	3.22	3.25

Source: This table is reproduced from Dunett, C. W. A multiple comparison procedure for comparing several treatments with a control. *J. Am. Stat. Assoc.* **50,** 1096–1121 (1955). Reprinted with permission from the Journal of the American Statistical Association.

APPENDIX C

Selected Governmental Sources of Biostatistical Data

Three types of data collections are described below: (1) a population census, (2) a vital statistics system, and (3) sample surveys. The sources of the data used in U.S. life tables are also described. To understand the data resulting from these collection mechanisms, it is essential to be familiar with some definitions and the organization of the data collection systems.

I. POPULATION CENSUS DATA

The *census* is a counting of the entire population at a specified time. In the United States, it occurs once every 10 years as required by the Constitution, and the latest census was taken on April 1, 1990. The U.S. census attempts to count people in the place where they spend most of their time. Most people are counted at their legal residence, but college students, military personnel, prison inmates, and residents of long-term institutions are assigned to the location of the institutions.

The information available from the U.S. census is derived from two types of questionnaires. The questions on the short form are intended for everybody in every housing unit and include such basic data items as age, sex, race, marital status, property value or rent, and number of rooms. The long form is intended for persons in sampled housing units and includes, in addition to the basic items, income, education, occupation, employment, and detailed housing characteristics. Data are tabulated for the nation and by two types of geographic areas: administrative areas (states, congressional districts, counties, cities, towns, etc.) and statistical areas (census regions, metropolitan areas, urbanized areas, census tracts, enumeration districts, block groups, etc.).

The tabulated census data are made available in several different forms: printed publications, computer tapes, microfiche, on-line information systems, laser disks, and flexible diskettes for microcomputers. To access the data, it is necessary to consult documentation for the data media of your choosing. In addition to the tabulated data, a 1 percent sample of individual records are available for demographic and other research.

The census data are used for a variety of purposes: by the federal, state, and local governments for political apportionment and allocation of federal funds for planning and management of public programs; by demographers to analyze population changes and the makeup of the nation's population; by social scientists to study social and economic characteristics of the nation's population; and by statisticians to design sample surveys for the nation and local communities. The census data, most importantly, provide the denominator data for the assessment of social and health events occurring in the population, for example, in calculating the birth and death rates.

II. VITAL STATISTICS

Vital statistics are produced from registered vital events including births, deaths, fetal deaths, marriages, and divorces. The scope and organization of the vital events registration system vary from one country to another. In the United States, the registration of vital events has been the responsibility of the states primarily and of a few cities. The federal government's involvement is to set reporting standards and to compile statistics for the nation.[1] Each state is divided into local registration districts (counties, cities, other civil divisions) and a local registrar is appointed for each district. The vital records are permanently filed primarily in the state vital statistics

[1]See History and organization of the vital statistics system (1954). *In* "Vital Statistics of the United States, 1950," Vol. 1, Chapter 1, pp. 2–19. Volume I: Analysis and Summary Tables with Supplemental Tables for Alaska, Hawaii, Puerto Rico, and Virgin Islands. United States Government Printing Office, Washington, D.C.

office. The local and state vital registration activities are usually housed in public health agencies. The National Center for Health Statistics (NCHS) receives processed data or microfilm copies of certificates to compile the national vital statistics.

Vital events are required to be registered with the registrar of the local district in which the event occurs. The reporting of births is the direct responsibility of the professional attendant at birth, generally a physician or midwife. Deaths are reported by funeral directors or person acting as such. Marriage licenses issued by town or county clerks and divorce and annulment records filed with the clerks or court official provide the data for marriage and divorce statistics. The data items on these legal certificates determine the contents of vital statistics reports. These certificates are revised periodically to reflect the changing needs of users of the vital statistics.

Vital statistics are compiled at the local, state, and federal levels. Data are available in printed reports and also on computer tapes. Data are tabulated either by place of occurrence or by place of residence; administrative areas are usually used and statistical areas are rarely used. Data by place of residence from the local vital statistics reports are often incomplete because the events for residents may have occurred outside the local registration districts and may not be included in the local data base.

What uses are made of vital statistics? In addition to calculating the birth and death rates, we obtain such well-known indicators of public health as the infant mortality rate and life expectancy from vital statistics. Much epidemiological research is based on an analysis of deaths classified by cause and contributing factors which comes from the vital statistics. Birth data are used by local health departments for planning and evaluation of immunization programs and by public health researchers to study trends in low-birth-weight infants, teenage pregnancy, midwife delivery, and prenatal care.

III. SAMPLE SURVEYS

To supplement the census and vital statistics, several important continuous *sample surveys* have been added to the statistics programs of the Census Bureau and the NCHS. Unlike the census and vital statistics, data are gathered from only a small sample of people. The sample is selected using a complex statistical design. To interpret the sample survey data appropriately, we must understand the sample design and the survey instrument.

The Current Population Survey (CPS) is a monthly survey conducted by the Census Bureau for the Department of Labor. It is the main source of current information on the labor force in the United States. The unemployment rate that is announced every month is estimated from this survey. In addition, it collects current information on many other population characteristics. The data from this survey are published in the Current Population Reports which include several series: Population Characteristics (P-20), Population Estimates and Projections (P-25), Consumer Income (P-60), and other subject matter areas. Public use tapes are also available.

The NCHS is responsible for two major national surveys: the National Health Interview Survey (NHIS) and the National Health and Nutrition Examination Survey (NHANES). The sampling design and the estimation procedures used in these surveys are similar to those of the CPS. Because of the complex sample design, analyses of data from these surveys are complicated. These two surveys are described below. Several other smaller surveys are conducted by the NCHS, including the National Survey of Family Growth, the National Hospital Discharge Survey, the National Ambulatory Medical Care Survey, the National Nursing Home Survey, and the National Natality and Mortality Surveys.[2]

The NHIS, conducted annually since 1960, is a principal source of information on the health of the noninstitutionalized civilian population of the United States. The data are obtained through personal interviews covering a wide range of topics: demographic characteristics, physician visits, acute and chronic health conditions, long-term limitation of physical activity, and short-stay hospitalization. Some specific health topics such as aging, health insurance, alcohol use, and dental care are included as supplements in different years of the NHIS. The data from this survey are published in the Vital and Health Statistics Reports (Series 10) and data tapes are also available.

The NHANES, conducted periodically, is a comprehensive examination of the health and nutrition status of the noninstitutionalized U.S. civilian population. The data are collected by interview as well as direct physical and dental examinations, tests, and measurements performed on the sample person. Among the many items included are anthropometric measurements, medical history, hearing test, vision test, blood test, and a dietary inventory. Several health examination surveys have been conducted since 1960; the two most recent surveys are the Hispanic HANES (conducted in 1982–1984 for three major Hispanic subgroups: Mexican-Americans in five southwestern states, Cubans in Dade County, Florida, and Puerto Ricans in the New York City area) and NHANES III (conducted in 1988–1994). The data from health examination surveys are published in the Vital and Health Statistics Reports (Series 11).

[2]For a list and summary of data tapes, see the National Center for Health Statistics. "Catalog of Electronic Data Products," Hyattsville, MD, DHHS Publication No. (PHS)92-1213, July 1992, or a later issue.

IV. LIFE TABLES

Life tables have been published periodically by the federal government since the mid-19th century. The first federally prepared life tables appeared in the report of the 1850 census. Life tables prior to 1900 were based on mortality and population statistics compiled from census enumerations. The accuracy of these life tables was questioned as mortality statistics derived chiefly from census enumeration were subject to considerable under-enumeration. The year 1900 is the first year in which the federal government began an annual collection of mortality statistics based on registered deaths. Since then life tables have been constructed based on registered deaths and the enumerated population. Prior to 1930, life tables were limited to those states that were included in the death registration area. Until 1946, the official life tables were prepared by the U.S. Bureau of the Census. All subsequent tables have been prepared by the U.S. Public Health Service (initially by the Nation's Office of Vital Statistics and later by the NCHS).

Life tables provide an essential tool in a variety of fields. Life insurance companies largely base their calculations of insurance premiums on life tables. Demographers rely on life tables in making population projections, in estimating the volume of net migration, and in computing certain fertility measures. In law cases involving compensation for injuries or deaths, life tables are used as a basis for adjudicating the monetary value of a life. Personnel managers and planners employ life tables to schedule retirement and pension programs and to predict probable needs for employee replacement. Applications are numerous in public health planning and management, clinical research, and studies dealing with survivorship.

Three series of life tables are prepared and published by the NCHS:

1. *Decennial life tables*. These are complete life tables, meaning that life-table values are computed for single years of age. These are based on decennial census data and the deaths occurring over 3 calendar years around the census year. The advantage of using a 3-year average of deaths is to reduce the possible abnormalities in mortality patterns that may exist in a single calendar year. The decennial life tables are prepared for the United States and for the 50 individual states and the District of Columbia. This series also includes life tables by major causes of death, which are known as multiple decrement life tables.

2. *Annual life tables*. These are abridged life tables, meaning that life-table values are computed for age intervals instead of single years of age, except for the first year of life. The age intervals used are 0–1, 1–5, 5–10, 10–15, . . . , 80–85, and 85 or over. The annual abridged tables are based on a complete count of the registered deaths and postcensal midyear population estimates. These are prepared for the total U.S. population by gen-

der and race. This series also contains summary tables showing life-table values for survivorship and expectation of life by single years of age, interpolated from the abridged tables.

3. *Provisional annual life tables*. These provisional abridged life tables are based on a 10 percent sample of registered deaths and population estimates and are prepared for the total U.S. population only. These are published in the Monthly Vital Statistics Report before the final annual life tables become available. This series has been published annually since 1958.

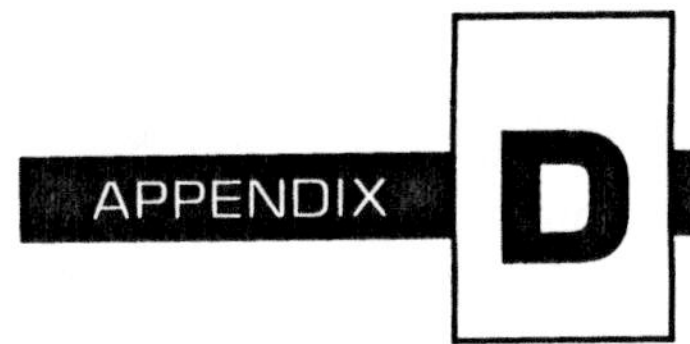

APPENDIX D

Solutions to Selected Exercises

CHAPTER 1

1.2. The change was made to protect the privacy of the adolescent in answering sensitive questions. The estimate of the proportion increased slightly immediately after the change, suggesting the earlier values were probably underestimated.

1.3. No. The difference in the infant mortality between Pennsylvania and Louisiana may be due to the difference in the racial/ethnic composition of the two states. The race-specific rates were indeed lower in Louisiana than in Pennsylvania. The proportion of blacks in Louisiana was sufficiently greater than that in Pennsylvania to make the overall rate higher than the overall rate in Pennsylvania.

CHAPTER 2

2.2. Not necessarily, as the choice of scale is dependent on the intended use of the variable. For example, we know that those completing high school have more economic opportunities than those that do not and the same is true for those completing college. Hence there is a greater difference between 11 and 12 years of education than between 10 and 11 years, and the same is true for the difference between 15 and 16 years compared with 13 and 14 or 14 and 15.

2.4. b. Counting the beats for 60 seconds may be considered too time consuming. On the other hand, counting for 20 seconds or 15 seconds and multiplying by 3 or 4 may be unreliable. Counting for 30 seconds and multiplying by 2 may be a good compromise.

2.6. Age recorded in the census is considered to be more accurate than that reported in the death certificate which was reported by grieving relatives and other informants. To alleviate some of these disagreements, the age-specific death rates are usually calculated by 5-year age groups.

CHAPTER 3

3.2. Read 25 four-digit random numbers and, if any random numbers are 2000 or greater, subtract a multiple of 2000 to obtain numbers less than 2000. Eliminate duplicates and draw additional random numbers to replace the number eliminated.

3.5. a. The population consists of all the pages in the book; the pages can be randomly sampled and number of words counted on the selected pages would constitute the data.
b. All moving passenger cars during the 1-week period can be considered as the population. The population can be framed in two dimensions: time and space. Passing cars can be observed at randomly selected locations at randomly selected times and the total number of cars and the number with only the driver can be observed.
c. The population consists of all the dogs in the county. Households in the county can be sampled in three stages: census tracts, blocks, and households. The number of the dogs found in the sample households and the number that have been vaccinated against rabies can then be recorded.

3.8. a. Some people have unlisted telephone numbers and others do not have telephones. People who have recently moved into the community are also not listed. Thus these groups are unrepresented in the sample. The advantage is that the frame, although incomplete, is already compiled.

CHAPTER 4

4.2. The actual expenditures increased, whereas the inflation-adjusted expenditures decreased. The trend in the inflated-adjusted expenditures would provide a more realistic assessment of the food stamp program.

4.5. b.

A/C	1	0	Total
1	8	20	28
0	8	10	18
Total	16	30	46

c. A (28) B(28) C(16)

4.8. As the total number of each type of hospital is not available, it is not possible to calculate the mean occupancy rate.

4.11. a. Mean = 747,000,000, CV = 344.1 percent.
b. Median = 10^5, geometric mean = 541,170.
c. The geometric mean, 5.4×10^5, seems to capture the sense of the data better than the mean or median.

4.14. b. Correlation = 0.094, adjusting for calories; the correlation of 0.648 is due to the fact that both protein and total fat are related to calories.

CHAPTER 5

5.2. a. 0.334
b. 0.524
c. $0.426 * 0.524 = 0.223$
d. $(0.372 - 0.223)/(1 - 0.524) = 0.313$

5.5. $1 - [1 - (1 - 0.99) * 0.2]^{120} = 0.24$

5.9. a. 79,590/96,334 = 0.826
b. (89,735 − 79,590)/98,223 = 0.103

CHAPTER 6

6.2. a. $1 - 0.8593 = 0.1407$
b. At least 10 persons with $p = 0.0226$
c. Virtually zero

6.5. 0.0146; 0.6057 (= 0.7073 − 0.1016)

6.8. Probability is 0.0116 (= 1 − 0.9884); would investigate further.

6.12. $z = -1.3441$; $\Pr(x < 7) = 0.0895$; yes, it is normal distributed; can be verified by a normal probability plot.

CHAPTER 7

7.1. a. Sample mean = 11.94; sample standard error = 1.75; 95 percent confidence interval = (8.42, 15.46).

c. Sample median = 8; 95 percent confidence interval = {3 (19th observation), 12 (32nd observation)}.

d. The 95 percent tolerance interval to cover 90 percent of observations, based on a normal distribution = 11.94 ± 1.992(12.5) = (0, 36.84; based on distribution-free method, the interval (0, 39) covers 89.6 percent of observations; the latter method is more appropriate, as the data are not distributed normally (data are skewed to the right).

7.4. Would expect a negative correlation because those states that have the higher workplace safety score should have the lower fatality rates. $r = -0.435$. As the data are based on population values, there is no need to calculate a confidence interval; however, if we viewed these data as a sample in time, then the formation of a confidence interval is appropriate. The 95 percent confidence interval is (−0.636, −0.178). A significant negative correlation exists, as the confidence interval does not include zero.

7.7. Correlation = 0.145. These data may be viewed as a sample in time; the 95 percent confidence interval is (−0.136, 0.404). No significant linear relationship exists, as the confidence interval includes zero. Region of the country, perhaps reflecting the unemployment levels, may play a role.

7.10. a. (0.052, 0.206)

b. (−0.082, 0.202); no difference, as the confidence interval includes zero.

7.13. Difference = −0.261. The 99 percent confidence interval is (−0.532, 0.010). No difference, as the confidence interval includes zero, although a 95 percent confidence interval would not include zero.

CHAPTER 8

8.2. a. Thirty classes can be randomly allocated to two curricula.

b. A simple random allocation of six teachers to two curricula may not be appropriate; instead, teachers can be matched based on teaching experience before randomly allocating one member of each pair to the new curriculum and the other member to the old curriculum.

8.5. a. Fewer subjects would be needed compared with the two-group comparison design.

b. The random assignment of subjects to the initial diet presumably balanced the sequencing effect but it might not be adequate because of the small sample size.

c. The carryover effect is ineffectively controlled by not allowing a washout period and the granting of a leave to some subjects.

8.7. a. Randomized block design

b. The effect of organizational and leadership types is not controlled effectively, although the matching may have reduced the effect of this confounder.

CHAPTER 9

9.3. The decision rule is to reject the null hypothesis when the number of pairs favoring diet 1 is 14 to 20 with $\alpha = 0.0577$ and $\beta = 0.0867$.

9.6. The confidence interval is (0.400, 0.852).

CHAPTER 10

10.1. Medians are 12.25 for group 1, 7.75 for group 2, and 5.80 for group 3; average ranks are 36.5 for group 1, 23.3 for group 2, and 16.9 for group 3. The Kruskal–Wallis test is appropriate to use. $H = 15.3$, df $= 2$, and $p = 0.001$, indicating the medians are significantly different.

10.4. Divide into three groups based on the toilet rate: 1–61, 133–276, and 385–749, with nine observations in group 1, six in group 2, and six in group 3. As $H = 6.67$ with $p = 0.036$, we reject H_0.

10.7. The results by the Wilcoxon signed rank test are consistent with those obtained by the sign test in Exercise 10.6, although the p value is slightly smaller with the Wilcoxon signed rank test than with the sign test.

CHAPTER 11

11.3. Note that there are 2 out of 10 cells with expected counts less than 5, but the smallest expected count (3.18) is greater than 1 [$= 5 * (2/10)$] and the chi-square test is valid. $X^2 = 6.66$, df $= 4$, $p = 0.1423$. We fail to reject the null hypothesis at the 0.05 significance level. This is a test of independence because it appears that the subjects were selected at random, not by degree of infiltration. By assigning scores of -1, 0, 1, 2, and 3, we calculate $X^2 = 6.67$, df $= 1$, $p = 0.0098$; we reject the null hypothesis of no trend. By assigning scores of -1, 0,

0.5, 1, and 1.5, we calculate $X^2 = 6.36$, df = 1, $p = 0.0117$; we again reject the null hypothesis of no trend.

11.5. $X^2 = 20.41$, df = 1, $p < 0.0001$. We reject the null hypothesis; the proportion of violation is nearly three times higher for the nonattendees (73.5 percent) than the attendees (24.3 percent). Without more information, we cannot draw any conclusion about the effect of attending the course. Our interpretation depends on whether the course was attended before or after the violation was found.

11.8. $X^2 = 103.3$, df = 1, $p < 0.0001$, ignoring the newspaper variable; significant. $X^2 = 24.65$, df = 1, $p < 0.0001$, ignoring the radio variable; significant. The radio variable seems to have the stronger association; however, it is difficult to recommend one media over the other, as these two media variables, in combination, appear to be related to the knowledge of cancer. Additionally, because people were not randomly assigned to the four levels of media, to use these results about knowledge of cancer, we must assume that the people in each of the four levels of the media initially had the same knowledge of cancer. Without such an assumption, it is difficult to attribute the status of cancer knowledge to the different media.

CHAPTER 12

12.2. a. For the group with serum creatinine concentrations of 2.00–2.49 mg/dl, the 5-year survival probability is 0.731 with a standard error of 0.050. For the group with serum creatinine concentrations of 2.5 mg/dl or greater, the 5-year survival probability is 0.583 with a standard error of 0.058.

b. Despite the considerable difference in the 5-year survival probabilities, the two survival distributions are not significantly different at the 0.01 level, with $X^2_{CMH} = 3.73$ and $p = 0.0535$, reflecting the small sample size.

12.5. Median for the fee-for-service group = 28.8 month; median for HMO = 29.5. The two survival distributions are not significantly different.

CHAPTER 13

13.1. The percent predicted FVC is used because it is adjusted for age, height, sex, and race. H_0: $\mu_1 = \mu_2$; H_a: $\mu_1 > \mu_2$. One-sided test is used to reflect the expected effect of asbestos on pulmonary function. Assuming equal population variances, pooled variance is used. $t = 30.08$; p value is virtually zero. Reject the null hypothesis, suggesting that those with less than 20 years of exposure have a significantly larger forced vital capacity than those with 20 or more years of exposure.

13.4. H_0: $\mu_d = 0$, H_a: $\mu_d < 0$, $t_d = -10.03$, which is smaller than $t_{23,0.01} = -2.50$. Reject the null hypothesis, suggesting that the weight reduction program worked.

13.7. H_0: $\pi = 0.06$, H_a: $\pi \neq 0.06$ (two-sided test); $z = 1.745$, which is not larger than $z_{0.975} = 1.96$. Fail to reject the null hypothesis, suggesting that there is no strong evidence for the community's attainment of the goal.

13.10. $r = -0.243$, H_0: $\rho = 0$, H_a: $\rho \neq 0$, $\lambda = -0.8224$, which is not smaller than $z_{0.05} = -1.645$. Fail to reject the null hypothesis, no evidence for nonzero correlation; $p = 0.21$.

13.13. H_0: $\mu = 190$; H_a: $\mu \neq 190$; $t = 3.039$, which is larger than $t_{14,0.01} = 2.6245$. Reject H_0.

CHAPTER 14

14.2. $F = 0.51$, $p = 0.479$, no significant difference. The test results are the same as those obtained using the t test, with the same p value. $F_{1,n-1,1-\alpha} = t^2_{n-1,1-2/\alpha}$.

14.5. Degrees of freedoms are 2, 2, 4, and 18 for smoking status, lighting conditions, interaction, and error, respectively. $F = 0.213$ for interaction, which is not significant. $F = 12.896$ for smoking status, significant. $F = 45.276$ for lighting conditions, significant.

CHAPTER 15

15.1. The zero value for the degree of stenosis in the 10th observation is suspicious; it appears to be a missing value rather than 0 percent stenosis. The zero value for the number of reactive nuclei at initial survey in the 12th observation is also suspicious, but it may well be a reasonable value, because there are other smaller numbers such as 1 and 2. The scatter plot seems to suggest that there is a very weak linear relationship. A regression analysis yields $\hat{\beta}_0 = 22.2$, $\hat{\beta}_1 = 2.90$, $F = 10.04$, $p = 0.007$. The 10th observation had the largest standardized residual and the 6th observation had the greatest leverage, almost three times greater than the average leverage. Eliminating the 6th observation, $\hat{\beta}_0 = 21.2$ and $\hat{\beta}_1 = 3.05$.

15.4. Eliminating the two largest blood pressure values (14th and 50th observations) and the two smallest values (22nd and 27th observations), $\hat{\beta}_0 = 63.1$, $\hat{\beta}_1 = 0.726$, and $R^2 = 23.4$ percent.

15.7. $r = -0.138$, $\hat{\beta}_0 = 8.53$, $\hat{\beta}_1 = -0.063$, $R^2 = 1.9$ percent, $F = 0.19$, and $p = 0.669$. By adding the new variable, $\hat{\beta}_0 = 8.13$, $\hat{\beta}_1 \doteq -0.067$, $\hat{\beta}_2 = 0.110$ (new variable), $R^2 = 37.8$ percent, $F = 2.73$, and $p = 0.118$. The new variable captured the nonlinear effect of BMI on serum cholesterol.

Index

A

D

K

L

M

Q

R

S

T

U

V